The Primate Visual System

The Primate Visual System

A Comparative Approach

Edited by

Jan Kremers

Novartis Institutes of Biomedical Research,
Novartis Pharma, Switzerland

John Wiley & Sons, Ltd

Telephone (+44) 1243 779777

Email (for orders and customer service enquiries): cs-books@wiley.co.uk
Visit our Home Page on www.wiley.com

Other Wiley Editorial Offices

John Wiley & Sons Inc., 111 River Street, Hoboken, NJ 07030, USA

Jossey-Bass, 989 Market Street, San Francisco, CA 94103-1741, USA

Wiley-VCH Verlag GmbH, Boschstr. 12, D-69469 Weinheim, Germany

John Wiley & Sons Australia Ltd, 33 Park Road, Milton, Queensland 4064, Australia

John Wiley & Sons (Asia) Pte Ltd, 2 Clementi Loop #02-01, Jin Xing Distripark, Singapore 129809

John Wiley & Sons Canada Ltd, 22 Worcester Road, Etobicoke, Ontario, Canada M9W 1L1

Wiley also publishes its books in a variety of electronic formats. Some content that appears in print may not be available in electronic books.

British Library Cataloguing in Publication Data

A catalogue record for this book is available from the British Library

ISBN-13 978-0-470-86809-0 (HB)
ISBN-10 0-470-86809-0 (HB)

Typeset in 10/12pt Times by Integra Software Services Pvt. Ltd, Pondicherry, India

This book is printed on acid-free paper responsibly manufactured from sustainable forestry in which at least two trees are planted for each one used for paper production.

Contents

Preface

We have Eyes like Twins
Where Your last Thought ends, my Next begins
Just one Heartbeat away
From everything, You meant to say
—Wilson Phillips

From Eyes Like Twins, Rupert Hine 1993, lyrics by Jeanette Obstoj

Primates are a highly fascinating group of animals partially because humans belong to them. From a Darwinistic viewpoint an interest in those individuals with whom we share a large part of our genome is natural. However, in my opinion what mainly distinguishes humans from animals is that humans have created new cultural and ethical laws besides, and overruling, the natural laws. Therefore for me it is more important that there is an ethical obligation to be interested in subjects which may be of relevance to our fellow human beings. Of course, the same ethical laws oblige us to evaluate whether the benefits of a study justify the consequences.

In addition to these anthropocentric viewpoints, primates deserve close study because they form a very heterogeneous group of animals with many fascinating and unique properties. The diversity in habitat, size and behaviour is extremely interesting to study.

This book is a tribute to all primates, human and non-human.

For the majority of primates, if not for all, vision is the most important sensory information for a normal daily life. For most non-human primates blindness is probably fatal. But for humans also, the loss of vision has a large impact on the quality of life. Many people, who lose their vision, face major problems in coping with and reorganizing their lives. It is thus very important to learn about and understand the anatomical and physiological properties of the visual system and the significance of these properties for visual perception. A better insight into the visual system may lead to better diagnoses and treatments of diseases and may help understand patients who suffer from them. The similarity between the visual systems of the different primate groups makes them the best model for those of humans. To quote the title of a song by Rupert Hine: primates have 'yes like Twins'.

The eye has fascinated theologists and biologists. Paley considered the exquisite organization of it as a proof for the existence of God because he could not imagine that it could be constructed without a plan. Even Darwin was puzzled by the eye and saw it as a challenge for his theory of evolution through natural selection. Today it is considered that the eye and the visual system underwent the same evolution as any other organ and that natural selection played the same shaping role. As Richard Dawkins would put it: the eye and the visual system are the result of a 'blind' process and not of a carefully planned design. Owing to natural selection, the visual system will have different anatomical and physiological properties in different species. But these differences are smallest in closely related species. Importantly, the investigation of the primate visual system will highly benefit from placing the available data in an evolutionary and ecological context using a comparative approach.

This book aims to give a comprehensive overview of what is known about the primate visual system. It tries to explore and correlate the anatomical, physiological and psychophysical properties and to place it in an evolutionary, developmental and ecological context. The book was not meant to go into all details. For that it might be better to go to the extensive original scientific literature which will probably fill libraries. It rather tries to help the reader to get a basic understanding and to find the way through the jungle of scientific literature. The book is a short trip through the visual system, starting with a discussion of the evolutionary, ecological and developmental constrains of the primate visual system. It then wanders through the visual system as the visual information would do, starting in the photopigments and ending at the cortex and in visual perception. Of course, there are many aspects which have been discussed only briefly but which deserve a much closer look. The clinically relevant work on the primate visual system is mentioned sporadically. The wealth of data on cortical and cognitive visual science is certainly much larger than those discussed in this book. Nevertheless, I am confident that the book might be a good start for obtaining an overview on most aspects of primate vision.

I am extremely proud that so many excellent scientists were willing to contribute to the book. The authors were asked to write for a relatively broad audience without being trivial and taking into account all scientific aspects, methods and theories. This though a nearly impossible task was managed admirably. Sometimes, I probably was a pain for them interfering with their texts, asking stupid questions and urging them to change the text here and there. But from my side it was a fantastic experience to collaborate with each one of them. I would like to thank them all. A few colleagues deserve special mention. Luiz Silveira, Paul Martin and Annette Werner were very helpful in bringing the book on its way. My colleagues in my former lab in Tübingen, Vladislav Kozyrev and Birgit Regelmann were extremely helpful. I would like to thank my new employer for having given me the opportunity to finish this project. I would also like to thank everybody at John Wiley and Sons, especially Rachael Ballard, who guided me through the process of book-making. I would like to thank my parents and my family for their constant support. This book is dedicated to Andrea and Leon who had to endure many of my uncertainties and doubts during the last couple of years and who kept me sane and on the road. We certainly have 'Eyes like Twins'.

List of Contributors

James K. Bowmaker Institute of Ophthalmology, University College London, Bath Street, London EC1V 9EL, UK

Vivien A. Casagrande Department of Cell and Developmental Biology, Vanderbilt Medical School, B2323 Medical Center North, Nashville, TN 37232–2175, USA

Jan Churan Generation Research Program, University of München, Prof. Max-Lange-Platz 11, 83646 Bad Tölz, Germany

Bogdan Dreher Anatomy and Histology, School of Medical Sciences and Institute for Biomedical Research, University of Sydney, Sydney, NSW 2006, Australia

Barbara L. Finlay Department of Psychology, Cornell University

Mark W. Greenlee Department of Experimental Psychology, University of Regensburg, 93053 Regensburg, Germany

Ulrike Grünert National Vision Research Institute of Australia and Department of Optometry and Vision Sciences, The University of Melbourne, Carlton, VIC 3053, Australia

David M. Hunt Institute of Ophthalmology, University College London, Bath Street, London EC1V 9EL, UK

Michael R. Ibbotson Visual Sciences, Research School of Biological Sciences, Australian National University, Canberra, ACT 2601, Australia

Uwe J. Ilg Oculomotor Lab, Department of Cognitive Neurology, Hertie-Institute for Clinical Brain Research, University of Tübingen, Germany, Hoppe-Seyler-Str. 3, 72076 Tübingen

Gerald H. Jacobs Neuroscience Research Institute and Department of Psychology, University of California, Santa Barbara 93106, USA

Jon H. Kaas Department of Psychology, Vanderbilt University, Nashville, TN 37240, USA

Jan Kremers Novartis Institutes for Biomedical Research, Novartis Pharma, Klybeckstrosse 141, 4057 Basel, Switzerland

Barry B. Lee Max Planck Institute for Biophysical Chemistry, Göttingen, Germany and SUNY College of Optometry, New York, USA

Robert D. Martin Academic Affairs, The Field Museum, 1400 S. Lake Shore Drive, Chicago, IL 60605-2496, USA

Paul R. Martin National Vision Research Institute of Australia and Department of Optometry and Vision Sciences, The University of Melbourne, Carlton, VIC 3053 Australia

D. Osorio School of Life Sciences, University of Sussex, Brighton. BN1 9QG, UK

Joel Pokorny Visual Science Laboratories, The University of Chicago, 940 East 57th Street, Chicago IL 60637, USA

Andreas Reichenbach Paul-Flechsig-Institut für Hirnforschung, University of Leipzig

Callum F. Ross Organismal Biology and Anatomy, The University of Chicago, 1029 East 57th Street, Chicago, IL 60637, USA

David W. Royal Department of Cell and Developmental Biology, Vanderbilt Medical School, B2323 Medical Center North, Nashville, TN 37232–2175, USA

Gyula Sáry Department of Cell and Developmental Biology, Vanderbilt Medical School, B2323 Medical Center North, Nashville, TN 37232–2175, USA

Stefan Schumann Anaesthesiologische Universitätsklinik, Sektion für Experimentelle Anaestesiologie, Hugstetter Straße 55, 79106 Freiburg

Luiz Carlos L. Silveira Department of Physiology, Biological Science Centre, Federal University of Pará, 66075-900 Belém, Pará, Brazil

Luiz Carlos de Lima Silveira Department of Physiology, Federal University of Pará

Vivianne C. Smith Visual Science Laboratories, The University of Chicago, 940 East 57th Street, Chicago IL 60637, USA

Samuel G. Solomon University Laboratory of Physiology, University of Oxford, UK

Arne Valberg Institute of Physics, Norwegian University of Science and Technology, 7491 Trondheim, Norway

M. Vorobyev Vision, Touch and Hearing Research Centre, School of Biomedical Sciences, The University of Queensland, Queensland 4072, Australia

Annette Werner Department of Experimental Ophthalmology, University Eye Hospital, Röntgenweg 11, 72076 Tübingen, Germany

1

The Evolutionary and Ecological Context of Primate Vision

Robert D. Martin and *Callum F. Ross*

1.1 Introduction

1.1.1 Overview of primates and their phylogenetic relationships

Excluding tree-shrews, now commonly relegated to the separate mammalian order Scandentia, approximately 350 species of modern primates can currently be recognized (Groves, 2001), most of them being arboreal inhabitants of tropical and subtropical forests. As has generally been recognized in some way in all major classifications, on morphological and biogeographical grounds these living primates fall fairly clearly into six 'natural groups' (Martin, 1990): (1) Madagascar lemurs (infraorder Lemuriformes); (2) lorises and bush babies (infraorder Lorisiformes); (3) tarsiers (infraorder Tarsiiformes); (4) New World monkeys (superfamily Ceboidea); (5) Old World monkeys (superfamily Cercopithecoidea); and (6) Old World apes and humans (superfamily Hominoidea). The last two groups (Old World monkeys; Old World apes and humans) are commonly combined in the infraorder Catarrhini, distinguishing them from the infraorder Platyrrhini established for New World monkeys. Because the first three natural groups (lemurs, lorisiforms, and tarsiers) have remained relatively primitive, they have often been labeled prosimians or lower primates, to distinguish them from the more advanced simians or higher primates (monkeys, apes, and humans), which are seen as having attained a higher grade of evolution. In traditional, grade-based classifications, it has accordingly been customary to allocate prosimians to the suborder Prosimii and simians to the suborder Anthropoidea. However, there is abundant evidence indicating that, although they have

The Primate Visual System: A Comparative Approach Edited by Jan Kremers

remained primitive in many respects, the tarsiers are in fact more closely related to the simians than to the other prosimians (Figure 1.1). In a cladistic classification (i.e. one designed to provide a direct reflection of inferred phylogenetic relationships), lemurs and lorises are hence allocated to the suborder Strepsirrhini, while tarsiers and simians are allocated to the suborder Haplorhini. Regardless of the choice of classification, which is a continuing matter of controversy, in some contexts it is convenient to distinguish between prosimians and simians in discussing primate features, whereas in others it is useful to contrast strepsirrhines with haplorhines.

Undoubted relatives of modern primates (euprimates or 'primates of modern aspect') first appear in the fossil record at the beginning of the Eocene epoch, approximately 55 mya (million years ago). The Plesiadapiformes, which were predominantly present during the preceding Palaeocene epoch (55–65 mya), have traditionally been allocated to the order Primates (e.g. Simons, 1972; Szalay and Delson, 1979). However, they lack many of the defining features of primates of modern aspect (Martin, 1968, 1990; Cartmill, 1972, 1974) and have been labeled 'archaic primates' to emphasize their distinctiveness. In fact, some authors (e.g. Martin, 1990, 1993; Kirk *et al.*, 2003) have questioned the supposed phylogenetic link between Plesiadapiformes and primates, and there have been alternative suggestions that plesiadapiforms are related to colugos (order Dermoptera) instead (Beard, 1990, 1993; Kay *et al.*, 1992). The recent discovery of a fairly complete skeleton of the plesiadapiform *Carpolestes*, which appears to show a primate-like grasping adaptation of the foot, has reopened the debate (Bloch and Boyer, 2002; but see Kirk *et al.*, 2003). Given continuing uncertainties about the status of Plesiadapiformes, the most prudent course is to treat them as a separate category in discussions of primate evolution. The present discussion will focus primarily on euprimates.

Fossil euprimates can be divided into two major categories. On the one hand, there are early Tertiary (predominantly Eocene) forms that can generally be allocated either to the infraorder Adapiformes or to the infraorder Omomyiformes. On the other hand, there are other, generally later, forms that can be linked more closely to the natural groups of modern primates. Direct relatives of the six modern natural groups of living primates tend to occur from the late Oligocene/early Miocene (i.e. from about 25 mya), although there are a few exceptions. For instance, the putative fossil tarsier *Tarsius eocaenus* is documented from the middle Eocene of China. Several authors have linked the Adapiformes to the strepsirrhine side of the primate tree and the Omomyiformes to the haplorhine side. Indeed, it has often been suggested that the Omomyiformes are specifically linked to tarsiers. However, it is possible that the Adapiformes and the Omomyiformes are not linked directly to strepsirrhines or haplorhines and that they constitute a separate adaptive radiation of early primates (Martin, 1993; Ross, 2003). It should be noted that Adapiformes and Omomyiformes have a northern continental distribution (Asia, Europe, and North America), while unquestioned strepsirrhines and haplorhines have a distribution that is predominantly confined to the southern landmasses.

1.1.2 Contributions of the comparative approach

With respect to vision, as in many other areas of biology, a comparative interspecific approach has much to offer with respect to evolutionary and functional interpretation. Among other things, it yields a different complementary perspective that can enhance

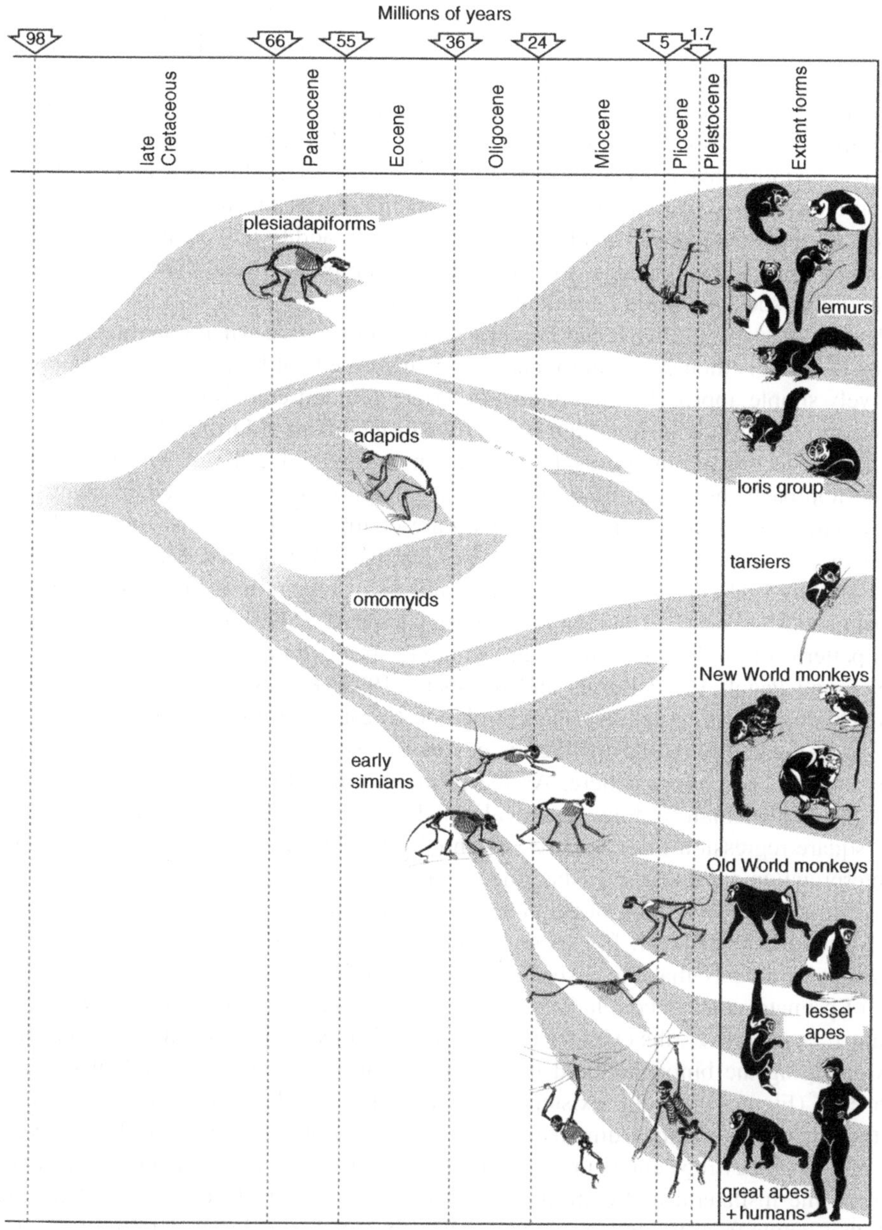

Figure 1.1 Outline phylogenetic tree of primates. Euprimates are represented by the six 'natural groups' of modern primates and various fossil relatives dating back to the beginning of the Eocene (55 mya). Note the initial subdivision between strepsirrhines and haplorhines in the evolution of euprimates. The tree also includes 'archaic primates' (Plesiadapiformes), which are of uncertain affinities. Original illustration by Lukrezia Bieler-Beerli. Reprinted by permission from Nature (Martin, R.D., vol. 363, pp. 223–234) copyright © (1993) Macmillan Journals Limited (http://www.nature.com/)

interpretation of results obtained from highly focussed studies of individual species. Furthermore, comparisons between species permit formulation of hypotheses and some degree of testing thereof. Indeed, a comparative approach is essential for the inclusion of fossil evidence and hence for any comprehensive phylogenetic interpretation, as is exemplified in this review. At the same time, it should be emphasized that comparisons can yield no more than correlations and indirect inferences. Although it is possible, given appropriate caution, to make certain functional inferences on the basis of interspecific comparative studies, causal analysis ideally requires additional inputs from testing of individual species. In an ideal world, interspecific comparisons and detailed investigation of individual species should be mutually enriching.

The comparative approach can be particularly informative when quantitative data are available, but it is then essential to take the scaling effects of body size into account. One relatively simple approach is to conduct bivariate allometric analysis, which requires the examination of trends in logarithmically transformed data (Martin, 1980, 1989, 1990). A basic model for allometric scaling is shown in Figure 1.2. Even with this relatively simple approach, however, there are several complex issues involved in analysis and interpretation. Bivariate allometric analyses are confronted by at least four fundamental problems: (1) choice of an appropriate best-fit line (Harvey and Mace, 1982; Martin and Barbour, 1989; Isler *et al.*, 2002); (2) successful recognition of grades in a dataset (Martin and MacLarnon, 1985; Martin, 1998; Isler *et al.*, 2002); (3) potential bias arising from patterns of phylogenetic relatedness among species in a sample (Felsenstein, 1985; Harvey and Pagel, 1991; Purvis and Rambaut, 1995; Purvis and Webster, 1999); and (4) questionable extrapolation from correlation to causation (Martin, 1998). Proper discussion of the complexities involved is beyond the scope of this review, so a pragmatic approach has been adopted here. As the primary aim is to identify general trends and principles, rather than to draw statistical conclusions from the data, bivariate plots with least-square regression lines are used essentially for description, and potential problems of interpretation are discussed where necessary.

In assessing the evolution of primate vision, the value of comparisons is evident even at a very superficial level. For example, comparison of primates with other mammals immediately reveals that the visual sense is of particular importance in primates. The eyes of primates tend to be relatively larger than in other mammals (Ross *et al.*, in press; Figure 1.3). Moreover, the eyes are rotated forward at least to some extent in all primates, thus enlarging the binocular field in which an object can be seen simultaneously with both eyes (Figure 1.4). The existence of a binocular field is an essential precondition for the emergence of three-dimensional vision, which also requires the development of appropriate processing centers in the brain. In all primates, there is an approximate balance between the ipsilateral and contralateral optic fibers passing to each side of the brain, such that inputs from the two eyes can be directly matched (Martin, 1990). Furthermore, it is a universal feature of primates that the lateral geniculate nucleus (an intermediate relay station in the visual system) shows clear lamination corresponding to these balanced inputs from the two eyes (Chapter 6). In fact, primates show an unusual condition that differentiates them from most or all other mammals with enhanced binocular vision. In other mammals, the entire binocular field is typically represented in the visual cortex on each side of the brain, whereas in primates only one half of the binocular field is represented on each side (Allman, 1982, 1999).

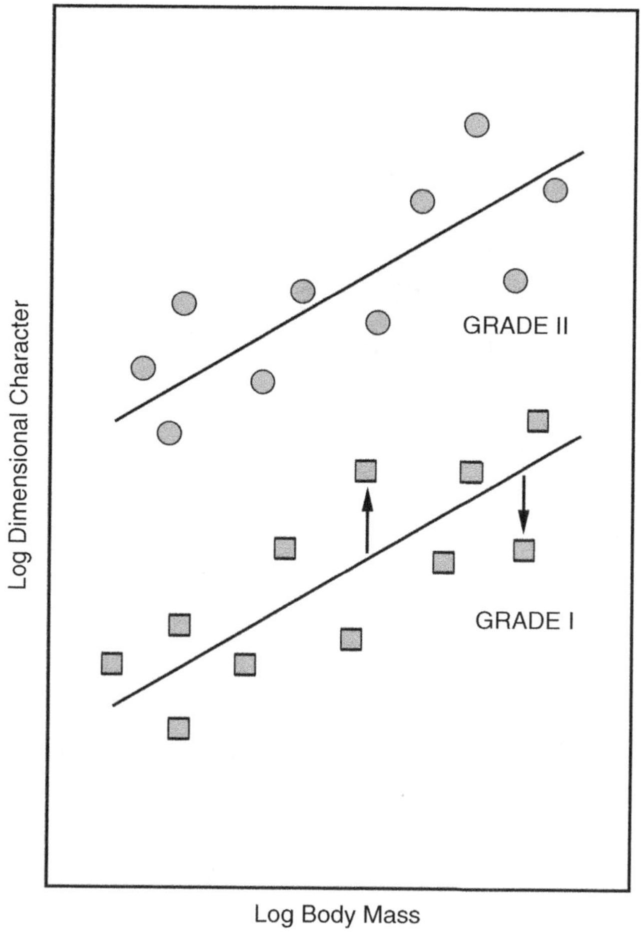

Figure 1.2 Illustration of basic principles of interspecific allometric scaling (after Martin, 1989). The standard allometric formula is $Y = k \cdot X^{\alpha}$, where X is some measure of body size, Y a dimensional character of interest, α the allometric exponent, and k the allometric coefficient. This potentially curvilinear relationship can be transformed into linear form by logarithmic conversion: $\log Y = \alpha \cdot \log X + \log k$. The allometric exponent (α) is indicated by the slope of the line, while the intercept ($\log k$) indicates the value of the allometric coefficient. For a given group of species (squares, representing average values for each species), a best-fit line can be determined as shown. The best-fit line indicates the idealized *scaling principle*, while positive or negative vertical deviations of individual species from the best-fit line (see arrows) – that is, their *residual values* – indicate special adaptations. In many cases, when a second group of species is taken (circles), a best-fit line of similar slope is obtained, but it is displaced vertically relative to the line determined for the first set. The two groups of species are then said to belong to distinct *allometric grades* (grade I; grade II). The vertical distance between the lines indicates the magnitude of the *grade shift* involved

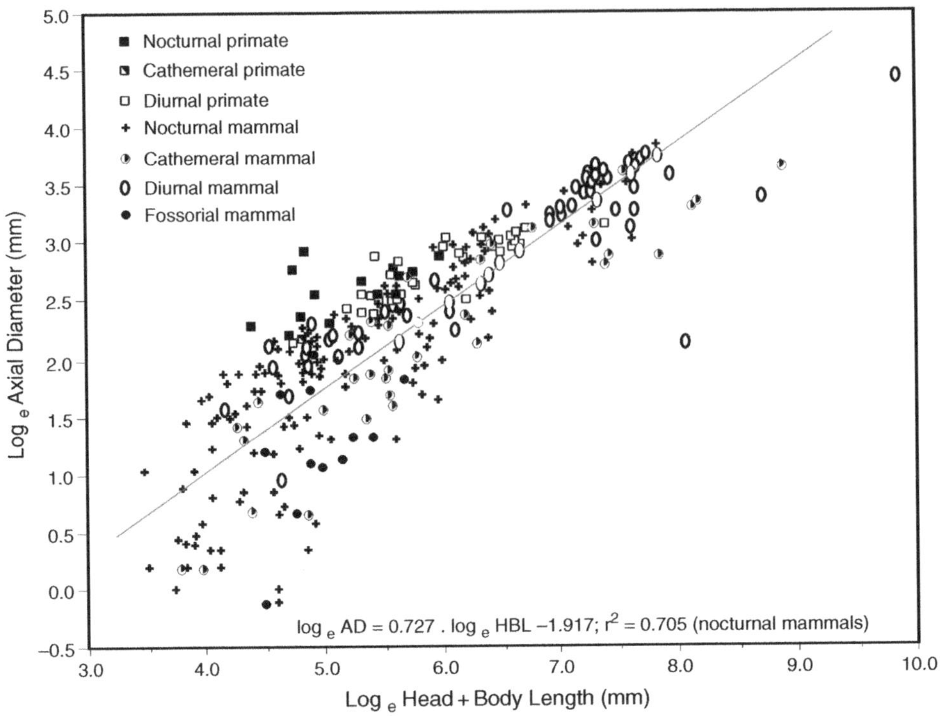

Figure 1.3 Bivariate plot of the axial diameter of the eye against head-and-body length for primates and other mammals, showing that both nocturnal primates and (to a lesser extent) diurnal primates tend to have relatively larger eyes in comparison to other mammals. (Adapted from Ross, 2000, incorporating data from Ritland, 1982.) Among non-primates, nocturnal species show a wide range of relative eye sizes. Some show comparatively large eyes, overlapping with values for primates, while others have very small eyes. Overall, in contrast to primates, among non-primates nocturnal species generally tend to have smaller eyes than diurnal species

The function of increased orbital convergence in euprimates has traditionally been linked to arboreality (Wood Jones, 1916; Elliot Smith, 1924; Le Gros Clark, 1959), but comparisons with other animals by Cartmill suggested that convergent orbits facilitate visual predation on insects on the fine branches of the shrub layer of tropical rainforests (Cartmill, 1970, 1972). 'Stereoptic integration of the two visual fields improves the accuracy of the final strike; increase in visual-field overlap facilitates compensation for evasive movements of the prey' (Cartmill, 1972, p. 113). Subsequent work by Pettigrew (cited by Allman, 1977, p. 29; Pettigrew, 1978) and Allman (1977) demonstrated that the dioptric benefits of orbital convergence are primarily obtained in nocturnal animals. The Allman–Pettigrew model posits that orbital convergence improves image quality by converging the optic axis (the axis of the dioptric apparatus of the eye; i.e. lens and cornea) with the visual axis, or 'physiological line of fixation' (Walls, 1942, p. 292). Another way to enhance retinal image quality is to decrease pupil diameter so as to restrict incoming images to the paraxial region of the dioptric apparatus. However, this option

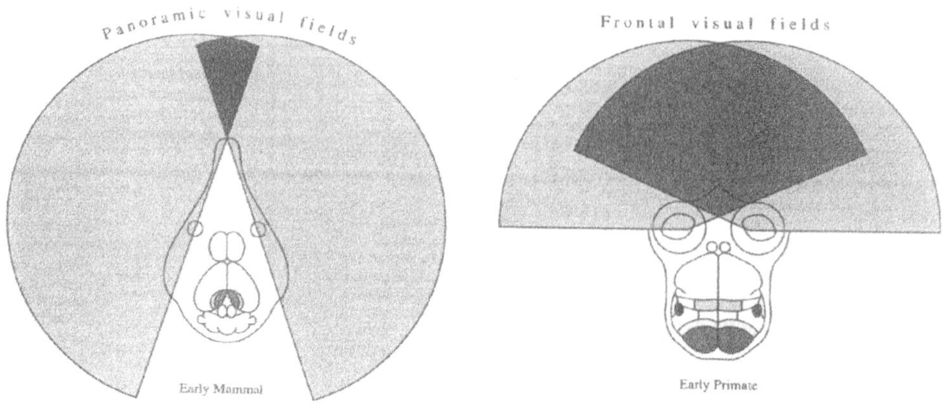

Figure 1.4 The primitive condition for mammals is to have relatively small, laterally oriented eyes that show very little convergence and hence only a small area of binocular overlap. In all euprimates (but not in plesiadapiforms), the eyes are relatively large and rotated forward at least to a moderate degree, such that there is a relatively large binocular field (adapted from Allman, 1999)

is not available to nocturnal animals, which must operate with enlarged pupils in order to maintain image brightness. Consequently, nocturnal animals improve image quality in the area of visual field overlap by increasing optic convergence. Nocturnal animals also benefit from increased visual field overlap because their eyes receive twice as many photons from any point in the binocular visual field as from points in the monocular visual fields, increasing their sensitivity to low-light levels.

Further significant characteristics of the primate visual system are revealed by comparative studies of the skull, notably with respect to the eye socket (orbit). One obvious feature that has long been seen as a defining feature of euprimates is that all of them possess at least a bony strut (postorbital bar) around the outer margin of the orbit (Figure 1.5). A simple bar is characteristic of all modern strepsirrhine primates and is also uniformly present in fossil adapiforms and omomyiforms. In modern haplorhine primates, by contrast, there is a bony partition (postorbital septum) closing the gap between the postorbital bar and the skull wall and effectively isolating the orbit from the temporal jaw musculature. This is one case in which tarsiers share an advanced feature with simian primates. Although it has been suggested that tarsiers and simians may have acquired the postorbital septum independently (Simons and Rasmussen, 1989), the fact that this feature is unique among mammals suggests that there was some basis in the common ancestor of haplorhine primates even if there may be differences in detail in the subsequent evolution of the postorbital partition (Cartmill, 1980; Ross, 1994).

Noble *et al.* (2000) reviewed three main hypotheses that have been proposed to explain the formation of the postorbital bar in primates: (1) resisting masticatory stresses (Greaves, 1985, 1995); (2) protecting the eye from injury (Prince, 1953; Simons, 1962); and (3) augmenting rigidity of the orbital margin to enhance visual acuity (Collins, 1921; Cartmill, 1970, 1972). A direct experimental test of the first hypothesis with the bush baby *Otolemur crassicaudatus* showed that strains arising from mastication are too low to account for development of the postorbital bar (Ravosa *et al.*, 2000). As the second

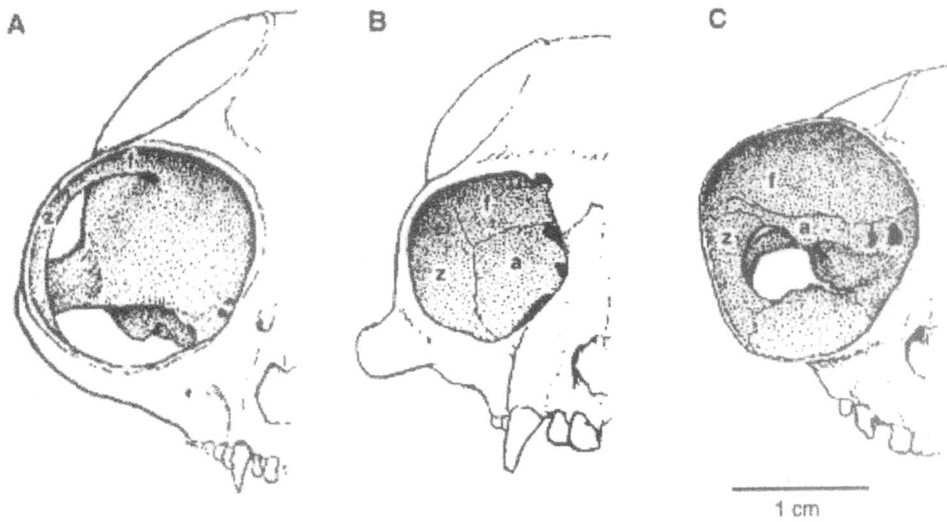

Figure 1.5 Illustration of the postorbital bar and the postorbital septum in primates. All strepsirrhine primates, such as *Nycticebus coucang* (A), possess a postorbital bar formed by contact between processes of the frontal (f) and zygomatic (z) bones. Uniquely among mammals, all haplorhine primates possess a postorbital septum formed from the frontal, alisphenoid (a), and zygomatic bones. This is illustrated for the New World simian *Saguinus* (B) and for *Tarsius* (C). Reproduced with permission from R.F. Kay *et al.*, *Science* **275**, 797–804 (1997). Copyright 1997 AAAS

hypothesis was effectively discounted by Cartmill, only enhancement of visual acuity remains as a promising hypothesis. It should be noted that the postorbital ligament, from which the bar arises, itself forms from the anteriormost free edge of the fascia of the temporal musculature, underscoring the close proximity between the eye and the main jaw muscles. This proximity means that any alterations in skull morphology that move the temporal fossa and the orbit out of the same plane put the eye in danger of disruption from movements in the temporal fossa (Heesy, 2003). These alterations might be due to increased orbital convergence, increased brain size, or some combination of the two (Cartmill, 1972; Noble *et al.*, 2000). Although most mammals do not possess a postorbital bar, this feature is by no means exclusive to primates. Tree-shrews possess a postorbital bar, and this has commonly been cited as a feature linking them to primates. However, postorbital bars are found in several carnivores and fruit-bats (Noble *et al.*, 2000), in horses and several artiodactyls, in some hyraxes, in a sea-cow and in some marsupials (Martin, 1990). Hence, possession of a postorbital bar is clearly not in itself sufficient to link tree-shrews to primates. Moreover, the presence of this feature in horses, artiodactyls, and a sea-cow reveals that it is not even linked to arboreal life as had been supposed by some authors. By contrast, given that possession of a postorbital bar (at least) is universally characteristic of all living and fossil euprimates, it is noteworthy that this feature is uniformly lacking in Plesiadapiformes. This is just one indication of the fact that any link between the Plesiadapiformes and euprimates is tenuous.

Ross (1995a, 2000) assesses possible functional reasons for the development of the postorbital septum in haplorhines, comparing the relative merits of two main hypotheses: (1) Cachel's (1979) proposal that the septum evolved to increase muscle attachment area in the anterior temporal fossa and (2) Cartmill's (1980) proposal that the septum evolved to insulate the eye from masticatory movements of muscles in the temporal fossa. Dissections of 55 primate species revealed that in all anthropoids the temporal muscles have their origins on the portion of the septum formed by the frontal bone and actually follow a curving path, thus indicating that the second hypothesis is more likely. Other alternative hypotheses (e.g. dissipation of masticatory stresses) have been falsified (Ross 1995a; Ross and Hylander, 1996). Thus, it would seem that both the development of the postorbital bar in ancestral euprimates and the subsequent development of a postorbital septum in haplorhines served to isolate the eye from the temporal jaw musculature. Such insulation is of particular importance in animals, like anthropoids, that have high visual acuity which they employ to search for prey or predators while chewing (Cartmill, 1980).

1.2 The phylogenetic background to primate vision

1.2.1 The nocturnal/diurnal divide

A key factor with respect to vision in primates is the distinction between nocturnal and diurnal species (Figure 1.6). Most primates are either clearly nocturnal (i.e. active between dusk and dawn) or clearly diurnal (i.e. active between dawn and dusk). A few lemurs (species of the genera *Eulemur* and *Hapalemur*) exhibit an unusual pattern involving a combination of nocturnality and diurnality that has been labeled 'cathemeral activity' (Tattersall, 1988). Most strepsirrhine primates are nocturnal. Lorisiforms are uniformly nocturnal, while among the Madagascar lemurs fully diurnal behavior is limited to certain members of two of the six families: Lemuridae (*Lemur* and *Varecia*) and Indridae (*Indri* and *Propithecus*). By contrast, the vast majority of haplorhine primates are diurnal. The only exceptions are species belonging to the two genera *Tarsius* and *Aotus*, which are typically nocturnal. It would seem that some populations of owl monkeys (*Aotus*) may also exhibit cathemeral rather than exclusively nocturnal activity. The basic division between nocturnal and diurnal primates is, of course, directly relevant to fundamental visual processes, as the former are adapted for activity in dim light intensities, whereas the latter are adapted for activity in relatively bright light (i.e. for scotopic and photopic conditions, respectively, using the terminology proposed by Walls, 1942; see also Chapters 5 and 6). Cathemeral primates are faced with the problem that they must cope with both scotopic and photopic conditions.

In connection with the nocturnal/diurnal divide, there is a pervasive distinction between strepsirrhines and haplorhines in that the former typically possess a reflecting tapetum lucidum, generally interpreted as an adaptation for vision under dim light conditions, whereas the latter do not. It is noteworthy that a well-developed tapetum is present not only in all nocturnal lorisiforms and lemurs but also in all diurnal lemurs (*Indri*, *Lemur*, *Propithecus*, and *Varecia*), whereas it is not well developed in lemurs with cathemeral

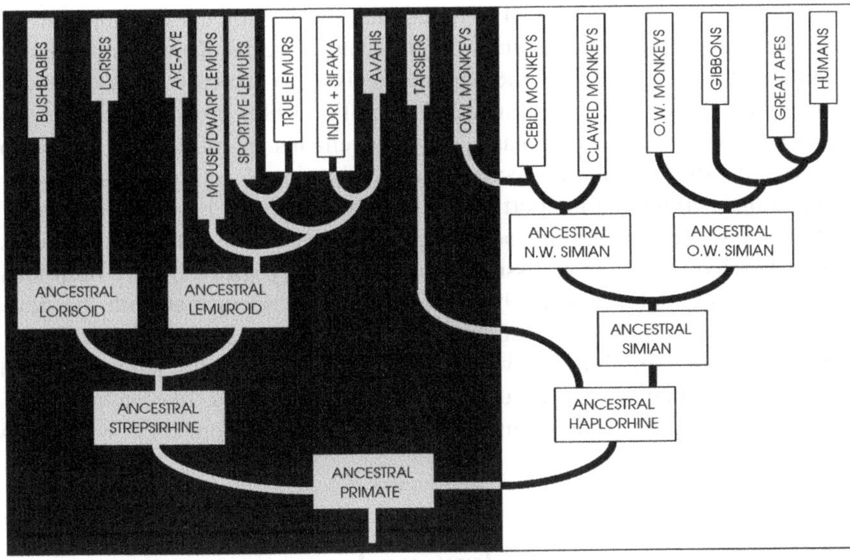

Figure 1.6 Outline phylogenetic tree of primates showing the distribution of nocturnal and diurnal habits among living representatives and inferred ancestral conditions. Note that some true lemurs (*Eulemur* and *Hapalemur*) show an unusual cathemeral pattern, involving both diurnal and nocturnal activity. Fossil forms have not been considered in this reconstruction, which is based exclusively on the characteristics of living primates. (Reproduced from the chapter by R.D. Martin in *Creatures of the Dark: The Nocturnal Prosimians* (1995), with permission from Kluwer Academic Publishers/Plenum Publishers.)

habits (*Eulemur* and *Hapalemur*). A notable feature of the tapetum in strepsirrhine primates is that, as far as is known, the active principle is a layer of plate-like riboflavin crystals (Pirie, 1959; Martin, 1990), which generates a conspicuous golden yellow eyeshine. Riboflavin, which belongs to the general class of flavins, is best known as the water-soluble vitamin B2, which serves a variety of biological roles (reviewed by Massey, 2000). Among other things, it plays a major part in aerobic metabolism, in photosynthesis, in light-dependent repair of DNA damage, in fetal development, and in regulation of biological clocks. However, the role of riboflavin as the active principle of a tapetum lucidum seems to be unique to strepsirrhine primates.

In fact, there are major differences between mammalian groups in the structure of the tapetum and in the active principle responsible for reflective properties (Schwab *et al.*, 2002). As a general rule, a tapetum serves to increase the amount of light absorbed by the photoreceptors, reflecting photons that were not initially absorbed on their first pass. It is always located behind the photoreceptors, but may occur either in the choroid (choroidal tapetum) or, more rarely, in the deep retina (retinal tapetum). Furthermore, two different kinds of choroidal tapeta can be recognized: (1) the tapetum fibrosum, which is the simplest type with stacks of densely packed collagen fibrils, and (2) the tapetum cellulosum, with reflecting cells stacked in a tile-like array. Strepsirrhine primates possess a tapetum cellulosum and therefore differ from several other mammal

groups that have either a tapetum fibrosum (e.g. elephants, horses, artiodactyls, cetaceans, some marsupials, and at least one rodent) or a retinal tapetum (e.g. some marsupials and megachiropteran bats). Certain mammals, such as carnivores and pinnipeds, superficially resemble strepsirrhine primates in possessing a tapetum cellulosum, but the active principle is different. The eyeshine of carnivores, for example, has a distinctive greenish hue, while it has been reported that the pen-tailed tree-shrew (*Ptilocercus*) – the only nocturnal scandentian – has a silvery eyeshine (Emmons, 2000). This wide variation in constitution indicates that tapeta have evolved several times independently in different mammalian lineages. This is, for instance, the most likely explanation for the fact that marsupials can possess either a tapetum fibrosum or a retinal tapetum. The development of a tapetum cellulosum with riboflavin as the active principle therefore seems to be a shared derived feature characterizing the common ancestor of strepsirrhine primates. This, in turn, strongly indicates that the common ancestor of strepsirrhines was nocturnal in habits. It is possible that a tapetum of some kind was already present in ancestral primates and was then lost in the common ancestry of haplorhines, as was suggested by Ross (2000); but it is also possible that ancestral primates did not possess a tapetum, or at least did not possess a tapetum with riboflavin as the active principle, and that this was first developed in the lineage leading to strepsirrhine primates (Martin, 1990).

The distinction between nocturnal, cathemeral, and diurnal habits among primates shows an obvious association with body mass (Figure 1.7). Nocturnal primates tend to be quite small, with a modal body mass of only 315 g, whereas diurnal primates are generally much larger, with a modal body mass that is more than 17 times greater (5.45 kg). The modal body mass of cathemeral primates (1.66 kg) is intermediate between these two main categories. The generalization that nocturnal species tend to be smaller than diurnal species applies across mammals, and Charles-Dominique (1975) proposed that this could be an outcome of competition with birds. Using data from tropical forest ecosystems in Africa and South America, he argued that mammals are typically nocturnal within the body size range covered by flying birds (up to about 5 kg). Diurnality is common only among mammals that exceed a body mass of 5 kg. In reasonable agreement with this, no nocturnal primate has a body mass greater than 2.5 kg, while 91 of 171 diurnal primate species have a body mass exceeding 5 kg. This is of crucial importance with respect to the overall phylogenetic tree of mammals (Figure 1.8). The precursors of mammals, the mammal-like reptiles (synapsids), were generally of moderate to large body size. By contrast, the first mammals, which appeared near the Jurassic/Triassic boundary some 200 mya, were relatively small, comparable in size to modern shrews or mice. This remained true for the next 135 my up until the end of the Cretaceous, about 65 mya (i.e. for the first two-thirds of mammalian evolutionary history). Partly for this reason, it has been generally accepted that early mammals were nocturnal and that diurnal habits did not emerge until the Tertiary, when a general trend toward increasing body size is seen in many lineages (Alroy, 1998). Accordingly, the late Jurassic/early Cretaceous common ancestors of marsupials and placentals were presumably both small and nocturnal. Against this background, it is most parsimonious to assume that, following their divergence from ancestral placental mammals, the ancestral primates had remained nocturnal (see Figure 1.6).

Whereas ancestral primates were seemingly nocturnal, there are several indications that the common ancestor of haplorhines was diurnal (Martin, 1990; Heesy and Ross,

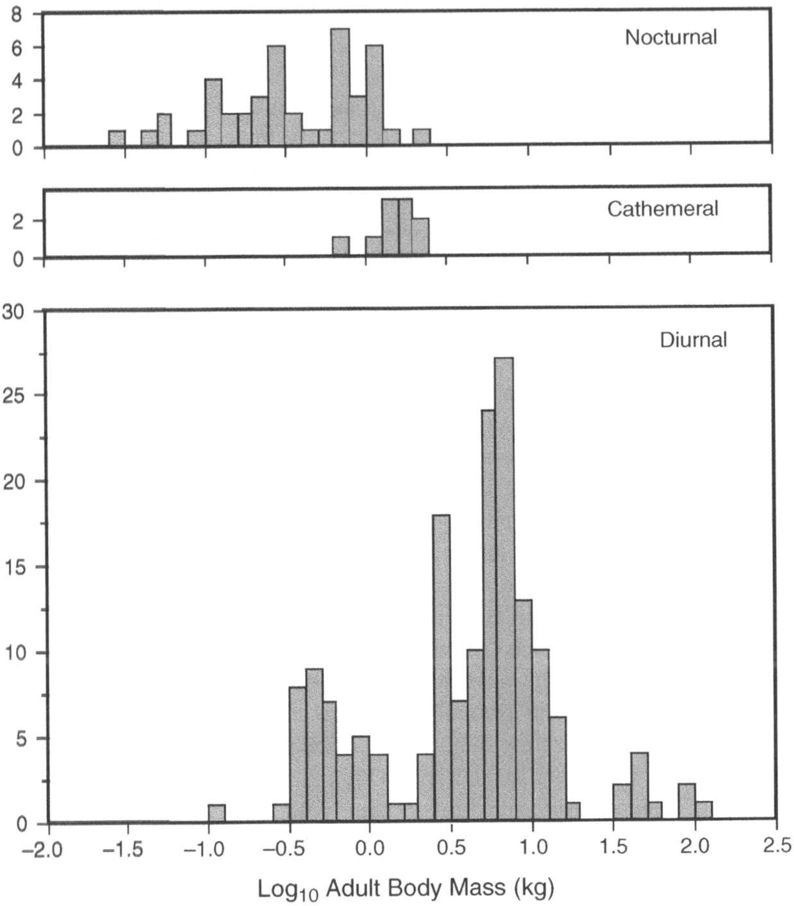

Figure 1.7 Histograms showing body mass distributions for nocturnal, cathemeral, and diurnal primates (data from Smith and Jungers, 1997). Nocturnal primates (mean value: 545 g; modal value: 315 g) are generally smaller than diurnal primates (mean value: 8.07 kg; modal value: 5.45 kg), with cathemeral primates being intermediate (mean value: 1.61 kg; modal value: 1.66 kg). However, there is considerable overlap, particularly because marmosets and tamarins are all diurnal despite their small body sizes

2001). In the only nocturnal haplorhines, *Tarsius* and *Aotus*, the retina shows a number of features indicative of a diurnal ancestry. In the first place, both *Tarsius* and *Aotus* lack a tapetum lucidum and have developed particularly large eyes that may compensate for this. More strikingly, a fovea (a pit with a higher density of receptors) is present in the central area of the retina of *Tarsius* (Polyak, 1957; Wolin and Massopust, 1970; Castenholz, 1984; Ross, 2000, 2004). Curiously, the presence of a fovea is variable in the nocturnal simian *Aotus* (Provis *et al.*, 1998), but there is definitely a foveal pit in some individuals. Haplorhine primates are the only mammals that consistently possess a

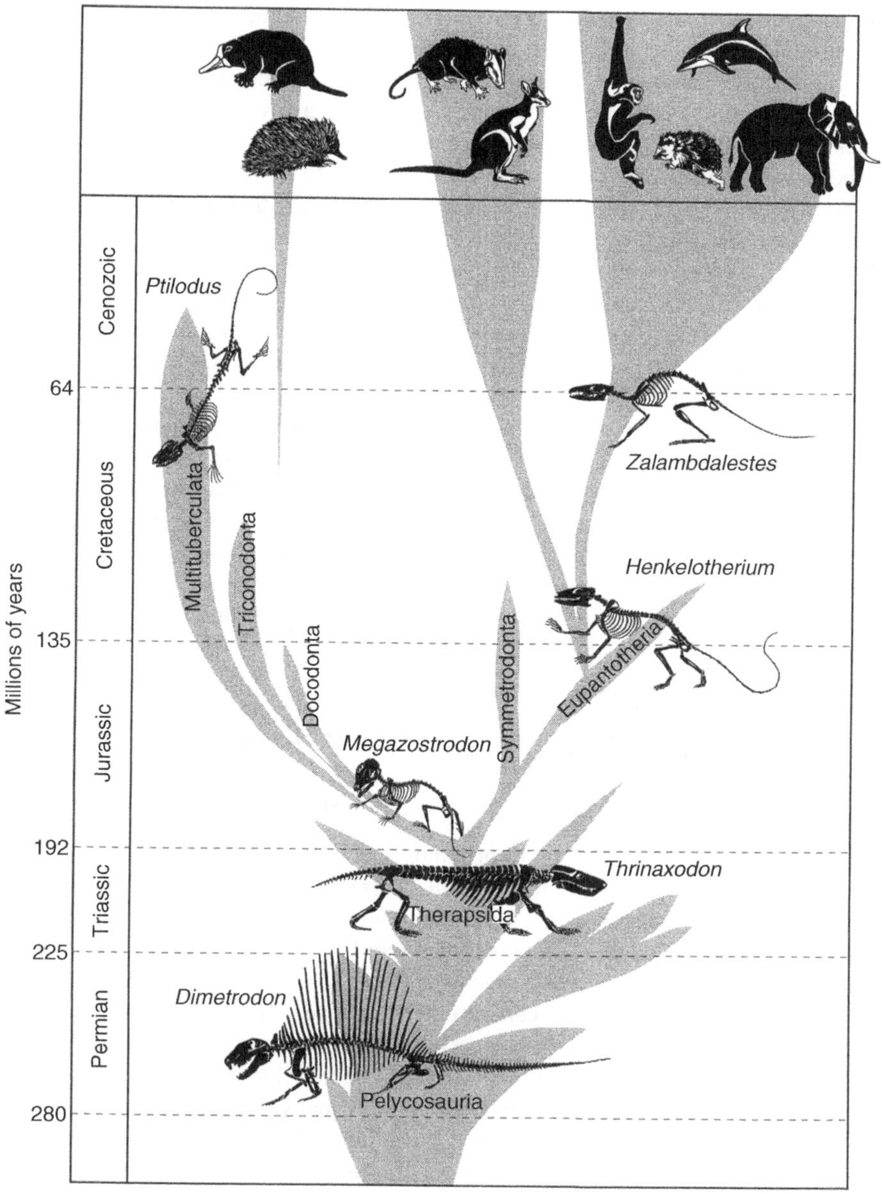

Figure 1.8 Outline phylogenetic tree for mammals. Modern mammals are divided into three groups: monotremes, marsupials, and placentals. Fossil mammals first appear in the record at the Triassic/Jurassic boundary, about 195 mya. They are derived from mammal-like reptiles (synapsids), which diverged from the diapsid reptiles leading to modern reptiles and birds at least 310 mya. For approximately two-thirds of their evolutionary history, between the Triassic/Jurassic boundary and the Cretaceous/Tertiary boundary 65 mya, the mammals remained very small and are widely presumed to have been nocturnal in habits. Original illustration by Lukrezia Bieler-Beerli

histologically identifiable retinal fovea, so this is another unique feature that characterizes this group. The fovea is interpreted as an adaptation for high visual acuity and it is typically associated with a high density of cones and hence with diurnal habits and some degree of color vision (see Chapter 5). Although a fovea is unique to haplorhine primates among mammals, it is a common feature in diurnal birds and lizards with well-developed visual acuity and color vision (Ross, 2004). In diurnal haplorhines, the fovea is consistently associated with a yellow spot (macula lutea) that has been interpreted as promoting high visual acuity (Kirk and Kay, 2004). Interestingly, in *Tarsius* (although not in *Aotus*) the fovea is surrounded by a faint yellow macular pigment (Hendrickson *et al.*, 2000). Overall, it is reasonable to conclude that both *Tarsius* and *Aotus* are derived from a diurnal ancestral haplorhine possessing both color vision and a fovea, and that there has been some subsequent reduction in the fovea that has for some reason progressed further in *Aotus* than in *Tarsius*.

Diurnal anthropoids also have a distinctive eye shape in comparison with other mammals, having a large axial diameter relative to cornea diameter (Ross, 2000). A larger axial diameter increases the focal length of the eye, thereby increasing image size and – all other things being equal – visual acuity as well. Just as other diurnal mammals do not share a retinal fovea and postorbital septum with anthropoids, they also lack the enlarged axial diameters relative to cornea size seen in anthropoids, along with their high visual acuity (Ross, 2000). Anthropoid primates are unusual among mammals in their adaptations for high visual acuity.

1.3 Comparative analyses of cranial dimensions

1.3.1 Allometric analysis of the eye and orbit

A basic guide to visual adaptations in primates and other mammals can be obtained by simply examining dimensions of the eyeball or of the bony socket (orbit) in which it resides. As has already been noted, a bivariate plot of eyeball diameter against head-and-body length for a representative sample of mammals (Figure 1.3) confirms the expectation that primates tend to have relatively large eyes (Ross, 2000). However, there is considerable overlap between primates and certain other mammals, and it is noteworthy that only some nocturnal primates (especially tarsiers) have exceptionally large eyes in comparison to all other mammals. Nevertheless, it should also be noted that most non-primate mammals are nocturnal, so diurnal primates overlap with many mammals that are nocturnal, while nocturnal primates commonly have larger eyes than other nocturnal mammals. Given that a larger retina can potentially accommodate more photoreceptors, it would seem that in nocturnal primates (but not in most other nocturnal mammals) the number of photoreceptors has generally increased markedly as an adaptation to high-acuity vision under scotopic conditions (Ross *et al.*, in press; see also Chapter 2 for differences between diurnal and nocturnal primates in the ontogenetic development of the retinae).

Data on the size of the eyeball are relatively scarce (Schultz, 1940; Rohen, 1962; Ritland, 1982) and are, of course, completely lacking for fossil species. It is therefore useful to have a substitute dimension on the skull that can be used as an indicator of eye size, and the size of the orbit has been widely used for this purpose (Kay and Cartmill, 1977; Martin, 1990; Kay and Kirk, 2000; Ross, 2000; Heesy and Ross, 2001).

It should at once be noted that there is a pervasive problem involved in taking the size of the orbit as an indicator of eye size. Although the size of the eye generally increases progressively with body size, the former does not keep pace with the latter (i.e. there is a negatively allometric relationship). This generalization, labeled 'Haller's Law', applies not only to primates and other mammals (Ritland, 1982; Martin, 1990; Kay and Kirk, 2000) but also to other terrestrial vertebrates such as birds (Brooke *et al.*, 1999). As a result, the ratio of eye size to skull size progressively declines as body size increases. One outcome of this is that the eye does not completely fill the orbit in large-bodied mammals, even in the relatively large-eyed primates (Martin, 1990; Kay and Kirk, 2000). Particular caution must therefore be exercised in inferring the size of the eye from the size of the orbit in large-bodied mammals. There is an additional problem in that body size must be indirectly inferred for fossil species, so a substitute measure must also be taken. A common approach is to take the maximum length of the skull (prosthion–inion length) as an indicator of overall body size. This has the advantage that a relatively complete skull will yield data indirectly indicating both eye size and body size. On the other hand, it is important to note that skull length itself can vary relative to body size and may hence be misleading as an indicator of overall body size. This applies particularly to the long-snouted baboons, which have undergone a secondary increase in muzzle length.

Despite the inherent limitations, a bivariate plot of orbit height against skull length for living and fossil primates (Figure 1.9) yields a number of valuable conclusions. In the first place, among living primates there is a fairly clear grade distinction between nocturnal and diurnal species. Tarsiers have the largest orbits relative to skull size, while nocturnal strepsirrhines (most lemurs and all lorises) and owl monkeys come next, generally having relatively larger orbits than diurnal primates (some lemurs and all simian primates except owl monkeys). However, although there is an overall grade separation between nocturnal strepsirrhines and diurnal primates, there is a considerable amount of overlap. As might be expected, cathemeral lemur species fall in this zone of overlap between nocturnal and diurnal species. For comparative purposes, especially for interpretation of data from fossil species, it is useful to examine positive and negative deviations (residual values) relative to some overall best-fit line. Unfortunately, because of the considerable scatter of data in Figure 1.9, the choice of an appropriate best-fit line to use as a standard for comparison is not immediately obvious. Following the pragmatic approach taken by Martin (1990), the least-squares regression line fitted to extant diurnal primates (excluding the long-snouted baboons) in Figure 1.9 can be taken as one option for calculation of residual values. The distributions of residual values (Figure 1.10) confirm the main conclusions from visual inspection of the bivariate plot. Nocturnal primates generally have relatively larger orbits than diurnal primates, with the nocturnal haplorhines (especially tarsiers) having conspicuously enlarged orbits. The distribution of residual values for relative orbit size in cathemeral primates is intermediate between those for nocturnal and diurnal primates. However, the histograms in Figure 1.10 also reveal that, despite the wide range of values for relative orbit size shown by each category, there is only partial overlap. The maximum logarithmic residual value for a diurnal primate is 0.123, found in the Old World leaf monkey *Trachypithecus cristatus*. This value in fact exceeds or equals those found in four nocturnal strepsirrhines (*Arctocebus calabarensis*, 0.080; *Perodicticus potto*, 0.090; *Cheirogaleus medius*, 0.095; *Microcebus murinus*, 0.123); but all other nocturnal primates show values exceeding that for *Trachypithecus*.

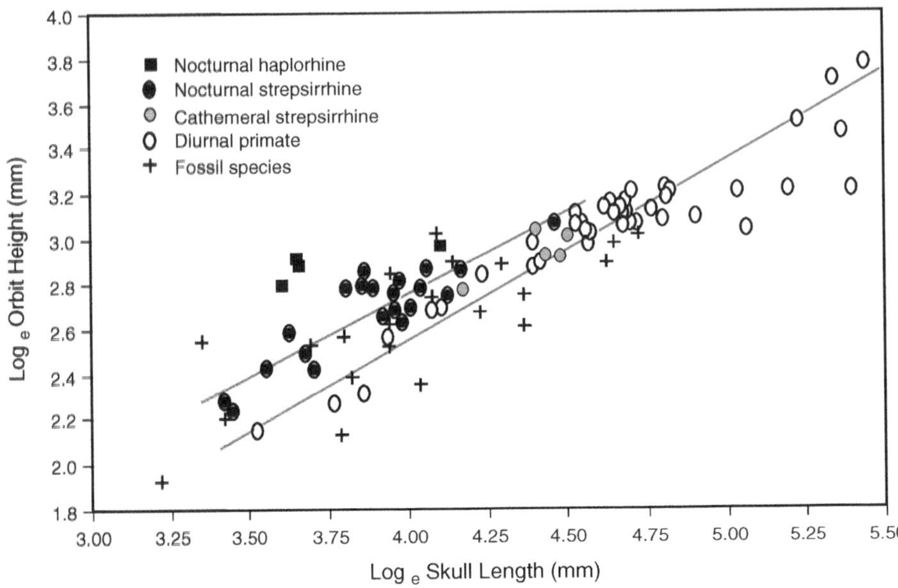

Figure 1.9 Scaling of orbit height (H) against maximal skull length (S) for 71 extant and 21 fossil primate species (data from Kay and Kirk, 2000, supplemented with data for 11 additional extant primate species measured by the first author). The two least-squares regression lines, for orientation only, are for nocturnal strepsirrhine primates (upper line; $n = 21$; $\log_e H = 0.736 \cdot \log_e S - 0.182$; $r^2 = 0.788$) and for diurnal primates, excluding the long-snouted baboons (lower line; $n = 37$; $\log_e H = 0.793 \cdot \log_e S - 0.616$; $r^2 = 0.948$). The lines are approximately parallel, indicating a grade distinction between these two categories of primates: extant nocturnal primates generally tend to have relatively larger orbits than extant diurnal primates. However, there is considerable overlap between members of the two grades. The scaling exponent is less than 1 for both lines, reflecting the fact that the size of the orbit does not keep pace with increasing skull size ('Haller's Law'). Fossil primates show considerable scatter for the relationship between orbit height and skull length. Note that the point for *Teilhardina asiatica* (minimal values for both skull length and orbit height) lies below the range of values for modern primates

In fossil primates, residual values indicating relative orbit height cover a wide range, extensively overlapping those for both nocturnal and diurnal species among extant primates. Relatively large orbits, indicated by residual values greater than for *Trachypithecus*, are found in some Eocene omomyiforms (*Microchoerus*, *Necrolemur*, *Shoshonius*), a few Eocene adapiforms (*Mahgarita*, *Pronycticebus*), an enigmatic Oligocene strepsirrhine (*Plesiopithecus*), and the Miocene lorisiform *Mioeoticus*. The Eocene omomyiform *Tetonius* has a value of 0.123, matching that of *Trachypithecus*, while all other fossil primates examined have smaller residual values lying within the range of extant diurnal primates. This applies to the Eocene omomyiform *Teilhardina*, several Eocene adapiforms (*Smilodectes*, *Leptadapis*, *Notharctus*, *Adapis*), the Oligocene prosimian *Rooneyia*, late Eocene/early Oligocene stem simians (*Catopithecus*, *Aegyptopithecus*, *Apidium* and *Proteopithecus*), and Miocene New World monkeys (*Tremacebus*, *Dolichocebus*). In fact,

Apidium, *Adapis*, and *Proteopithecus* all have strikingly small orbits. Among living primates, only the long-snouted baboons (*Mandrillus*, *Papio*, and *Theropithecus*) and the patas monkey (*Erythrocebus*) have comparably small orbits relative to skull length.

Because early fossil euprimates include species with both large and small orbits relative to skull size, it is not immediately apparent from this source of evidence whether ancestral euprimates would have been nocturnal or diurnal. However, a phylogenetic analysis conducted by Heesy and Ross (2001) yielded as the most parsimonious solution the conclusion that ancestral euprimates were probably both nocturnal and dichromatic. This inference was subsequently challenged by Ni *et al.* (2004) on the basis of the small relative size of the orbits in a newly discovered skull of the early omomyiform *Teilhardina asiatica*. It is, indeed, true that for *Teilhardina* the residual value for orbit size in relation to skull length falls among values for extant diurnal primates. Probably because *Teilhardina asiatica* seems to occupy a very basal position in the phylogenetic tree for primates, when Ni *et al.* (2004) repeated the phylogenetic analysis conducted by Heesy and Ross (2001), they came to the opposing conclusion that the ancestral primates were diurnal. Thus, this new analysis seemingly challenges the long-standing interpretation that ancestral primates had retained nocturnal habits from their mammalian heritage. Before abandoning this well-established view, however, it is necessary to consider a number of special factors in the interpretation of relative orbit size of *Teilhardina asiatica* (Martin, 2004). In the first place, it should be noted that the inferred body mass of this early fossil primate is less than 30 g, which is smaller than that of any extant primate species. Hence, calculation of the residual value for the size of the orbit relative to skull length requires extrapolation below the range of variation found among modern primates (Figure 1.9). On statistical grounds alone, such extrapolation is questionable. In addition, as is evident from Figure 1.3, the relationship between eye and body size is clearly curvilinear, being positively allometric at small body size and becoming negatively allometric at large body size. Ni and coworker's extension of the relationship seen in extant primates to the smaller fossil does not take this curvilinearity into account. More importantly, caution is necessary in another kind of extrapolation, namely in interpreting the relative size of the orbit in fossil primates directly on the basis of relationships determined for modern primates. Given the very wide range of variation in relative size of the orbit found among fossil primates, it seems likely that both nocturnal and diurnal habits were represented. In particular, it would seem likely that species with very large orbits (e.g. the Eocene omomyiforms *Microchoerus*, *Necrolemur*, and *Shoshonius*) were nocturnal, while those with very small orbits (e.g. the adapiform *Adapis* and the early simians *Apidium* and *Proteopithecus*) were diurnal. But for species with orbits of intermediate size (including *Teilhardina*, which lies close to the borderline between modern diurnal and nocturnal species; see Figure 1.10), it would be unwise to draw any firm conclusions in the absence of additional evidence.

It should also be noted that there is a marked phylogenetic influence on the relationship between orbit size and body size in primates. In a plot of orbit diameter against body mass for prosimians (Figure 1.11), it can be seen that lorisiforms tend to have relatively larger orbits than lemuriforms, while the largest orbits of all are found in tarsiers. Furthermore, all lemur species conform closely to a single best-fit line despite the fact that they may exhibit nocturnal, cathemeral, or diurnal habits. Hence, for lemurs the scaling of orbital diameter to body size reveals no clear separation according to activity pattern. It is

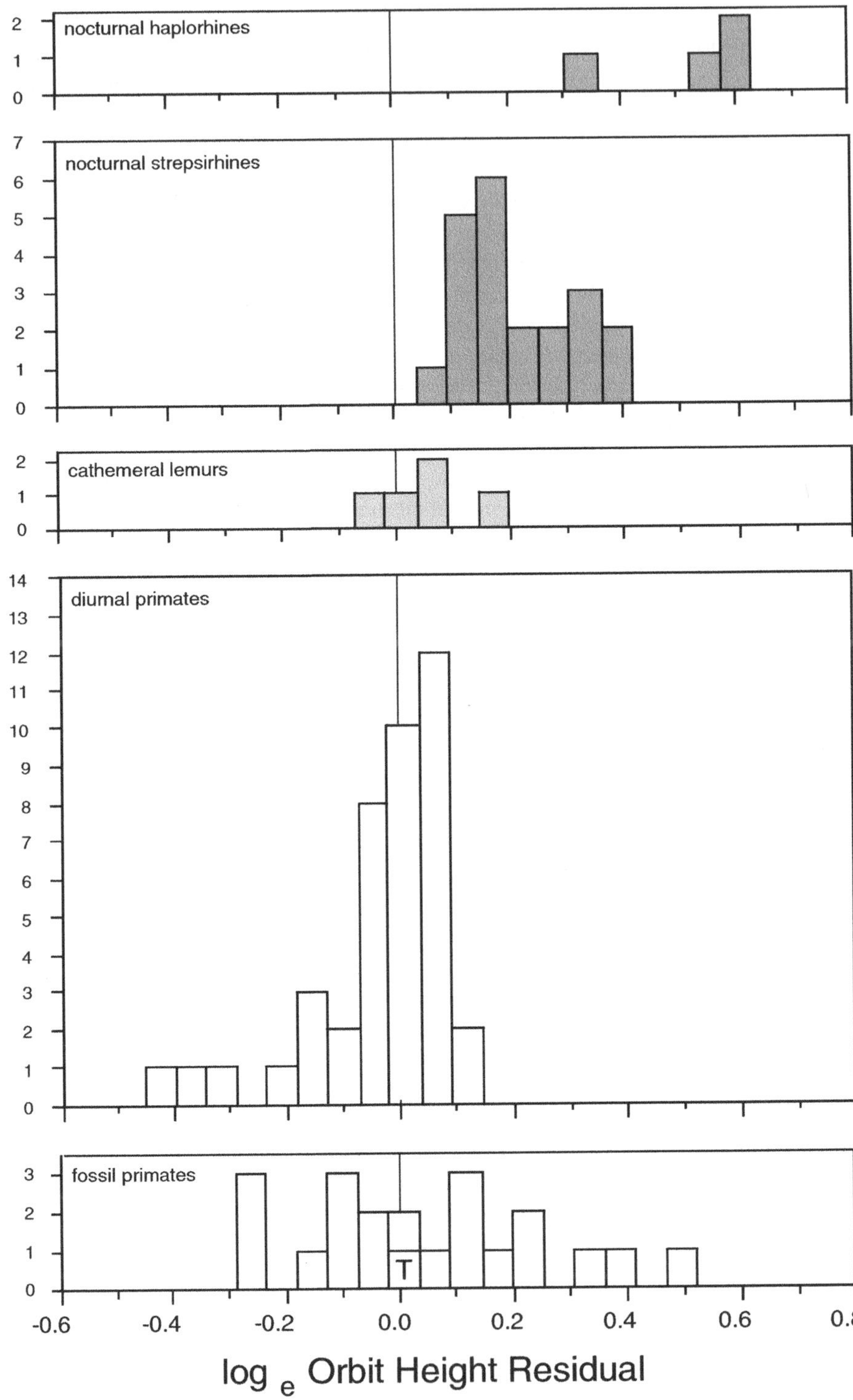
nocturnal haplorhines
nocturnal strepsirhines
cathemeral lemurs
diurnal primates
fossil primates
T
-0.6
-0.4
-0.2
0.0
0.2
0.4
0.6
0.8
log e Orbit Height Residual

also noteworthy that the two African lorisid genera (*Arctocebus* and *Perodicticus*) both possess relatively small orbits in comparison with other lorisiforms. Given this apparent phylogenetic influence, the interpretation of relative orbit size in isolation should be treated with appropriate caution. A reconstruction of nocturnal habits in basal primates is also more parsimonious when the appropriate outgroups are added to the primate phylogeny of Ni *et al.* (2004). Most importantly, the pen-tailed tree-shrew (*Ptilocercus*) and the flying lemurs are all nocturnal, firmly fixing nocturnality at the basal primate node, regardless of the activity pattern inferred for *Teilhardina asiatica*.

Measurements conducted on the skull can also yield valuable information on the orientation of the orbit and, by inference, of the eyes themselves. Orientation of the orbit includes both convergence and frontation (Figure 1.12). Orbital convergence is the degree of forward rotation of the orbit, measured as the dihedral angle between the mid-sagittal plane and the plane of the orbital margin (Cartmill, 1972). Forward rotation of the orbit directly influences the angular magnitude of the binocular field and hence the extent of three-dimensional vision. The primitive condition in mammals is for the orbits to be oriented almost entirely laterally (degree of convergence not much above 0°), such that there is only a relatively small binocular field (Figure 1.4; Ross, 2000; Heesy, 2004). Accordingly, progressive increase in the level of orbital convergence to approach the maximum value of 90° (orbits facing directly forwards) is an advanced feature corresponding to an increase in the magnitude of the binocular field. Frontation is the degree of verticality of the orbital margin, measured as the plane angle between the nasion–inion line and the chord along which the orbital and mid-sagittal planes intersect (Cartmill, 1972). Primitively, the plane of the orbital margin is oriented obliquely upwards in primates and the degree of frontation can be as low as 40° in some extant prosimian primates. As an advanced feature among primates, the orbital plane becomes more vertical in orientation and can in fact overshoot the vertical (i.e. 90°) to reach a value of up to 110°.

Convergence and frontation of the orbits in primates were originally studied in detail by Cartmill (1970, 1972), with an emphasis on prosimians. In a later study using the same methods, Ross (1995b) examined a larger sample of primate species and focussed more on the particularly advanced condition found in simian primates. Simians differ quite markedly from prosimians in the range of values for both convergence and frontation, which are generally higher in simians and show relatively little overlap with the values for prosimians. Thus, in simians the orbits are generally not only more forward-oriented

---

Figure 1.10 Histograms showing residual values for orbit height for modern and fossil primates, derived from the bivariate plot in Figure 1.9. The least-squares regression line for diurnal primates (excluding the long-snouted baboons) was taken as the baseline for calculations. Modern nocturnal species (dark bars) generally have relatively larger orbits than modern diurnal species (white bars), while cathemeral species (hatched bars) are intermediate. However, there is some overlap between the three categories. Fossil primates cover almost the entire range of relative orbit sizes found among modern primates and presumably included nocturnal, diurnal, and perhaps cathemeral species. The value for the early fossil omomyiform *Teilhardina* (T) falls very close to the boundary between modern nocturnal and diurnal species

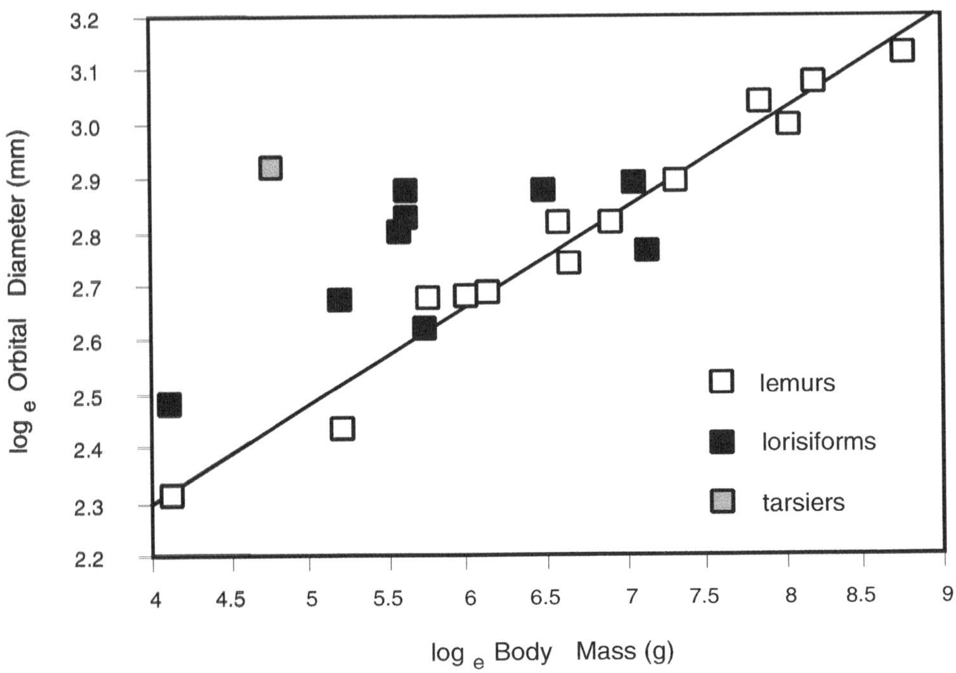

Figure 1.11 Scaling of orbit diameter (D) against body mass (M) for prosimian primates. The least-squares regression line, for orientation only, is for lemurs alone ($n = 13$; $\log_e H = 0.182 \cdot \log_e M + 1.569$; $r^2 = 0.973$). Although different lemur species exhibit nocturnal, cathemeral, or diurnal habits, they all conform closely to a single best-fit line. Lorisiforms, all of which are nocturnal, tend to have larger orbits than lemurs, although the African lorisids *Arctocebus* and *Perodicticus* are exceptional in possessing relatively small orbits

but also more vertical. The results are best summarized by a combined plot of frontation against convergence, which reveals a clear separation between modern prosimians and simians (Figure 1.13). Among simian primates, increases in convergence are accompanied by increases in frontation, whereas in prosimians increases in convergence are accompanied by decreases in frontation. Simians converge their orbits to the fronts of the heads, prosimians to the top. It is noteworthy that the tarsiers in this case lie firmly within the distribution for strepsirrhine primates and do not show the more advanced condition that typifies the skull of simians. Most fossil primates in the sample examined by Ross (1995b) also fall within the strepsirrhine range, although there are a few exceptions. In addition to a number of Eocene fossil prosimians (*Adapis*, *Leptadapis*, and *Microchoerus*), two large-bodied subfossil lemurs (*Megaladapis* and *Palaeopropithecus*) fall within the distribution for modern prosimian primates. The Oligocene 'omomyiform' *Rooneyia* falls on the boundary between modern prosimians and simians, while two medium-sized subfossil lemurs (*Archaeolemur* and *Mesopropithecus*) and the early Oligocene simian *Aegyptopithecus* clearly fall among the simians. In this context, it should be noted that one implication of 'Haller's Law' is that a high degree of convergence may be easier

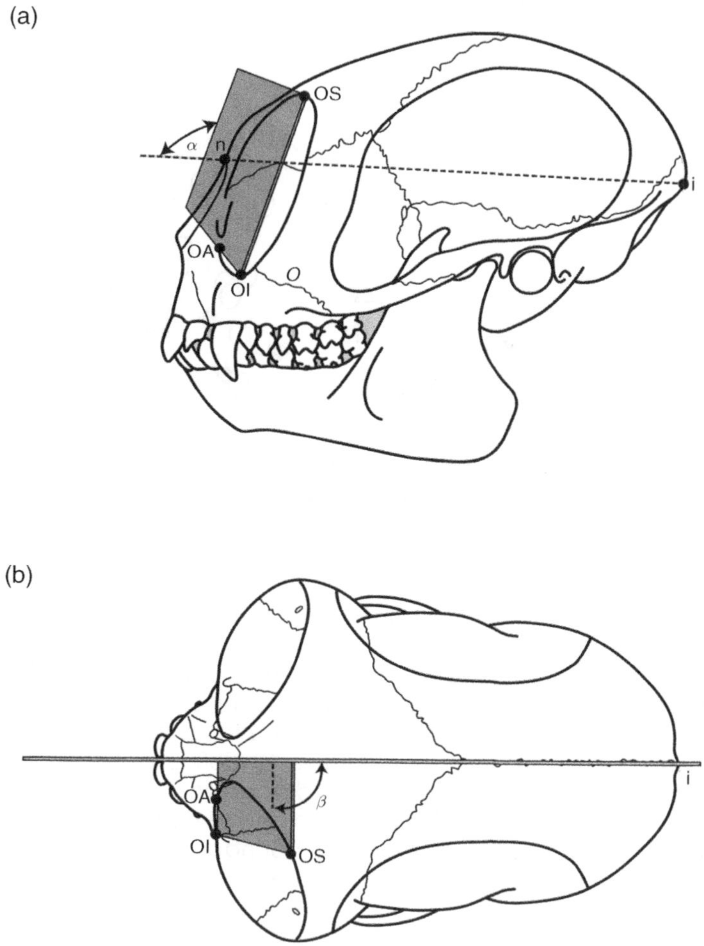

Figure 1.12 Illustrations of degrees of convergence and frontation in the primate skull, taking the example of *Aotus*. (A) Frontation (lateral view): the sagittal plane is shown with light shading, while the plane of the orbital margin is heavily shaded. The sagittal plane passes through nasion (n) and inion (i). The plane of the orbital margin passes through orbitale inferius (OI), orbitale superius (OS), and orbitale anterius (OA). The angle of frontation is equal to $180° - \alpha$. (B) Convergence (superior view): the sagittal plane is indicated by the shaded line. The angle of convergence is equal to β. Reproduced from Ross (1995b), with kind permission from Elsevier, Inc.

to acquire in larger-bodied species because the eyes are smaller in relation to skull size. Hence, it is conceivable that the generally greater degrees of convergence found in simian primates are attributable, at least in part, to the fact that the average body size of simians is considerably greater than that of prosimians. However, this does not fit well with the observation that large-bodied subfossil lemurs fall into the prosimian range with respect to orbital convergence whereas medium-sized subfossil lemurs overlap

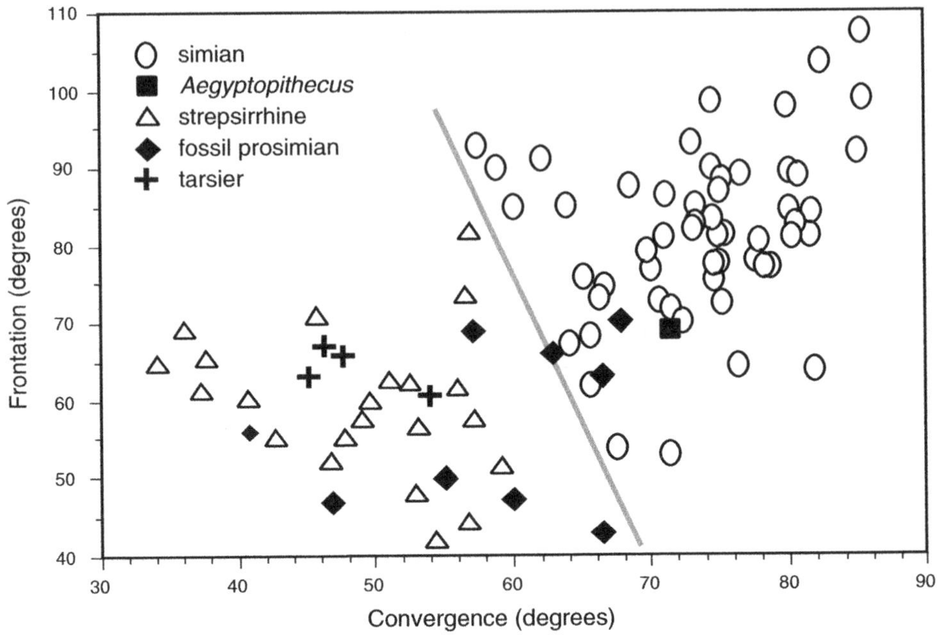

Figure 1.13 Plot of frontation against convergence for modern and fossil primates (data from Ross, 1995b). Modern prosimians (strepsirrhines and tarsiers) show moderate degrees of convergence and frontation, while modern simians generally show higher values for both measures. It is possible to define a boundary that separates all prosimians from all simians (oblique dashed line). Whereas most fossil primates in the sample fall into the distribution for modern prosimians, the enigmatic Oligocene prosimian *Rooneyia* falls onto the boundary, while two subfossil lemurs (*Archaeolemur* and *Mesopropithecus*) fall at the lower end of the simian range

with simians. Cartmill (1970, 1972) proposed that convergence is primarily influenced by relative orbit diameter and relative interorbital breadth. However, comparisons with other mammals (carnivores and fruit-bats) throw some doubt on this conclusion. Noble *et al.* (2000) found that data from felid carnivores and fruit-bats provided limited support for Cartmill's hypothesis, while data from herpestid carnivores provided no support at all. Hence, it would seem that additional factors might be involved in the evolution of orbital convergence in primates. Noble and coworkers found evidence that brain size (which tends to be larger in primates than in other mammals and is reflected in a relatively larger braincase) may be an important contributory factor.

1.3.2 Allometric analysis of optic canal size

Valuable additional information for interpreting visual adaptations can be obtained from the skull of living and fossil primates by examining the relative size of the optic foramen, through which the optic nerve passes on its way to the brain. A key factor in the processing of photoreceptor signals is the trade-off between sensitivity and visual acuity. In dim light, increased sensitivity can be achieved through summation of the inputs from photoreceptors, but this necessarily entails a reduction in visual acuity, which is enhanced by maximizing inputs from small numbers of photoreceptors (Chapters 4–6). In line with this, it is a general rule that rods show more summation than cones. Hence, it is only to be expected that nocturnal primate species should show a greater degree of summation than diurnal species, and it was indeed demonstrated by Hassler (1967) and Stephan *et al.* (1984) that nocturnal primates typically have much smaller optic nerves than diurnal primates. This was further confirmed by a study conducted by Kay and Kirk (2000), who also reviewed data confirming that diurnal haplorhines perform considerably better on tests of visual acuity than all other mammals. After demonstrating with a sample of living primates that data for the size of the optic foramen yield very similar results to those provided by information on the size of the optic nerve itself, Kay and Kirk proceeded to investigate scaling of the optic foramen in extant and fossil primates. In accordance with expectation, it was found that nocturnal primates showed the smallest values for relative size of the optic foramen, whereas diurnal simians showed the largest values. Interestingly, although the relative size of the optic foramen was found to be significantly larger in diurnal lemurs than in nocturnal primates, the difference was relatively small and far more limited than for diurnal simians. Thus, it would seem that in diurnal lemurs visual sensitivity is only slightly decreased in comparison with nocturnal primates, while visual acuity is only marginally greater. This fits well with the observation that there is very little distinction between nocturnal and diurnal lemurs with respect to the relative size of the orbit (Figure 1.11).

In fact, re-analysis of some of the data collected by Kay and Kirk (2000) provides an additional perspective on scaling of the optic foramen (Figure 1.14). When the diameter of the foramen is plotted against orbit height (Figure 1.14A), there is a relatively clear grade separation between modern strepsirrhine primates and modern diurnal simians. However, *Tarsius* species are found to lie very close to the line for diurnal simians, whereas *Aotus* lies among the modern strepsirrhines. When optic foramen diameter is plotted against skull length (Figure 1.14B), an even clearer grade separation between modern strepsirrhines and modern diurnal simians is found. In this case, however, both *Tarsius* species and *Aotus* lie clearly within the distribution for modern strepsirrhines. The interpretation of this would seem to be that in *Tarsius* (but not in *Aotus*) the very large size of the eyes is associated with a relatively large diameter of the optic nerve, matching that in diurnal simians. In other words, it seems that *Tarsius* may be able to combine a high degree of visual sensitivity with a relatively high level of visual acuity simply by possessing a particularly large number of photoreceptors for its body size. The difference between *Tarsius* and *Aotus* in this respect may explain why *Tarsius* has consistently retained a clearly developed fovea whereas *Aotus* has not. It should also be noted that there is no obvious distinction between diurnal lemurs and other strepsirrhines in either Figure 1.14A or B, once again emphasizing the fact that there is no strong indication of diurnal habits in lemurs from scaling of orbital dimensions.

As noted by Kay and Kirk (2000), the presence of a relatively small optic foramen in the Eocene omomyiforms *Microchoerus* and *Necrolemur*, in the Eocene adapiform *Pronycticebus*, and in the Oligocene strepsirrhine *Plesiopithecus* indicates that their eyes were similar to those of modern nocturnal primates, with rod-dominated retinas showing a marked degree of summation. By contrast, it can be inferred that the Eocene adapiform *Leptadapis* and the enigmatic Oligocene prosimian *Rooneyia* resembled diurnal lemurs,

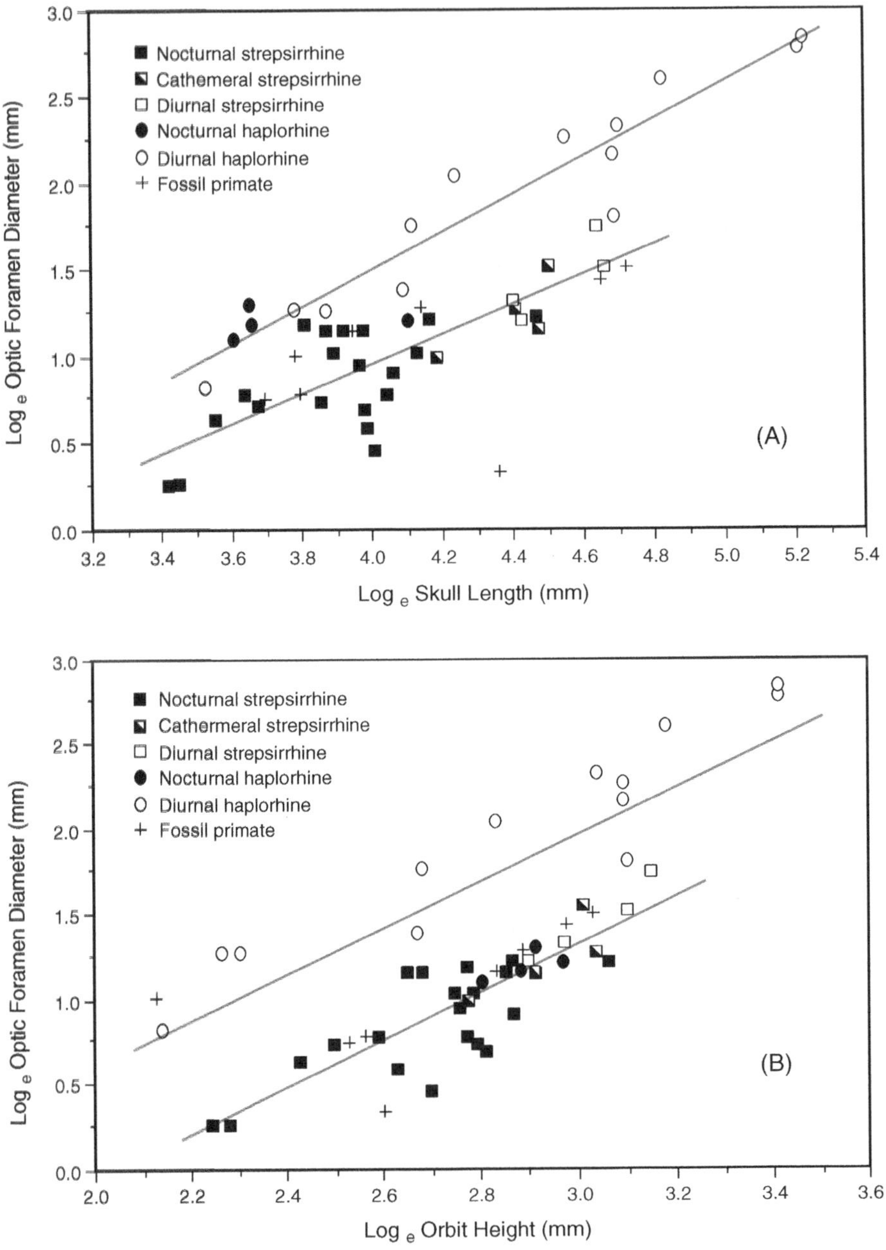

Figure 1.14 (A) Scaling of optic foramen diameter (D) against maximal skull length (S) for extant and fossil primates (data from Kay and Kirk, 2000). The two least-squares regression lines, for orientation only, are for modern haplorhine primates (upper line; $n = 17$; $\log_e D = 1.089 \cdot \log_e S - 2.865$; $r^2 = 0.900$) and for modern strepsirrhine primates (lower line; $n = 28$; $\log_e D = 0.866 \cdot \log_e S - 2.516$; $r^2 = 0.665$). In this plot, modern tarsiers and diurnal simians together represent a distinct grade in relation to other

probably possessing moderate retinal summation and a larger cone:rod ratio than in nocturnal primates. In these cases, information from the relative size of the optic foramen reinforces inferences from the relative size of the orbit in indicating nocturnal or diurnal habits for those particular fossil primates. On one hand, it should be noted that the optic foramen is unusually small relative to skull length in the Eocene adapiform *Adapis*, suggesting a particularly high degree of summation (Figure 1.14A). On the other hand, when the size of the optic foramen in *Adapis* is considered in relation to orbital height, the deviation from other prosimian primates is far less marked (Figure 1.14B). In other words, the very small orbits of *Adapis* are matched by a very small size of the optic foramen. However, the comparatively high degree of summation that is indicated even by Figure 1.14B does not fit well with the common inference based on the small size of its orbits that *Adapis* was diurnal. This provides one very good example in which inference of the activity pattern through direct comparison with modern primates can be problematic.

1.4 Evolution of color vision

1.4.1 Occurrence of color vision among mammals

Very useful general reviews of the evolutionary background to color vision in mammals are provided by Jacobs (1993) and Ahnelt and Kolb (2000). Light-sensitive cells (photoreceptors) in the vertebrate retina can typically be divided into rods, adapted for monochromatic (black-and-white) vision under scotopic conditions, and cones, adapted for color vision under photopic (normal daylight) conditions. No mammal is known to have more than one kind of rod receptor, but most mammals have two types of cones, in principle providing a basis for dichromatic color vision. Rods are more sensitive to light than cones, so in dim light only the rods are functional. Conversely, signals from rods are saturated at the higher light intensities at which color vision operates well, so a fairly clear functional separation between rods and cones is to be expected. It is generally accepted that cones provide the basis for color vision and that two cone types are the minimal requirement. However, there are several indications (e.g. from work with humans) that rod signals can influence color vision. In fact, it has been shown that apparent behavioral evidence for trichromatic color vision in certain lemur species (*Eulemur fulvus* and *Lemur catta*) is attributable to a combination of signals from two cone types

---

Figure 1.14 *(continued)*
modern primates. (B) Scaling of optic foramen diameter (*D*) against orbital height (*H*) for extant and fossil primates (data from Kay and Kirk, 2000). The two least-squares regression lines, for orientation only, are for modern haplorhine primates (upper line; $n = 17$; $\log_e D = 1.367 \cdot \log_e H - 2.143$; $r^2 = 0.629$) and for modern strepsirrhine primates (lower line; $n = 28$; $\log_e D = 1.392 \cdot \log_e S - 2.864$; $r^2 = 0.734$). In this plot, modern diurnal simians together represent a distinct grade in relation to all other modern primates. Note the extreme negative outlier among fossil primates in both plots is the Eocene adapiform *Adapis*, in which the foramen is strikingly small in relation to skull length and also quite small in relation to orbit height

with signals from rods (Jacobs and Deegan, 1993). However, some primates – notably the Old World simians – have three types of cones, which is unique among mammals (Jacobs, 1993; Kremers *et al.*, 1999; Dominy *et al.*, 2003; see Chapters 3 and 4).

The original distinction between rods and cones was based on morphological differences, with rods being typically long and thin and cones being short and squat. Because no clear morphological distinction was evident between different classes of photoreceptors, several investigators (notably Walls, 1942) concluded that cones were completely lacking from the retina of many nocturnal mammals. As it was also widely believed that the ancestral mammals were nocturnal, it was only logical to infer that cones had been completely lost in the ancestral stock, leaving an all-rod retina, and that they had to be redeveloped in those mammals that subsequently became diurnal in habits. Conversely, a comparable absence of any clear morphological distinction between photoreceptors led to the view that some diurnal mammals, especially squirrels and tree-shrews, have all-cone retinas. However, the distinction between rods and cones has been increasingly refined with additional criteria: cytological, neuroanatomical, physiological, immunological, and molecular. In the wake of this, it has emerged that nocturnal mammals – including monotremes, marsupials, and placentals – generally do not seem to have completely cone-free retinas. Furthermore, it is now known that diurnal mammals, including squirrels and tree-shrews, generally have at least some rods. In the retina of nocturnal mammals, cones are usually present, albeit at a relatively low frequency, and there are commonly two cone types with different photopigments. The retention of two cone classes (which presumably function only under photopic conditions) in the retina of nocturnal mammals is an enigma that requires some explanation. Be that as it may, it is now clear that the distinction between nocturnal and diurnal habits is reflected by a shift in the ratio between rods and cones, rather than by complete absence of one photoreceptor type. It should also be noted that even in fully diurnal mammals the rods often heavily outnumber the cones in the retina. In the human retina, for example, there are just 6 million cones compared with 120 million rods, a ratio of only 1:20. Tree-shrews and squirrels, originally thought to have pure cone retinas, are exceptional in having a high cone:rod ratio. In *Tupaia*, for instance, 95 percent of the photoreceptors are cones (Müller and Peichl, 1989).

Despite the fact that nocturnal mammals generally seem to have retained a small proportion of cones in the retina, there are several compelling reasons to accept Walls' hypothesis that ancestral mammals were nocturnal and had lost diurnal visual features found in other vertebrate groups. In the first place, most mammals (especially nocturnal species) lack vivid coloration of the pelage, indicating that color signals were generally not favored by selection. Further, as noted by Ahnelt and Kolb (2000, p. 715): 'Compared to the diversity of photoreceptors in groups such as teleost fish or reptilian families such as geckoes . . . mammalian photoreceptors are uninterestingly uniform and difficult to study.' This very uniformity, along with the associated difficulties in distinguishing rods from cones in many mammal species, suggests that in the course of their evolution mammals passed through an extended period of nocturnal adaptation. This interpretation is reinforced by the fact that mammalian cones are relatively simple compared with those of other vertebrates (fish, amphibians, birds, and reptiles). In the latter, a colored oil droplet (ellipsosome) is commonly present between the inner and outer segments of each cone (Kremers *et al.*, 1999). The droplets absorb light of particulars wavelengths and hence narrow the spectral sensitivity of each of the four cone photopigments that are present in the typical tetrachromatic visual system. In all placental mammals and some marsupials and monotremes oil droplets are completely lacking. Although certain marsupials and monotremes have retained oil droplets in some cones, they are colorless and their original function has presumably been diminished if not abolished. Walls (1942) proposed that colored oil droplets appeared at a very early stage of vertebrate evolution and that a shift to nocturnal habits in mammals was accompanied by loss of color from the droplets and eventually by their complete loss. This hypothesis was rejected as unparsimonious by some authors, but subsequent demonstration of colored oil droplets in the cones of lungfish has increased the probability that they were present, possibly in conjunction with tetrachromatic vision, in the common ancestor of lungfish and land vertebrates (Robinson, 1994).

In the vertebrate retina, the outer segment of each photoreceptor is composed of a stack of membranes in which the photosensitive visual pigments (photopigments) are embedded. Different photopigments are maximally sensitive to different parts of the spectrum (spectral tuning). The photon-capturing properties of vertebrate photopigments depend essentially on apoproteins known as opsins, each containing approximately 350 amino acids, which are integral membrane proteins. Each photopigment is formed by the binding of an opsin with the chromophore retinal, derived from vitamin A. Each type of opsin corresponds to a different gene. Accordingly, differential spectral tuning of photopigments is determined by amino acid differences between the different opsin proteins, and those differences have turned out to be relatively few in number (Hunt, 2001; see Chapter 3). The amino acid substitutions responsible occur at particular sites in the transmembrane portion of the pigment molecule.

Recent evidence indicates that five opsin types are commonly present in the retina of vertebrates other than mammals, namely fish, amphibians, reptiles, and birds (Ahnelt and Kolb, 2000; Hunt, 2001; see Chapter 3). In addition to the rhodopsin present in rods, there are four cone opsins that are maximally sensitive to different regions of the spectrum: long-wave sensitive (L-) cones (peak sensitivity at 530–570 nm), medium-wave sensitive (M-) cones (peak sensitivity at 480–520 nm), short-wave sensitive I (S-I) cones (peak sensitivity at 440–460 nm), and short-wave sensitive II (S-II) violet to ultraviolet cones (peak sensitivity at 355–430 nm). Mammals typically have just three of these opsins – rod rhodopsin and two cone opsins (an M-/L-opsin and an S-I-opsin) – and it is now generally agreed that the basic conformation of the mammalian retina is dichromatic (Ahnelt and Kolb, 2000). The gene for long-wave sensitive red-opsin is located on the X-chromosome, while the gene for short-wave sensitive blue-opsin is located on an autosomal chromosome. Some primates are unique among mammals in having three kinds of cone opsins – L-opsin, M-opsin, and S-I-opsin – and in all cases the gene for M-opsin (which is also located on the X-chromosome) is secondarily derived in some way from the gene for L-opsin. Valuable information has been provided by molecular comparisons of opsin genes (Jacobs, 1993; Yokoyama, 2000). DNA sequence comparisons indicate that divergence between the ancestral rod gene and a common ancestor for the cone pigment genes occurred very early in vertebrate evolution. As cones typically develop before rods in the vertebrate retina, it is possible that the ancestral photoreceptor was adapted for photopic, rather than scotopic, conditions. This inference is reinforced by the fact that rod signals converge on the cone system prior to reaching the ganglion cells (Kremers *et al.*, 1999). In the evolution of cones, a divergence in the cone pigment gene, probably at some time during the Permian or Triassic, yielded one gene producing a pigment with maximal sensitivity at short wavelengths (S) and another producing a pigment with maximal sensitivity at long wavelengths (L). This ancient divergence is reflected by the fact that there is now only about 40 percent sequence similarity between the S-opsin gene and the M-/L-opsin genes in primates (Chapter 3). Given that sequence comparisons suggest lineal continuity in the evolution of S and L genes in all mammals, it seems highly likely that ancestral mammals did, indeed, possess a duplex retina (one containing both rods and cones). Moreover, it would seem that two spectrally distinct types of cone photopigment were present and provided an ancestral basis for the typical condition of dichromacy in mammals.

In recent years, through a combination of new methods (e.g. electroretinography; labeling of photoreceptors with antibodies) and sequencing of opsin genes, a relatively clear picture of the evolutionary history of color vision in primates has emerged. It has been confirmed that nocturnal primates have heavily rod-dominated retinas and essentially monochromatic vision. Although it was originally thought that nocturnal primates possess cone-free retinas, this conclusion has proved to be erroneous across the board, as was anticipated by Martin (1990). For example, Dartnall *et al.* (1965) reported that *Galago crassicaudatus* has a pure rod retina, but labeling with a cone-specific antibody showed that *Galago garnettii* has a rod:cone ratio that varies from 100:1 to 30:1 across the retina (Wikler and Rakic, 1990). It was subsequently confirmed that 1–3 percent of the photoreceptors in the related species *Galago crassicaudatus* are cones (Jacobs *et al.*, 1995). The cones were all found to be of a single type, with an absorption peak corresponding to human L-cones. In fact, DNA sequencing revealed the presence of a gene for another cone type (corresponding to human S-cones), but is non-functional because of mutational modification (see also Zhou *et al.*, 1997). It therefore seems that *Galago crassicaudatus* and *Galago garnettii* lack color vision despite the possession of some cones. It was also originally believed that the retina of tarsiers exclusively contains closely packed rods (Castenholz, 1984), but it has now been clearly demonstrated that the retina of *Tarsius* does contain a small proportion of cones. Morphologically, the cones are not easily distinguishable from rods, but their outer segments are only half as long. Immunocytochemical labeling revealed the presence of two cone types in the retina of *Tarsius spectrum*, one with an M-/L-opsin and the other with an S-opsin (Hendrickson *et al.*, 2000). Thus, *Tarsius* does seem to possess a limited basis for dichromatic vision. A small proportion of cones have also been found in the retina of *Aotus*, which was originally believed to possess an all-rod retina (Jacobs, 1993). However, only one type of cone was found, with a sensitivity peak again corresponding to human L-cones. Furthermore, unlike many diurnal New World monkeys, *Aotus* shows no evidence of photopigment polymorphism in its L-cones. A limited degree of color discrimination that has been identified for this nocturnal species is probably attributable to interaction between signals from cones and rods (Jacobs, 1993).

Corresponding to behavioral evidence indicating limited powers of color discrimination, the retina of diurnal lemurs generally possesses at least a basis for dichromacy. For example, both *Lemur catta* and *Eulemur fulvus* have two cone types corresponding to human S-cones and L-cones, respectively. Indeed, it is almost certain that these diurnal lemurs use rods in combination with the two cone types to achieve some degree of trichromatic vision (Jacobs, 1993). However, an analysis of opsin genes on the

X-chromosome revealed that there is a polymorphism involving M- and L-alleles in *Propithecus verreauxi coquereli* and *Varecia variegata* (Tan and Li, 1999). The presence of an M-/L-opsin polymorphism in *Propithecus verreauxi coquereli* was subsequently confirmed with electroretinographic evidence (Jacobs *et al.*, 2002). It would therefore seem that female *Propithecus* and *Varecia* with different opsin alleles on their X-chromosomes possess a basis for trichromacy.

It now seems likely that some basis for trichromacy was already present in ancestral simians (Kremers *et al.*, 1999). At one time, all New World monkeys were generally thought to possess dichromatic color vision, with an autosomal gene for an S-opsin and a gene for an L-opsin on the X-chromosome. However, it then progressively emerged that several platyrrhine species possess a polymorphism of the opsin gene on the X-chromosome similar to that subsequently discovered in *Propithecus* and *Varecia* (Jacobs, 1993). Here, too, females with different opsin alleles on their X-chromosomes possess a basis for trichromacy. This condition increasingly seemed to be typical for diurnal New World monkeys, until it was discovered that the howler monkey (*Alouatta*) possesses three separate opsin genes, one autosomal and the other two on the X-chromosome. Electroretinograms revealed that *Alouatta* possesses trichromatic color vision with three photopigments in males as well as females (Jacobs *et al.*, 1996). This closely resembles the system for trichromatic color vision that is universally found in catarrhine primates (Old World monkeys, apes, and humans). The genetic basis is very similar to that of catarrhines; the M- and L-opsins on the X-chromosome have comparable spectral peaks.

Trichromatic color vision based on the presence of three opsin genes is a consistent feature of all catarrhine primates, so it seems highly likely that this was an ancestral feature of Old World simians. All catarrhines possess an autosomal gene coding for an S-opsin and two genes on the X-chromosome that code for an M-opsin and an L-opsin (Jacobs, 1993). The high degree of sequence similarity between the two opsin genes on the X-chromosome (approximately 97 percent) indicates that they arose through duplication of a single ancestral locus (Kremers *et al.*, 1999). The two genes lie very close to one another on the X-chromosome in a head-to-tail array (Hunt, 2001; see Chapter 3).

It is accordingly reasonable to suggest that color vision in diurnal primates evolved in stages. The first stage was development of a dichromatic system with an increased proportion of cones with two photopigments, an S-opsin coded by an autosomal gene and an M-/L-opsin coded by a gene on the X-chromosome. Such a dichromatic system was undoubtedly present as a minimum in the common ancestor of simian primates and it is possible that it was present in ancestral primates. It is a moot point whether the dichromatic system of diurnal lemurs was retained from the ancestral primates or whether it was redeveloped following derivation from an ancestral lemur that possessed only one cone photopigment. In any event, polymorphism of the opsin gene on the X-chromosome presumably developed separately in some diurnal lemurs and in numerous New World monkeys (or perhaps in the common platyrrhine ancestor), yielding a basis for trichromatic vision in heterozygous females. (It should be noted that trichromacy in female New World monkeys is well established. On the other hand, no trichromatic lemurs have been found. The data only showed the possibility of the presence of trichromacy.) This was possibly an intermediate stage in the development of full trichromatic vision in both sexes. The final stage in the evolution of trichromatic color vision in primates, which was achieved independently by at least one New World monkey (*Alouatta*) and by the common ancestor of the catarrhine primates, was duplication of the opsin gene on the X-chromosome to produce an M-opsin in addition to the ancestral L-opsin.

This scenario was questioned by Tan and Li (1999), who examined X-linked opsin genes in 20 prosimian species. In addition to finding evidence for a polymorphism of the opsin gene on the X-chromosome in the diurnal lemurs *Propithecus* and *Varecia*, they also reported a similar polymorphism in *Cheirogaleus major*. Furthermore, they reported that in other prosimians examined the gene on the X-chromosome may code either for an M-opsin or for an L-opsin. The former was found most often among the species examined (despite the fact that it is presumably derived from an ancestral L-opsin), but the latter was found in nocturnal mouse lemurs (*Mirza coquereli* and *Microcebus murinus*) and in the cathemeral bamboo lemur (*Hapalemur griseus*). It was also reported that *Tarsius syrichta* has an L-opsin, whereas *Tarsius bancanus* has an M-opsin. On this basis, Tan and Li (1999) concluded that the common ancestor of tarsiers and strepsirrhines (i.e. the ancestral primate) might have been trichromatic, such that trichromacy originated much earlier than is commonly believed. However, this interpretation was challenged by Heesy and Ross (2001) because it conflicts with a large body of morphological, phylogenetic, and behavioral data. In fact, in a footnote added in press, these authors reported a personal communication from Tan to the effect that the polymorphism attributed to *Cheirogaleus major* had proven to be unfounded. Hence, polymorphic trichromacy has so far been demonstrated only for two diurnal lemurs (*Propithecus* and *Varecia*), and this provides no convincing basis for inferring that ancestral primates were trichromatic.

Regardless of whether the ancestral primates were nocturnal or diurnal, the fact remains that the majority of modern prosimians are nocturnal. It now seems to be clearly established that all nocturnal primates have a small proportion of cones scattered among the rods in the retina. As far as is known, however, there is usually only one type of cone in the retina of any nocturnal strepsirrhine species. So far, no S-cones have ever been identified in the retina of a nocturnal strepsirrhine and the single cone type demonstrated has been found to belong to the M/L class (Jacobs, 1996; Tan and Li, 1999). It may initially seem surprising that nocturnal primates possess cones at all. However, it is perfectly possible that some residual capacity for color discrimination may be advantageous (e.g. when a nocturnal primate is disturbed by a predator during the daytime). Furthermore, it has been suggested that possession of some cones may facilitate detection of the transition from daytime (photopic) to night-time (scotopic) conditions. This suggestion was prompted by the observation that the onset of activity in various nocturnal strepsirrhines typically coincides with the time at dusk when human cones cease to function (Martin, 1990).

The most likely hypothesis, given present evidence, seems to be that the ancestral primates were nocturnal and that the acquisition of diurnal habits represents a secondary development in three lineages of diurnal primates (Ross *et al.*, unpublished data). However, there has been some support for the alternative suggestion that ancestral primates were diurnal (Ni *et al.*, 2004). It certainly seems likely that the ancestral primate possessed two types of cone (S-cones and M-/L-cones), as sequence comparisons indicate lineal continuity for these two types of cone in terrestrial vertebrates including primates and other mammals (Yokoyama, 2000). Incidentally, such lineal continuity indicates that mammals retained S-cone and M-/L-cone opsins throughout the Jurassic and Cretaceous periods (a total of 135 my) when they were presumably consistently nocturnal. It would therefore seem that inactivation of S-cones occurred in one or more nocturnal strepsirrhine lineages, as diurnal lemurs are at least dichromatic and have both S-cones and M-/L-cones. Independent inactivation of S-cones has also occurred in the lineage leading to nocturnal *Aotus*. Retention of two types of cone in the retina throughout the first two-thirds of mammalian evolution and apparently in the ancestral primate is somewhat surprising and is not explicable on the basis of detection of the switch from photopic to scotopic conditions, as a single cone type would presumably suffice for this.

Among modern primates, diurnal behavior seems to be a prerequisite for effective color discrimination. Diurnal primates are at least dichromatic, but trichromacy has emerged a number of times, either through polymorphism of a single opsin gene on the X-chromosome (possibly some diurnal lemurs; many New World monkeys) or by duplication and sequence divergence of that gene (howler monkeys and all catarrhine primates). As dichromacy would seem to be the basic condition for mammals (Jacobs, 1993; Ahnelt and Kolb, 2000), this does not require any special explanation within the context of primate evolution. However, trichromatic color vision is unique among mammals and does need some explanation. The primary hypothesis that has been considered is that trichromacy is particularly advantageous for detection of yellow or orange fruits against a background of green foliage (Osorio and Vorobyev, 1996; Regan *et al.*, 2001; see Chapter 4). This fits well with the fact that fruits typically constitute a significant part of the diet in primates. Furthermore, the two diurnal lemurs that have been found to have a polymorphic basis for trichromacy (*Propithecus*, *Varecia*) are predominantly frugivorous, and the same applies to New World monkeys that have been shown to have the same development of the visual system. However, it should be noted that in Old World monkeys and apes (catarrhine primates), which all possess full trichromacy in both sexes, folivory is quite common (which is not the case among the predominantly frugivorous New World monkeys). It is also noteworthy that *Alouatta*, the only New World monkey so far shown to possess full trichromacy paralleling that of catarrhines, is one of only two platyrrhines that includes a large proportion of leaves in its diet (the other being *Brachyteles*, whose visual adaptations have not yet been explored). In fact, a field study of four catarrhine species in Kibale Forest (Uganda) showed that they eat immature leaves which are discriminated from mature leaves only on the basis of red, as opposed to green, coloration. Red coloration of leaves correlated with high protein content and low toughness. By contrast, fruits are discriminated by a much broader range of visual cues (Dominy and Lucas, 2001). In a subsequent paper, Dominy *et al.* (2003) proposed a two-stage explanation of the evolution of primate color vision. Initially, historical biogeography of figs and arborescent palms accounted for patterns of variation in primate color vision. With respect to polymorphic systems in which some females are trichromatic but other females and all males are dichromatic, it was suggested that foraging groups with mixed capabilities for chromatic distinction gained a selective advantage in relation to the abundance and inconspicuous coloration of figs and palms. In the second stage, following regional extinction of palms and probably figs because of climatic change, the evolution of routine trichromatic vision permitted primates to exploit protein-rich immature leaves as a substitute resource. Hence, it seems quite possible that the evolution of color vision was connected both with fruits and with leaves, but that full trichromacy is more strongly linked to folivory (see also Chapter 4).

As this brief survey illustrates, satisfactory explanations for the distribution of many aspects of the primate visual system require that their points of origin be precisely located on the primate evolutionary tree. Not only do primate visual systems differ, but visual systems differ even among simian primates ('monkeys') and prosimians, often quite radically. An evolutionary and comparative approach to the primate visual system is therefore needed that invokes specific sequences of evolutionary events. This must take into account not only the distribution of visual adaptations in extant primates, but also the inferred visual adaptations of fossil primates. Fossils can exhibit combinations of characteristics that are not seen in living primates, falsifying hypotheses based on extant animals, and enriching our understanding of the diversity of primate visual adaptations.

Acknowledgements

RDM is grateful to Dr Jan Kremers and his team for the invitation to attend the conference entitled 'Structure, Function and Evolution of the Primate Visual System: Primavision', held in Tübingen, Germany, on 6–9 July 2003. This well-organized and highly stimulating conference provided the initial impetus and inspiration for preparation of this chapter. Input from Chris Heesy is gratefully acknowledged. Thanks are also due to Jill Seagard for skilled assistance with some of the illustrations.

References

Ahnelt, P.K. and Kolb, H., 2000. The mammalian photoreceptor mosaic-adaptive design. *Prog. Retin. Eye Res.* **19**, 711–777.

Allman, J.M., 1977. Evolution of the visual system in the early primates, in *Progress in Psychobiology and Physiological Psychology,* Vol. 7 (eds J.M. Sprague and A.N. Epstein), Academic Press, New York, pp. 1–53.

Allman, J.M., 1982. Reconstructing the evolution of the brain in primates through the use of comparative neurophysiological and neuroanatomical data, in *Primate Brain Evolution* (eds E. Armstrong, and D. Falk), Plenum Press, New York, pp. 13–28.

Allman, J.M., 1999. *Evolving Brains*, W.H. Freeman/Scientific American, New York.

Alroy, J., 1998. Cope's rule and the dynamics of body mass evolution in North American fossil mammals. *Science* **280**, 731–733.

Beard, K.C., 1990. Gliding behaviour and palaeoecology of the alleged primate family Paromomyidae (Mammalia, Dermoptera). *Nature (Lond.)* **345**, 340–341.

Beard, K.C., 1993. Phylogenetic systematics of the Primatomorpha, with special reference to Dermoptera, in *Mammal Phylogeny: Placentals* (eds F.S. Szalay, M.J. Novacek, and M.C. McKenna), Springer-Verlag, New York, pp. 129–150.

Bloch, J.I. and Boyer, D.M., 2002. Grasping primate origins. *Science* **298**, 1606–1610.

Brooke, M.d.L., Hanley, S., and Laughlin, S.B., 1999. The scaling of eye mass with body size in birds. *Proc. R. Soc. Lond. B* **266**, 405–412.

Cachel, S.M., 1979. A functional analysis of the primate masticatory system and the origin of the anthropoid post-orbital septum. *Am. J. Phys. Anthropol.* **50**, 1–18.

Cartmill, M., 1970. *The Orbits of Arboreal Mammals: A Reassessment of the Arboreal Theory of Primate Evolution*, Ph.D. Thesis, University of Chicago.

Cartmill, M., 1972. Arboreal adaptations and the origin of the order Primates, in *The Functional and Evolutionary Biology of Primates* (ed R.H. Tuttle), Aldine-Atherton, Chicago, pp. 97–122.

Cartmill, M., 1974. Rethinking primate origins. *Science* **184**, 436–443.

Cartmill, M., 1980. Morphology, function, and evolution of the anthropoid postorbital septum, in *Evolutionary Biology of the New World Monkeys and Continental Drift* (eds R.L. Ciochon and A.B. Chiarelli), Plenum Press, New York, pp. 243–274.

Castenholz, A., 1984. The eye of *Tarsius*, in *Biology of Tarsiers* (ed C. Niemitz), Gustav Fischer, Stuttgart, pp. 303–318.

Charles-Dominique, P., 1975. Nocturnality and diurnality: An ecological interpretation of these two modes of life by an analysis of the higher vertebrate fauna in tropical forest ecosystems, in *Phylogeny of the Primates* (eds W.P. Luckett and F.S. Szalay), Plenum Press, New York, pp. 69–88.

Collins, E.T., 1921. Changes in the visual organs correlated with the adoption of arboreal life and with the assumption of the erect posture. *Trans. Ophthalmol. Soc. UK* **41**, 10–90.

Dartnall, H.J.A., Arden, G.B., Ikeda, H. *et al.*, 1965. Anatomical, electrophysiological and pigmentary aspects of vision in the bush baby: An interpretive study. *Vision Res.* **5**, 399–424.

Dominy, N.J. and Lucas, P.W., 2001. Ecological importance of trichromatic vision to primates. *Nature (Lond.)* **410**, 363–366.

Dominy, N.J., Svenning, J.C., and Li, W.H., 2003. Historical contingency in the evolution of primate color vision. *J. Hum. Evol.* **44**, 25–45.

Elliot Smith, G., 1924. *The Evolution of Man*, Oxford University Press, Oxford.

Emmons, L.H., 2000. *Tupai: A Field Study of Bornean Treeshrews*, University of California Press, Berkeley.

Felsenstein, J., 1985. Phylogenies and the comparative method. *Am. Nat.* **125**, 1–15.

Greaves, W.S., 1985. The mammalian postorbital bar as a torsion-resisting helical strut. *J. Zool., Lond.* **207**, 125–136.

Greaves, W.S., 1995. Functional predictions from theoretical models of the skull and jaws in reptiles and mammals, in *Functional Morphology in Vertebrate Paleontology* (ed J.J. Thomason), Cambridge University Press, Cambridge, pp. 99–115.

Groves, C.P., 2001. *Primate Taxonomy*, Smithsonian Institution Press, Washington, DC.

Harvey, P.H. and Mace, G.M., 1982. Comparison between taxa and adaptive trends: Problems of methodology, in *Current Problems in Sociobiology* (eds King's College Sociobiology Group), Cambridge University Press, Cambridge, pp. 343–361.

Harvey, P.H. and Pagel, M.D., 1991. *The Comparative Method in Evolutionary Biology*, Oxford University Press, Oxford.

Hassler, R., 1967. Comparative anatomy of the central visual structures in day- and night-active primates, in *Evolution of the Forebrain* (eds R. Hassler and H. Stephan), Plenum Press, New York, pp. 419–434.

Heesy, C.P., 2003. *The Evolution of Orbit Orientation in Mammals and the Function of the Primate Postorbital Bar*, Ph.D. Thesis, Stony Brook University.

Heesy, C.P., 2004. On the relationship between orbit orientation and binocular visual field overlap in mammals. Anat. Rec. 281A:1104–1110.

Heesy, C.P. and Ross, C.F., 2001. Evolution of activity patterns and chromatic vision in primates: Morphometrics, genetics and cladistics. *J. Hum. Evol.* **40**, 111–149.

Hendrickson, A., Djajadi, H.R., Nakamura, L. *et al.*, 2000. Nocturnal tarsier retina has both short and long/medium-wavelength cones in an unusual topography. *J. Comp. Neurol.* **424**, 718–730.

Hunt, D.M., 2001. Seeing rainbows. *Biologist* **48**, 67–71.

Isler, K., Barbour, A.D., and Martin, R.D., 2002. Line-fitting by rotation: A nonparametric method for bivariate allometric analysis. *Biometr. J.* **44**, 289–304.

Jacobs, G.H., 1993. The distribution and nature of colour vision among the mammals. *Biol. Rev.* **68**, 413–471.

Jacobs, G.H., 1996. Primate photopigments and primate color vision. *Proc. Natl. Acad. Sci. USA* **93**, 577–581.

Jacobs, G.H. and Deegan, J.F., 1993. Photopigments underlying color vision in ringtail lemurs (*Lemur catta*) and brown lemurs (*Eulemur fulvus*). *Am. J. Primatol.* **30**, 243–256.

Jacobs, G.H., Neitz, M., and Neitz, J., 1995. Why bushbabies lack color vision. *Am. J. Primatol.* **36**, 130.

Jacobs, G.H., Neitz, M., Deegan, J.F. *et al.*, 1996. Trichromatic colour vision in New World monkeys. *Nature (Lond.)* **382**, 156–158.

Jacobs, G.H., Deegan, J.F., Tan, Y. *et al.*, 2002. Opsin gene and photopigment polymorphism in a prosimian primate. *Vision Res.* **42**, 11–18.

Kay, R.F. and Cartmill, M., 1977. Cranial morphology and adaptation of *Palaechthon nacimienti* and other Paromomyidae (Plesiadapoidea, Primates), with a description of a new genus and species. *J. Hum. Evol.* **6**, 19–53.

Kay, R.F., Ross, C.F., and Williams, B.A., 1997. Anthropoid origins. *Science* **275**, 797–804.

Kay, R.F. and Kirk, E.C., 2000. Osteological evidence for the evolution of activity pattern and visual acuity in primates. *Am. J. Phys. Anthropol.* **113**, 235–262.

Kay, R.F., Thewissen, J.G.M., and Yoder, A.D., 1992. Cranial anatomy of *Ignacius graybullianus* and the affinities of the Plesiadapiformes. *Am. J. Phys. Anthropol.* **89**, 477–498.

Kirk, E.C. and Kay, R.F., 2004. The evolution of high visual acuity in the Anthropoidea, in *Anthropoid Origins: New Visions* (eds C.F. Ross and R.F. Kay), Kluwer Academic/Plenum Publishers, New York, pp. 539–602.

Kirk, E.C., Cartmill, M., Kay, R.F. *et al.*, 2003. Comment on "grasping primate origins". *Science* **300**, 741.

Kremers, J., Silveira, L.C.L., Yamada, E.S. *et al.*, 1999. The ecology and evolution of primate color vision, in *Color Vision: From Genes to Perception* (eds K.R. Gegenfurtner and L.T. Sharpe), Cambridge University Press, Cambridge, pp. 123–142.

Le Gros Clark, W.E., 1959. *The Antecedents of Man*, Edinburgh University Press, Edinburgh.

Martin, R.D., 1968. Towards a new definition of primates. *Man* **3**, 377–401.

Martin, R.D., 1980. Adaptation and body size in primates. *Z. Morphol. Anthropol.* **71**, 115–124.

Martin, R.D., 1989. Size, shape and evolution, in *Evolutionary Studies – A Centenary Celebration of the Life of Julian Huxley* (ed M. Keynes), Eugenics Society, London, pp. 96–141.

Martin, R.D., 1990. *Primate Origins and Evolution: A Phylogenetic Reconstruction*, Princeton University Press, New Jersey.

Martin, R.D., 1993. Primate origins: Plugging the gaps. *Nature (Lond.)* **363**, 223–234.

Martin, R.D., 1995. Prosimians: From obscurity to extinction? in *Creatures of the Dark: The Nocturnal Prosimians* (eds L. Alterman, G.A. Doyle, and M.K. Izard), Plenum Press, New York, pp. 535–563.

Martin, R.D., 1998. Comparative aspects of human brain evolution: Scaling, energy costs and confounding variables, in *The Origin and Diversification of Language* (eds N.G. Jablonski and L.C. Aiello), California Academy of Sciences, San Francisco, pp. 35–68.

Martin, R.D., 2004. Chinese lantern for early primates. *Nature (Lond.)* **427**, 22–23.

Martin, R.D. and Barbour, A.D., 1989. Aspects of line-fitting in bivariate allometric analyses. *Folia Primatol.* **53**, 65–81.

Martin, R.D. and MacLarnon, A.M., 1985. Gestation period, neonatal size and maternal investment in placental mammals. *Nature, (Lond.)* **313**, 220–223.

Massey, V., 2000. The chemical and biological versatility of riboflavin. *Biochem. Soc. Trans.* **28**, 283–296.

Müller, B. and Peichl, L., 1989. Topography of cones and rods in the tree shrew retina. *J. Comp. Neurol.* **282**, 581–594.

Ni, X., Wang, Y., Hu, Y., and Li, C., 2004. A euprimate skull from the early Eocene of China. *Nature (Lond.)* **427**, 65–68.

Noble, V.E., Kowalski, E.M., and Ravosa, M.J., 2000. Orbit orientation and the function of the mammalian postorbital bar. *J. Zool., Lond.* **250**, 405–418.

Osorio, D. and Vorobyev, M., 1996. Colour vision as an adaptation to frugivory in primates. *Proc. R. Soc. Lond. B* **263**, 593–599.

Pettigrew, J.D., 1978. Comparison of the retinotopic organization of the visual wulst in nocturnal and diurnal raptors, with a note on the evolution of frontal vision, in *Frontiers of Visual Science* (eds S.J. Cool and E.L. Smith), Springer-Verlag, New York, pp. 328–335.

Pirie, A., 1959. Riboflavin in the tapetum of the bushbaby. *Nature (Lond.)* **183**, 985.

Polyak, S., 1957. *The Vertebrate Visual System*, University of Chicago Press, Chicago.

Prince, J.H., 1953. Comparative anatomy of the orbit. *Br. J. Physiol. Optics* **10**, 144–154.

Provis, J.M., Diaz, C.M., and Dreher, B., 1998. Ontogeny of the primate fovea: A central issue in retinal development. *Prog. Neurobiol.* **54**, 549–580.

Purvis, A. and Rambaut, A., 1995. Comparative analysis by independent contrasts (CAIC): An Apple Macintosh application for analysing comparative data. *CABIOS* **11**, 247–251.

Purvis, A. and Webster, A.J., 1999. Phylogenetically independent comparisons and primate phylogeny, in *Comparative Primate Socioecology* (ed P.C. Lee), Cambridge University Press, Cambridge, pp. 44–70.

Ravosa, M.J., Johnson, K.R., and Hylander, W.L., 2000. Strain in the galago facial skull. *J. Morphol.* **245**, 51–66.

Regan, B.C., Julliot, C., Simmen, B. *et al.*, 2001. Fruits, foliage and the evolution of primate colour vision. *Philos. Trans. R. Soc. Lond. B* **356**, 229–283.

Ritland, S., 1982. *The Allometry of the Vertebrate Eye*, Ph.D. Thesis, University of Chicago.

Robinson, S.R., 1994. Early vertebrate colour vision. *Nature (Lond.)* **367**, 121.

Rohen, J.W., 1962. Sehorgan, in *Primatologia*, Vol. 2, Part 6 (eds H. Hofer, A.H. Schultz, and D. Starck), S. Karger, Basel, pp. 1–210.

Ross, C.F., 1994. The craniofacial evidence for anthropoid and tarsier relationships, in *Anthropoid Origins* (eds J.G. Fleagle and R.F. Kay), Plenum Press, New York, pp. 469–547.

Ross, C.F., 1995a. Muscular and osseous anatomy of the primate anterior temporal fossa and the functions of the postorbital septum. *Am. J. Phys. Anthropol.* **98**, 275–306.

Ross, C.F., 1995b. Allometric and functional influences on primate orbit orientation and the origins of the Anthropoidea. *J. Hum. Evol.* **29**, 201–227.

Ross, C.F., 2000. Into the light: The origin of Anthropoidea. *Annu. Rev. Anthropol.* **29**, 147–194.

Ross, C.F., 2003. Review of *The Primate Fossil Record* (ed Walter C. Hartwig). *J. Hum. Evol.* **45**, 195–201.

Ross, C.F., 2004. The tarsier fovea: Functionless vestige or nocturnal adaptation? in *Anthropoid Origins: New Visions* (eds C.F. Ross and R.F. Kay), Kluwer Academic/Plenum Publishers, New York, pp. 477–537.

Ross, C.F. and Hylander, W.L., 1996. In vivo and in vitro bone strain in the owl monkey circumorbital region and the function of the postorbital septum. *Am. J. Phys. Anthropol.* **101**, 183–215.

Ross C.F., Hall, M.I., and Heesy, C.P., in press. Were basal primates nocturnal? Evidence from eye and orbit shape, in *Primate Origins and Adaptations* (eds M.J. Ravosa and M. Dagosto), Kluwer Academic/Plenum Publishers, New York.

Schultz, A.H., 1940. The size of the orbit and of the eye in primates. *Am. J. Phys. Anthropol.* **26**, 389–408.

Schwab, I.R., Yuen, C.K., Buyukmihci, N.C. *et al.*, 2002. Evolution of the tapetum. *Trans. Am. Ophthalmol. Soc.* **100**, 187–200.

Simons, E.L., 1962. Fossil evidence relating to the early evolution of primate behavior. *Ann. N.Y. Acad. Sci.* **102**, 282–295.

Simons, E.L., 1972. *Primate Evolution: An Introduction to Man's Place in Nature*, Macmillan, New York.

Simons, E.L. and Rasmussen, D.T., 1989. Cranial morphology of *Aegyptopithecus* and *Tarsius* and the question of the tarsier-anthropoidean clade. *Am. J. Phys. Anthropol.* **79**, 1–23.

Smith, R.J. and Jungers, W.L., 1997. Body mass in comparative primatology. *J. Hum. Evol.* **32**, 523–559.

Stephan, H., Frahm, H., and Baron, G., 1984. Comparison of brain structure volumes in primates. VI. Non-cortical visual structures. *J. Hirnforsch.* **25**, 385–403.

Szalay, F.S. and Delson, E., 1979. *Evolutionary History of the Primates*, Academic Press, New York.

Tan, Y. and Li, W.-H., 1999. Trichromatic vision in prosimians. *Nature (Lond.)* **402**, 36.

Tattersall, I., 1988. Cathemeral activity in primates: A definition. *Folia Primatol.* **49**, 200–202.

Walls, G.L., 1942. *The Vertebrate Eye and its Adaptive Radiation*, Cranbrook Press, Michigan.

Wikler, K.C. and Rakic, P., 1990. Distribution of photoreceptor subtypes in the retina of diurnal and nocturnal primates. *J. Neurosci.* **10**, 3390–3401.

Wolin, L.R. and Massopust, L.C., 1970. Morphology of the primate retina, in *The Primate Brain* (eds C.R. Noback and W. Montagna), Appleton-Century-Crofts, New York, pp. 1–27.

Wood Jones, F., 1916. *Arboreal Man*, Edward Arnold, London.
Yokoyama, S., 2000. Molecular evolution of vertebrate visual pigments. *Prog. Retin. Eye Res.* **19**, 385–419.
Zhou, Y.-H., Hewett-Emmett, D., Ward, J.P. *et al.*, 1997. Unexpected conservation of the X-linked color vision gene in nocturnal prosimians: Evidence from two bush babies. *J. Mol. Evol.* **45**, 610–618.

the case where competition favors the survival of animals of larger body size. If individuals varied in the degree to which their developmental programs coordinate the size and structure of body parts, in this case the organization of the retina, and vision was seriously compromised by eye enlargement consequent to body enlargement in a number of animals (e.g. as we will discuss later, their visual sensitivity became poor in the dark), these animals will not fare well. Only the animals with good coordinating programs will be with us today. Scaling up and scaling down are commonplace in evolutionary time, and it is thus reasonable to examine present-day developmental mechanisms for the details of their scaling properties, stabilized by repeated challenges of the kind just described.

Similarly, every present-day primate is the descendant of species which at minimum has made the transitions from a diurnal niche to a nocturnal niche and again back to a diurnal niche, beginning from the first vertebrates. A second structural feature of evolution to consider in contexts like these is how the genome is translated to variable morphology. Duplications and accretions characterize evolving developmental networks, as opposed to substitutions of new pathways (Gerhart and Kirschner, 1997; Wilkins, 2001). These two phenomena together – the repetition of movement between nocturnal and diurnal niches, and genetic change that build on redundancy and duplication – suggest we should see the evidence of successful transitions in building eyes suited to both low- and high-light levels in the developmental programs of present-day primates.

In the present comparative analysis of visual system development in primates, we will compare two things – present primates to non-primates, and primates to each other. We will first describe the commonalities of fundamental retinogenesis, and the development of ocular morphology of primates compared with other vertebrates. We will then describe variations within primates in retinal size, color vision, and nocturnality and diurnality, and what is known about their development. This review will concentrate on the retina and the eye, though we will note a few features of central nervous system organization of particular developmental interest in primates, such as the pattern of crossing of the optic nerves at the optic chiasm. Finally, we will discuss the development of a feature unique to primates (among mammals), the fovea, with its high central concentration of photoreceptors coupled with centripetal displacement of photoreceptor cell bodies and the rest of its associated retinal processing architecture.

2.2 Fundamental organization and development of the retina

The basic organization of the vertebrate retina has been fixed at least since the first divergence of the jawless vertebrates, over 400 million years ago, insofar as we can judge from the retinas of the current representatives of the major vertebrate radiations, a rather remarkable fact. 'Basic organization' in fact refers to virtually all functional and structural features of importance, including the image-forming eye with its components of cornea, lens, and retinal conformation; the method of phototransduction, the presence of rod and cone photoreceptors; multiplication of opsin types allowing the possibility

2

Comparative Aspects of Visual System Development

Barbara L. Finlay, Luiz Carlos de Lima Silveira, and *Andreas Reichenbach*

2.1 Introduction

Teleological thinking in evolution tends to worsen the closer the animals we compare come to ourselves. It is not difficult to objectify a shark or a bat – to examine the niche they inhabit and the problems they must solve, and begin the exploration of how their nervous systems combine species-general neural structure with species-typical adaptations. When we examine other primates, however, we often fail to view them in the large evolutionary context that makes sense of their design features, viewing them as the 'model systems' so favored by biomedical funding agencies. In this chapter, we will attempt to place primate visual system development in the larger context of vertebrate visual system evolution. We will argue that only an evolutionary context makes sense out of the developmental mechanisms that are deployed to generate eyes and connect them up.

A concept useful for understanding the nature of developmental mechanisms found in all present-day animals is 'evolvability' (this concept borrowed and modified from the sense used by Gerhart and Kirschner (1997) in *Cells, Embryos and Evolution*). Every extant animal is the descendant of animals going back to the beginning of evolutionary time, the successful survivor of multiple mass extinctions, climatic shifts, niche invasions, diseases, disasters, and mishaps. These forces alone predispose to powerful, redundantly organized developmental strategies. Not only disastrous change, however, but also persistent kinds of changes seen over evolutionary time should be considered in the context of evolvability. Changes in body size are a change of this kind. Consider

The Primate Visual System: A Comparative Approach Edited by Jan Kremers

of color vision; the remaining classes of retinal cells, including horizontal cells, bipolar and amacrine cells, and retinal ganglion cells and their pattern of layering; lateral inhibition as realized both in the photoreceptor layer and in interactions between photoreceptors, horizontal cells, and bipolars, and again in the interactions of bipolar and amacrine cells with retinal ganglion cells; many transretinal classes of neuromodulators, and finally oculomotor organization relevant to eye stabilization (Ramón y Cajal, 1972 (translation date); Rodieck, 1973; Braun, 1996). A discussion of the evolution of the vertebrate retina, and secondarily, the primate retina, is about the adaptation of the eye to the broad nature of the environment (e.g. eye-in-water versus eye-in-air), to size (eye diameters range over approximately a hundredfold, from the smallest fish to the largest whale eye), and to niche (nocturnal versus diurnal, predator versus prey, and so on). Niche adaptations may have many facets, for example, the characteristic chromaticity of the environment, the specialization of the eye for central vision and associated ocular motility, characteristic patterns of whole-body motion, the use of the eyes as social signaling devices, and so on. Resulting variable features of the eye and visual system include the number and spectral features of photopigments (Chapters 3 and 4), the number and topological arrangement of photoreceptors, bipolar cells and ganglion cells (Chapter 5); the presence of neural machinery capable of producing eye movement (capable of overriding ancestral eye stabilization mechanisms under vestibular control; see Chapter 10) and various mechanisms for the central modulation of retinal function.

In broad strokes, the development of the eye is similarly conserved, and in fact, the fundamental mechanism of the embryonic positioning of the eye through the action of the conserved developmental patterning gene, *PAX6*, antedates vertebrates. The nature of this conserved function is the subject of much current debate (Callaerts *et al.*, 1997; Fernald, 2000). We will not discuss here the early morphogenesis of the eye (Robinson, 1991), though we will consider some aspects of early gene expression when we discuss the development of the primate fovea. At this point, we will discuss retinal development following from the time that the primordium of the eye has evaginated, contacted the tissue that will give rise to the lens, and folded back in to give the bilaminar cup-shaped organ that will give rise to the retina in the ventricular lamina adjacent to the prospective lens, and the choroid and supporting tissue in the ventricular region forming the outside of the cup.

2.3 Neurogenesis

As in all neural tissue, the retina arises from the cells of the ventricular zone directly adjoining the ventricle. The onset of retinogenesis is signaled by the interaction of ‘symmetry-breaking’ cell surface molecules and receptors that take the population of undifferentiated neural precursor cells, and drive some proportion of them toward their terminal, differentiating division. After cells exit the precursor pool, a second step determines what sort of cell the postmitotic neuroblast becomes (Cepko, 1999; Dyer and Cepko, 2001). The order of retinogenesis is conserved in all mammals (Clancy *et al.*, 2001), though not in all vertebrates (Beazley *et al.*, 1989). Ganglion cells are produced first, then cones, then horizontal cells. A slight decrement in cell production follows, and then amacrine, Müller cells, and bipolar cells, and finally rods are produced. This pattern of neurogenesis as it is seen in the macaque retina, schematized for retinal ganglion cell, cone, bipolar, and rod cell classes from LaVail *et al.* (1991), is shown in Figure 2.1. In this representation, the amount of each cell class that is produced on each postnatal day is represented as a percentage of its own total population, with area under each curve roughly approximated; in subsequent representations, the absolute numbers of cells of each type will be emphasized instead. This pattern of neurogenesis has many intrinsic

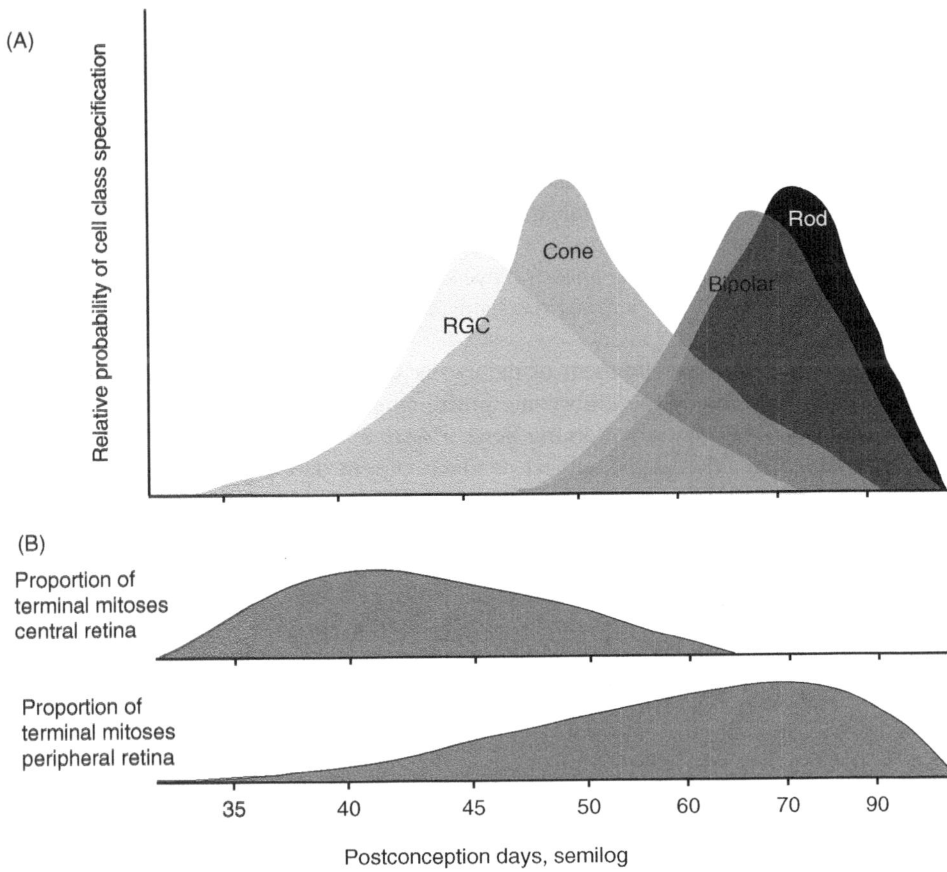

Figure 2.1 (A) Order of production of retinal ganglion cells, cones, bipolar cells, and rods in the retina of rhesus macaque. For the two types of neurons omitted from this map, horizontal cells virtually overlap cones; amacrine cells lead bipolars slightly. The ordinate shows the postconceptional day of terminal cell division, the abscissa percent of each cell class produced on that day. (B) The temporal gradient in the probability of specification of cell of a particular class from an undifferentiated cell (top graph) versus the spatial gradient in the probability of exit of precursor cells in central retina (middle graph) versus peripheral retina (bottom graph). Replotted and schematized from Rapaport *et al.*, 1996

features of interest for understanding retinal variation in primates – scaling of the retina to different eye sizes and for nocturnality – and we will return to it several times more. At this point, however, we will turn to general features of retinal maturation.

2.3.1 Gradients of maturation

The mammalian retina differs from the non-mammalian retinas that have been the most studied – the goldfish (Raymond, 1991), the cichlid fish (Fernald, 1989, 1991), and the 'workhorse' of developmental biology, the zebra fish (Easter and Nicola, 1996), in that the mammalian retina is produced in a single bout in early embryogenesis, while most teleost retinas have indeterminate growth, setting up a basic retinal structure, but adding cells to it throughout the lifespan. Retinas with indeterminate growth must essentially solve scaling problems on the fly as they grow. As in mammals, a separate strategy for the genesis of cones versus rods exists, to solve the distinct scaling required to maintain acuity in high-light conditions versus sensitivity in low-light conditions (Fernald, 1991). The solution to this same problem appears in a different form in the mammalian retina, which we will discuss in some detail.

Primates, compared generally with mammals, have long periods of retinogenesis (Clancy *et al.*, 2001) and a relatively large retina (Ross, 2000). This extended time and large space allow probably the most distinctive feature of neurogenesis in primates to be seen, which is the presence of major gradients in the onset and cessation of neurogenesis across the retinal surface (LaVail *et al.*, 1991; Rapaport *et al.*, 1992, 1996). The nature of the retinal gradient with respect to control of cell specification is somewhat counterintuitive. Cells exit the precursor pool first in the central retina to begin their terminal maturation, and the specification of cell type is determined by a complex interaction of cell lineage and the extracellular environment (Alexiades and Cepko, 1997; Livesey and Cepko, 2001; Ohnuma *et al.*, 2002). One could imagine that this process could overall be best described by each retinal location in turn proceeding through the cascade of events that first statistically favors retinal ganglion cells, then cones, and so forth. The process, however, seems better described by two gradients, one spatial and one temporal (Figure 2.1A and B). The exit of the precursor population to terminal neurogenesis has a central-to-peripheral spatial gradient (Figure 2.1B), while cell specification appears best described as a transretinal clock (of unknown kind) without a strong spatial gradient (Figure 2.1A). Experimental support for this conjecture comes from the manipulation of time of exit from the cell cycle in the rat (Austin *et al.*, 1995; Bao and Cepko, 1997) and in *Xenopus* (Chang and Harris, 1998; Ohnuma *et al.*, 2002). Production of excessive terminal cells in early retinogenesis produces larger numbers of retinal ganglion cells and cones, while exit of precursor cells later produces mostly rods – if a local count of divisions advanced the specification clock, we would not see this pattern. Extrapolating these findings to the particular case of the primate retina, it is likely that at the time that the majority of cells in the central retina become postmitotic, cell specification mechanisms produce principally retinal ganglion cells and cones and their attendant horizontal cells and some bipolars, producing the very low convergence ratio from photoreceptors to ganglion cells seen in the central primate retina. By the time most peripheral retina cells enter terminal mitoses, principally bipolars and rods are specified, but fewer ganglion cells, automatically producing the high convergence ratio of the peripheral retina. In smaller retinas with flatter gradients, the disparities in cell type between central and peripheral retinas are not so pronounced. The statistical probabilities of producing each cell class are broad enough that no retinal location totally lacks any cell type, with one exception, the primate fovea (hypotheses for the development of this primate feature will be discussed subsequently). A possible cause of the absence of rods in the fovea is that neurogenesis is terminated, or central precursor pools are exhausted, before the progression of cell-type specification reaches the period when rods are specified, leaving a central rod-free zone.

Across species, this gradient of maturation appears to have immediate consequences for the rest of retinal maturation, as the gradient becomes realized in the mechanical properties of the retina. The retina grows in diameter after initial cytogenesis principally by passive stretch – the 'balloon' model (Coulombre, 1957; Mastronarde *et al.*, 1984; Kelling *et al.*, 1989; Robinson *et al.*, 1989; Reichenbach *et al.*, 1991). The early 'area centralis' becomes relatively inelastic compared with the rest of the retina and remains relatively fixed at its embryonic dimensions and, most important, embryonic cell density. The cause of the lesser central elasticity could simply be the direct result of features of early maturation, i.e. greater thickness, absence of cell interpolation, and growth of connective processes between cells, or might involve additional directed changes in the cytoskeleton or cell adhesion. Thus, the peripheral retina reduces cell density per unit visual angle substantially during early development, even as it adds cells, due to its greater elasticity during the period when ocular dimensions are growing most rapidly; in several species, including humans, there is evidence for excess cell loss in the peripheral retina as well (Sengelaub and Finlay, 1982; Provis, 1987). Extrapolating from several species of which the relationship of embryonic cell density, elasticity, and retinal stretch have been studied (chick, cat, and rabbit), it is likely that the unusually steep spatial gradients of neurogenesis in the primate produce very steep gradients of cell density by this biomechanical process.

2.4 Topology and specification of cell-type subcategories

The prior discussion has lumped all variation in cell categorization to the seven large retinal cell classes. There is a second aspect of cell deployment, which is the spatial arrangement of cell subclasses across the tangential aspect of the retina. These include the different subclasses of photoreceptors and their associated bipolars; ON- and OFF-bipolars and ganglion cells; the very wide diversity of amacrine cells; and the various subtypes of ganglion cells. With two exceptions, there are few studies in primates that would suggest any unusual variations on vertebrate-general mechanisms that order the array of subclasses of cells across the retinal surface. Research to date suggests that any identifiable cell class 'tiles' the retina completely (Cook and Chalupa, 2000). Work on ON- and OFF-bipolar cells and the dendritic development of retinal ganglion cells suggests that the initial distribution of cells is more random than the terminal distribution, which becomes hyperdispersed due to the combined effects of competition with like cells for synaptogenesis, some cell death, and biomechanical factors (Reese and Galli-Resta, 2002).

2.4.1 Cone photoreceptors

All primates, with the exception of two cases of nocturnal cone monochromats, the New World monkey *Aotus*, the owl monkey, and the bush baby *Otolemur crassicaudatus* (Jacobs *et al.*, 1996) possess at least two cone types, a short- and a middle-wavelength sensitive cone (Ross, 2000; Heesy and Ross, 2001; Figure 2.2). A large fraction, including all Old World monkeys and the great apes (Jacobs and Deegan, 1999), the New World howler monkey *Alouatta* (Jacobs and Deegan, 2001), and female New World monkeys of the tamarin and cebid groups (Jacobs, 1998), further differentiates the middle-wavelength cone into long- and medium-wavelength-sensitive classes (i.e. are trichromatic). We will first discuss the disposition of the first two cell classes in dichromats, those that express the 'blue' (S) opsin versus those that express any one of the family of intermediate opsins (historically named R/G or M/L), which can be distinguished immunocytochemically. Work in fetal rhesus monkey shows that cones are initially laid out in a regular array, particularly one precocious cone that leads its neighbor cones in the expression of opsins, an arrangement that appears similar to the early arrangement of photoreceptor mosaics in *Drosophila* (Wikler and Rakic, 1994). Eventually, the S and M/L groups form independent mosaics, with about 10 times as many M-/L- cones as the S class; different primate species differ distinctly in the regularity of their cone mosaics, but the significance of this observation is unclear (Wikler and Rakic, 1996). In addition to the opsin expressed, these two cone photoreceptors (S versus. M/L, Figure 2.2) differ from each other in a number of dimensions, including number, retinal distribution, size of cell body, and connectivity at maturity.

The duplication and differentiation of two opsins (M, L) from the single M-/L-opsin appear to have occurred at least three times in the primate lineage, producing obligatory trichromacy in all Old World monkeys and their descendants, and at least once in the ateline line of New World monkeys (demonstrated only in *Alouatta*, the howler monkey). In other New World monkeys, non-obligate trichromacy appears in females only, as the gene for opsin in the yellow range takes two forms and is linked to the sex chromosome (Jacobs, 1998; Jacobs and Deegan, 1999, 2001). In only those females who happen to carry two different alleles of the gene, trichromacy is possible, as only one opsin will be expressed per photoreceptor, the other silenced at random (Chapters 3 and 4). In all these cases, however, the only difference between the L- and M-opsin-expressing photoreceptors lies in the opsins themselves, in the substitution of a few amino acid groups that cannot be distinguished immunocytochemically. In fact, there is no known feature of cell type – size, distribution, or dendritic appearance – that can as yet distinguish the L and M photoreceptors. Recent work shows highly variable foveal distributions of these photoreceptors, in no way mosaic in appearance (Roorda and Williams, 1999; Roorda *et al.*, 2001). The likelihood is that in the absence of other cell-type specification or intrinsic mosaic organization, it is only the one-to-one (or better) photoreceptor to ganglion cell convergence in central retina that allows the L/M chromatic distinction to be conveyed to the central nervous system (Sjostrand *et al.*, 1999; see Chapter 5). The generic activity-dependent organizational capabilities of the

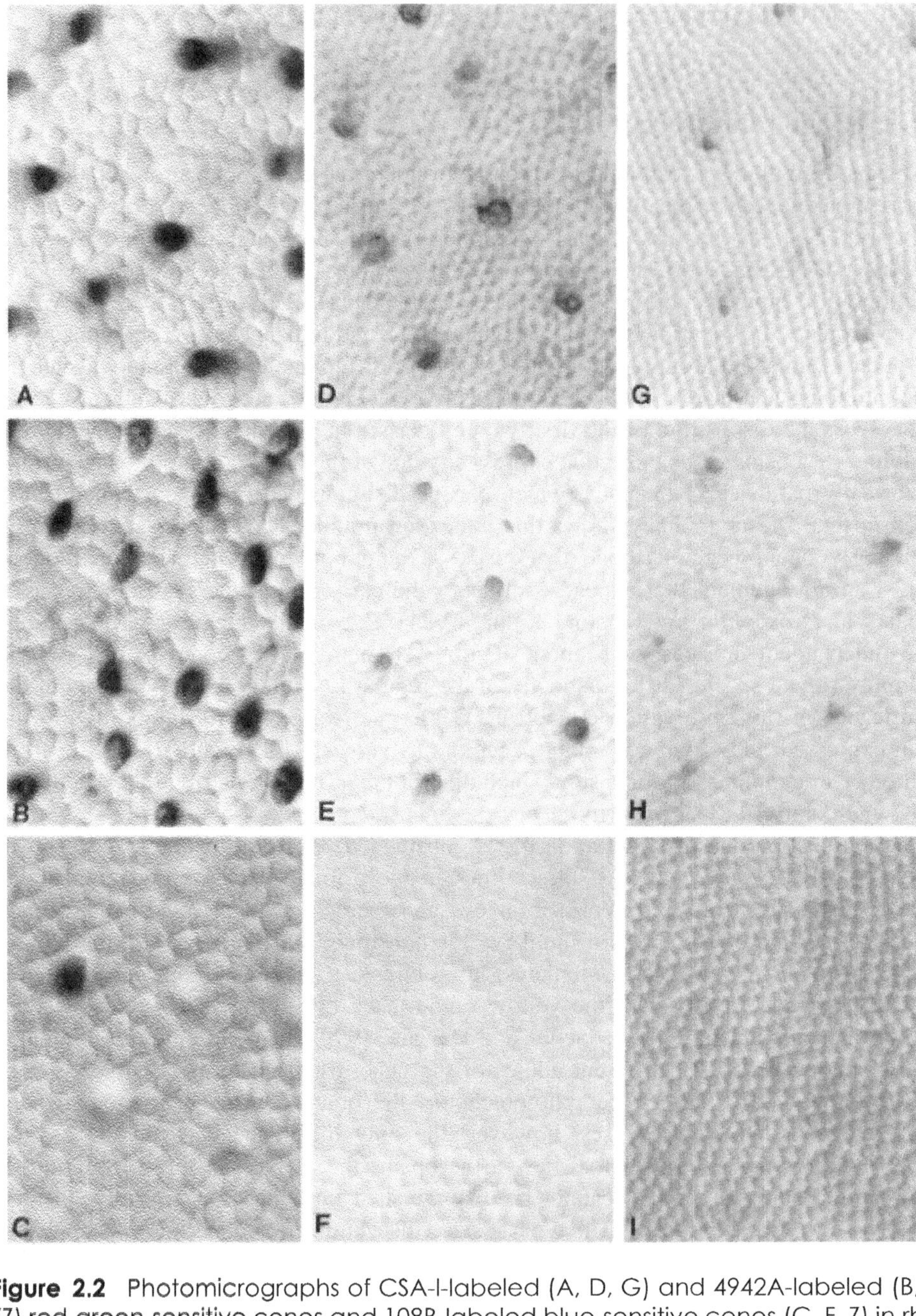

Figure 2.2 Photomicrographs of CSA-I-labeled (A, D, G) and 4942A-labeled (B, E, ZZ) red-green-sensitive cones and 108B-labeled blue-sensitive cones (C, F, Z) in rhesus monkey (first column), owl monkey (middle column), and bush baby retina (third column) (Wikler and Rakic, 1990). Note the regular and periodic absence of labeling of putative blue-sensitive cones in the CSA-1- and 4942A-labeled rhesus monkey retinas. Magnification: 1900× for E, F, H, and I; 1350× for A, B, C, D, and G. Reproduced by permission of The Society for Neuroscience

central nervous system rather than molecularly mediated cell recognition mechanisms may then permit the development of a three-dimensional color space. This Hebbian 'fire together, wire together' mechanism is already employed in such basic developmental visual functions as sharpening topographic maps and organizing ocular dominance layers and columns, and could also be employed in the segregation of central chromatic pathways. In addition to Hebbian mechanisms, existing dichromatic color vision has mechanisms available potentially useful to implement trichromatic vision. In particular, the ordinary contextual dependency of color computations, whether the context is immediate ambient illumination, or the relative number of photoreceptors of a particular type, could be used to extract a signal from the apparently random expression of L- versus M-opsins for useful perceptual distinctions (Golz and MacLeod, 2002; Neitz *et al.*, 2002).

2.4.2 Retinal ganglion cells

The retinal ganglion cells of both New and Old World monkeys come in several subclasses, 'PC' ('midget', or the small-celled classes that project to the parvocellular layers of the lateral geniculate nucleus (LGN)), 'MC' (also parasol, projecting to the magnocellular layers of the LGN), and several other smaller groups (Silveira *et al.*, 2003; see Chapters 5 and 6). In development, the PC class is produced first, and MC second (Rapaport *et al.*, 1992). The relative order of the final precursor divisions producing these two cell types, and the spatial distribution of rods and cones and the order of their final divisions may allow convergence to remain proportional in eyes of different sizes. Though both cell classes get rod input, the greater rod input is to MC cells; in addition, there is greater convergence of all bipolar input, cone and rod, to MC cells compared with PC cells. As will be discussed in the following text, the similar position of MC cells and rods in the relative order of neurogenesis will allow these functional pathways to scale similarly in eyes of different sizes.

2.4.3 Other dimensions of cell-type specification within broad classes

One dominating feature of the historical retinal literature is the classification of cells into types (e.g. nine types of bipolar cells in rats; multiple classes of amacrine cells), where the process is usually carried to maximal distinctiveness, sorting neurons on every possible dimension of function (ON/OFF; temporal and spatial selectivity; chromatic response) and morphology (size of arbor; branching pattern and lamination pattern – e.g. Wässle *et al.*, 1987, 1991). A developmental classification system might produce a more condensed taxonomy. Considering activity-dependent organization and other aspects of competition in early development, many aspects of 'cell type' are revealed to be epigenetic rather than intrinsic, depending on the pattern of interaction of the cell with its neighbors, or the relative densities of cells (Bodnarenko and Chalupa, 1993; Bodnarenko *et al.*, 1995; Reese and Galli-Resta, 2002). The rapid growth of genomic approaches to cell specification should ultimately give us a better taxonomy of 'type' – how many rules exist for arbor growth, receptor expression, and so forth. For now, however, the understanding of the details of adult typologies in terms of specific developmental growth rules is only beginning to be established.

2.5 Lamination; synaptogenesis; axon outgrowth; and cell death

Extension of processes, and expression of the metabolic processes associated with the production and degradation of neurotransmitters, begins rapidly following the final division of precursors producing retinal neurons and photoreceptors. The first evidence of processes are found in the inner plexiform lamina nearest the retinal ganglion cells (Robinson, 1987). Development of processes spreads to the peripheral retina following the gradient of cessation of neurogenesis (Robinson, 1991). Slightly later differentiation of the outer plexiform layer begins and the connections between photoreceptors, bipolar cells, and horizontal cells develop in a similar manner. The lamination in the adult retina is highly precise, with bipolars and ganglion cells, respectively, sorting themselves into laminae according to their ON or OFF response to light, to their chromatic responses (Rodieck, 1973). While no experimental demonstrations of the activity dependence of development of cell processes have been done in primates, from work with cats and ferrets, we may infer three possible organizing roles or manifestations of activity dependence in early retinal organization. If ganglion cell spiking activity is experimentally eliminated, the normal developmental course of dendritic spine reduction of ganglion cells is altered (Wong *et al.*, 1991). In initial development, ganglion cells extend dendrites in all laminae, while bipolar axons are restricted in their lamination. In cats, if retinal activity is blocked by glutamate blockers, the ganglion cells retain their overlapping and uniform arborization to adulthood (Bodnarenko *et al.*, 1995), much in the way the axons of the lateral geniculate neurons representing the right and left eyes retain their overlapping and uniform distribution if their activity is blocked. However, it should be noted that virtually all activity-dependent retraction effects that have been described in the central nervous system involve the retraction of axons to stabilize their pattern of connectivity on dendrites, while in the case of the retina, the *dendrites* of retinal ganglion cells segregate according to the lamination of the bipolar 'axons' which appears to be fixed by some other source of information, yet unknown (Katz and Shatz, 1996).

Finally, in a number of species (Wong, 1999) in the time preceding eye opening, organized activity is produced in the retina in the form of retinal waves, which sweep across the retina in spatially organized fronts. The spatial correlation of neuronal activity that these wavefronts represent may provide some of the statistical structure required for activity-dependent organization of retinal projections in the central targets of the retina before actual visual experience. The locally correlated activity in each eye could initiate the organization of a number of important aspects of visual system circuitry, for example, the oriented receptive fields of the visual cortex, the topographic mapping of retinal projections on any central array, or the sorting of projections that are decorrelated, as the two eyes are before they begin to view the external world.

2.5.1 Connecting with central targets

In the rhesus monkey, axons appear in the optic stalk at approximately postconception day 35 (PCD 35), and reach the chiasm about five days later, which occurs before the peak of retinal ganglion cell genesis occurs in the retina, reminder of the ongoing and overlapping nature of cytogenesis and development of connectivity in retinal development. The optic nerve reaches its principal central targets (LGN and superior colliculus) at about PCD 48 and progressively invades them over the next 10 days (Clancy *et al.*, 2001). The peak of optic axon nerve number is reached at PCD 65, and is reduced by more than half by PCD 100, due to the apoptotic death of retinal ganglion cells (Rakic, 1983; Provis, 1987). For comparison, eye opening occurs in rhesus at PCD 126 and birth at PCD 165. Comparative studies of the timing of such events showed that the order of these events is highly conserved and predictable across species including primates (Clancy *et al.*, 2001).

At the optic chiasm, the hemi-decussating pattern of contralateral and ipsilateral retinal projections in primates is quite precise, and while it follows the general vertebrate strategy of aligning congruent representations of the visual field in the central targets of the retina, some interesting details vary (Godement *et al.*, 1990; Chalupa and Dreher, 1991; Chalupa and Lia, 1991). In particular, in most mammals at least one cell class in the retina projects contralaterally from all parts of the retina, temporal and nasal, while in primates all cells from the temporal retina project contralaterally and cells from the nasal retina project contralaterally. The molecules mediating midline pathway choice, including the optic chiasm, appear to be highly conserved across invertebrates and vertebrates (Erskine *et al.*, 2000). The question of how some retinal neurons and their axons from the nasal retina come to carry different labels to dictate their midline choice is not known.

2.5.2 Cell death

Cell death in the retina has been the subject of many past reviews that have summarized its incidence and topographic distribution (Finlay, 1992), and current work concentrates upon the developmental context of cell survival (de la Rosa and de Pablo, 2000), the characterization of the molecular communication between interconnecting populations of neurons, and the nature of the pathways that initiate the orchestrated cell death termed apoptosis (Nijhawan *et al.*, 2000). Cell death occurs in all retinal layers (Georges *et al.*, 1999; Linden *et al.*, 1999; de la Rosa and de Pablo, 2000). In nervous system development, two-way signaling between most interconnected populations, like motoneuron and muscle, or retinal ganglion cells and their targets, is required for the growth and often survival of both populations. Not all apoptotic cell death is regulated by such trophic interactions between neurons – sometimes a period of neuronal apoptosis will occur in development entirely independent of the connections of cells. Retinal ganglion cells, however, which are highly dependent on trophic support for their survival, are the best-studied cells of the visual system and thus we know the most about this aspect of early cell death. Retinal ganglion cell survival requires that axons establish connections, so there is an absolute requirement for trophic support mediated by target contact (Finlay and Pallas, 1989). The absolute amount of cell loss in regions of the retinal ganglion cell layer can be extremely high in normal development (e.g. 90 percent in the human retina (Provis, 1987)) and varies between species. The pattern of cell loss also is variable with topographic position in the retina, with excess ganglion cell loss in the retinal periphery, and is thus a potential contributor to retinal organization and topography.

2.6 Emmetropization

In primates, much retinal maturation is still to occur after birth, though all cells have been generated, all central connections are generally stabilized, and lamination is adult-like. The eye grows substantially in size from birth to adulthood with proportional growth of the retina, principally by stretching in the non-uniform manner described previously. The fovea is beginning its development, which will continue in the first year of life, to be described later in this chapter. In early development, the growth of the eye is in part under the control of experience. 'Deprivation myopia' was first described in primates (Wiesel and Raviola, 1977), noting the unusual growth of the eye after early eye suture in infant macaques; and early 'runaway myopia' is a well-known clinical syndrome. While most of the detailed work on mechanism has been done in the chick (Troilo, 1992), work in marmosets (Troilo and Judge, 1993), rhesus monkey (Hung *et al.*, 2000), and human clinical syndromes make extrapolation from the work in chick retina quite plausible (Wallman *et al.*, 1987a).

'Emmetropization' refers to the process of matching the length of the eye to the power of the eye's optics (principally contributed by lens and cornea). While most primates are born with a general match achieved, early acuity is very poor compared with the adult, and there may be very substantial astigmatic errors of refraction – that is, the refracting power of the eye may be quite different on its horizontal and vertical axes (Howland and Sayles, 1985). During the first year of life, the optics undergo much modification, under the control of eye activity. Defocus can be measured directly by the retina and alter retinal growth, even if the optic nerve is severed (Troilo *et al.*, 1987). 'Defocus' can be represented crudely as the relative amount of retinal activity, as a blurred image is a poor stimulus to photoreceptors, and other features of image movement can be used to infer sign of defocus (Schaeffel and Diether, 1999). In chicks, a high-contrast image signals that the length of eye and optics match, and growth of the eye is checked. A blurred image is taken as evidence that the eye is too small, and the eye continues to grow, which can result in the positive feedback condition of 'runaway myopia' if the eye has made the wrong guess about the direction of defocus and the eye is already too long (Wallman, 1995). This activity must be initially mediated by cones, though it is thought that the signal that modulates eye growth directly probably goes through the amacrine pathway, causing changes in choroidal thickness and scleral elasticity (Wallman *et al.*, 1994). In particular, the glucagonergic amacrine cells have been shown to respond to defocus in retinal image and even to its sign, by changed expression of the immediate-early gene product, ZENK (Feldkaemper and Schaeffel, 2002).

Anthropocentrism may have caused us to somewhat mischaracterize this process, as what was described above is the sequence of events necessary to properly focus an eye in the day. A nocturnal animal will want a different control regime – although it is always useful for the image to be focused on the retina, linking focus to inhibition of growth is a poor strategy for a nocturnal animal, as a much larger eye will produce a

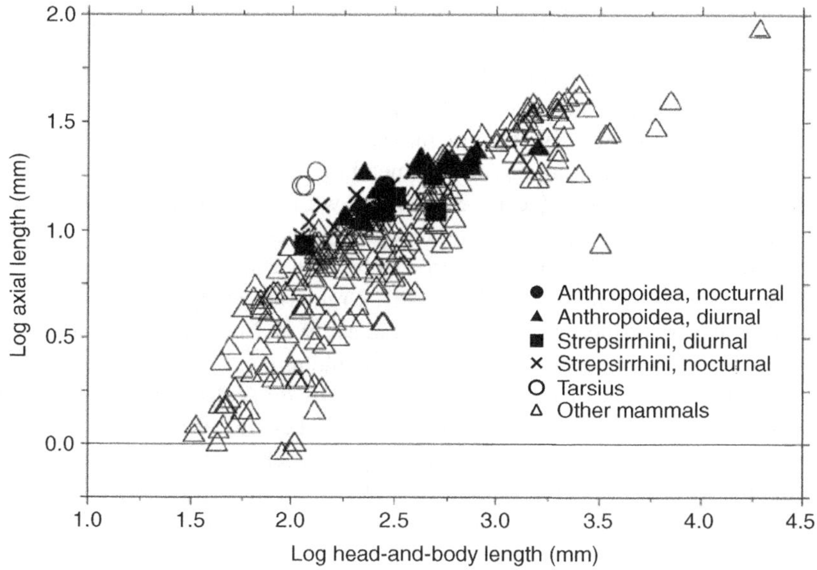

Figure 2.3 Log–log plot of eyeball axial diameter against head-and-body length (HBL). Redrawn from Ross (2000). Reduced major axis slope = 0.815. Best-fit line is axial diameter $= -2.353 + 1.97$ HBL -0.244 HBL^2; $R^2 = 0.775$

much greater photon capture. Most nocturnal animals do have larger eyes than diurnal ones; such is the case, for example, for the owl monkey compared to other New World monkeys of similar body size, as well as other nocturnal primates (Heesy and Ross, 2001; Figures 2.3 and 1.3). We hypothesize that mammalian eyes may have two competing control regimes available, one rod-dominated and one cone-dominated. In fact, it is known that the 'emmetropization' signal is gated by circadian rhythms in the very diurnal chick (Schaeffel *et al.*, 1995) – in a diurnal animal, the absence of cone activity at night should not be taken as evidence of inadequate focus. However, if there is a great deal of rod activity at night, or an absence of cone activity during the day, this could conceivably be transduced as a signal to produce the optimally large nocturnal eye, an adaptation, not a 'failure of emmetropization'. Later, we will consider how employing this proposed developmental route might produce the coordinated changes seen in the conformation of the eye in the owl monkey, and in part account for the curious pattern of cone monchromacy they have evolved.

2.7 Scaling the eye

The eyes of diurnal primates are absolutely large compared with most mammals and scale allometrically with body size (Ross, 2000; see also Chapter 1). Curiously, however, given the fact that diurnal primates are thought to be visual specialists, the eye of diurnal primates scales with a rather flat slope, particularly notable if you consider the human case – if we scaled at the general mammalian slope, our eye would be considerably larger in diameter than it is (Figure 2.3). We have proposed that one unusual, non-scaling feature of the primate eye could account for it. In all primates, regardless of overall eye size, the fovea is approximately 0.5 mm in diameter, and it is possible that it can be no larger due to various physiological and metabolic constraints (Franco *et al.*, 2000). The high metabolic activity of the fovea coupled with the absence of vasculature, and the drawn-out fibers of Henle from photoreceptor processes to cell body may limit its size to the size observed. As the fovea probably first arose in a primate of relatively small eye and body size (Heesy and Ross, 2001), retention of the fovea would have put a substantial brake on further enlargement of the eye with brain and body size, so as not to have the fovea subtend an excessively small visual angle.

Even with the constraint of the fovea, primate eyes do scale with body size, ranging from around 10 mm in diameter in the smallest monkeys to around 30 mm in various anthropoid apes (Heesy and Ross, 2001). Scaling an organ (made of cells of constrained size) which has several geometrical features under different constraints is an interesting construction puzzle – not all parts of the eye may scale the same and retain function. The overall conformation of the optics of the eye scale up linearly – for example, the eye of the rat and the eye of the mouse, appropriately scaled, are superimposable (Remtulla and Hallett, 1985). Within the eye, however, retinal thickness may not vary much, due to the constraints of perfusion and light passage, and stays close to a thickness of some 200 μm or so; as we have already discussed, the size of the fovea appears to be constrained to approximately 0.5 mm.

Rods and cones must scale at different slopes with eye size, in order to hold constant their particular functions. If an eye becomes twice as large in diameter, no change is necessary in the number of cones to retain the same visual acuity – since the retina is flooded with photons in diurnal vision, a single cone will have no difficulty encountering a photon in the visual angle it represents regardless of the angle the cone itself subtends. More cones could of course be added, to improve acuity, but we are discussing here what is required to maintain equivalent, not improved, function over different eye sizes.

The same solution will not work for rods – working at low-light levels and low photon numbers, a single rod located in a larger absolute retinal area even if retinal angle is unchanged will detect proportionately fewer photons, even allowing for biologically plausible increases in the size of a single rod. Rods must tile the surface of the retina to maintain sensitivity, increasing in number approximately at the square of change in retinal diameter. The observed scaling of rods and cones in diurnal primates conforms closely to this functional necessity, where cones increase in number by less than a factor of 2 between marmosets (*Callithrix jacchus*) and humans, while rods increase by more than a factor of 10 (Figure 2.4).

How is this consistent within- and across-species scaling necessity executed in the schedule of neurogenesis of the retina? While the precise kinetics remain to be worked out, the schedule of neurogenesis in the retina is arranged such that extension of the period of embryogenesis (the principal way mammals get bigger) should automatically produce the desired differential scaling. In much prior work, for the central nervous system in general, we have determined that components of the nervous system scale very

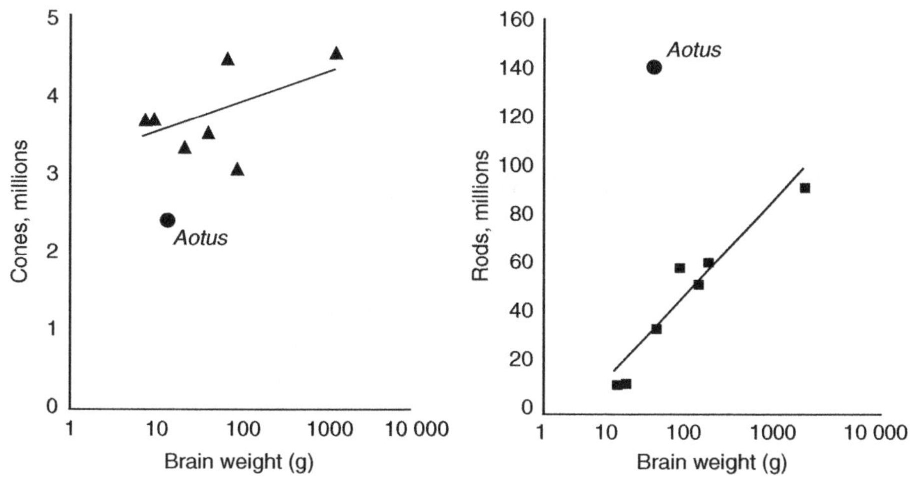

Figure 2.4 Rod and cone numbers versus brain weight in seven diurnal primates (in order, *Callithrix jacchus*, *Saguinus midas niger*, *Saimiri sciureus*, *Alouatta caraya*, *Cebus apella*, *Macaca mulatta*, and *Homo sapiens*) and one nocturnal primate (*Aotus azarae*). Data for *Macaca* and *Homo* from Curcio *et al.* (1987, 1990) and Curcio and Hendrickson (1992) and the remainder from Snow *et al.* (1997), Franco *et al.* (2000), Kaskan *et al.* (2005). Regression equation, cones, diurnal primates only, $y = 0.167 \ln(x) + 3.159$; $R^2 = 0.27$; regression equation rods, diurnal primates only, $y = 0.16.137 \ln(x) - 16.96$; $R^2 = 0.92$

predictably with brain size, but with different slopes – for example, as mammalian brains become larger, they are composed of predictably and proportionately more neocortex, but proportionately less medulla. The pattern of neurogenesis, conserved across mammals, accounts for which structures will grow disproportionately large in neuron number – 'late equals large' (Finlay and Darlington, 1995; Finlay *et al.*, 2001). That is, if a schedule of neurogenesis is extended, for example, from 15 days in one species to 30 days in another, groups of neurons go into terminal neurogenesis at roughly proportionate times within the entire period. However, the precursor pools from which these neurons are derived are increasing at an exponential rate in absolute time, and the last-differentiating cell groups (like the cortex and cerebellum) will be launched from a disproportionately larger pool. Such is the case for the relative timing of cone and rod neurogenesis in the retina, as modeled for marmoset versus human (Figure 2.5). Those cell types that must change in number exponentially with eye diameter, rods and their attendant bipolar cells, are located last in order of differentiation, and those that need not change are produced first.

This obligatory, coordinated scaling of retinal cell classes to match functional requirements is an example of the idea of 'evolvability' with which we began this chapter. Those potential ancestors with perhaps a reversed order of neurogenesis in the retina who might have had a selective advantage at a larger body size, but who unfortunately became blind in the dark as a result, presumably would enjoy less reproductive success. Common selection points, such as body size, will naturally select on coordinating features

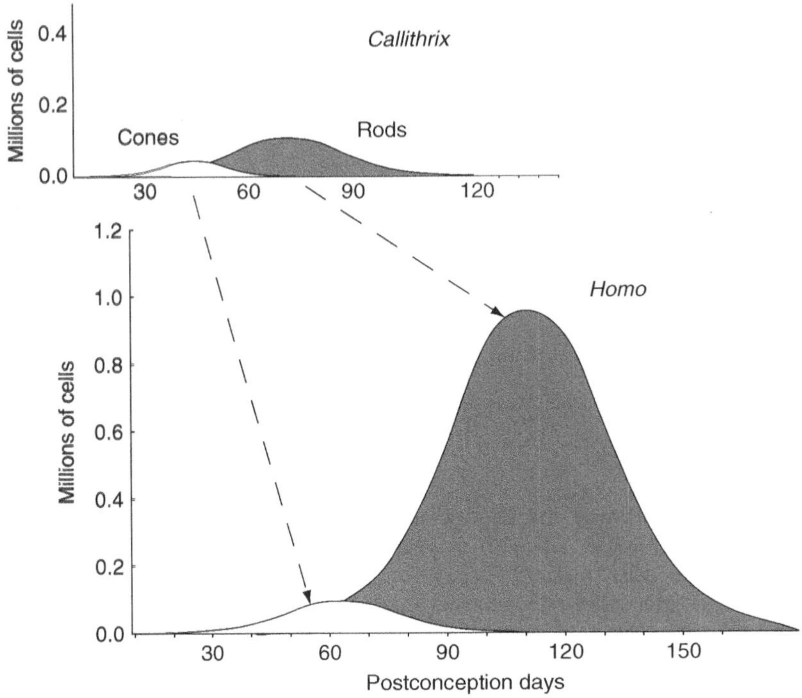

Figure 2.5 Schematic to demonstrate how extension of the period of retina neurogenesis may disproportionately increase the numbers of later-generated cell groups by allowing disproportionate increase in the precursor pool from which later cell groups are drawn. Neuron numbers and developmental durations drawn from Finlay *et al.* (2001) and Clancy *et al.* (2001)

in embryogenesis, and the retina is a nice example of order of neurogenesis adapted to two contrasting functional constraints.

2.8 Producing the nocturnal eye

Nocturnal eyes, overall, have a number of features that are different from diurnal eyes. Underscored at the outset should be the fact that most eyes are duplex, with the ability to function at both night and day, but most eyes have features that optimize one or the other niche. Eyes specialized for nocturnal vision have design features which maximize photon capture – they are often frontal to double light catch by focusing two eyes on the same scene, each eye is larger in size, with larger pupils, relatively more rods, and an absence of central specializations for high-acuity vision which would be wasted, and often nocturnal eyes have a reflective tapetum behind the retina to double again the chances of photon capture. Diurnal eyes often contain central specializations where cones are packed at the maximal density possible, an area often accorded special chromatic sensitivity or special processing, in combination with a more spatially distributed set of lower density over the rest of the retina (Walls, 1963).

We will discuss the particular case of the owl monkey, *Aotus*, the only nocturnal representative of the anthropoids, with some additional observations from nocturnal strepsirrhine primates and the tarsier. The owl monkey, thought to be derived from originally diurnal monkeys, possesses all of the specializations listed above with the exception of a tapetum and has a variable partial appearance of a fovea, usually only an increased cone density containing many rods (Figure 2.6D). Its retinal diameter is strikingly large, and it has fewer cones and many more rods that would be expected compared to diurnal primates of similar brain and body size (Figures 2.3 and 2.4). Interestingly, it is a cone monochromat – the photopigment which is found in the S cones is not expressed (Figure 2.2; Wikler and Rakic, 1990). How is development altered to produce these new constellations of features? We propose a hypothesis here whose predictions we are presently investigating.

As described earlier, the production of retinal photoreceptors and retinal neurons is a two-stage process: First, the uniform population of multiplying precursor cells is driven to a terminal division by the symmetry-breaking actions of Notch/Delta signaling. Second, cells are specified as to type with respect to some as-yet-unidentified feature of the retinal environment in combination with lineage (Cepko, 1999). A single biasing event could shift the numbers of terminally differentiating neurons early in development, when cones and retinal ganglion cells are being specified, or later, when rods are produced (Figure 2.7). Such an alteration may be produced in a developing rat retina in tissue culture; it is also essentially the same mechanism already described to produce the different cell constituents of the central and peripheral retina (Austin *et al.*, 1995; Alexiades and Cepko, 1997). This single change in developmental timing alters the relative proportions of all retinal cells, and the production of rods in central retina may disable the means by which the fovea is normally produced, which is to be discussed later.

The larger eye of the owl monkey may be a secondary consequence of alteration of the diurnal mechanism of 'emmetropization' – recall that a high-contrast image, which must be transduced through the cone pathway in the daylight hours, is what checks the growth of the eye. In the owl monkey, the dose of the growth-limiting cone signal is reduced in two ways – not only is the overall production of cells during the cone-generating period limited, but also the S photopigment is not expressed, as described earlier (Figure 2.2). One explanation for the lack of cones is to make maximal room for rods, optimizing sensitivity. However, the removal of the cone signal may have an active developmental function as well, to allow the rapid growth of the large nocturnal eye; the fact that the cone lost is the one widely distributed across the retina would support this argument, as the growth of the eye appears to be controlled locally, not globally (Wallman *et al.*, 1987b). Interestingly, another nocturnal prosimian, the bush baby *Otolemur crassicaudatus*, is also a cone monochromat (Figure 2.2; Jacobs *et al.*, 1996).

The tarsier is a particularly interesting, and confusing, case. Tarsiers are classed with Haplorrhini (the group that includes monkeys and apes, but not lemurs and bush babies), a separate branch from the New and Old World monkeys we have been discussing. The stem haplorrhine primate is thought to have been diurnal, and the tarsier has a peculiar mixture of diurnal and nocturnal features, perhaps having become secondarily nocturnal like the owl monkey (Ross, 2000; see Chapter 1). It has a fovea, variable in morphology,

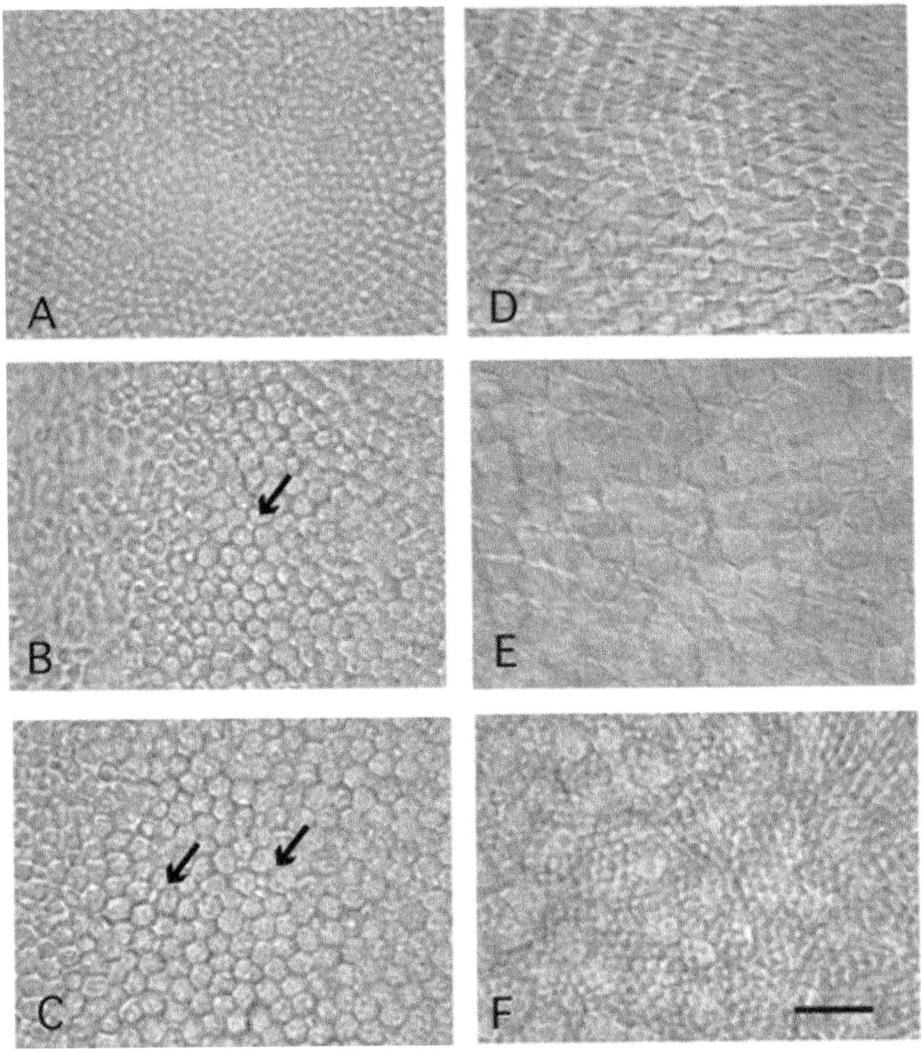

Figure 2.6 Foveae in *Alouatta caraya*, the howler monkey (A, B, C); *Cebus apella* (D, E), and *Aotus azarae* (D), showing the very reduced cone size and high cone density of *Alouatta* compared with *Cebus* (whose fovea is typical of most New and Old World monkeys) and central rods in *Aotus*. Scale bar 10 μm for all photomicrographs. (A) Center of fovea of *Alouatta caraya*. (B) Location 0.123 mm from foveal center of *Alouatta*; arrows point to two of a number of rods interposed among the larger cones. (C) Location 0.168 mm from foveal center of *Alouatta*; arrows point to two of a number of rods interposed among the larger cones. Other morphological and distributional evidence that these are rods are outlined by Franco *et al.* (2000). (D) Center of fovea, *Cebus apella*. (E) Outer fovea, *Cebus apella*. (F) Center of 'area centralis' in *Aotus azarae* focused on the internal segments of cones; principally rods are visible

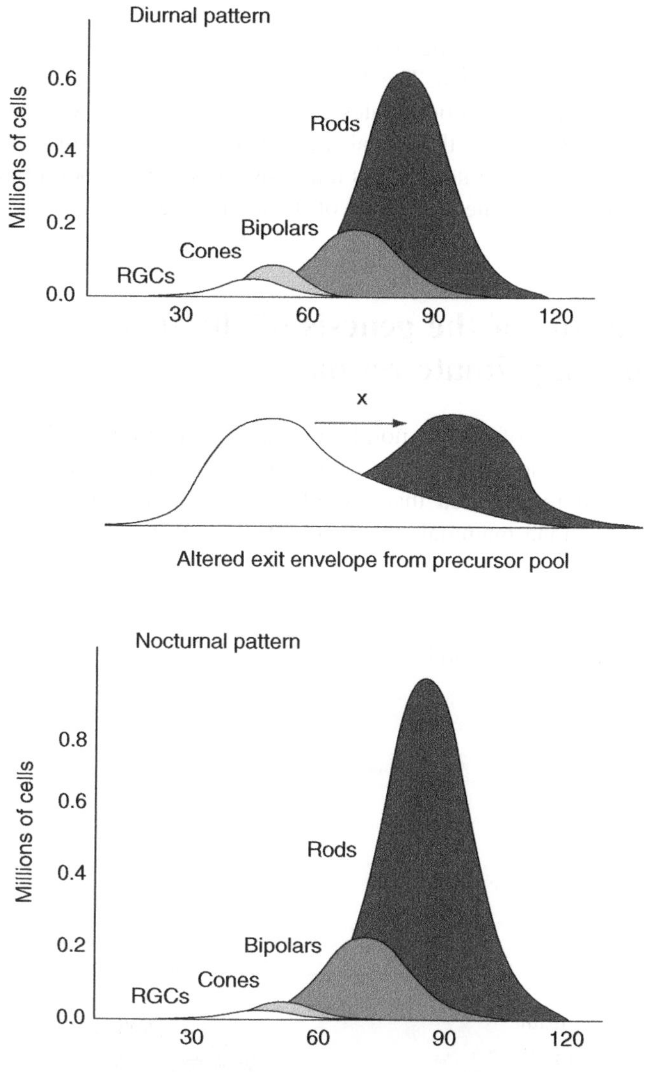

Figure 2.7 Schematic of the hypothesis of how precursor-pool exit may be biased to select principally the period of diurnal cell specification, cones and ganglion cells (diurnal precursor pool) or nocturnal cell specification (rods, and fewer ganglion cells improving convergence). Rod and cone numbers and durations from Franco *et al.* (2000) and Clancy *et al.* (2001)

but also has the largest eyes per body size seen in mammals (Figure 2.3). Nothing is known about the development of its retina, which might reveal another mechanism of decoupling eye growth from cone production.

The order of neurogenesis in the retina seems optimized for the production of a coordinated day-to-night niche transition, poised for 'exaptation' through its fundamental permissive organization into a number of adaptive, niche-specific varieties (Gould and

Vrba, 1982). The eye of the owl monkey elaborates on a feature of variability already present in primate eyes between the central and peripheral retinas – essentially, the owl monkey has an all-peripheral retina. The dose of cones (either activity or gene expression) may be the functional link that coordinates eye morphology with receptor complement, and possibly, as we will discuss, the foveal specialization. This coordinating linkage thus minimizes the number of separate features that must be selected upon to produce an eye appropriately configured for the light level of its environment.

2.9 Mechanisms of the genesis of the fovea centralis in primate retina

We now turn to a feature of ocular morphology unique to primates. It would be highly desirable to understand the ontogenetic mechanisms, which generate the fovea centralis, the small but crucial retinal area that provides the vast majority of our visual input (Figure 2.8). Since among mammals the occurrence of a fovea is restricted to primates,

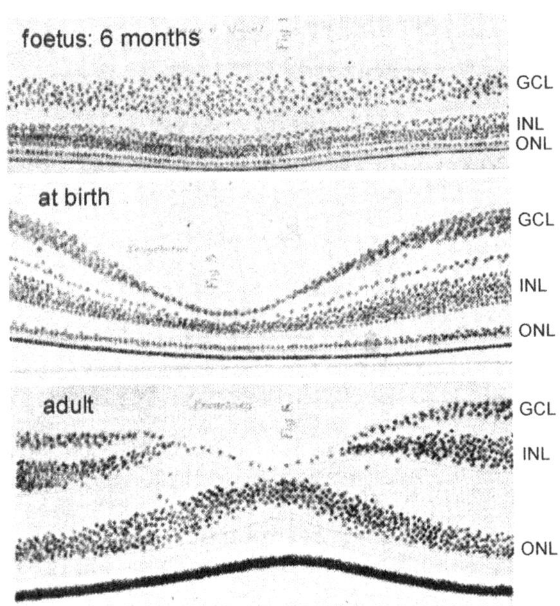

Figure 2.8 Ontogenetic development of the human fovea (modified after Bach and Seefelder, 1914; from Mann, 1950). Note that at 6th month of gestation, the ganglion cell layer (GCL) of the future fovea is even thicker than that of the neighboring retina, whereas it is completely missing in the mature fovea; by contrast, the foveal outer nuclear layer (ONL) consists of only one row of cone nuclei at 6th month but displays several layers in the adult. The inner nuclear layer (INL) undergoes similar developmental changes as the GCL

the amount of available data on its ontogenetic development is limited, and experimental approaches are hardly possible. Moreover, it should be pointed out that although foveae (meaning 'pit') may occur in the retinas of all vertebrate classes (with the probable exception of the amphibians), the structure of the primate fovea is very distinct, and differs from that of other known foveae (Figure 2.9) so that non-mammals cannot serve as model systems. Thus, our ideas about the make up of the primate fovea (Provis *et al.*, 1998) largely originate from 'experiments of nature' (such as albinism) and from mechanistic interpretations of histological data (Springer, 1999). This section is aimed at an attempt to critically review the available data, and to draw some cautious conclusions about the underlying mechanisms.

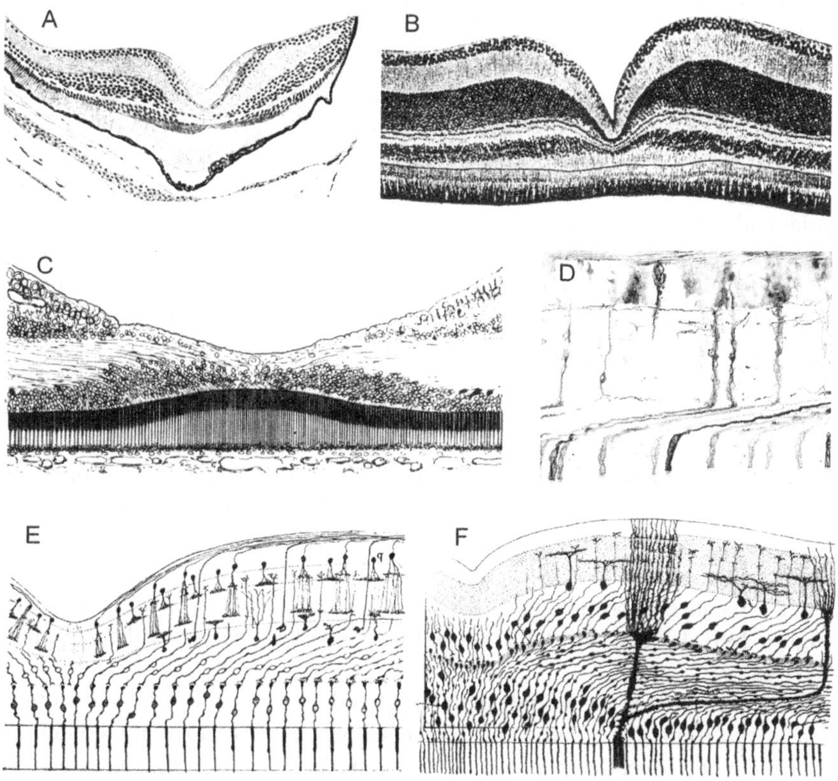

Figure 2.9 Comparison between primate (C, D) and non-mammalian foveae (A, B, E, F). (A–C) General structure of the fovea; (A) fish retina (*Evermanella indica*, Scopolidae; modified after Brauer from Franz, 1913); (B) avian retina (European bank swallow; modified after Rochon-Duvignaud from Walls, 1963); (C) human retina (modified after Polyak, 1941). (D–F) Retinal cells; (D) original microphotograph of a Golgi-labeled rhesus monkey retina (preparation of B.B. Boycott, courtesy of H. Wässle); (E) drawing of a Golgi-labeled Greenfinch retina (*Fringilla chloris*; modified after Ramón y Cajal, 1972); (F) drawing of a Golgi-labeled chameleon retina (*Chamaeleo vulgaris*; modified after Ramón y Cajal, 1972)

It has been shown that the morphogenesis of the primate fovea begins rather late in ontogenesis, i.e. *after* cytogenesis has ceased, and that the future foveal area at this time differs greatly from the mature conditions (Bach and Seefelder, 1914). The main differences are that in the immature or future fovea (i) the inner retinal layers are thicker rather than thinner than the adjacent retina, and (ii) the outer nuclear layer consists of a single layer of nuclei of cone photoreceptor cells which are thick and short rather than long and thin as in the adults (Figure 2.8). This means that – as no new cells are generated, and only a few cells die (Robinson, 1991; Georges *et al.*, 1999) – the available cells must undergo considerable changes in their shape and even in their location in order to achieve the adult configuration. These mechanisms may differ from those involved in the generation of other vertebrate foveae. As shown in Figure 2.9B, the foveae of birds may be smaller in diameter but their walls may be much steeper (i.e. they possess a 'convexiclivate' shape, enabling them to act as 'magnification lenses'). This difference prompted Walls (1963) to differentiate between the 'well-developed' avian fovea (Figure 2.9B) and the 'poorly developed' human fovea (Figure 2.9C). However, the foveae of other vertebrates, including birds, usually contain continuous inner nuclear and ganglion cell layers within their center (Figure 2.9A, B, E, F) where these layers are missing in the foveola of humans (Figure 2.9C) and most monkeys; thus, it should be kept in mind that the lateral displacement between a foveolar cone and its circuit partners (i.e. bipolar and ganglion cells) is much larger in the 'flat' primate fovea (up to $>300\,\mu m$ in humans: Yuodelis and Hendrickson, 1986; Reichenbach, unpublished data) than in the 'steep' avian fovea (up to $40\,\mu m$, as estimated from drawings of Ramón y Cajal, 1972). This means that unusual mechanisms may be required to generate the primate design. In particular, a characteristic feature of the typical primate fovea is the 'Z course' of the photoreceptor cell axons (the so-called Henle fibers) and the accompanying Müller cell processes (Figure 2.9D), suggesting that much of the re-location of retinal cells during foveogenesis occurs by an 'en block' counter-shift of the inner retina (inner part of the outer plexiform layer up to ganglion cell layer) against the outer retina (outer nuclear layer), using the (outer part of the) outer plexiform layer, together with the developing Henle fiber layer, as a 'compliant zone'. This 'Z course' is much less distinct (if at all present) in the foveae of lower vertebrates (where much of the transversal cone-ganglion cell distance is bridged by obliquely oriented bipolar cells in the inner nuclear layer: Figure 2.9E, F), and will thus be used here as a key feature of the typical mature primate fovea.

In order to search for an initial signal required to switch on the development of the fovea at the right place and time, a promising way is to look for conditions that are associated with foveal hypoplasia. It has long been known that several forms of albinism are accompanied by foveal hypoplasia or even aplasia (Elschnig, 1913; Naumann *et al.*, 1976; Fulton *et al.*, 1978). More recently it has been discovered that missense mutations of the 'major ocular control gene', *PAX6* (Azuma *et al.*, 1996, 2003), as well as Trisomy 2p(p23→pter) (Al-Saffar *et al.*, 2000), and other hitherto unknown conditions (Oliver *et al.*, 1987) may also be involved in foveal hypoplasia.

At least for albinism-associated foveal hypoplasia it has been shown that the central retina (where the fovea is supposed to be located) is not devoid of rod photoreceptor cells (Naumann *et al.*, 1976; Fulton *et al.*, 1978) as is always the case in the normal foveolar region from the very beginning of cytogenesis (Mann, 1950; Hollenberg and

Spira, 1973; Rhodes, 1979; Yuodelis and Hendrickson, 1986). Furthermore, the retina of *Aotus*, which is the only monkey species without a fovea (Kolmer, 1930), neither displays a central rod-free area (Odgen, 1975; Wikler and Rakic, 1990; Franco *et al.*, 2000; Figure 2.6). This suggests that the presence of a rod-free area in the embryonic retina may be necessary to trigger (at least, some aspects of) the generation of the fovea. This is of particular interest since it may be causally related to albinism as a well-established factor in foveal hypoplasia. In many forms of albinism, the underlying gene defect causes impairments of the availability of dopamine and related metabolites. One of these metabolites, 3,4-dihydroxyphenyl-alanine (DOPA), has been shown to stimulate the proliferation of early retinal progenitor cells but to inhibit the proliferative activity of late progenitor cells (Jeffery, 1997; Jeffery *et al.*, 1997). As described previously, early progenitor cells produce cone photoreceptor cells, retinal ganglion cells, and horizontal cells, and the late progenitor cells predominantly generate rod photoreceptor cells (for reviews, see Robinson, 1991; Reichenbach and Robinson, 1995). A deficiency in dopamine and/or DOPA may prevent both the enrichment of cones and ganglion cells and the lack of rods which are characteristic (and probably essential) for the future fovea (Figure 2.10). In other terms, a growth factor-induced block of late retinal cell generation may be a precondition to generate a rod-free area, and, thus, a fovea, well before any fovea in the morphological sense is present. As discussed earlier when describing the severe gradients of neurogenesis, one possibility raised for the absence of cones was the shifting of the time course of exit for terminal neurogenesis with respect to the clock of specification of cell determination in the retina, such that the precursor pool for the area of the retina producing the fovea was depleted before the time period for specification of rods was reached.

Both possibilities must result in an absence of cones centrally, and may be discriminated by the examination of gene activation patterns and transcription factors present in the central region at the time when the above-mentioned distinct growth pattern is generated at the place of the future fovea but nowhere else. Homeobox genes could play a crucial role in this determination, and in fact, missense mutations of *PAX6* were found to be associated with foveal hypoplasia (Azuma *et al.*, 1996, 2003). There are quite a few signaling molecules that have been shown to display a locally restricted expression pattern in the embryonic retina (for review, see Reichenbach and Pritz-Hohmeier, 1995); for example, there is a nasal-to-temporal gradient in the expression of the homeodomain factors, SOHo-1 and GH6 (Deitcher *et al.*, 1994; Schulte and Cepko, 2000), and the homeodomain transcription factor *VAX2* was shown to control the patterning of the eye dorso-ventral axis (Barbieri *et al.*, 1999). A particular combination of these, or another hitherto unidentified key molecule, may determine the location of the future fovea from the very beginning of retinal development (Figure 2.11). Once this location has been specified, the same signal and/or its downstream signals must trigger (at least) two fovea-specific events, viz. (i) either the failure to reach or a direct block of rod cell generation, and (ii) a repelling activity directed onto the growing axons of the retinal ganglion cells which thus grow *away* from the center of the future fovea even if this means (with the exception of those located at the side of the fovea directed toward the optic nerve head) that they also initially grow away from the optic nerve head (Figure 2.12, top); this latter event may be indirectly triggered by a high local density of retinal ganglion cells (Leventhal *et al.*, 1989). It may be of interest for a comparative view that both features

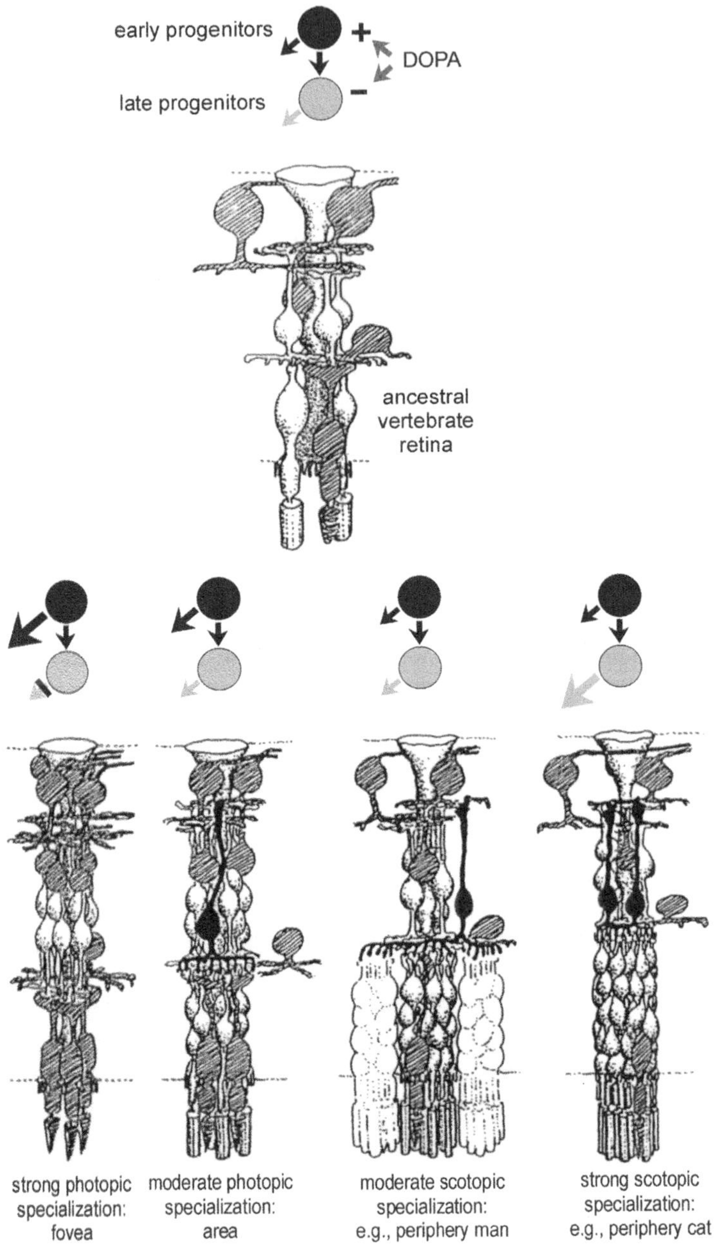

Figure 2.10 Hypothesis of the generation of (local) photopic/scotopic retina specialization by differential activation of early and late progenitor cells. Modified after Reichenbach and Robinson (1995). Early progenitors (black circles and arrows) mainly generate cells of the photopic pathways (hatched cells in the drawings: cones, ganglion cells, horizontal cells, and a subpopulation of amacrine cells) and the late progenitors (grey circles and arrows) then generate the rest (white cells in the drawings: rods, bipolar cells, another subpopulation of amacrine

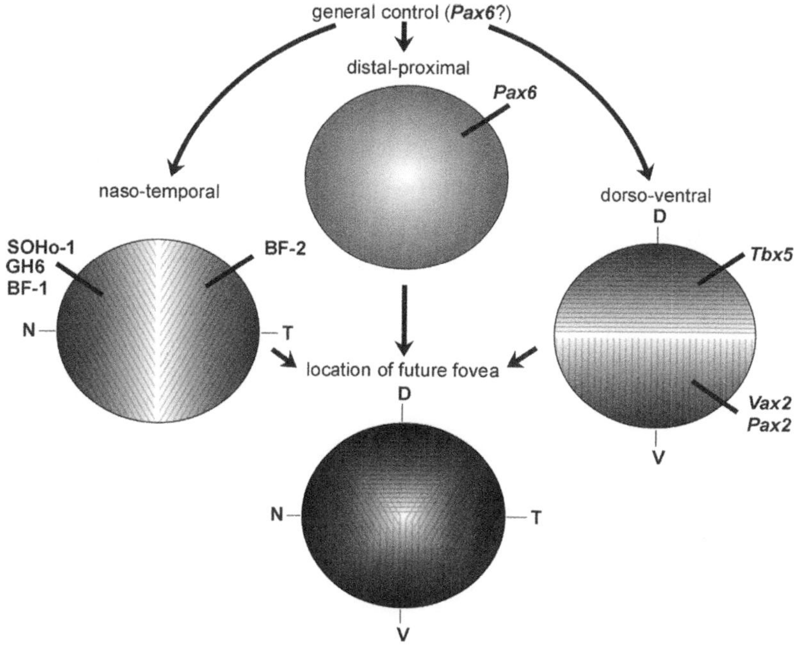

Figure 2.11 Possible mechanisms of the determination of the (site of the) future fovea by homeobox genes. The 'major ocular control gene', *PAX6*, may determine the future retina as a whole (for review, see e.g. Reichenbach and Pritz-Hohmeier, 1995), organize a peripheral-to-central pattern directly (Bäumer *et al.*, 2002), and be (indirectly?) involved in the organization of naso-temporal and dorso-ventral patterns (Bäumer *et al.*, 2002), and thus, finally, in the location of the future fovea (Azuma *et al.*, 1996, 2003). The establishment of naso-temporal and dorso-ventral patterns involves several other homeodomain transcription factors and related molecules (Deitcher *et al.*, 1994; Barbieri *et al.*, 1999; Schulte and Cepko, 2000; Bäumer *et al.*, 2002; and references therein)

Figure 2.10 (*continued*)
cells, and one type of Müller cell). A more or less equal proliferative activity of both types of progenitors will generate what is called here the 'ancestral vertebrate retina', typical e.g. for lampreys and most amphibians, and apparently capable of both scotopic and photopic vision. The preconditions for a future foveal specialization may be achieved by an enhanced proliferation of the early progenitors (generating e.g. a high density of cones and ganglion cells) and a decreased proliferation of the late progenitors (resulting in a local lack of rods). By contrast, the strong scotopic specialization of the retina of nocturnal mammals may be generated by an enhanced activity of the late progenitors, leading to a high density of rods. It is noteworthy in this context that DOPA was shown to stimulate the proliferation of early progenitors but to inhibit that of the late progenitors (Jeffery, 1997; Jeffery *et al.*, 1997; for review, see Reichenbach *et al.*, 1998). The black cells represent rod bipolar cells

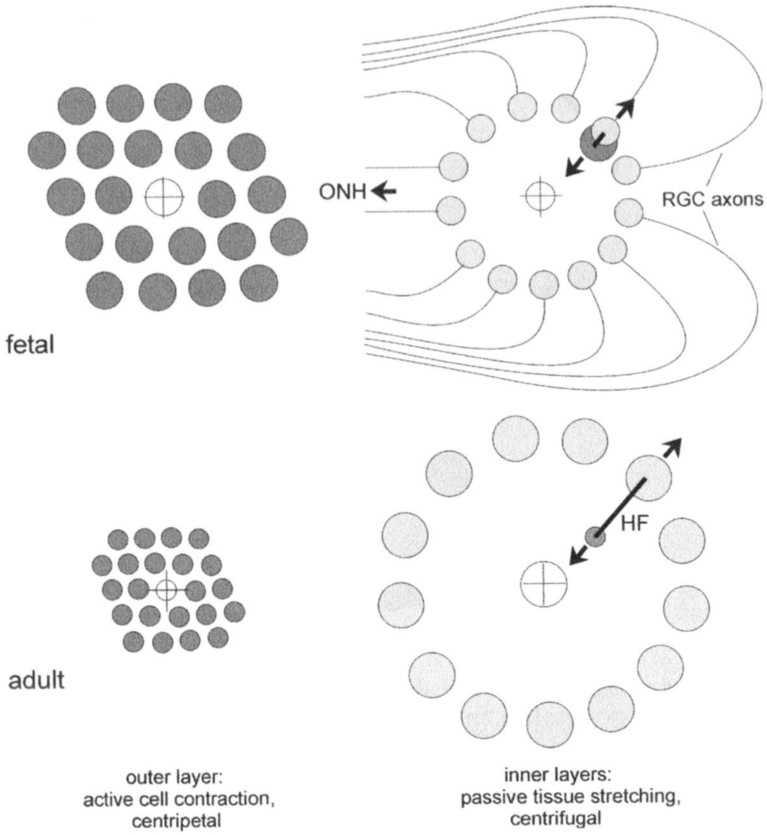

Figure 2.12 Schematic presentation of the differential developmental dislocation of cells in the inner and outer retinas during foveogenesis. Original. In the fetal retina, the thick cone cells (dark) are not very densely arranged, and retinal ganglion cells (RGC, bright) are still present in the future fovea, such that the (bipolar cell-mediated) coupling between a cone and 'its' ganglion cell(s) is arranged straightforward (i.e. 'vertical' if viewed from inner retinal surface), as exemplified by the couple of cells at right top. During further development, the cone cells become more slender and move up closer, whereas the ganglion cells (together with their partners in the inner nuclear layer) are drawn apart from foveal center along (and by?) their axons. This generates a dislocation within the 'cone-ganglion cell pairs' that must be bridged by the Henle fibers (HF) which elongate parallel to the retinal surface (i.e. 'horizontal' rather than 'vertical'), as exemplified by the couple of cells at right bottom. ONH, optic nerve head

may not apply to the fovea of the fish retina, as some deep sea fish possess a pure rod fovea (Walls, 1963), and as many nerve fibers are present within the fish foveal pit (Figure 2.9A).

The first of these 'secondary fovea-specific events' (i.e. the generation of a rod-free area) may stimulate morphological alterations of the local cone cells (perhaps, via extensive cone–cone contact signaling, by missing near-field signals secreted by rods, or by activation of the fibroblast growth factor receptor, FGFR-4, on cones (Cornish *et al.*, 2004)). Anyway, the most probable response of the central cones to their pure cone environment seems to be a contraction of their actin filaments such that the cells become more slender; within the next 11–15 months of development in humans, the diameter of the central cone cells is reduced from about 7.5 to 2 μm (Yuodelis and Hendrickson, 1986; Curcio *et al.*, 1990). As the cones (and their accompanying Müller cells) are tightly glued together by zonulae adherentes, a wave of contraction compacts the cones within the prospective fovea (Diaz-Araya and Provis, 1992). As a result of this 'snuggling up', the local density of cones increases from about 20 000 cones/mm^2 in the neonate to up to more than 150 000 cones/mm^2 in the adult (Yuodelis and Hendrickson, 1986; Curcio *et al.*, 1990). Among other morphological changes, this leads to a characteristic stacking of the cone cell nuclei (Ahnelt *et al.*, 2004) which eventually form several rows in the adult after they have been arranged in a monolayer in the fetus (Figure 2.8).

It is remarkable that the inner retinal layers fail to follow this centripetal movement; rather, they undergo a centrifugal dislocation. This latter dislocation is the reason for the generation of the central pit, which is (almost) devoid of secondary and tertiary retinal neurons (and which is responsible for the appropriateness of the term 'fovea'). The mechanism of this centrifugal dislocation is difficult to understand. There are two reasons to assume that there is no active migration of ganglion cells: (i) these cells establish very complex dendritic trees and synaptic contacts before this dislocation (one can hardly imagine these complex 'adnexes' to be dragged behind), and, more importantly, (ii) there is no apparent movement of the ganglion cells relative to their local environmental 'landmarks' such as their neuronal partners in the inner nuclear layer or the neighboring Müller cells (Figures 2.9D and 2.12). This suggests the action of forces generated outside the foveal cells. The most probable of these is the 'passive growth' or expansion of the retinal as a whole, driven by the inner ocular pressure (and following the enlargement of the sclera; Mastronarde *et al.*, 1984; Kelling *et al.*, 1989; Reichenbach *et al.*, 1991). In mammalian retinas without a fovea such as those of cat and rabbit, this mechanism causes a flattening and areal enlargement of the retinal tissue which is more pronounced in the periphery than in the center of the retina. Owing to different mechanical properties of the tissue, the (developmentally more advanced, and thicker) central retina displays more resistance against tangential stretching than the periphery (Kelling *et al.*, 1989; Reichenbach *et al.*, 1991). The mechanical properties of the fetal (future) foveal retina have not yet been measured; probably, however, its resistance against tangential stretching forces is relatively low rather than high since there are no nerve fibers traversing the inner surface of the tissue, at this place (all ganglion cell axons grow away from the future foveolar center such that there exists a central nerve fiber-free area (Figure 2.12)). During further development and thinning of the future foveal tissue (Figure 2.8), its mechanical resistance must even further decrease, as the thickness of both the tissue and, particularly, the nerve fiber layer are important determinors of retinal tensility (Reichenbach *et al.*, 1991). It may thus be concluded that the future foveal area is subjected to much tangential stress and stretching during the period of 'passive retinal growth'.

These 'passive stretching forces' all seem to act centrifugally possibly mediated by the specific centrifugal arrangement of the ganglion cells axons. The 'turning points' of the axons move together with the surrounding retinal tissue such that the force is transmitted to the ganglion cells (and probably to their synaptic partners, as well) in a strictly centrifugal direction (Figure 2.12). Because there is no tangentially connecting network of neurites in the center of the future foveola (the dendritic trees of central neurons are small – e.g. Figure 2.9E – and ganglion cell axons are missing), the cells of this area are then drawn away from the center where eventually a cell-free area arises. This hypothesis requires, in the case of the primate fovea, that each ganglion cell forms a rather stiff 'tissue bloc' together with its synaptic partners in the inner nuclear layer, such that small columns of neural tissue are dragged behind the stretched ganglion cell axons. Although this hypothesis remains to be proven, it is consistent with the developmental changes that can be observed when the Müller (radial glial) cells are used as markers of retinal 'columnar units' (Reichenbach *et al.*, 1993; Reichenbach and Robinson, 1995; Germer *et al.*, 1997). Figure 2.13 shows unpublished data obtained from a neonatal and an adult baboon retina. In the neonate, the fovea was not yet established. Along the straight central Müller cells, specific clusters of retinal neurons were identified for central and peripheral regions of the retina. Virtually identical clusters were found in the adult, but now the (para-) foveolar Müller cells displayed the typical 'Z- course' which indicated the course of the Henle fibers and, thus, of the 'wiring' of the retinal columnar units.

Following the above argumentation the genesis of the fovea may be the result of a centripetal movement of the outer nuclear layer and a centrifugal movement of the inner retina including all tissue from nerve fiber layer down to the inner part of their outer plexiform layer. In order to enable such a counter-movement, a 'compliant zone' must be interposed between the two layers. Such a zone should be characterized by initially a very low resistance against shear forces and subsequently an elongation of the fibers. What do we know about the border between the outer nuclear and outer plexiform layers (beginning) and the Henle fiber layer (later) which constitutes this zone? It consists of a mixture of only two elements, the axons of the cone photoreceptor cells, and the outer processes of Müller cells. In general, axons are known to be able to grow behind the surrounding growing embryonic/fetal tissue (for instance, after the connection between spinal nerves and a muscle has been established, the nerve may grow from a few micrometers up to a length of about a meter). In the case of rabbit Müller cells, it has been shown that they express cytoskeletal elements along much of their length but not in that part of the outer stem process which is located within the outer plexiform layer (Magalhães and Coimbra, 1973; Reichenbach *et al.*, 1988; Reichenbach, 1989). Although this feature remains to be demonstrated on fetal primate Müller cells, Figure 2.14 shows how it may contribute to the function of the outer plexiform layer as a 'compliant zone'.

2.9.1 The unusual fovea of the howler monkey

The retina of *Alouatta* presents a very interesting deviation from the foveae described for nearly all anthropoids, and may eventually provide a clue to the cellular mechanisms of foveal production. As the retina of other Atelidae have not been systematically examined for this feature yet, we do not know whether the howler monkey fovea is a singularity, or a characteristic of the entire family, which includes spider and woolly monkeys. Since *Alouatta* appears to be a singularity among New World primates in its obligatory trichromacy (Chapters 3 and 4), however, it may also be a singularity in foveal organization.

The fovea of *Alouatta* is of the same size as all other primate foveae, about 0.5 mm in diameter, but has twice the number of cones, which are reduced in diameter (Figure 2.6A,

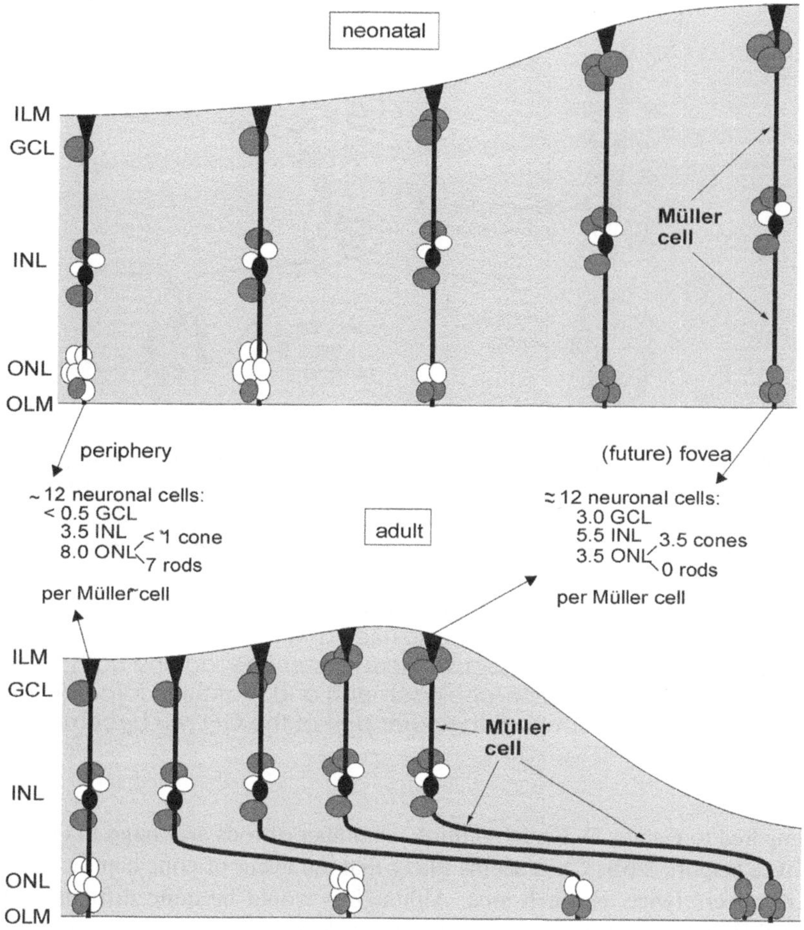

Figure 2.13 Comparison of a newborn and an adult baboon retina. Schematic drawing of the columnar units as estimated from cell counts (Müller cells were labeled by vimentin immunocytochemistry, and cell nuclei were counter-stained by hematoxylin; local densities of Müller cells and neuronal cell nuclei were counted along the course of Müller cell processes). The grey cells represent early born neurons, whereas white cells are the progeny of the late progenitors; Müller cells are drawn in black. Original (H. Kuhrt, unpublished results). Note that (i) the cellular constituents of foveal and peripheral columnar units differ although a total of 12 neurons per Müller cell was counted at both places, and (ii) the number and specification of cells within the foveal units remain unchanged during the large morphological alterations between neonatal and adult stages. The foveal data represent units whose cones are located outside the very center of the foveola. ILM, inner limiting membrane; GCL, (cells within the) ganglion cell layer; INL, (cells within the) inner nuclear layer; ONL, (cells within the) outer nuclear layer; OLM, outer limiting 'membrane'

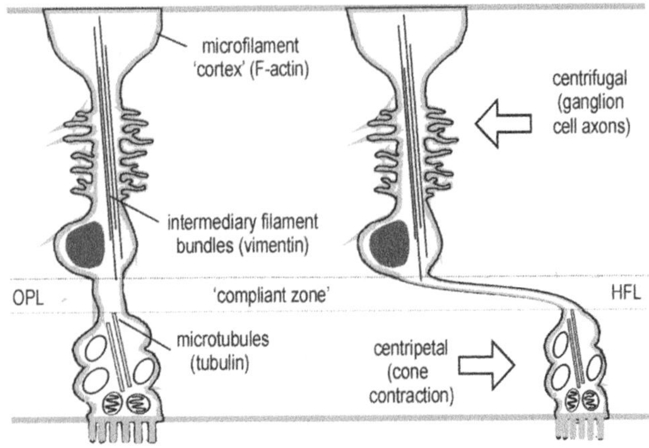

Figure 2.14 Hypothetic view of the contribution of the subcellular specifications of the Müller cell cytoskeleton to the function of the outer part of the outer plexiform layer (OPL) as a 'compliant zone' during foveogenesis. Original. In their inner stem process and soma (i.e. the 'inner part') the Müller cells contain densely packed bundles of intermediate filaments, and the distal outer Müller cell process contains microtubuli; only in the proximal outer process (located within the OPL) longitudinal cytoskeletal elements are missing (Magelhães and Coimbra, 1972; Reichenbach *et al.*, 1988). This may generate a '*locus minoris resistentiae*' against both shear forces (beginning cone cell contraction/displacement) and stretching forces (during later stages of development) such that the outer part of the OPL will become the Henle fiber layer (HFL)

B, C, compared to *Cebus*, D, E). In addition, a number of rods are found in the outer half of this fovea (Figure 2.6B, C). It seems likely that the event of cone contraction is twice as long or covers twice as much area. Although it would be quite difficult to procure retinas in this species during the time of foveal development, it is possible that close inspection of the details of foveal architecture with respect to remaining cell classes may illuminate the early events that produce this unusual cone density.

2.10 Summary

In this chapter, we have attempted to place the development of the primate retina in the larger context of vertebrate retinal evolution. We have also described four interesting dimensions of primate variation: scaling of the entire retina with respect to the different scaling requirements of rods and cones, the development of trichromacy in some primates, adaptation of the retina for nocturnal and diurnal niches, and finally, the unique case of the primate fovea. In all these cases, it is interesting to see how a very few genetic changes in timing or expression of opsin variation could be coordinated by existing developmental programs to produce adaptive variations in the eye and retina to produce a large number of secondary, epigenetic alterations. In the case of retinal scaling, extension of neurogenesis to produce a larger eye may automatically have the feature of generating the required many more rods than cone. In the development of trichromacy, it is possible that the sole change is the alteration of several amino acids in a duplicated opsin; generic mechanisms for silencing the expression of more than one opsin per cell and generic features of the nervous system allow this signal to be employed for perceptual decisions. To produce a nocturnal eye, it may only be necessary to shift the envelope of precursor-pool exit later in development to produce more rods and fewer cones, and after that, the loss of the cone signal may allow the eye to become larger, and halt the production of a fovea. Finally, for the fovea, creation of an all-cone location and contraction of the cone segments can set in motion a series of events which get a free ride from the normal balloon expansion of the eye, moving cell bodies away from the foveal pit.

Particularly unusual in the case of the retina is the coordination of genetic changes, generic features in information-processing and biomechanical qualities of the eye. All serve to seat new primate adaptations into cross-vertebrate functional constraints. Evolutionary and developmental approaches have so far been very successfully integrated to understand the organization of the basic vertebrate plans, and we have argued in this chapter that we may profit by extending the approach to the design of complex organs and complex information-processing problems.

Acknowledgements

This work was supported by a joint NSF/CNPq #910149/96-98 to LCLS and BLF, and NSF IBN-0138113 to BLF, LCLS, C. Cepko, and M. Dyer. We thank the Centro Nacional de Primatas (Ananindeua, PA, Brazil) for provision of many of the monkeys described in this chapter.

References

Ahnelt, P.K., Schubert, C., and Anger, E., 2004. Foveal cone mosaic organization: Modular sub-structuring – a consistent feature along all cone elements, in *The Macula: Diagnosis, Treatment and Future Trends* (ed S. Binder), Springer Wien, New York, pp. 1–8.

Alexiades, M.R. and Cepko, C.L., 1997. Subsets of retinal progenitors display temporally regulated and distinct biases in the fates of their progeny. *Development* **124**, 1119–1131.

Al-Saffar, M., Lemyre, E., Koenekoop, R. *et al.*, 2000. Phenotype of a patient with pure partial trisomy 2p(p23 → pter). *Am. J. Med. Genet.* **94**, 428–432.

Austin, C.P., Feldman, D.E., Ida, J.A. *et al.*, 1995. Vertebrate retinal ganglion cells are selected from competent progenitors by the action of notch. *Development* **121**, 3637–3650.

Azuma, N., Nishina, S., Yanagisawa, H. *et al.*, 1996. *PAX6* missense mutation in isolated foveal hypolasia. *Nat. Genet.* **13**, 141–142.

Azuma, N., Yamaguchi, Y., Handa, H. *et al.*, 2003. Mutation of the *PAX6* gene detected in patients with a variety of optic-nerve malformations. *Am. J. Hum. Genet.* **72**, 1565–1570.

Bach, L. and Seefelder, R., 1914. *Atlas zur Entwicklungsgeschichte des menschlichen Auges*, Verlag Wilhelm Engelmann, Leipzig und Berlin (plate XLIX).

Bao, Z-Z. and Cepko, C.L., 1997. The expression and function of Notch pathway genes in the developing rat eye. *J. Neurosci.* **17**, 1425–1434.

Barbieri, A.M., Lupo, G., Bulfone, A. *et al.*, 1999. A homeobox gene, *vax2*, controls the patterning of the eye dorsoventral axis. *Proc. Natl. Acad. Sci. USA* **96**, 10729–10734.

Bäumer, N., Marquardt, T., Stoykova, A. *et al.*, 2002. *Pax6* is required for establishing naso-temporal and dorsal characteristics of the optic vesicle. *Development* **129**, 4535–4545.

Beazley, L.D., Dunlop, S.A., Harman, A.M. *et al.*, 1989. Development of ganglion cell gradients in the retinal ganglion cell layer of amphibians and marsupials: Two solutions to one problem, in *Development of the Vertebrate Retina* (eds B.L. Finlay and D.R. Sengelaub), Plenum Press, New York, pp. 221–226.

Bodnarenko, S.R. and Chalupa, L.M., 1993. Stratification of ON and OFF ganglion cell dendrites depends on glutamate-mediated afferent activity in the developing retina. *Nature* **364**, 144–146.

Bodnarenko, S.R., Jeyarasasingam, G., and Chalupa, L.M., 1995. Development and regulation of dendritic stratification in retinal ganglion cells by glutamate-mediated afferent activity. *J. Neurosci.* **15**, 7037–7045.

Braun, C.B., 1996. The sensory biology of the living jawless fishes: A phylogenetic assessment. *Brain Behav. Evol.* **48**, 262–276.

Callaerts, P., Halder, G., and Gehring, W.J., 1997. *Pax-6* in development and evolution. *Annu. Rev. Neurosci.* **20**, 483–532.

Cepko, C.L., 1999. The roles of intrinsic and extrinsic cues and bHLH genes in the determination of retinal cell fates. *Curr. Opin. Neurobiol.* **9**, 37–46.

Chalupa, L.M. and Dreher, B., 1991. High precision systems require high precision blueprints – a new view regarding the formation of connections in the mammalian visual system. *J. Cog. Neurosci.* **3**, 209–219.

Chalupa, L.M. and Lia, B., 1991. The nasotemporal division of retinal ganglion cells with crossed and uncrossed projections in fetal rhesus monkey. *J. Neurosci.* **11**, 191–202.

Chang, W.S. and Harris, W.A., 1998. Sequential genesis and determination of cone and rod photoreceptors in *Xenopus*. *J. Neurobiol.* **35**, 227–244.

Clancy, B., Darlington, R.B., and Finlay, B.L., 2001. Translating developmental time across mammalian species. *Neuroscience* **105**, 7–17.

Cook, J.E. and Chalupa, L.M., 2000. Retinal mosaics: New insights into an old concept. *Trends Neurosci.* **23**, 26–34.

Cornish, E.E., Natoli, R.C., Hendrickson, A. *et al.*, 2004. Differential distribution of fibroblast growth factor receptors (FGFRs) on foveal cones: FGFR-4 is an early marker of cone photoreceptors. *Mol. Vis.* **10**, 1–14.

Coulombre, A., 1957. The role of intraocular pressure in the development of the chick eye. II. Control of corneal size. *AMA Arch. Ophthalmol.* **57**, 250–253.

Curcio, C.A. and Hendrickson, A.E., 1992. Organization and development of the primate photoreceptor mosaic. *Prog. Retin. Eye Res.* **10**, 89–120.

Curcio, C.A., Sloan, K.R., Packer, O. *et al.*, 1987. Distribution of cones in human and monkey retina: Individual variability and radial asymmetry. *Science* **236**, 579–582.

Curcio, C.E., Sloan, K.R., Kalina, R.E. *et al.*, 1990. Human photoreceptor topography. *J. Comp. Neurol.* **292**, 497–523.

Deitcher, D.L., Fekete, D.M., and Cepko, C.L., 1994. Asymmetric expression of a novel homeobox gene in vertebrate sensory organs. *J. Neurosci.* **14**, 486–498.

de la Rosa, E. and de Pablo, F., 2000. Cell death in early neural development: Beyond the neurotrophic theory. *Trends Neurosci.* **23**, 455–458.

Diaz-Araya, C.M. and Provis, J.M., 1992. Evidence of photoreceptor migration during early foveal development: A quantitative analysis of human fetal retinae. *Vis. Neurosci.* **8**, 505–514.

Dyer, M.A. and Cepko, C.L., 2001. Regulating proliferation during retinal development. *Nat. Rev. Neurosci.* **2**, 333–341.

Easter, S.S. and Nicola, G.N., 1996. Development of vision in the Zebrafish (*Danio rerio*). *Dev. Biol.* **180**, 646–663.

Elschnig, A.Z., 1913. Zur Anatomie des menschlichen Albionoauges. *Graefes Arch. Ophthalmol.* **84**, 401–419.

Erskine, L., Williams, S.E., Brose, K. *et al.*, 2000. Retinal ganglion cell axon guidance in the mouse optic chiasm: Expression and function of Robos and Slits. *J. Neurosci.* **20**, 4975–4982.

Feldkaemper, M.P. and Schaeffel, F., 2002. Evidence for a potential role of glucagon during eye growth regulation in chicks. *Vis. Neurosci.* **19**, 755–766.

Fernald, R.D., 1989. Seeing through a growing eye, in *Perspectives in Neural Systems and Behavior* (eds T.J. Carew and D.B. Kelly), Liss, New York, pp. 151–174.

Fernald, R.D., 1991. Retinal rod neurogenesis, in *Development of the Vertebrate Retina* (eds B.L. Finlay and D.R. Sengelaub), Plenum Press, New York, pp. 31–42.

Fernald, R.D., 2000. Evolution of eyes. *Curr. Opin. Neurobiol.* **10**, 444–450.

Finlay, B.L., 1992. Cell death and the creation of regional differences in neuronal numbers. *J. Neurobiol.* **23**, 1159–1171.

Finlay, B.L. and Darlington, R.B., 1995. Linked regularities in the development and evolution of mammalian brains. *Science* **268**, 1578–1584.

Finlay, B.L. and Pallas, S.L., 1989. Control of cell number in the developing visual system. *Prog. Neurobiol.* **32**, 207–234.

Finlay, B.L., Darlington, R.B., and Nicastro, N., 2001. Developmental structure in brain evolution. *Behav. Brain Sci.* **24**, 263–307.

Franco, E.C.S., Finlay, B.L., Silveira, L.C.L. *et al.*, 2000. Conservation of absolute foveal area in new world monkeys – A constraint on eye size and conformation. *Brain Behav. Evol.* **56**, 276–286.

Franz, V., 1913. *Sehorgan. Teil 7 des Lehrbuchs der vergleichenden mikroskopischen Anatomie der Wirbeltiere*, Gustav Fischer Verlag, Jena.

Fulton, A.B., Albert, D.M., and Craft, J.L., 1978. Human albinism: Light and electron microscopy study. *Arch. Ophthalmol.* **96**, 305–310.

Georges, P., Madigan, M.C., and Provis, J.M., 1999. Apoptosis during development of the human retina: Relationship to foveal development and retinal synaptogenesis. *J. Comp. Neurol.* **413**, 198–208.

Gerhart, J. and Kirschner, M., 1997. *Cells, Embryos and Evolution*, Blackwell Science, Malden, Massachusetts.

Germer, A., Kühnel, K., Grosche, J. *et al.*, 1997. Development of the neonatal rabbit retina in organ culture. 1. Comparison with histogenesis in vivo, and the effect of a gliotoxin (a-aminoadipic acid). *Anat. Embryol.* **196**, 67–79.

Godement, P., Salaun, J., and Mason, C.J., 1990. Retinal axon pathfinding in the optic chiasm: Divergence of crossed and uncrossed fibers. *Neuron* **5**, 173–186.

Golz, J. and MacLeod, D.I.A., 2002. Influence of scene statistics on colour constancy. *Nature* **415**, 637–640.

Gould, S.J. and Vrba, E.S., 1982. Exaptation – a missing term in the science of form. *Paleobiology* **8**, 4–15.

Heesy, C.P. and Ross, C.F., 2001. Evolution of activity patterns and chromatic vision in primates: Morphometrics, genetics and cladistics. *J. Hum. Evol.* **40**, 111–149.

Hollenberg, M.J. and Spira, A.W., 1973. Human retinal development: Ultrastructure of the outer retina. *Am. J. Anat.* **137**, 357–385.

Howland, H.C. and Sayles, N., 1985. Photokeratometric and photorefractive measurements of astigmatism in infants and young children. *Vision Res.* **25**, 73–81.

Hung, L.-F., Wallman, J., and Smith, E.L., 2000. Vision-dependent changes in the choroidal thickness of macaque monkeys. *Invest. Ophthalmol. Vis. Sci.* **41**, 1259–1269.

Jacobs, G.H., 1998. A perspective on color vision in platyrrhine monkeys. *Vision Res.* **38**, 3307–3313.

Jacobs, G.H. and Deegan, J.F., 1999. Uniformity of colour vision in Old World monkeys. *Proc. R. Soc. Lond. Ser. B* **266**, 2023–2028.

Jacobs, G.H. and Deegan, J.F., 2001. Photopigments and colour vision in New World monkeys from the family Atelidae. *Proc. R. Soc. Lond. B* **268**, 695–702.

Jacobs, G.H., Neitz, M., and Neitz, J., 1996. Mutations in S-cone pigment genes and absence of color vision in two species of nocturnal primate. *Proc. R. Soc. Lond. B* **263**, 705–710.

Jeffery, G., 1997. The albino retina: An abnormality that provides insight into normal retinal development. *Trends Neurosci.* **20**, 165–169.

Jeffery, G., Brem, G., and Montoliu, L., 1997. Correction of retinal abnormalities found in albinism by introduction of a functional tyrosinase gene in transgenic mice and rabbits. *Dev. Brain Res.* **99**, 95–102.

Kaskan, P.M., Franco, E.C.S., Yamada, E.S. *et al.*, 2005. Peripheral variability and central constancy in mammalian visual system evolution. *Proc. R. Soc. Biol. Sci.* **272**, 91-100.

Katz, L.C. and Shatz, C.J., 1996. Synaptic activity and the construction of cortical circuits. *Science* **274**, 1133–1138.

Kelling, S.T., Sengelaub, D.R., Wikler, K.C. *et al.*, 1989. Differential elasticity of the immature retina: A contribution to the development of the area centralis? *Vis. Neurosci.* **2**, 117–120.

Kolmer, W., 1930. Zur Kenntnis des Auges der Primaten. *Z. Anat. Entwicklungsgesch.* **93**, 679–722.

LaVail, M.M., Rapaport, D.H., and Rakic, P., 1991. Cytogenesis in the monkey retina. *J. Comp. Neurol.* **309**, 86–114.

Leventhal, A.G., Ault, S.J., Viztek, D.J. *et al.*, 1989. Extrinsic determinants of retinal ganglion cell development in primates. *J. Comp. Neurol.* **286**, 170–189.

Linden, R., Rehen, S.K., and Chiarini, L.B., 1999. Apoptosis in developing retinal tissue. *Prog. Retin. Eye Res.* **18**, 133–165.

Livesey, F.J. and Cepko, C.L., 2001. Vertebrate neural cell-fate determination: Lessons from the retina. *Nat. Rev. Neurosci.* **2**, 109–118.

Magalhães, M.M. and Coimbra, A., 1972. The rabbit retina Müller cell: A fine structural and cytochemical study. *Journal of Ultrastructural Research*, **39**, 10–26.

Magalhães, M.M. and Coimbra, A., 1973. Electron microscopic autoradiographic detection of sites of protein synthesis in the rabbit retina Müller cells. *J. Cell. Biol.* **57**, 77–87.

Mann, I., 1950. *The Development of the Human Eye*, 2nd edn, Grune and Stratton, New York.

Mastronarde, D.N., Thibeault, M.A., and Dubin, M.W., 1984. Non-uniform postnatal growth of the cat retina. *J. Comp. Neurol.* **228**, 598–608.

Naumann, G.O.H., Lerche, W., and Schroeder, W., 1976. Foveola-Aplasie bei Tyrosinase-positivem oculocutanen Albinismus. *Graefes Arch. Clin. Exp. Ophthalmol.* **200**, 39–50.

Neitz, J., Carroll, J., Yamauchi, Y. *et al.*, 2002. Color perception is mediated by a plastic neural mechanism that is adjustable in adults. *Neuron* **35**, 783–792.

Nijhawan, D., Honarpour, N., and Wang, X.D., 2000. Apoptosis in neural development and disease. *Annu. Rev. Neurosci.* **23**, 73–87.

Odgen, T.E., 1975. The receptor mosaic of *Aotes trivirgatus*: Distribution of rods and cones. *J. Comp. Neurol.* **163**, 193–202.

Ohnuma, S., Hopper, S., Wang, K.C. *et al.*, 2002. Co-ordinating retinal histogenesis: Early cell cycle exit enhances early cell fate determination in the *Xenopus* retina. *Development* **129**, 2435–2446.

Oliver, M.D., Dotan, S.A., Chemke, J. *et al.*, 1987. Isolated foveal hypoplasia. *Br. J. Ophthalmol.* **71**, 926–930.

Polyak, S.L., 1941. *The Retina*, University of Chicago Press, Chicago.

Provis, J.M., 1987. Patterns of cell death in the ganglion cell layer of the human fetal retina. *J. Comp. Neurol.* **259**, 237–246.

Provis, J.M., Diaz, C.M., and Dreher, B., 1998. Ontogeny of the primate fovea: A central issue in retinal development. *Prog. Neurobiol.* **54**, 549–581.

Rakic, P., 1983. Overproduction and elimination of retinal axons in the fetal rhesus monkey. *Science* **219**, 1441–1444.

Ramón y Cajal, S., 1972. *The Structure of the Retina* (translated by S.A. Thorpe and M. Glickstein), Charles C. Thomas Publishers, Springfield, Illinois.

Rapaport, D.H., Fletcher, J.T., LaVail, M.M. *et al.*, 1992. Genesis of neurons in the retinal ganglion cell layer of the monkey. *J. Comp. Neurol.* **322**, 577–588.

Rapaport, D.H., Rakic, P., and LaVail, M.M., 1996. Spatiotemporal gradients of cell genesis in the primate retina. *Perspect. Dev. Neurobiol.* **3**, 147–159.

Raymond, P.A., 1991. Cell determination and positional cures in the teleost retina: Development of photoreceptors and horizontal cells, in *Development of the Visual System* (eds D.M.-K. Lam and C.J. Shatz), MIT Press, Cambridge, MA, pp. 37–58.

Reese, B.E. and Galli-Resta, L., 2002. The role of tangential dispersion in retinal mosaic formation. *Prog. Retin. Eye Res.* **21**, 153–168.

Reichenbach, A., 1989. Organelle-free cytoplasmic volume fraction of rabbit retinal Müller (glial) cells. *J. Hirnforsch.* **30**, 513–516.

Reichenbach, A. and Pritz-Hohmeier, S., 1995. Normal and disturbed development of the eye anlagen. *Prog. Retin. Res.* **14**, 1–46.

Reichenbach, A. and Robinson, S.R., 1995. Phylogenetic constraints on retinal organization and development: An Haeckelian perspective. *Prog. Retin. Res.* **15**, 139–171.

Reichenbach, A., Hagen, E., Schippel, K. *et al.*, 1988. Quantitative electron microscopy of rabbit Müller (glial) cells in dependence of retinal topography. *Z. mikroskop.-anat. Forsch.* **102**, 721–755.

Reichenbach, A., Eberhardt, W., Scheibe, R. *et al.*, 1991. Development of the rabbit retina. IV. Tissue tensility and elasticity in dependence on topographic specializations. *Exp. Eye Res.* **53**, 241–251.

Reichenbach, A., Stolzenburg, J.-U., Eberhardt, W. *et al.*, 1993. What do retinal Müller (glial) cells do for their neuronal "small siblings". *J. Neurochem. Anat.* **6**, 201–213.

Reichenbach, A., Ziegert, M., Schnitzer, J. *et al.*, 1994. Development of the rabbit retina. V. The question of columnar units. *Dev. Brain Res.* **79**, 72–84.

Reichenbach, A., Germer, A., Bringmann, A. *et al.*, 1998. Glio-Neuronal interactions in retinal development, in *Development and Organization of the Retina from Molecules to Function*, Vol. 299 (eds L.M. Chalupa and B.L. Finlay), NATO ASI Series A, Life Sciences, Plenum Press, New York, pp. 121–146.

Remtulla, S. and Hallett, P.E., 1985. A schematic eye for the mouse and comparisons with the rat. *Vision Res.* **25**, 21–31.

Rhodes, R.H., 1979. A light microscopic study of the developing human neural retina. *Am. J. Anat.* **154**, 195–210.

Robinson, S.L., 1987. Ontogeny of the area centralis in the cat. *J. Comp. Neurol.* **255**, 50–67.

Robinson, S.R., 1991. Development of the mammalian retina, in *Neuroanatomy of the Visual Pathways and their Development* (eds B. Dreher and S.R. Robinson), Macmillan, London, pp. 69–128.

Robinson, S.R., Dreher, B., and McCall, M.J., 1989. Nonuniform retinal expansion during the formation of the rabbit's visual streak: Implications for the ontogeny of mammalian retinal topography. *Vis. Neurosci.* **2**, 201–219.

Rodieck, R.W., 1973. *The Vertebrate Retina: Principles of Structure and Function*, W.H. Freeman, San Francisco.

Roorda, A. and Williams, D.R., 1999. The arrangement of the three cone classes in the living human eye. *Nature* **397**, 520–522.

Roorda, A., Metha, A.B., and Lennie, P. *et al.*, 2001. Packing arrangement of the three cone classes in primate retina. *Vision Res.* **41**, 1291–1306.

Ross, C.F., 2000. Into the light: The origin of Anthropoidea. *Annu. Rev. Anthropol.* **29**, 147–194.

Schaeffel, F. and Diether. S., 1999. The growing eye: An autofocus system that works on very poor images. *Vision Res.* **39**, 1585–1589.

Schaeffel, F., Bartmann, M., Hagel, G. *et al.*, 1995. Studies on the rule of the retinal dopamine/melatonin system in experimental refractive errors in chickens. *Vision Res.* **35**, 1247–1264.

Schulte, D. and Cepko, C.L., 2000. Two homeobox genes define the domain of EphA3 expression in the developing chick retina. *Development* **127**, 5033–5045.

Sengelaub, D.R. and Finlay, B.L., 1982. Cell death in the mammalian visual system during normal development. I. Retinal ganglion cells. *J. Comp. Neurol.* **204**, 311–317.

Silveira, L.C.L., Saito, C.A., Lee, B.B. *et al.*, 2003. Morphology and physiology of primate M- and P-cells. *Prog. Brain Res.* **144**, 21–46.

Sjostrand, J., Olsson, V., Popovic, Z. *et al.*, 1999. Quantitative estimations of foveal and extra-foveal retinal circuitry in humans. *Vision Res.* **39**, 2987–2998.

Snow, R.L., Nelson, A., Driscoll, L.L. *et al.*, 1997. Scaling of the visual system, photoreceptors to extrastriate cortex, emphasizing primates. *Soc. Neurosci. Abs.* **23**, 1308.

Springer, A., 1999. A virtual biomechanical model of foveal development. *Vis. Neurosci.* **16**, 629–636.

Troilo, D., 1992. Neonatal eye growth and emmetropization: A literature review. *Eye* **6**, 154–160.

Troilo, D. and Judge, S.J., 1993. Ocular development and visual deprivation myopia in the common marmoset (*Callithrix -jacchus*). *Vision Res.* **33**, 1311–1324.

Troilo, D., Gottlieb, M.D., and Wallman, J., 1987. Visual deprivation causes myopia in chicks with optic nerve section. *Curr. Eye Res.* **6**, 993–999.

Wallman, J., 1995. How many myopias? *Vision et Adaptation* **6**, 140–150.

Wallman, J., Gottlieb, M.D., Rajaram, V. *et al.*, 1987a. Retinal control of myopia and eye growth. *Science* **237**, 73–77.
Wallman, J., Gottlieb, M.D., Rajaram, V. *et al.*, 1987b. Local retinal regions control local eye growth and myopia. *Science* **237**, 73–76.
Wallman, J., Wildsoet, C., Xu, A.M. *et al.*, 1994. Moving the retina: Choroidal modulation of refractive state. *Vision Res.* **35**, 37–50.
Walls, G.L., 1963. *The Vertebrate Eye and its Adaptive Radiation*, Hafner Publ. Co., New York and London.
Wässle, H., Hoon Chun, M., and Müller, F., 1987. Amacrine cells in the ganglion cell layer of the cat retina. *J. Comp. Neurol.* **265**, 391–408.
Wässle, H., Yamashita, M., Greferath, U. *et al.*, 1991. The rod bipolar cell of the mammalian retina. *Vis. Neurosci.* **7**, 99–112.
Wiesel, T.N. and Raviola, E., 1977. Myopia and eye enlargement after neonatal lid fusion in monkeys. *Nature* **266**, 66–68.
Wikler, K.C. and Rakic, P., 1990. Distribution of photoreceptor subtypes in the retina of diurnal and nocturnal primates. *J. Neurosci.* **10**, 3390–3401.
Wikler, K.C. and Rakic, P., 1994. An array of early differentiating cones precedes the emergence of the photoreceptor mosaic in the fetal monkey retina. *Proc. Natl. Acad. Sci. USA* **91**, 6534–6538.
Wikler, K.C. and Rakic, P., 1996. Development of photoreceptor mosaics in the primate retina. *Perspect. Dev. Neurobiol.* **3**, 161.
Wilkins, A.S., 2001. *The Evolution of Developmental Pathways*, Sinauer Associates, Sunderland, Massachusetts.
Wong, R.O.L., 1999. Retinal waves and visual system development. *Annu. Rev. Neurosci.* **22**, 29–47.
Wong, R.O.L., Hermann, K., and Shatz, C.J., 1991. Remodeling of retinal ganglion cell dendrites in the absence of action potential activity. *J. Neurobiol.* **22**, 685–697.
Yuodelis, C. and Hendrickson, A., 1986. A qualitative and quantitative analysis of the human fovea during development. *Vision Res.* **26**, 847–855.

3

The Genetics and Evolution of Primate Visual Pigments

David M. Hunt, Gerald H. Jacobs, and *James K. Bowmaker*

3.1 Introduction

In vertebrates, four different cone visual pigments form the ancestral complement that first appeared at the base of the vertebrate lineage around 540 million years ago (mya) (Collin *et al.*, 2003). Amongst eutherian mammals, however, this complement has been reduced to two and it is only in the primates that routine trichromacy has evolved to partially reverse this loss. Color vision in primates is superior therefore to that in other eutherian mammals but still falls short of the tetrachromacy that is not uncommon in other vertebrate classes.

3.2 Structure of visual pigments

Visual pigments are formed by a seven-transmembrane (TM) opsin protein and a chromophore that is covalently attached to a lysine residue via a Schiff base. In mammals, the chromophore is invariably 11-*cis*-retinal derived from vitamin A1, so the peak spectral absorption (λ_{max}) of a visual pigment is determined not by the chromophore but by the amino acid sequence of the opsin protein, with certain residues tuning the pigment to particular spectral locations. Opsin proteins are members of the super-family of G protein-coupled receptors that function through the activation of a guanine nucleotide binding protein (G protein) and an effector enzyme that changes the level of a second messenger in the cell cytoplasm.

The Primate Visual System: A Comparative Approach Edited by Jan Kremers

In vertebrates, sensitivity to dim light (scotopic vision) is generally achieved by a single class of rod photoreceptors in the retina that contains a pigment with a λ_{max} around 500 nm. In contrast, cones are responsible for vision at higher light levels (photopic vision), and for color vision. In all vertebrate taxa except eutherian mammals, up to four different cone visual pigments may be present. The four cone classes are distinguished on the basis of the amino acid sequence of their respective opsins and roughly correlate with spectral sensitivity as follows: a long wave-sensitive (LWS) class with λ_{max} of 500–570 nm; a middle wave-sensitive (MWS) class with λ_{max} of 480–520 nm; and two short wave-sensitive (SWS) classes, SWS2 with λ_{max} of 415–470 nm and SWS1 with λ_{max} of 355–435 nm. In eutherian mammals, this complement has been reduced to only two classes, LWS and SWS1, an event that is believed to have resulted from a nocturnal life style that mammals went through during their evolution. As a result of this loss, most mammals are dichromats although there is evidence that some marsupials express additional pigments and could possess trichromatic color vision (Arrese *et al.*, 2002). In anthropoid or simian primates, this reduction in the number of cone pigments has been partially reversed, thereby achieving trichromacy. The evolutionary mechanism by which this has arisen differs, however, in the two major simian groups, the Old World (catarrhines) primates from Africa and Asia and the New World (platyrrhine) primates from Central and South America. The major driving force behind the evolution of trichromacy with its improved color discrimination in the red/green region of the spectrum is argued to be the detection and evaluation of ripe fruits (Mollon, 1989; Osorio and Vorobyev, 1996; Sumner and Mollon, 2000; Regan *et al.*, 2001) or young nutritious leaves (Dominy and Lucas, 2001) against the green foliage of the rainforest.

3.3 Visual pigment genes in primates

3.3.1 Rod opsin

In all primates there is single rod opsin gene (on human chromosome 3) that encodes a pigment with a λ_{max} around 500 nm. The gene is composed of five coding exons. In humans, missense mutations in each exon are responsible for the disease of retinitis pigmentosa (RP), a generally progressive loss of vision that is often inherited in a dominant fashion (see review by al-Maghtheh *et al.*, 1993) and, more rarely, congenital stationary night blindness (Dryja *et al.*, 1993; Sieving *et al.*, 1995; al-Jandal *et al.*, 1999). The cause of blindness in RP is a loss of rod photoreceptors by apoptosis that is most likely triggered by the accumulation of mutant opsin protein.

3.3.2 Cone opsins

Photopic vision is subserved by an autosomal SWS1 gene (on chromosome 7 in humans) and one or more X-linked genes.

SWS1 opsin

The SWS1 gene is composed of five exons. Mutations in this gene give rise to a rare form of dichromacy called *tritanopia*; four missense mutations have been reported in the human gene, Gly79Arg, Ser214Pro, Pro264Ser (Weitz *et al.*, 1992a,b), and Pro56Leu (Gunther *et al.*, 2003), each resulting in a dominant mode of disease inheritance.

The SWS1 class of vertebrate visual pigments vary in λ_{max} from around 355 nm in the ultraviolet (UV-sensitive, UVS) to 435 nm and above in the violet (violet-sensitive, VS) region of the spectrum. In all teleost fish species examined so far, the SWS1 pigment is invariably UVS (Johnson *et al.*, 1993; Hisatomi *et al.*, 1997; Carleton *et al.*, 2000; Allison *et al.*, 2003) and a UVS pigment is also present in the lamprey, a member of the jawless vertebrates that separated from the main vertebrate lineage around 540 mya (Collin *et al.*, 2003). There is little doubt therefore that the ancestral vertebrate SWS1 pigment was UVS (Hunt *et al.*, 2001) with shifts to VS occurring many times during evolution.

Primates are amongst the many species in which a VS pigment with a λ_{max} of 415–420 nm has replaced a UVS pigment (Hunt *et al.*, 2001). However, even with this long-wavelength shift, spectral sensitivity in primates would extend into the UV if primate lenses did not strongly absorb light in this region of the spectrum. Transmittance of UV light generates two problems. Firstly, the refractive index for blue light is sufficiently different from red and green that when they are in focus, the blue is slightly out of focus. This would be exaggerated with UV light in the relatively large eyes of primates. Secondly, UV light produces high levels of reactive free radicals that are damaging to the cornea, lens, and retina and this may be a particular problem for animals that are long-lived. For both of these reasons, it is generally thought that the loss of UV sensitivity in primates is adaptive.

The tuning of primate VS pigments has been examined in detail by Shi *et al.* (2001). A chimeric opsin comprising TM 1-3 from human VS and TM 4-7 from mouse UVS pigments, when expressed and regenerated with 11-*cis*-retinal, was shown to produce a pigment with λ_{max} very close to the native human pigment (Figure 3.1A). This indicates that the same region of the opsin protein (TM 1-3) is important for violet spectral shifts in primates as in other vertebrates. Moreover, the simultaneous replacement of the residues present at seven sites (Phe46Thr, Phe49Leu, Thr52Phe, Phe86Leu, Thr93Pro, Ala114Gly, and Ser118Thr) in mouse UVS with those in human VS pigment (Shi *et al.*, 2001) produced a shift from 359 to 411 nm (Figure 3.1B), and the reverse substitutions shifted the λ_{max} of human VS to 360 nm. Overall therefore, these substitutions achieve the spectral shift between mouse UVS and human VS pigments, although their effects must be synergistic since single substitutions had no effect on λ_{max}. Additionally, the residue present at site 114 is not conserved across other primate species so is unlikely to be critically important for spectral tuning (Hunt *et al.*, 2004).

L and M opsins

The LWS gene in mammals is on the X-chromosome. In primates, two different forms of the gene are present that have arisen from a duplication of the ancestral gene (Nathans *et al.*, 1986). These variants show close identity with each other (97 percent nucleotide) and encode L and M pigments with λ_{max} values of around 563 and 532 nm, respectively. Since all Old World primates appear to possess this duplication (Ibbotson *et al.*, 1992), it must have occurred at the base of the catarrhine lineage around 30 mya. The L and M genes each comprise six exons and extend in humans over 15.2 and 13.3 kb of DNA, respectively, with the difference largely attributable to a 'long' intron 1 in the L gene. This long intron is rarely present (1 percent) in the M gene of Caucasians but is polymorphic in the M gene of African–Americans at a frequency of 35 percent (Jorgensen *et al.*, 1990). The genes are organized into a head-to-tail array on the X-chromosome with the L gene at the 5′ or upstream side and the M gene at the 3′ or downstream side of the array, as shown in Figure 3.2. The number of genes within the array does vary, with duplicate copies of the

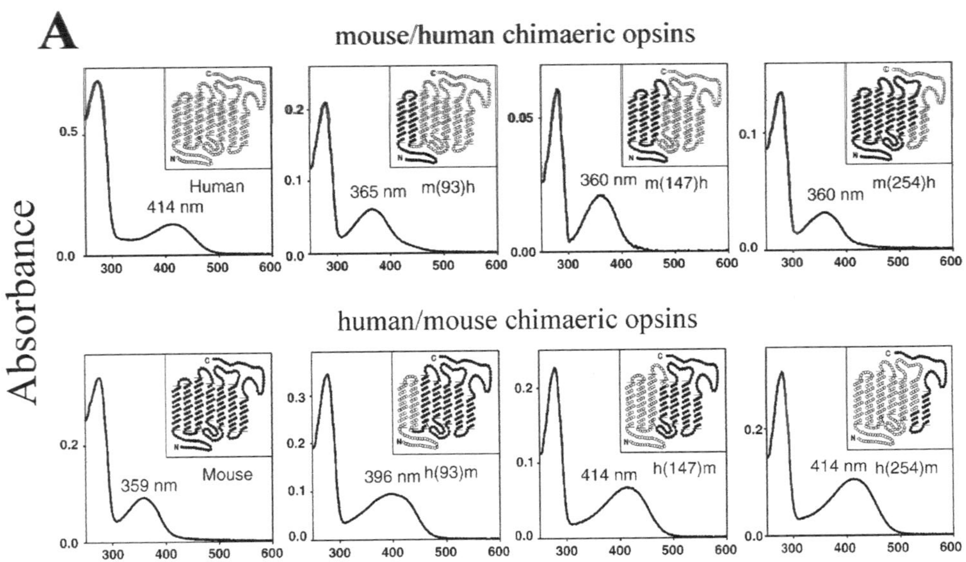

B

Amino acid sites							λ_{max} nm
46	**49**	**52**	**86**	**93**	**114**	**118**	
Phe	Phe	Thr	Phe	Thr	Ala	Ser	359
Thr	Phe	Thr	Phe	Thr	Ala	Ser	359
Phe	**Leu**	Thr	Phe	Thr	Ala	Ser	359
Phe	Phe	**Phe**	Phe	Thr	Ala	Ser	359
Phe	Phe	Thr	**Leu**	Thr	Ala	Ser	359
Phe	Phe	Thr	Phe	**Pro**	Ala	Ser	359
Phe	Phe	Thr	Phe	Thr	**Gly**	Ser	359
Phe	Phe	Thr	Phe	Thr	Ala	**Thr**	
Mutagenesis at 2 sites							
Phe	Phe	Thr	**Leu**	**Pro**	Ala	Ser	378
Mutagenesis at 4 sites							
Phe	Phe	Thr	**Leu**	**Pro**	**Gly**	**Thr**	399
Mutagenesis at all sites							
Thr	**Leu**	**Phe**	**Leu**	**Pro**	**Gly**	**Thr**	411

Figure 3.1 Spectral tuning of the violet-sensitive human SWS1 pigment. (A) Spectra of mouse/human chimeric pigments. Modified from Shi *et al.* (2001) and reproduced by permission. (B) Amino acid substitutions at seven sites in mouse UVS pigment. Top row shows residues present in the mouse UVS pigment. Subsequent rows show single, double, or multiple substitutions (boxed) of the residues present in human VS pigment. Data from Shi *et al.* (2001)

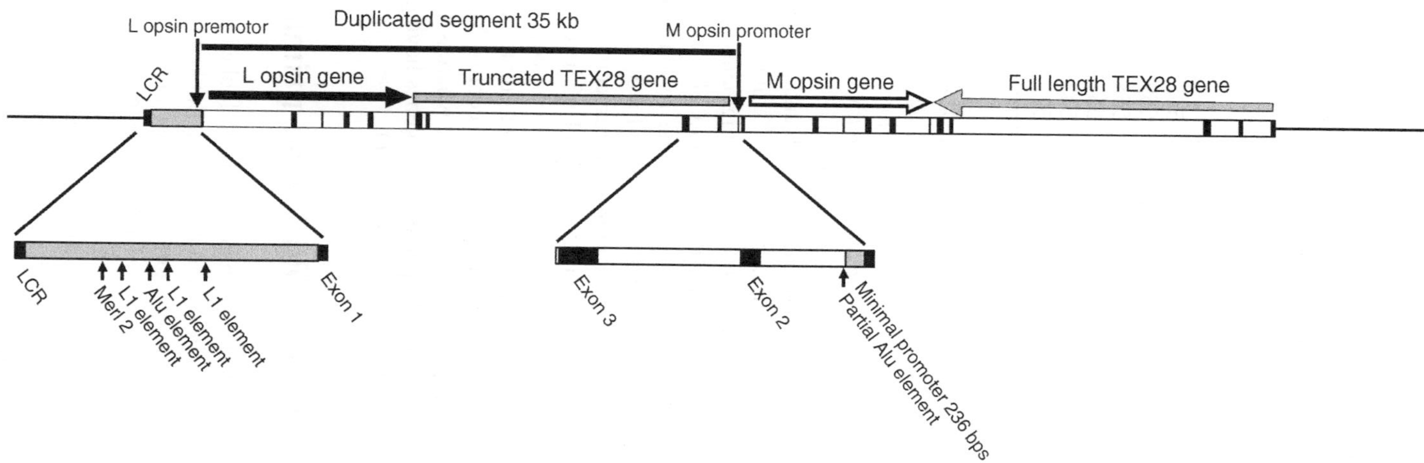

Figure 3.2 Origin of opsin gene duplication on the X-chromosome of Old World primates. The exons for the opsin and *TEX28* genes are shown as black bars. Arrows indicate the direction of transcription. The positions of the LCR, the minimal promoter region, and various repeat elements are shown

Table 3.1 Site-directed mutagenesis of L and M pigments showing the relative importance of substitutions at sites 116, 180, 230, 277, and 285 on spectral tuning. The amino acid substitutions were introduced sequentially into a human L pigment sequence. Data from Asenjo *et al.* (1994)

Amino acid sites					λ_{max} (nm)
116	180	230	277	285	
Ser	Ser	Ile	Tyr	Thr	563 (L pigment)
Tyr	Ser	Ile	Tyr	Thr	559
Tyr	Ala	Ile	Tyr	Thr	555
Tyr	Ala	Thr	Tyr	Thr	551
Tyr	Ala	Thr	Phe	Thr	543
Tyr	Ala	Thr	Phe	Ala	532 (M pigment)

M gene common in humans (Drummond-Borg *et al.*, 1989). The array is bounded on the upstream side by a so-called locus control region (LCR) (Wang *et al.*, 1992), the presence of which is critical for the expression of either gene. Deletions of the LCR give rise to another rare color vision defect called *blue cone monochromacy* where no L or M pigment is produced, even though both genes may be present and fully intact (Nathans *et al.*, 1989).

The human L and M opsin genes were first cloned and sequenced by Nathans *et al.* (1986). They encode proteins that differ by only 15 amino acids. Of these, five are involved in the spectral shift between the L and M pigments (Table 3.1). Differences at only three sites are, however, responsible for most of the spectral shift; these are 180, 277, and 285 (Asenjo *et al.*, 1994), with the polar residues Ser, Tyr, and Thr present at the respective sites in the L opsin and accounting for 3–7, 6–10, and 10–16 nm of the 29–30 nm difference between L and M pigments, respectively. Minor shifts are associated with Ser/Tyr116 encoded by exon 2 and Ile/Thr230 encoded by exon 4. Site 180 is polymorphic in both the L and M genes, with a consequent shift in the λ_{max} of the pigment. The frequencies vary, however, in different ethnic groups; e.g. Ala180 is encoded by 38 percent of L genes and Ser180 by 15 percent of M genes in a cohort of Causcasian individuals (Winderickx *et al.*, 1992, 1993), whereas the frequencies are lower for a Japanese population with only 21 percent of L genes having Ala180 and only 8 percent of M genes having Ser180 (Hayashi *et al.*, 2001).

3.4 Origin of duplication in Old World primates

The opsin gene duplication in Old World primates has been looked at in detail by Dulai *et al.* (1999). The opsin gene array on the X-chromosome is flanked on the upstream side by the LCR, with the L gene transcription start site approximately 3.5 kb downstream. Within this 3.5-kb flanking region, a number of DNA sequences called *repeat elements* are found. Repeat elements have limited function but, as their name implies, are found scattered throughout the primate genome. The different types of repeat element present in the flanking region include an Alu element at −1991, three L1 elements at −1360, −1781, and −2261, and a MER12 element at −2438. The M gene is a further 25.5 kb downstream of the L gene (Feil *et al.*, 1990) and sequence comparison of the upstream flanking region of the M gene demonstrates that homology with the flanking sequence of the L gene is restricted to the first 236 bp (Figure 3.2).

The non-homologous region between the L and M genes was shown by Hanna *et al.* (1997) to contain a truncated copy of a gene called *TEX28* that is expressed only in the testes. Truncated copies of this gene are present immediately upstream of all copies of the M gene in the array and it is only the final opsin gene at the 3′ end of the array that is flanked by a complete copy of *TEX28*. As shown in Figure 3.2, the truncated copy of *TEX28* was generated by the duplication; the absence of the first exon of the *TEX28* gene places one of the duplication breakpoints in intron 1 of *TEX28*. Overall, the duplicated region encompasses 35 kb of DNA which includes all six exons of the opsin gene, 236 bp of upstream flanking sequence, and a truncated copy of the *TEX28* gene on the downstream side. The whole of this duplicated region was inserted either just upstream of the original opsin gene or into the first intron of the *TEX28* gene.

A striking feature of the upstream region of the human M gene beyond the region of homology is the presence of Alu elements, a partial element of 120 bp at position −234, the point of insertion of the duplicated segment, and three more complete elements further upstream (Dulai *et al.*, 1999). Limited sequence data from the opsin array of the diana monkey, *Cercopithecus diana*, indicate that these elements are present in other catarrhines. The juxtapositioning therefore of Alu elements with the M opsin gene must have arisen from the gene duplication event rather than from a more recent insertion. Moreover, since Alu repeat elements have been implicated in the insertion event that resulted in the ‘long’ intron 1 of the human L opsin gene (Meagher *et al.*, 1996) and in unequal crossing-over within a number of other genes (see e.g. Campbell *et al.*, 1995; Rudiger *et al.*, 1995; Harteveld *et al.*, 1997; Levran *et al.*, 1998), the presence of a half Alu element exactly at the point of insertion of the original duplication suggests that this element may have been involved in promoting insertion at this position. These sequences may also be important in promoting unequal crossing-over within the opsin gene array that leads to either the complete loss of a gene, the generation of additional copies of the M gene (Drummond-Borg *et al.*, 1989; Ibbotson *et al.*, 1992), or the generation of L/M gene hybrids (Neitz *et al.*, 1995), as discussed in section 3.5.1. Such changes are responsible for inherited defects in red/green color vision in humans (Nathans *et al.*, 1986; Deeb *et al.*, 1992; Neitz and Neitz, 2000).

The close proximity of the duplication insertion breakpoint to the transcription start site of the M gene would imply that the promoter which regulates the expression of the M gene is contained within the 236 bp of homologous flanking sequence. This has been confirmed by an *in vitro* study of the activity of this region in driving the expression of the human L and M genes in transfected WERI Rb1 cells, a retinoblastoma cell line that is known to express cone opsins (Shaaban and Deeb, 1998). The minimal promoter region, numbered from immediately upstream of the first codon, was defined in this study as the region from −64 to −1, with positive elements that up-regulate expression within the region from −83 to −171, and negative elements that down-regulate expression from −171 to −240. This study was unable to provide, however, any information on the mechanism of selective gene expression that ensures that only one or other but not both of the L and M opsin genes is activated per cone photoreceptor, although differences in the activity of the two promoters were clearly apparent.

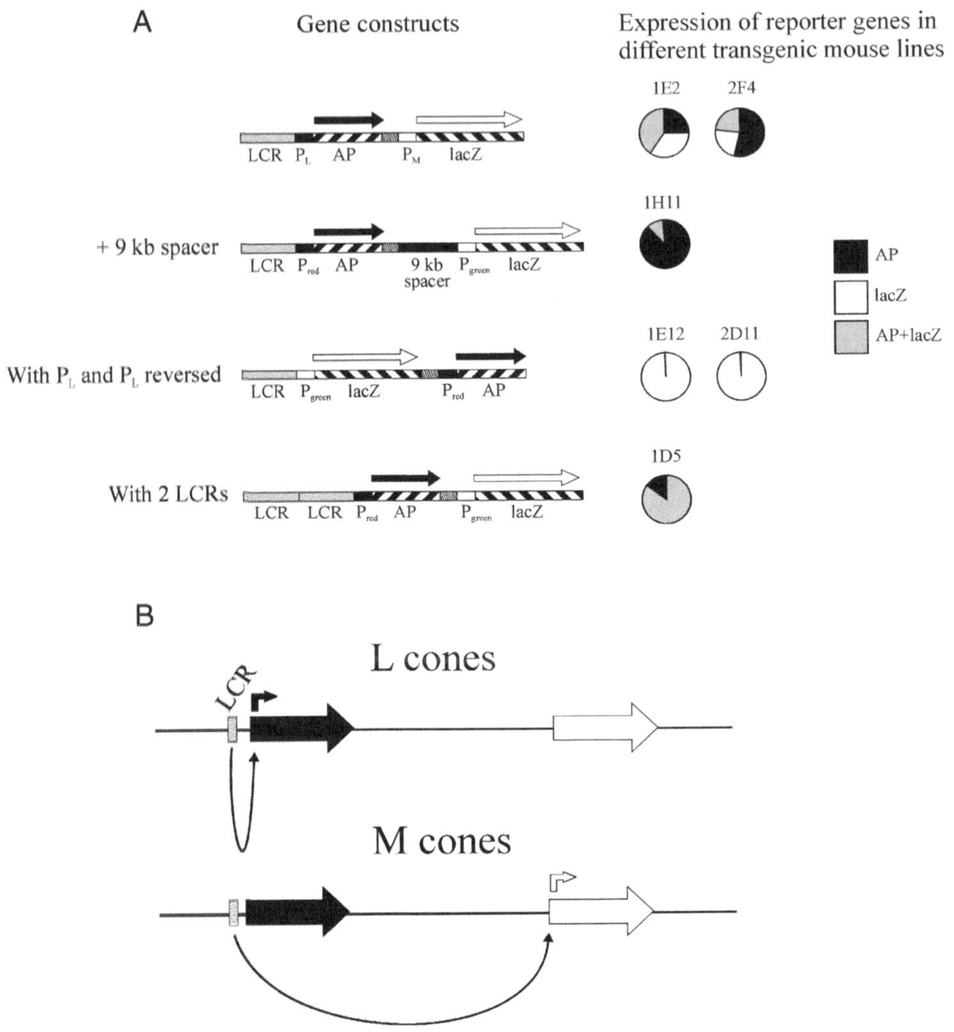

Figure 3.3 The effect of the LCR on reporter gene expression from L and M opsin gene promoters. (A) The relative activity of the L and M promoters in four different transgene arrays was assessed in a number of transgenic mouse lines by scoring individual cone cells for the expression of the reporter genes. The direction of transcription is shown by arrows. P_L, human L pigment gene promoter; P_M, human M pigment gene promoter; AP, human placental alkaline phosphatase; lacZ, *E. coli* β-galactosidase. Re-drawn from Smallwood *et al.* (2002). (B) Potential mechanism for the selective expression of L or M pigment gene via LCR–promoter pairing

The role of the LCR in L and M gene expression is evident from the failure of expression of either gene if the LCR is deleted, as in some forms of blue cone monochromacy (Nathans *et al.*, 1989). In females with two X-chromosomes, the process of X-chromosome inactivation will ensure that only one opsin gene array is potentially active in any given cell, but another mechanism must exist that ensures that only one gene, either L or M, is switched on. This may be the role of the LCR. L and M cones express a common set of phototransduction genes, and the close sequence identity between the L and M opsin genes suggests that there are no functional differences between L and M cone types other than spectral sensitivity. The selection of one or other gene for expression may be therefore the single determinant for the production of either an L or M cone. In such a circumstance, it is unlikely that gene-specific transcription factors are involved. More likely, it depends on a stochastic process for the selection of one of two stable and mutually exclusive states via *cis*-acting DNA sequences and identical *trans*-acting factors. The interaction of the LCR with the minimal promoter region of either the L or M gene could provide such a mechanism. This has been directly examined by Smallwood *et al.* (2002) through the generation of transgenic mice with a transgene construct that included the LCR and the L and M promoters driving different reporter genes (alkaline phosphatase and β-galactosidase). As shown in Figure 3.3, the activity of the promoter nearest to the LCR was highest while a 9-kb spacer interposed between the L and M reporter genes reduced the expression of the downstream gene even further. These data are consistent with the LCR interacting with either the L or M promoter to stably activate the corresponding opsin gene. The likelihood of interaction with either the L or M promoter may also depend on certain key regions of the promoter that affect the efficiency of this interaction, with different promoters showing stronger or weaker affinities. Promoter sequence differences may be responsible therefore for differences in the average L to M cone ratio of 2:1 in humans as compared with 1:1 in cercopithecoid monkeys (Bowmaker *et al.*, 1991).

Individual human males also show widely differing L to M cone ratios; McMahon *et al.* (2004) have examined the promoter regions of the L and M genes in humans for variation which might underlie differences in the strength of the promoter/LCR interaction. Two subjects with promoter mutations were found from a total of 73 individuals in which the cone ratio had been determined, and both had unusual cone ratios. However, this cannot be the only mechanism for determining L and M cone numbers since other individuals in this cohort also had unusual cone ratios but lacked any changes in the promoter regions.

3.5 L and M gene variation in Old World primates

3.5.1 Red/green color blindness

The head-to-tail arrangement of the L and M opsin genes in the array and their close sequence identity (even the intron sequences show relatively little divergence as discussed in section 3.5.2) would appear to promote a high level of mispairing and unequal crossing-over at meiosis (Figure 3.4). Depending on the exact position of the cross-over, this results in either the gain of a gene on one chromosome and the loss on the other, or the production of a hybrid gene. Gene gain or loss is responsible for the numerical polymorphism in opsin genes that is a feature of the opsin gene array, with 66 percent of human X-chromosomes having more than two genes within the array (Drummond-Borg *et al.*, 1989). The mean number of opsin gene copies in the array is, however, less certain;

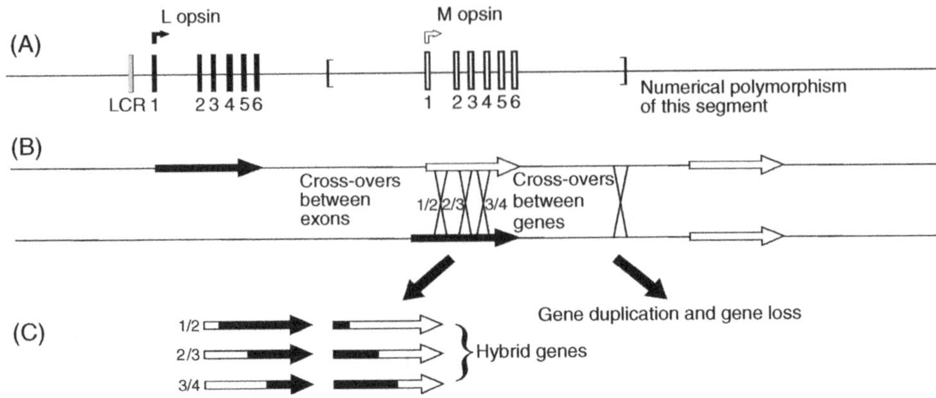

Figure 3.4 Crossing-over within the L and M opsin gene array. (A) Genomic organization of the L and M genes. (B) Mispairing of the L and M genes. (C) Products of crossing-over within and between mispaired genes

studies based on the ratio of restriction enzyme fragments (Drummond-Borg *et al.*, 1989) or on array size (Macke and Nathans, 1997) indicated an average of three copies per array, whereas a PCR-based study indicated a higher average number of more than four copies per array (Neitz and Neitz, 1995).

Up to 10 percent of human males have some form of red/green color blindness arising from changes in the opsin gene array (Neitz and Neitz, 2000). Full dichromacy arises when there is only a single functional gene in the array and, depending on whether it encodes for an L or M pigment, the dichromacy is further defined as deuteranopia or protanopia, respectively. In most cases, either the L or M gene has been lost by the mechanism described above. Less commonly, the L and M genes are both present but one is made non-functional by the presence of a mutation. The most common mutation is a Cys203Arg substitution that disrupts the structure of the opsin protein (Karnik *et al.*, 1988; Kazmi *et al.*, 1997) although three other mutations, Asn94Lys, Arg330Gln, and Gly338Glu, in L or M opsin genes have been reported (Ueyama *et al.*, 2002, 2003). Recently, it has been shown that in the retina of a dichromat with a color vision defect based on a mutant M opsin gene that presumably encodes a non-functional pigment, there is complete loss of M cones (Carroll *et al.*, 2004). In this instance therefore, it would appear that the production of mutant protein leads in due course to cone photoreceptor death. There are clear parallels therefore between this mechanism of disease and that arising from rod opsin mutations in RP. The dominant mode of inheritance seen for tritanopia may also result from the death of S cones arising from the accumulation of mutant SWS1 opsin.

The structure of the opsin gene array is almost certainly the same in other Old World primates as in humans and the gene sequences are no less identical (Ibbotson *et al.*, 1992), so it is not surprising to find that color vision defects are also present in other Old World primate species. A protanomalous male chimpanzee, *Pan troglodytes*, with a hybrid gene combining M exons 1–4 and L exons 5–6 and a normal M gene, has recently been identified by Saito *et al.* (2003), and in a cohort of 3153 macaque monkeys, three dichromats were found, all from a group of crab-eating macaques, *Macaca fascicularis* (Onishi *et al.*, 1999). All three lacked an L gene and, in one of these, it was confirmed that the single gene present was an L/M hybrid that combined exons 1–4 of the L gene and 5 and 6 of the M gene (Onishi *et al.*, 1999, 2002). The overall frequency of dichromats in crab-eating macaques was 0.4 percent compared with 2 percent obtained from a survey of 30 000 Caucasian male humans (Pokorny *et al.*, 1979). This value is similar to the maximum frequency of 0.65 percent for dichromatism in Old World monkeys estimated by Jacobs and Williams (2001) from the total number of animals examined across a number of studies and judged to be trichromats. The frequency of dichromacy amongst Old World monkeys is therefore significantly lower than in humans. A numerical polymorphism is also present in monkeys with an average ratio of 2M:1L genes found in a cohort of talapoin monkeys (*Miopithecus talapoin*) by Ibbotson *et al.* (1992). A lower frequency was found by Onishi *et al.* (2002) in macaques with only 4.5 percent of opsin gene arrays containing more than two genes. These latter animals were all from the same region of South Thailand, suggesting that the distribution of arrays with more than two genes may be geographically limited.

A potentially less severe form of red/green color blindness arises when the λ_{max} of one of the pigments is shifted toward the other such that color discrimination in the red/green region of the spectrum is compromised. Spectrally shifted pigments arise from the generation of hybrid genes (Figure 3.5) and, depending on the exonic composition of the hybrid, a reduced spectral separation of the hybrid and normal pigment may give rise to color vision defects that range from mild through to very severe.

3.5.2 Gene conversion in Old World L and M opsin genes

Gene conversion, a process whereby the sequence of one gene is converted to that of another, was first noted in exons 4 and 5 of the L and M opsin genes of a group of cercopithecoid monkeys by Ibbotson *et al.* (1992); contrary to expectation for duplicated

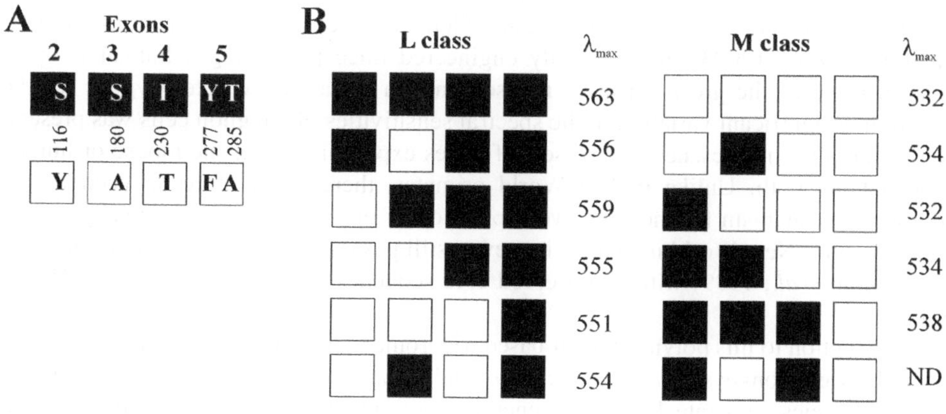

Figure 3.5 Spectral tuning of L and M pigments. (A) Major spectral tuning sites in exons 2–5. (B) L, M, and hybrid genes and λ_{max} values for corresponding pigments. Data from Asenjo *et al.* (1994)

genes, the exon sequences of the two genes are more alike within species and between genes than vice versa. Similar observations have been made for the intronic regions of these genes (Shyue *et al.*, 1995; Zhao *et al.*, 1998). Non-coding introns are generally found to diverge more rapidly than the coding exons, but this is not the case for the L and M opsin genes where, for example, intron 4 of 1519 bp shows no differences at all between genes. This implies that there is a process of sequence homogenization in play that limits sequence diversification. This may be a specific process, but is more likely a consequence of the ongoing process of unequal crossing-over within the array and the production of hybrid genes as discussed above.

3.6 Color vision in platyrrhines and prosimians

3.6.1 New World primates

Red/green color vision is much more variable in New World primates (Table 3.2). The nocturnal owl monkey, *Aotus trivirgatus*, lacks a functional SWS1 gene; Wikler and Rakic (1990) noted the absence of S cones in the owl monkey retina and Jacobs *et al.* (1996b) identified a number of insertions and deletions in exon 1 of the SWS1 gene that introduce a premature stop codon. The owl monkey possesses therefore only a single class of cones and is monochromatic. Most other New World species exhibit a trichromacy that is based on only two opsin genes, an autosomal SWS1 gene as in Old World primates, and a polymorphic X-linked L/M gene with multiple allelic forms that encode pigments with differing λ_{max} values (Mollon *et al.*, 1984; Neitz *et al.*, 1991; Williams *et al.*, 1992). Platyrrhines thus lack the routine trichromacy of Old World primates, since male monkeys can combine the SWS1 gene with just one of the different allelic forms of the X-linked gene and are therefore dichromats. In contrast, those females that inherit a different form of the LWS (M or L) gene from each parent possess trichromatic vision, since X-inactivation will ensure that only one allele is expressed per cell (Mollon *et al.*, 1984). The ability of the X-inactivation system to generate cone photoreceptors expressing just one of the two opsin gene alleles has been elegantly demonstrated by Smallwood *et al.* (2003) in genetically engineered mice possessing a human L opsin gene in place of the green-sensitive mouse gene. In female mice heterozygous for the mutation, a significant variation in the spectral sensitivities of ganglion cells was present, consistent with the presence of two sets of cones expressing either the mouse or human gene but not both. Unlike in Old World primates, there is therefore no need for an additional mechanism for the selective expression of only one L/M opsin gene per cone photoreceptor. New World monkeys, however, still possess an LCR upstream of the L/M gene (Dulai *et al.*, 1999) with a conserved core sequence identical to that in Old World primates.

An exception to this polymorphism-based trichromacy in New World primates is found in the howler monkeys, *Alouatta seniculus* and *Alouatta caraya* (Jacobs *et al.*, 1996a). In these species, separate L and M genes are present, thus conferring routine trichromacy in both males and females. The duplication is not present in two other Atelid genera, the spider monkey (*Ateles* sp.) and woolly monkey (*Lagothrix* sp.), with both possessing a polymorphic L/M gene (Jacobs and Deegan, 2001). Fluorescent *in situ*

Table 3.2 Variation in LWS genes and alleles in New World primates. All λ_{max} values shown have been determined experimentally. Data from Jacobs and Deegan (2001, 2003), Mollon *et al.* (1984), Saito *et al.* (2004), and Travis *et al.* (1988)

Infraorder: Platyrrhini		Superfamily: Ceboidea		L/M opsin genes	Number of variants per gene	λ_{max} of pigments
Family	Subfamily	Genus	Common name			
Atelidae	Atelinae	*Alouatta*	Howler monkey	Two genes	1	560, 530
		Ateles	Spider monkey	Single gene	2	563, 550
		Lagothrix	Woolly monkey	Single gene	2	
Pitheciidae	Pitheciinae	*Callicebus*	Titi monkey	Single gene	5	
		Pithecia	Saki monkey	Single gene	3	
Cebidae	Cebinae	*Cebus*	Capuchin monkey	Single gene	3	563, 549, 535
		Samiri	Squirrel monkey	Single gene	3	564, 550, 536
	Aotinae	*Aoutus*	Owl monkey	Single gene	1	545
	Callitrichinae	*Leontopithecus*	Lion tamarin	Single gene	3	563, 555, 543
			Saddle back tamarin	Single gene	3	563, 557, 545
		Callithrix	Marmoset	Single gene	3	565, 559, 543

hybridization (FISH) analysis places the duplicate copies in the same region of the howler X-chromosome (Hunt, Stanyon, Halford, and von Dornum, unpublished results) but, unlike the arrangement in Old World primates, each gene is associated with a separate LCR (Dulai *et al.*, 1999; Figure 3.6). Nevertheless, microspectrophotometric data demonstrate that the L and M pigments are present in separate cone photoreceptors, confirming the differential expression of these genes and trichromacy in the howler monkey. The role of the LCR in gene regulation is therefore unclear in this species. As mentioned above, the LCR is also present in other New World primates (Dulai *et al.*, 1999) where the single polymorphic X-linked gene does not require any selection process other than X-inactivation to limit expression to a single gene copy per photoreceptor. The role of the LCR in these species needs therefore to be re-examined.

Tuning of L and M pigments

New World primates use amino acid substitutions at sites 180, 277, and 285 to tune the L/M pigments. In this regard they are identical to Old World monkeys. However, sites 116 and 230, which produce small shifts in the L and M pigments of Old World primates, do not show consistent substitution across the different New World pigments and are unlikely therefore to be involved in the tuning process (Table 3.3). In addition, site 277 does not vary in species from the family Callitrichidae; the long and short-wave pigments differ spectrally therefore by only 19–20 nm, compared with the approximately 27 nm seen in Old World primates and in members of the other major New World family, the Cebidae, where site 277 is used. Intermediate alleles are also found with different combinations of amino acids at these sites, thereby yielding pigments with a λ_{max} at 556 or 550 nm, depending on family. The howler monkey pigments also show substitutions at all three sites (Jacobs *et al.*, 1996a), consistent with the λ_{max} of the two pigments at approximately 530 and 558 nm (Saito *et al.*, 2004).

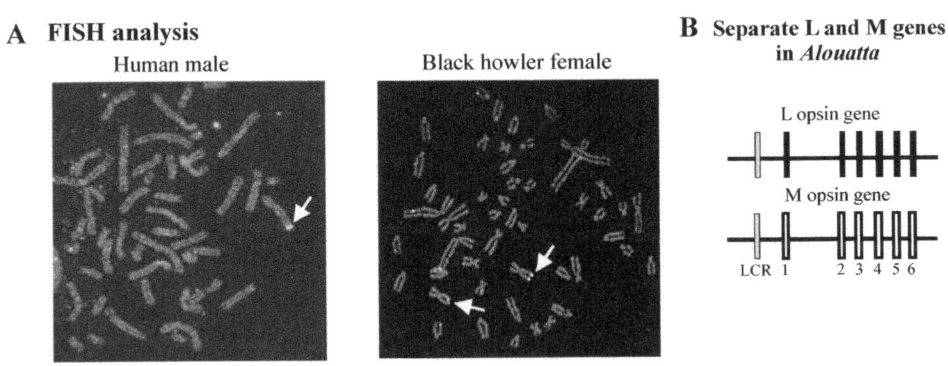

Figure 3.6 Chromosomal localization of the L and M opsin genes in human and howler monkey by FISH. (A) Metaphase chromosome preparations were probed with a human L opsin cosmid probe that included the 5′ flanking region labeled with biotin-dUTP or digoxigenin-dUTP. Arrows indicate position of hybridization on human and howler monkey X-chromosomes. (B) Duplication of coding region and LCR in howler monkey. The FISH analysis was carried out by Dr R Stanyon, Laboratory of Genomic Diversity, National Cancer Institute, Frederick, Maryland, USA

Table 3.3 Amino acid substitutions involved in the spectral tuning of LWS pigments in New World primates

	Amino acid sites				
	116	180	230	277	285
Capuchin monkey					
P563	Tyr	Ser	Leu	Tyr	Thr
P550	His	Ala	Leu	Phe	Thr
P530	His	Ala	Leu	Phe	Ala
Squirrel monkey					
P563	Tyr	Ser	Leu	Tyr	Thr
P550	Tyr	Ala	Leu	Phe	Thr
P530	Tyr	Ala	Leu	Phe	Ala
Marmoset					
P563	His	Ser	Leu	Tyr	Thr
P556	Tyr	Ala	Leu	Tyr	Thr
P543	Tyr	Ala	Leu	Tyr	Ala
Tamarin					
P563	His	Ser	Leu	Tyr	Thr
P556	His	Ala	Leu	Tyr	Thr
P543	His	Ala	Ile	Tyr	Ala
Howler monkey					
L pigment		Ser	Leu	Tyr	Thr
M pigment		Ala	Leu	Phe	Ala

3.6.2 Prosimians

Conventional systematics places the entire platyrrhine radiation subsequent to the separation of the Old and New World lineages. The different trichromatic systems present in modern day species in the two lineages may have evolved therefore from a polymorphism-based trichromacy present in the ancestral primate. Implicit in this hypothesis is that the mechanism of spectral tuning is essentially the same in New World and Old World primates, despite the sexual dimorphism of trichromacy in platyrrhines. As described above, this has largely turned out to be the case (Neitz *et al.*, 1991; Ibbotson *et al.*, 1992; Williams *et al.*, 1992). One hypothesis is that the two-gene system of Old World primates originated by unequal crossing-over between two chromosomes of the ancestral primate carrying different L/M alleles, thereby placing non-identical copies of the gene onto a single X-chromosome and generating in a single step the genetic basis for routine trichromacy. This hypothesis requires that the evolution of the polymorphic system pre-dates the separation of the Old and New World primate lineages. Recent studies on the prosimians lend some support to this concept. A polymorphic L/M gene has been reported in three species by Tan and Li (1999) from three different strepsirhine families, Coquerel's sifaka (*Propithecus verreauxi coquereli*), the greater dwarf lemur (*Cheirogaleus major*), and the red ruffed lemur (*Varecia variegata rubra*). The alleles were identified by DNA sequencing and classified as L or M on the basis of the amino acids encoded at the major tuning sites. In all three species, only two alleles were identified and they differed only at site 277, with either Ala or Thr present. The λ_{max} values for the two pigments were estimated by Tan and Li (1999) from sequence comparison with New World primates to be at 543 and 558 nm, and this has been confirmed by flicker photometry in one species, Coquerel's sifaka, where λ_{max} values around 545 and 558 nm were obtained (Jacobs *et al.*, 2002). Trichromacy is not, however, a common feature of prosimians: only a single allele was identified in a further 13 species where more than one X-chromosome was examined (Tan and Li, 1999), although a second allele may have been missed since the sample size of individuals for each species was small. Of these latter species, the majority have an M pigment. L pigments are, however, present in three species, Coquerel's dwarf lemur (*Microcebus coquereli*), the grey mouse lemur (*Microcebus murinus*), and the Phillippine tarsier (*Tarsius syrichta*). It is striking therefore that different species of lemurs have genes that encode either an L or M pigment and that an L pigment gene is found in the tarsier; trichromacy may have been present in the common ancestor of the tarsiers and strepsirhines but only retained by a sub-set of modern day species. Its original presence is, however, still reflected in the L or M genes present in different species.

An unexpected feature of color vision in the prosimians is the observation that a functional SWS1 pigment is absent from all members of the Lorisiformes so far examined, but present in the Lemuriformes. The Lorisiformes comprise two subfamilies, the Lorisinae, which includes the lorises, and the Galaginae, which includes the galagos. For the greater galago, *Otelemur crassicaudatus*, sequencing of exon 4 of the SWS1 gene identified a 2-bp insertion and 1-bp deletion that result in the introduction of a stop codon (Jacobs *et al.*, 1996b). A more extensive study of the SWS1 gene in the lesser galago, *Galago senegalensis*, and the slow loris, *Nycticebus coucang*, identified in addition to the 2-bp insert in exon 4 described above, a 46-bp deletion in exon 1, and two nonsense mutations in the codon interrupted by intron 2 (Kawamura and Kubotera, 2004). The presence of the 2-bp deletion in exon 4 in all three species and the shared mutations in the slow loris and lesser galago (which may also be present in the greater galago) imply a common ancestry for the loss of a functional SWS1 gene and this may extend to all members of the Lorisiformes. In contrast, Kawamura and Kubotera (2004) found that two members of the Lemuriformes, the brown lemur, *Eulemur fulvus*, and the western tarsier, *Tarsius bancanus*, have a fully intact SWS1 gene. All three members of the Lorisiformes are nocturnal, as is the western tarsier, but the brown lemur is defined as cathemeral (irregular bursts of activity throughout a 24-h cycle). Moreover, the SWS1 gene in the tarsier and lemur shows evidence of constraint in the acquisition of nucleotide substitutions (Kawamura and Kubotera, 2004), indicating that the gene is under positive selection pressure in these species.

As mentioned previously, the nocturnal owl monkey also has a mutant non-functional SWS1 gene (Jacobs *et al.*, 1996b). The loss of S cones most likely occurred in this species as a result of a reduced dependence on color vision; S cones would appear to be preferentially lost as they contribute little toward spatial or temporal resolution and their loss does not have major structural implications for the retina. The same would appear to be the case for the nocturnal lorises and galagos but not for the nocturnal lemurs and tarsiers. In the former group, the SWS1 has become inactivated by mutation, whereas in the latter group, the gene is not only retained intact but there is also evidence for purifying selection (Kawamura and Kubotera, 2004).

3.7 Evolution of trichromacy

3.7.1 Gene duplication in the howler monkey

The duplication and generation of two spectrally distinct pigment genes in the howler monkey and in Old World primates is an example of convergent evolution. Sequence differences in intron 4 are informative about evolutionary origin, with both genes in the howler monkey having 13 indels (insertions or deletions) and an Alu element in common with other New World primates (Boissinot *et al.*, 1997). The duplicated genes of the howler monkey are closer in homology therefore to other New World primate sequences than to human. Surprisingly, however, the duplicated genes of the howler monkey show less divergence over exons 3, 4, and 5 than the P562 and P543/P535 alleles of four species of New World primates, with only eight differences (if the three key tuning changes are discounted) compared to an average of 14.25 changes between the allelic forms (Boissinot *et al.*, 1997; Hunt *et al.*, 1998). The normal process of recombination between alleles would be expected to limit divergence but separate genes would not be exposed to this process. The lower variability of the howler L and M genes would suggest therefore that they were not derived from two different allelic forms of the ancestral gene. In fact, only two sites support an origin from different alleles, at positions 685 and 697 in exon 4 where the same nucleotide dimorphism is seen both in the separate howler monkey genes and in the polymorphic alleles of other New World primates. The interpretation of such differences is complicated, however, by the presence of gene conversion in the howler monkey genes, as demonstrated by the uneven distribution of variation along intron 4, with the 5′ and 3′ ends of the intron showing a greater level of homogenization than the central region (Boissinot *et al.*, 1997; Surridge and Mundy, 2002). It is possible therefore that other dimorphisms have been eliminated from the howler monkey L and M genes and that the two genes are derived from different allelic forms, as indicated by the phylogenetic analysis discussed in the next section.

3.7.2 Origin of different allelic forms in New World primates

Phylogenetic analysis of nucleotide substitutions in exons 3, 4, and 5 of the polymorphic gene from six species of New World primate shows that the clustering of sequences does not follow a species pattern (Boissinot *et al.*, 1998; Hunt *et al.*, 1998). Rather, they mostly group according to the spectral characteristics of the corresponding pigments (Figure 3.7). In particular, all of the P562 pigments form a single clade. Some subdivision along subfamily lines is present with the other alleles, with the tamarin and marmoset P556 and P543 sequences grouping together and away from the P550 and P535 sequences of the capuchin, squirrel, and saki monkeys. The topology of the tree does not support a multiorigin hypothesis for the alleles in the different genera. Rather, it implies a common origin of the P562 alleles, with the possibility that the P556 and P543 alleles of the Callitrichinae and the P550 and P535 alleles of the Cebinae and Pitheciinae arose separately. The utilization of site 277 only in the latter group is consistent with this interpretation. Within the tree, the howler monkey L sequence is placed within the P562

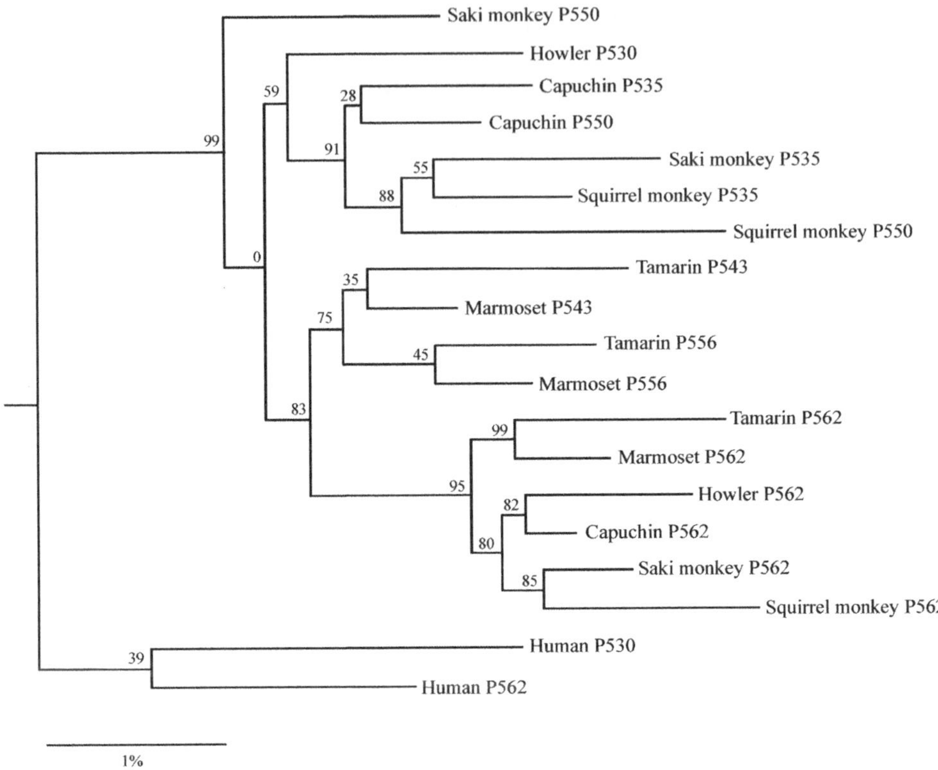

Figure 3.7 Phylogenetic analysis of exons 3–5 of New World primate alleles. Neighbor-joining tree was generated from the nucleotide sequences using weights of 80 and 20 percent for non-synonymous and synonymous substitutions, respectively. The number at each node denotes the percentage of 500 bootstrap replicates that supported the branch point. Re-drawn from Boissinot *et al.* (1998)

clade and the M sequence within the Cebid P550 and P535 clade, indicating an origin of the duplicated genes from the different allelic forms of an ancestral polymorphic gene.

3.7.3 Routine trichromacy in Old World primates

A number of lines of evidence bear on the question of the evolution of trichromacy in Old World primates as follows:

1. The presence of a polymorphic opsin gene in some species of prosimians suggests that this system of trichromacy is not just confined to the New World primates, and may pre-date the Old World/New World split. It is possible therefore that the duplicated genes in Old World primates arose from different allelic forms of an ancestral polymorphic gene, although the utilization of only site 285 for spectral tuning in modern day prosimians indicates that the evolution of the full system may not

have occurred until the advent of the platyrrhine lineage. Since the entire platyrrhine radiation is thought to have occurred subsequent to the Old World/New World split, this would tend to argue against a common origin for the Old World and New World systems. The utilization of the same amino acid substitutions at sites 180, 277, and 285 in the duplicated genes in Old World primates and the different allelic forms of New World primates is also not in itself convincing evidence for a common origin since these sites are frequently used for the tuning of a range of visual pigments. For example, the L and M duplicates of the LWS gene in the cave fish, *Astyanax fasciatus*, use the same three amino acid substitutions to generate red- and green-sensitive pigments (Yokoyama and Yokoyama, 1990) and site 277 is used for the tuning of rod and blue-sensitive (SWS2) cone pigments in cottoid fish from Lake Baikal (Hunt *et al.*, 1996; Cowing *et al.*, 2002). It is possible therefore that substitution at these sites is the only way to generate the spectral variants of the LWS pigment; convergent evolution between the Old World L and M genes and the New World alleles cannot therefore be ruled out.

2. The nucleotide divergence between the allelic forms of the L/M opsin gene in the marmoset and capuchin monkey ranges from 1.9 to 3.5 percent, with an average value of 2.6 percent (Table 3.4; Hunt *et al.*, 1998). This is considerably less than the divergence between the Old World primate genes that show an average of 6.1 percent. Taken at face value, this would imply that the New World polymorphism is more recent than the Old World gene duplication. An unquantifiable factor, however, when

Table 3.4 Divergence of exons 3, 4, and 5 of L/M opsin genes in Old and New World primates. The values are shown as percentage nucleotide differences. The Old World gene differences are between the L gene and the M gene of the same species. Data from Hunt *et al.* (1998)

		New World primates				Old World primates
		Capuchin		Marmoset		
	λ_{max}	550	535	556	543	M
New World primates						
	563	2.8	2.8			
Capuchin						
	550		1.9			
	563			2.3	3.5	
Marmoset						
	556				2.3	
Old World primates						
Chimpanzee	L					8.3
Human	L					4.7
Gorilla	L					5.6
Diana	L					5.2
Macaque	L					7.1
Talapoin	L					5.9

considering divergence between alleles is the extent of homogenization that has arisen from inter-allelic recombination; the polymorphism may therefore be much older and pre-date the Old and New World primate split. Sequence homogenization within the New World genes is certainly present and has been shown to extend to the introns (Boissinot *et al.*, 1997, 1998).

3. Except for the major tuning sites, there are few examples in the Old World genes and New World alleles of a nucleotide dimorphism between the L and M genes that is also present in the polymorphic L/M gene. A comparison of the sequences of exons 3, 4, and 5 identifies just three such sites (Table 3.5), and for two of these, the nucleotide present also differs between the L and M genes in different Old World species. Therefore, either these are the only sites that have escaped gene conversion or the separate L and M genes did not arise directly from different allelic forms of a polymorphic gene. Also, even if the nucleotide differences at the three major tuning sites are discounted, there is still a total of 21 sites across exons 3, 4, and 5 that differ in a consistent way between the separate L and M genes of Old World primates but do not differ in the different forms of the New World primate gene, so conversion has not resulted in the complete homogenization of the L and M sequences.

Table 3.5 Sequence dimorphisms in Old World L and M genes and New World L/M alleles. These are the only sites that show the same nucleotide differences between the Old World L and M genes and the New World L/M alleles. Data from Boissinot *et al.* (1997), Deeb *et al.* (1994), Dulai *et al.* (1994), Hunt *et al.* (1998), Ibbotson *et al.* (1992), and Nathans *et al.* (1986)

	Pigments	Exon 3	Exon 4	
		552	685	697
Great Apes	L	A	T/A	G
	M	A	A	A
Old World monkeys	L	G/A	A	G
	M	A	A	A
Squirrel monkey	P562	G	T	G
	P550	A	A	A
	P535	G	A	A
Capuchin	P562	G	T	G
	P550	A	A	A
	P535	A	A	A
Marmoset	P562	G	T	G
	P556	A	T	A
	P543	A	A	A
Tamarin	P562	G	T	G
	P556	G	T	A
	P543	G	A	A
Howler monkey	L	G	T	G
	M	G	A	A

For these reasons, the origin of the gene duplicates in Old World primates remains uncertain.

3.8 Summary and conclusions

Phylogenetic analysis of the New World primate polymorphic L/M opsin alleles indicates a common origin for this system. Despite this, differences do exist: members of the Callitrichinae (marmosets and tamarins) utilize substitutions at only two sites, 180 and 285, whereas members of the Pitheciinae (saki monkey) and Cebinae (capuchin and squirrel monkey) additionally utilize substitutions at site 277 to give a wider spectral separation between pigments. A polymorphism-based trichromacy is also present in a sub-set of prosimians, based in this case on substitution at site 285 only, to give a relatively small spectral separation of around 13 nm between the L and M pigments. The presence of a polymorphic gene in prosimians indicates that a system for trichromacy evolved in a more limited form before the separation of the Old World and New World primate lineages, so it is possible that the duplicated genes in Old World primates may have originated from different forms of a polymorphic gene in the ancestral primate. Unfortunately, the presence of extensive gene conversion between the L and M genes in Old World primates arising from the high rate of mispairing and crossing-over between these genes has served to homogenize sequences and thereby largely to obliterate the molecular evidence that would have clarified the situation. Nevertheless, there are differences between the L and M genes that have escaped conversion that are not present in the New World alleles. This would indicate that the evolution of trichromacy in Old World primates arose from the production of two identical copies of the L/M gene which have diverged over time and under the pressure of adaptive selection to produce L and M genes that encode spectrally distinct pigments.

There would appear to be little doubt that the howler monkey duplication arose quite separately from that in Old World primates and phylogenetic analysis indicates that the duplicated gene copies arose directly from different allelic forms of the polymorphic L/M gene present in an ancestral member of the Atelidae. However, since this duplication is not found in the spider monkey or woolly monkey, it must have occurred in the howler monkey lineage after the separation of these other lineages. Both genes are in the same region of the X-chromosome but the duplication differs from that in Old World primates in the presence of an LCR upstream of both copies. The organization of the two genes is otherwise unknown, so the role, if any, for the LCR in selecting either the L or M gene for expression remains uncertain. The LCR has been implicated in this process in Old World primates. However, since it is present upstream of the polymorphic gene in New World primates, this cannot be its primary role.

In contrast, monochromacy is present in the owl monkey and in one of the major groups of prosimians, the lorises and galagos. In all cases, S cones have been lost and this is associated with a nocturnal life style where the dependence on color vision is greatly reduced. The sparse distribution of S cones in the primate retina means that their loss would not compromise spatial and temporal vision, nor greatly affect retinal structure. The loss of S cones is not, however, an inevitable consequence of nocturnality, since lemurs, many of which are nocturnal, retain S cones and have a fully functional SWS1 gene.

References

al-Jandal, N., Farrar, G.J., Kiang, A.S. *et al.*, 1999. A novel mutation within the rhodopsin gene (Thr-94-Ile) causing autosomal dominant congenital stationary night blindness. *Hum. Mutat.* **13**, 75–81.

Allison, W.T., Dann, S.G., Helvik, J.V. *et al.*, 2003. Ontogeny of ultraviolet-sensitive cones in the retina of rainbow trout (*Oncorhynchus mykiss*). *J. Comp. Neurol.* **461**, 294–306.

al-Maghtheh, M., Gregory, C., Inglehearn, C. *et al.*, 1993. Rhodopsin mutations in autosomal dominant retinitis pigmentosa. *Hum. Mutat.* **2**, 249–255.

Arrese, C.A., Hart, N.S., Thomas, N. *et al.*, 2002. Trichromacy in Australian marsupials. *Curr. Biol.* **12**, 657–660.

Asenjo, A.B., Rim, J., and Oprian, D.D., 1994. Molecular determinants of human red/green color discrimination. *Neuron* **12**, 1131–1138.

Boissinot, S., Zhou, Y.-H., Qui, L. *et al.*, 1997. Origins and molecular evolution of the X-linked duplicate color vision genes in howler monkeys. *Zool. Studies* **36**, 360–369.

Boissinot, S., Tan, Y., Shyue, S.K. *et al.*, 1998. Origins and antiquity of X-linked triallelic color vision systems in New World monkeys. *Proc. Natl. Acad. Sci. USA* **95**, 13749–13754.

Bowmaker, J.K., Astell, S., Hunt, D.M. *et al.*, 1991. Photosensitive and photostable pigments in the retinae of Old World monkeys. *J. Exp. Biol.* **156**, 1–19.

Campbell, C., Marondel, I., Montgomery, K. *et al.*, 1995. Unequal homologous recombination of human DNA on a yeast artificial chromosome. *Nucleic Acids Res.* **23**, 3691–3695.

Carleton, K.L., Harosi, F.I., and Kocher, T.D., 2000. Visual pigments of African cichlid fishes: Evidence for ultraviolet vision from microspectrophotometry and DNA sequences. *Vision Res.* **40**, 879–890.

Carroll, J., Neitz, M., Hofer, H. *et al.*, 2004. Functional photoreceptor loss revealed with adaptive optics: An alternate cause of color blindness. *Proc. Natl. Acad. Sci. USA* **101**, 8461–8466.

Collin, S.P., Knight, M.A., Davies, W.L. *et al.*, 2003. Ancient colour vision: Multiple opsin genes in the ancestral vertebrates. *Curr. Biol.* **13**, R864–R865.

Cowing, J.A., Poopalasundaram, S., Wilkie, S.E. *et al.*, 2002. Spectral tuning and evolution of short wave-sensitive cone pigments in cottoid fish from Lake Baikal. *Biochemistry* **41**, 6019–6025.

Deeb, S.S., Lindsey, D.T., Hibiya, Y. *et al.*, 1992. Genotype-phenotype relationships in human red/green color-vision defects: Molecular and psychophysical studies. *Am. J. Hum. Genet.* **51**, 687–700.

Deeb, S.S., Jorgensen, A.L., Battisti, L. *et al.*, 1994. Sequence divergence of the red and green visual pigments in great apes and humans. *Proc. Natl. Acad. Sci. USA* **91**, 7262–7266.

Dominy, N.J. and Lucas, P.W., 2001. Ecological importance of trichromatic vision to primates. *Nature* **410**, 363–366.

Drummond-Borg, M., Deeb, S.S., and Motulsky, A.G., 1989. Molecular patterns of X chromosome-linked color vision genes among 134 men of European ancestry. *Proc. Natl. Acad. Sci. USA* **86**, 983–987.

Dryja, T.P., Berson, E.L., Rao, V.R. *et al.*, 1993. Heterozygous missense mutation in the rhodopsin gene as a cause of congenital stationary night blindness. *Nat. Genet.* **4**, 280–283.

Dulai, K.S., Bowmaker, J.K., Mollon, J.D. *et al.*, 1994. Sequence divergence, polymorphism and evolution of the middle-wave and long-wave visual pigment genes of great apes and Old World monkeys. *Vision Res.* **34**, 2483–2491.

Dulai, K.S., von Dornum, M., Mollon, J.D. *et al.*, 1999. The evolution of trichromatic color vision by opsin gene duplication in New World and Old World primates. *Genome Res.* **9**, 629–638.

Feil, R., Aubourg, P., Heilig, R. *et al.*, 1990. A 195-kb cosmid walk encompassing the human Xq28 color vision pigment genes. *Genomics* **6**, 367–373.

Gunther, K.L., Neitz, J., and Neitz, M., 2003. A novel missense mutation in the S cone photopigment in a male who made tritan errors on the Neitz test of color vision. *Invest. Ophthalmol. Vis. Sci.* **44**, E-Abstract 1907.

Hanna, M.C., Platts, J.T., and Kirkness, E.F., 1997. Identification of a gene within the tandem array of red and green color pigment genes. *Genomics* **43**, 384–386.

Harteveld, K.L., Losekoot, M., Fodde, R. *et al.*, 1997. The involvement of Alu repeats in recombination events at the alpha-globin gene cluster: Characterization of two alphazero-thalassaemia deletion breakpoints. *Hum. Genet.* **99**, 528–534.

Hayashi, S., Ueyama, H., Tanabe, S. *et al.*, 2001. Number and variations of the red and green visual pigment genes in Japanese men with normal color vision. *Jpn. J. Ophthalmol.* **45**, 60–67.

Hisatomi, O., Satoh, T., and Tokunaga, F., 1997. The primary structure and distribution of killifish visual pigments. *Vision Res.* **37**, 3089–3096.

Hunt, D.M., Fitzgibbon, J., Slobodyanyuk, S.J. *et al.*, 1996. Spectral tuning and molecular evolution of rod visual pigments in the species flock of cottoid fish in Lake Baikal. *Vision Res.* **36**, 1217–1224.

Hunt, D.M., Dulai, K.S., Cowing, J.A. *et al.*, 1998. Molecular evolution of trichromacy in primates. *Vision Res.* **38**, 3299–3306.

Hunt, D.M., Wilkie, S.E., Bowmaker, J.K. *et al.*, 2001. Vision in the ultraviolet. *Cell. Mol. Life Sci.* **58**, 1583–1598.

Hunt, D.M., Cowing, J.A., Wilkie, S.E. *et al.*, 2004. Divergent mechanisms for the tuning of shortwave sensitive visual pigments in vertebrates. *Photochem. Photobiol. Sci.* **3**, 713–720.

Ibbotson, R.E., Hunt, D.M., Bowmaker, J.K. *et al.*, 1992. Sequence divergence and copy number of the middle- and long-wave photopigment genes in Old World monkeys. *Proc. R. Soc. Lond. B* **247**, 145–154.

Jacobs, G.H. and Deegan, J.F., 2nd. 2001. Photopigments and colour vision in New World monkeys from the family Atelidae. *Proc. R. Soc. Lond. B Biol. Sci.* **268**, 695–702.

Jacobs, G.H. and Deegan, J.F., 2nd. 2003. Cone pigment variations in four genera of new world monkeys. *Vision Res.* **43**, 227–236.

Jacobs, G.H. and Williams, G.A., 2001. The prevalence of defective color vision in Old World monkeys and apes. *Col. Res. Appl.* **26**, S123–S127.

Jacobs, G.H., Neitz, M., Deegan, J.F. *et al.*, 1996a. Trichromatic colour vision in New World monkeys. *Nature* **382**, 156–158.

Jacobs, G.H., Neitz, M., and Neitz, J., 1996b. Mutations in S-cone pigment genes and the absence of colour vision in two species of nocturnal primate. *Proc. R. Soc. Lond. B* **263**, 705–710.

Jacobs, G.H., Deegan, J.F., 2nd, Tan, Y. *et al.*, 2002. Opsin gene and photopigment polymorphism in a prosimian primate. *Vision Res.* **42**, 11–18.

Johnson, R.L., Grant, K.B., Zankel, T.C. *et al.*, 1993. Cloning and expression of goldfish opsin sequences. *Biochemistry* **32**, 208–214.

Jorgensen, A.L., Deeb, S.S., and Motulsky, A.G., 1990. Molecular genetics of X chromosome-linked color vision among populations of African and Japanese ancestry: High frequency of a shortened red pigment gene among Afro-Americans. *Proc. Natl. Acad. Sci. USA* **87**, 6512–6516.

Karnik, S.S., Sakmar, T.P., Chen, H.B. *et al.*, 1988. Cysteine residues 110 and 187 are essential for the formation of correct structure in bovine rhodopsin. *Proc. Natl. Acad. Sci. USA* **85**, 8459–8463.

Kawamura, S. and Kubotera, N., 2004. Ancestral loss of short wave-sensitive cone visual pigment in lorisiform prosimians, contrasting with its strict conservation in other prosimians. *J. Mol. Evol.* **58**, 314–321.

Kazmi, M.A., Sakmar, T.P., and Ostrer, H., 1997. Mutation of a conserved cysteine in the X-linked cone opsins causes color vision deficiencies by disrupting protein folding and stability. *Invest. Ophthalmol. Vis. Sci.* **38**, 1074–1081.

Levran, O., Doggett, N.A., and Auerbach, A.D., 1998. Identification of Alu-mediated deletions in the Fanconi anemia gene FAA. *Hum. Mutat.* **12**, 145–152.

Macke, J.P. and Nathans, J., 1997. Individual variation in size of the human red and green visual pigment gene array. *Invest. Ophthalmol. Vis. Sci.* **38**, 1040–1043.

McMahon, C., Neitz, J., and Neitz, M., 2004. Evaluating the human X-chromosome pigment gene promoter sequences as predictors of L:M cone ratio variation. *J. Vis.* **4**, 203–208.

Meagher, M.J., Jorgensen, A.L., and Deeb, S.S., 1996. Sequence and evolutionary history of the length polymorphism in intron 1 of the human red photopigment gene. *J. Mol. Evol.* **43**, 622–630.

Mollon, J.D., 1989. 'Tho' she kneel'd in that place where they grew "..." The uses and origins of primate colour vision. *J. Exp. Biol.* **146**, 21–38.

Mollon, J.D., Bowmaker, J.K., and Jacobs, G.H., 1984. Variations of colour vision in a New World primate can be explained by polymorphism of retinal photopigments. *Proc. R. Soc. Lond. B Biol. Sci.* **222**, 373–399.

Nathans, J., Thomas, D., and Hogness, D.S., 1986. Molecular genetics of human color vision: The genes encoding blue, green, and red pigments. *Science* **232**, 193–202.

Nathans, J., Davenport, C.M., Maumenee, I.H. *et al.*, 1989. Molecular genetics of human blue cone monochromacy. *Science* **245**, 831–838.

Neitz, M. and Neitz, J., 1995. Numbers and ratios of visual pigment genes for normal red-green color vision. *Science* **267**, 1013–1016.

Neitz, M. and Neitz, J., 2000. Molecular genetics of color vision and color vision defects. *Arch. Ophthalmol.* **118**, 691–700.

Neitz, M., Neitz, J., and Jacobs, G.H., 1991. Spectral tuning of pigments underlying red-green color vision. *Science* **252**, 971–974.

Neitz, M., Neitz, J., and Jacobs, G.H., 1995. Genetic basis of photopigment variations in human dichromats. *Vision Res.* **35**, 2095–2103.

Onishi, A., Koike, S., Ida, M. *et al.*, 1999. Dichromatism in macaque monkeys. *Nature* **402**, 139–140.

Onishi, A., Koike, S., Ida-Hosonuma, M. *et al.*, 2002. Variations in long- and middle-wavelength-sensitive opsin gene loci in crab-eating monkeys. *Vision Res.* **42**, 281–292.

Osorio, D. and Vorobyev, M., 1996. Colour vision as an adaptation to frugivory in primates. *Proc. R. Soc. Lond. B Biol. Sci.* **263**, 593–599.

Pokorny, J., Smith, V., Verriest, G. *et al.*, 1979. *Congenital and Acquired Colour Vision Defects*, Grune Stratton, New York.

Regan, B.C., Julliot, C., Simmen, B. *et al.*, 2001. Fruits, foliage and the evolution of primate colour vision. *Philos. Trans. R. Soc. Lond. B Biol. Sci.* **356**, 229–283.

Rudiger, N.S., Gregersen, N., and Kielland-Brandt, M.C., 1995. One short well conserved region of Alu-sequences is involved in human gene rearrangements and has homology with prokaryotic chi. *Nucleic Acids Res.* **23**, 256–260.

Saito, A., Mikami, A., Hasegawa, T. *et al.*, 2003. Behavioral evidence of color vision deficiency in a protanomalia chimpanzee (*Pan troglodytes*). *Primates* **44**, 171–176.

Saito, C.A., da Silva Filho, M., Lee, B.B. *et al.*, 2004. Alouatta trichromatic color vision – single-unit recording from retinal ganglion cells and microspectrophotometry. *Invest. Ophthalmol. Vis. Sci.* **45**, E-Abstract 4276.

Shaaban, S.A. and Deeb, S.S., 1998. Functional analysis of the promoters of the human red and green visual pigment genes. *Invest. Ophthalmol. Vis. Sci.* **39**, 885–896.

Shi, Y., Radlwimmer, F.B., and Yokoyama, S., 2001. Molecular genetics and the evolution of ultraviolet vision in vertebrates. *Proc. Natl. Acad. Sci. USA* **98**, 11731–11736.

Shyue, S.K., Hewett-Emmett, D., Sperling, H.G. *et al.*, 1995. Adaptive evolution of color vision genes in higher primates. *Science* **269**, 1265–1267.

Sieving, P.A., Richards, J.E., Naarendorp, F. *et al.*, 1995. Dark-light: Model for nightblindness from the human rhodopsin Gly-90-Asp mutation. *Proc. Natl. Acad. Sci. USA* **92**, 880–884.

Smallwood, P.M., Wang, Y., and Nathans, J., 2002. Role of a locus control region in the mutually exclusive expression of human red and green cone pigment genes. *Proc. Natl. Acad. Sci. USA* **99**, 1008–1011.

Smallwood, P.M., Olveczky, B.P., Williams, G.L. *et al.*, 2003. Genetically engineered mice with an additional class of cone photoreceptors: Implications for the evolution of color vision. *Proc. Natl. Acad. Sci. USA* **100**, 11706–11711.

Sumner, P. and Mollon, J.D., 2000. Catarrhine photopigments are optimized for detecting targets against a foliage background. *J. Exp. Biol.* **203 Pt 13**, 1963–1986.

Surridge, A.K. and Mundy, N.I., 2002. Trans-specific evolution of opsin alleles and the maintenance of trichromatic colour vision in Callitrichine primates. *Mol. Ecol.* **11**, 2157–2169.

Tan, Y. and Li, W.H., 1999. Trichromatic vision in prosimians. *Nature* **402**, 36.

Travis, D.S., Bowmaker, J.K., and Mollon, J.D., 1988. Polymorphism of visual pigments in a callitrichid monkey. *Vision Res.* **28**, 481–490.

Ueyama, H., Kuwayama, S., Imai, H. *et al.*, 2002. Novel missense mutations in red/green opsin genes in congenital color-vision deficiencies. *Biochem. Biophys. Res. Commun.* **294**, 205–209.

Ueyama, H., Li, Y.H., Fu, G.L. *et al.*, 2003. An A-71C substitution in a green gene at the second position in the red/green visual-pigment gene array is associated with deutan color-vision deficiency. *Proc. Natl. Acad. Sci. USA* **100**, 3357–3362.

Wang, Y., Macke, J.P., Merbs, S.L. *et al.*, 1992. A locus control region adjacent to the human red and green visual pigment genes. *Neuron* **9**, 429–440.

Weitz, C.J., Miyake, Y., Shinzato, K. *et al.*, 1992a. Human tritanopia associated with two amino acid substitutions in the blue-sensitive opsin. *Am. J. Hum. Genet.* **50**, 498–507.

Weitz, C.J., Went, L.N., and Nathans, J., 1992b. Human tritanopia associated with a third amino acid substitution in the blue-sensitive visual pigment [letter]. *Am. J. Hum. Genet.* **51**, 444–446.

Wikler, K.C. and Rakic, P., 1990. Distribution of photoreceptor subtypes in the retina of diurnal and nocturnal primates. *J. Neurosci.* **10**, 3390–3401.

Williams, A.J., Hunt, D.M., Bowmaker, J.K. *et al.*, 1992. The polymorphic photopigments of the marmoset: Spectral tuning and genetic basis. *EMBO J.* **11**, 2039–2045.

Winderickx, J., Lindsey, D.T., Sanocki, E. *et al.*, 1992. Polymorphism in red photopigment underlies variation in colour matching. *Nature* **356**, 431–433.

Winderickx, J., Battisti, L., Hibiya, Y. *et al.*, 1993. Haplotype diversity in the human red and green opsin genes: Evidence for frequent sequence exchange in exon 3. *Hum. Mol. Genet.* **2**, 1413–1421.

Yokoyama, R. and Yokoyama, S., 1990. Convergent evolution of the red- and green-like visual pigment genes in fish, *Astyanax fasciatus*, and human. *Proc. Natl. Acad. Sci. USA* **87**, 9315–9318.

Zhao, Z., Hewett-Emmett, D., and Li, W.H., 1998. Frequent gene conversion between human red and green opsin genes. *J. Mol. Evol.* **46**, 494–496.

4

The Ecology of the Primate Eye: Retinal Sampling and Color Vision

D. Osorio, M. Vorobyev, and *G.H. Jacobs*

4.1 Introduction: sampling and retinal specialization

'To any vision must be brought an eye adapted to what is to be seen, and bearing some likeness to it.' (Plotinus, Section 9, 6th tractate 1st Ennaid; 3rd Century C.E.)

Because natural selection cannot take place in a perfect world, Darwin saw the perfection of human optics as a challenge to his theory of evolution and was, apparently, relieved when Helmholtz found that eyes suffer from chromatic aberration (Cronin, 1993). Nonetheless, the notion that natural selection can optimize vision may be instructive. We might infer what an animal is adapted to see by discovering what it sees best. Constraints imposed by the physics of light and metabolic costs mean that performance in one task generally comes at the expense of another. For instance, compromises between spatial resolution and absolute sensitivity, or between chromatic and luminance coding, can suggest the signals or behaviors that influence the evolution of vision.

This chapter outlines how primate eyes perform as devices for sampling natural signals. Animals make behavioral decisions based on imperfect sense data, so that perception resembles statistical inference (Rao *et al.*, 2002). Understanding of eye design (as well as later stages of sensory processing and perception) has therefore been informed by statistical models that specify how to sample data, how to use prior knowledge, and the magnitudes of errors. To illustrate this approach the first sections deal with spatial

The Primate Visual System: A Comparative Approach Edited by Jan Kremers

sampling, and we then look in more detail at *color*, where efforts to explain the spectral tuning of primate cone pigments have led to a range of adaptive accounts.

In eyes an array of photoreceptors samples an optical image. As in cameras, the quality of the picture encoded depends upon the brightness and resolution of the optical image, and on the sensitivity and 'grain' (i.e. sampling density) of the receptor array. Photoreceptor arrays are, however, different from camera film or CCDs (Hughes, 1977; Roorda and Williams, 1999) in that artificial devices sample the image uniformly, whereas the density of photoreceptors varies across the visual field (Figure 4.1). This retinal topography often reflects habitat and lifestyle (Hughes, 1977). Animals that inhabit open country typically have an 'acute zone' of high cell densities in the streak of retina that

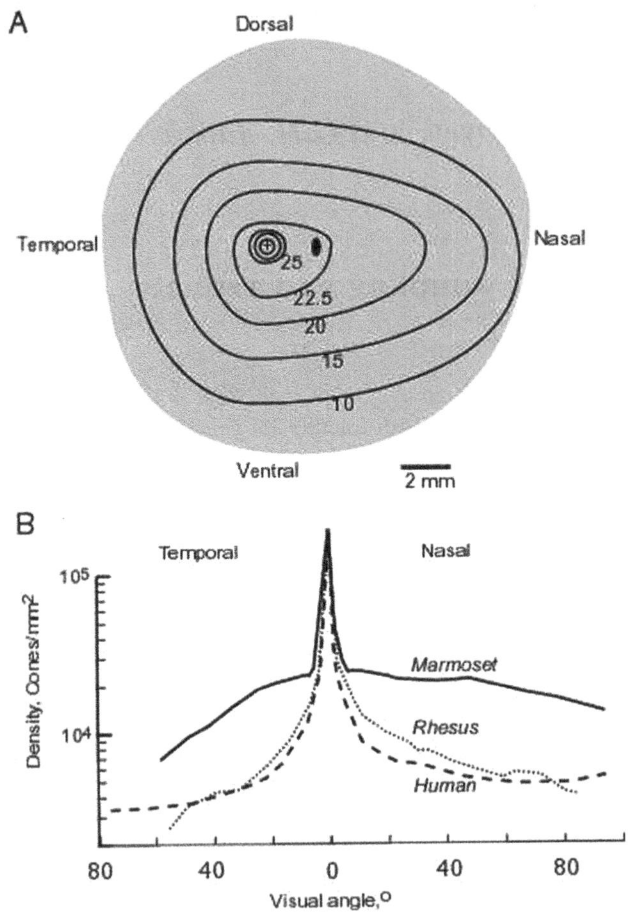

Figure 4.1 Retinal topography. (A) Density of cones in a marmoset retina, and (B) comparison between cone densities of three primates with widely different eye sizes plotted as a function of visual angle. Note that the maximum densities are the same, but marmoset has an appreciably higher density of cones in the periphery (see also Table 4.1; from Troilo *et al.*, 1993; Reproduced by permission of Elsevier.)

views the horizon; presumably because this is where important events occur, or objects of interest are small. By comparison arboreal species have circularly symmetrical retinal topographies, with high acuity in the frontal visual field. Primates have approximately circularly symmetrical topography (Figure 4.1), although there is often a slightly elevated cone density along the horizon (Curcio *et al.*, 1987, 1990). In addition, a foveal pit containing an area centralis with very high cone density is present in all haplorrhine primates including nocturnal species (Table 4.1), and to a lesser extent in strepsirrhines (Stone and Johnston, 1981; Chapter 1). Foveas are also found in birds, but not other mammals; they are likely to reflect the use of mobile eyes that can be directed to any point of interest.

The fact that eyes do not sample uniformly implies that the economics of vision are different from those of manufactured devices. Good glass lenses are costly, so in photography it is desirable to sample up to limits imposed by lens quality, and hence make full use of the optical image. By comparison, as Darwin was aware, physiological optics are generally high-quality, and aside from chromatic aberration they can be perfect – limited only by diffraction (Land and Nilsson, 2002). Instead, the need to maintain ionic concentration gradients in photoreceptors makes the retina amongst the most metabolically active of all tissues (Laughlin, 2001a,b). As a result phototransduction – the conversion of light to electrical signals – is a significant part of the energy-budget of many animals. The high cost of phototransduction, as opposed to image formation, means that retinas should

Table 4.1 Cone densities and theoretical resolution of some primate eyes (see also Franco *et al.*, 2000). Resolution is estimated from the cone density and the optical magnification of the eye, according to the formulae given by Snyder and Miller (1977); this is an upper limit. The cited publications listed give details of retinal topography outside the area centralis

Species	Eye size/PND* (mm)	Ret magn. μm/deg	Peak cone density/ 10^3 mm^2	Theoretical resolution cpd	Reference
Microcebus murinus	9.65/5.00	87	8	4.2	Dkhissi-Benyahya *et al.*, 2001
Galago sp.	18.26/9.65	165	8.5	8.2	Dkhissi-Benyahya *et al.*, 2001
Aotus sp.	–	200	16.3	13.7	Dkhissi-Benyahya *et al.*, 2001
Tarsier spectrum	15/–	150*	50	18	Hendrickson *et al.*, 2000
Human	24/16.7	280	199	67	Curcio *et al.*, 1987, 1990
Macaca fascicularis	17.12/12.17	211	100	36	Lapuerta and Schein, 1995
Callithrix jacchus	10.9/7.63	128	190	30	Troilo *et al.*, 1993
Cebus apella	18.4/11.3	197	169	44	Andrade da Costa and Hokoc, 2000

*Estimate PND, posterior nodal distance.

sample an optical image to make efficient use of costly photoreceptors and neurons (Figure 4.2; Hughes, 1996; Laughlin, 2001a,b). The need for efficiency accounts for the varied retinal topography of different animals, and more generally may be an important principle in eye design.

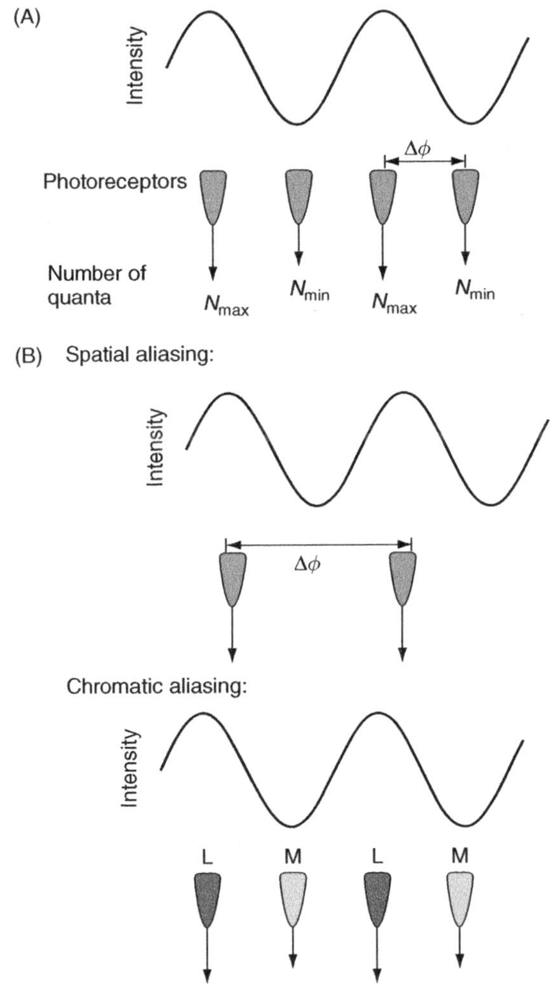

Figure 4.2 Spatial sampling and aliasing. (A) The optimal separation of receptors ($\Delta\phi$) depends upon optical resolution of the eye, and the difference in quantum catch ($N_{max} - N_{min}$) between neighboring receptors (Snyder and Miller, 1977). (B) Spatial aliasing can occur if $\Delta\phi$ exceeds the optical resolution, so that high spatial frequencies are represented as a lower spatial-frequency signal. Chromatic aliasing can occur when outputs of two different spectral types of receptor are used for spatial and color vision. In this case fine spatial patterns might be confused with a color that excites L receptors more than M receptors

4.2 Spatial sampling: signals, noise and image statistics

Resolution of a lens is limited because a point source is not focused to a point, but a blurred spot whose intensity distribution is known as the point spread function (Land and Nilsson, 2002). The extent of the blur determines the smallest separation that allows two points to be distinguished from one. With aberration-free optics, the point spread function is set by diffraction at the aperture, forming a so-called Airy disc. The half-width of this disc, w, is given by: $w = \lambda/D$ rad, where λ is the wavelength of light and D the diameter of the aperture. For a human 2-mm pupil and $\lambda = 550\,\text{nm}$, $w = 0.016°$ (approx. $1'$ arc). When diffraction-limited optical resolution is raised only by increasing the aperture (e.g. pupil diameter). In humans and other primates aberrations exceed diffraction limits with large pupil sizes, but it is not clear how optical aberrations affect vision at the relatively low intensities when the pupil diameter exceeds about 3 mm (Laughlin, 1992), because they may be masked by photon noise (section 4.2.2). Off-axis aberrations may in fact be beneficial because they limit the effects of spatial aliasing by the relatively coarse cone array (Williams *et al.*, 1996; see following section).

Optimal receptor spacing depends upon image quality. There is no benefit in sampling beyond the limit imposed by optical resolution because intensities registered by neighboring receptors become indistinguishable. Conventional principles of optical design suggest that the smallest useful receptor separation is approximately the half-width of the point spread function (Figure 4.2; Snyder and Miller, 1977). This arrangement is known as matched (or Nyquist) sampling, and allows recovery of the finest details transmitted by the optics. More widely spaced detectors are said to undersample; this is undesirable, not only because details are missed, but also fine patterns (i.e. high spatial frequencies) are 'aliased' to corrupt lower spatial frequencies (Figure 4.2; Snyder and Miller, 1977; Williams *et al.*, 1996; Williams and Hofer, 2004). Aliasing resembles Moiré effects, which occur when two fine gratings are superimposed to give an artifactual coarser pattern. Despite the potential disadvantages, most eyes have a receptor density that is a factor of 2 or more lower than needed for matched sampling (Snyder *et al.*, 1977, 1986). Foveas of primates and birds of prey are unusual in sampling close to the Nyquist limit (Snyder *et al.*, 1986; Williams and Hofer, 2004).

Given that receptor and neural densities seem to vary adaptively it seems unlikely that undersampling is an evolutionary imperfection. It can be shown that in the presence of noise it is indeed efficient to undersample when photoreceptors are energetically costly (Snyder *et al.*, 1977; Hughes, 1996; Laughlin, 2001a,b). To take account of the loss of information through aliasing, Snyder *et al.* (1977) suggested that the eye should maximize the number of discriminable (natural) images, and showed that the optimal sampling rate depends on spatial correlation in both visual signals and receptor noise.

4.2.1 Cone densities in primate retinas

In diurnal primates the area centralis is a regular hexagonal array (Hirsch and Hylton, 1984), with cone densities ranging from 1.0 to 2.5×10^5 cones/mm^2 (Table 4.1; Franco *et al.*, 2000 report a value of 4×10^5 cones/mm^2 for a howler monkey). Although there is substantial variation within species, the maximum cone density seems not to differ systematically between species, even where eye sizes are as different as marmosets and humans (Samy and Hirsch, 1989; Troilo *et al.*, 1993). This uniformity might seem surprising given that retinal magnification, image brightness, and resolution all vary with eye size and/or pupil diameter. However, as Snyder and Miller (1977) pointed out, for a receptor array at the Nyquist limit the amount of light captured by each receptor is independent of pupil diameter. Moreover, the linear dimensions of the Airy disc depend only on the f-number (i.e. focal length/aperture). Therefore, as long as the f-number is independent of eye size, the relationship between the receptor width and the blur-circle diameter is also independent of eye size. This means that if performance is photon-noise limited (following section), for a given f-number a given cone spacing will result in a fixed sampling efficiency; that is the signal-to-noise ratio is constant relative to the diffraction limit. Thus the uniformity across species suggests that diurnal primates share similar sampling strategies, at least in the fovea, where the relationship between the optical quality of the image and the receptor density is fixed, and differences between species are dependent only on eye size. (An additional constraint here may be wave-guide properties of photoreceptors, which limit the outer segment diameter to a minimum of about 2 μm.) Less easy to understand by sampling theory are the findings that the absolute area of the fovea, and hence its overall 'bandwidth', varies little with total eye size (Franco *et al.*, 2000), and that the total number of cones in the entire eye is approximately the same in humans, macaques, and marmosets (Figure 4.1; Troilo *et al.*, 1993).

Compared with diurnal anthropoid primates amongst prosimians there is considerable variation in cone densities, and there is less obvious specialization in the area centralis. Tarsiers (like the anthropoid owl monkeys) have a fovea, with maximum cone densities of about 50 000 cones/mm^2, whilst the nocturnal strepsirrhines have lower cone densities (see also below). There is little data on lemurs, but the peak cone density in the diurnal species *Propithecus* reaches about 34 000 cones/mm^2 (Peichl *et al.*, 2001).

4.2.2 Photoreceptor noise

Noise sets detection thresholds and consequently has a crucial influence on sensory coding (Barlow, 1964, 1977; Cohn, 2004). Noise is introduced at each stage of the sensory pathway, so that it may be difficult to attribute a threshold to any particular mechanism – for example, receptors, retinal neurons, or cortical cells. Nonetheless, given that phototransduction is metabolically costly, it is *a priori* reasonable to suppose that in certain evolutionarily important tasks performance is limited by receptor noise. Otherwise selection should simply favor more economical eyes or photoreceptors, whose noise levels match those in higher-level mechanisms.

Historically it was thought that internal processes set thresholds, and the finding by Hecht *et al.* (1942) that dark-adapted humans can detect lights containing five or fewer photons was a key discovery because it showed that psychophysical performance might be set by fundamental physical constraints (Barlow, 1964). What sets this absolute threshold, and what sets thresholds at higher intensities? Photoreceptors suffer three main types of noise, each dominating at different light levels (Figure 4.3; Barlow, 1964): (i) the presence of a fixed threshold (>1 photon) is attributed to a light-independent process called 'dark noise'; (ii) at intermediate intensities thresholds are set by variations in photon catch; and (iii) at the highest intensities the main source of noise is from metabolic processes of

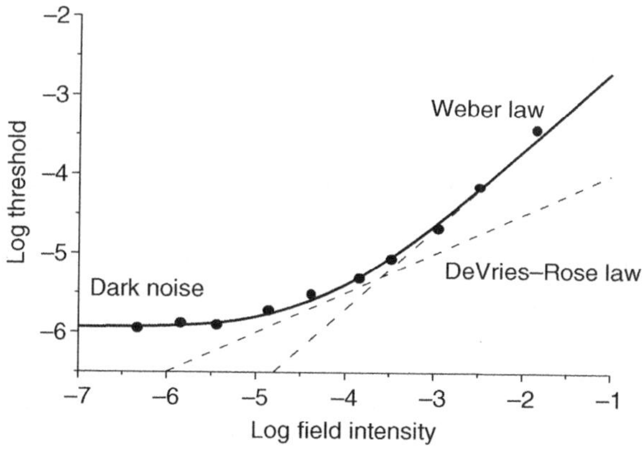

Figure 4.3 Threshold versus intensity curve is explained by the noise in a cone mechanism
Note: Points indicate psychophysical thresholds for 1°, 0.06 s test stimulus imaged 5° from the fovea. Wavelengths of test and field stimuli are 580 and 500 nm, respectively (Wyszecki and Stiles, 1982, p. 128). A theoretical dependence of threshold on the field intensity is shown as a solid line. The model predicts the effects of (i) dark noise, (ii) noise due to fluctuations of the number of absorbed quanta (DeVries–Rose law), and (iii) neural noise that is proportional to the number of absorbed quanta (Weber law). The dark noise $d = 126$ quanta per receptive field, per integration time, the Weber fraction $\omega = 0.02$, the scaling factor $s = 9.88 \times 10^6$ photons per unit radiance. Description of the model: Let M be the field radiance, then the corresponding number of absorbed quanta per receptive field and integration time is $x = sM$, where s is a scaling factor. The increment threshold, N, is set as the inaccuracy of the estimate of the number of absorbed quanta, δx. Hence, $N = s\delta x$. Let d be the dark light, ω be a Weber fraction of a cone mechanism, then the variance of x is given by: $(\delta x)^2 = (d+x) + \omega^2(d+x)^2$ where the first term corresponds to fluctuations of the number of absorbed quanta (variance is proportional to the signal) and the second term corresponds to the neural noise that obeys the Weber law (variance is proportional to the square of the signal). This equation can be rewritten in terms of field radiance and thresholds as: $N = \dfrac{\sqrt{(d+sM) + \omega^2(d+sM)^2}}{s}$

phototransduction and transmitter release. The level of this internal noise is proportional to light intensity.

Spontaneous isomerization of photopigment molecules is sometimes called dark-light (Barlow, 1964). For humans Donner (1992) estimated a rate of 0.01 spontaneous isomerizations per rod per second. By comparison, cones have a much higher level of dark noise. Early work (Barlow, 1964) and a physiological study (Schneeweis and Schnapf, 1999) indicate the dark noise in cones is equivalent to about 3000 photopigment isomerization events per cone per second, whereas Donner (1992) estimated the rate to be about 110 events per cone per second for human foveal vision. To investigate the origins of this difference Kefalov *et al.* (2003) expressed human and salamander L cone photopigments in *Xenopus* rods and human rod pigment in *Xenopus* cones, but did not otherwise alter the transduction pathways. They found that dark noise in cones is indeed due to photoisomerization-like events in the photopigments, and that it occurs at over 10 000 times the rate for rod pigment. The reasons why cone pigments have such a high level of dark noise remain unclear. It has been suggested that the longer wavelength sensitivity maximum of cone photopigments accounts for their elevated rate of thermal isomerization (Barlow, 1964), but although there is a relationship between spectral tuning and dark noise this does not explain the difference between rods and cones. Recent work (Ala-Laurila *et al.*, 2004) suggests that the lower energy barrier for cone pigment isomerization is due to the chomophore pocket being 'looser', perhaps to allow faster responses or recovery from bleaching. Thus cones may trade-off dark noise against response speed. Regardless of its origins, dark noise favors maximizing light flux per photopigment molecule (e.g. image brightness), rather than total light capture, and hence a low f-number, rather than a large eye (Barlow, 1964).

Above absolute threshold, photon noise (i.e. variation in photon catch) may be important. Random quantum process means that the number of photons absorbed from a fixed source varies with Poisson statistics, i.e. the variance equals the mean. As noise is proportional to the standard deviation (the square root of the variance), detection thresholds are proportional to the square root of intensity, and thresholds are said to follow DeVries–Rose law, named after the discoverers of this relationship (Figure 4.3; Barlow, 1964; Cohn, 2004). When photon noise is limiting there is pressure to increase both light capture and image brightness, which favors large pupil apertures and large eyes.

Photon noise is not dominant at high intensities because it is exceeded by fluctuations associated with the biochemical process of transduction and electrical properties of photoreceptors, and perhaps subsequent neural mechanisms. This noise is (approximately) a fixed proportion of the mean light level, and as a result the contrast threshold is independent of the mean intensity, hence following Weber's law. In a recent study Rovamo *et al.* (2001) found that for chromatic signals the transition from a photon-noise limit to Weber's law occurs at about 160 trolands (i.e. approx. 2700 photons per cone per second; Makous, 1997), which corresponds to moderate daylight (and is also close to the level of dark noise). Interestingly, although chromatic thresholds are dependent upon the size of the target, the intensity at which this transition from photon-limited to Weber law behavior occurs is independent of spatial scale, and hence spatial pooling of receptor signals. This scale independence suggests that even where Weber's law holds, noise originates prior to spatial pooling, e.g. in the cones.

4.2.3 Information and image statistics

For natural images outputs of neighboring receptors tend to be highly correlated, so that the signal at any given location can be predicted from other information (e.g. its neighbors). The resulting statistical redundancy means that the information can be encoded more compactly. When parallel channels, such as individual axons in the optic nerve, transmit a message the redundancy is minimized when their signals are statistically independent, which for Gaussian signal distributions is achieved by decorrelation. In practice, even though statistical distributions of real messages may not be Gaussian, the reduction of statistical correlation generally leads to reduction of redundancy. Barlow (1961) proposed that the lateral inhibition in the retina, which produces center-surround receptive fields, operates to reduce spatial redundancy. Subsequently, a general theoretical framework developed by workers such as van Hateren (1992, 1993a) and Atick (1992) showed how retinal processing could maximize information capacity in natural conditions.

Removal of redundancy is, however, only one of several possible functions of center-surround receptive fields (Barlow, 1961). An alternative is that retinal neurons are tuned to particular features of interest, such as flies, worms, or colored fruit (Lettvin *et al.*, 1959; Ewert, 1997; Parraga *et al.*, 2001). Although the design principles seem quite different, distinguishing between models is not easy, and requires data on neural signals and physiological noise. In a key experimental study Srinivasan *et al.* (1982) observed that a receptive field 'surround' represents a weighted mean of the receptor signals in the neighborhood of the center, and hence is a statistical prediction of the signal at the center. They measured image statistics in a photograph of a natural scene, and found that for fly visual neurons the surround size and its changes with light intensity were in quantitative agreement with a method known as 'predictive coding', which is used to reduce redundancy in image processing. They concluded that the neural code was adapted to maximize general coding efficiency, as suggested by Barlow (1961), rather than to detect any particular type of object.

Since the late 1980s digital imaging has simplified recording of natural images, and modeling or direct recording of neural responses to them. For spatial sampling a key statistic is the spatial (auto) correlation function, which specifies the statistical similarity of pairs of points in an image as a function of their separation – and hence redundancy in the outputs of the receptor array. A key observation (Field, 1987) is that spatial statistics from a wide range of natural and artificial scenes are much alike, and can be described by simple mathematical functions[1] (see also Ruderman and Bialek, 1994; Baddeley, 1997). For example, Baddeley (1997) found that spatial correlation declines logarithmically with distance, according to the function: $\rho = a \log \theta + b$, where ρ is the correlation between two points in an image, θ their separation (>0), and a, b are constants.

Given that images are composed of many different kinds of objects – pebbles, trees, mountains, animals, etc. – with many sizes it was a surprise that their spatial correlation functions should follow a common function. One possible interpretation is that images are composed of discrete (more or less) uniform surfaces that occlude one another (Baddeley, 1997), which gives the characteristic spatial statistics seen in natural images. Regardless of its origins, this consistency in image statistics is of great value both in evolutionary terms and for researchers, because it means that a general-purpose visual coding strategy will suit a wide range of natural environments.

4.2.4 Retinal coding

The principle of coding efficiency is increasingly influential in neuroscience (Dan *et al.*, 1996; Simoncelli and Olshausen, 2001), but open questions remain about the retinal processing in primates. An interesting problem is to account for the range of responses produced by the parasol and midget ganglion cells that project to the lateral geniculate nucleus (LGN), and, respectively, form the inputs to the magnocellular (MC) and parvocellular (PC) pathways (Chapters 5 and 6).

[1] Field (and others) described image statistics in terms of spatial-frequency power spectra, and observed that power $\alpha 1/\text{frequency}^2$. Here we refer to spatial correlation which is mathematically equivalent.

The differences between PC and MC pathways are well known. MC cell r.f. centers pool the outputs of several (>5) cones; they have relatively high contrast sensitivity and fast responses. PC cell centers are driven by a single cone, at least in the central visual field, and have relatively low contrast sensitivity and sluggish responses. PC neurons transmit the red-green signal to the brain, whereas the MC pathway is color-blind, and represents the luminance signal (Chapters 5 and 6).

Midget ganglion cells with single cone centers are found in all known primates, but are unknown in other mammals (including fully diurnal types such as arboreal squirrels; Kremers, 1998; Yamada *et al.*, 1998). Mollon (1989) suggested that the PC system initially evolved for spatial resolution, and was co-opted for color vision (see also Chapter 6). Whereas parasol cells pool the outputs of L and M cones unselectively (Diller *et al.*, 2004), the PC pathway transmits them separately up the optic nerve. If cortical mechanisms segregate L and M signals on the basis of their correlated activity – for example by a Hebbian mechanism – trichromacy may be possible without the need to evolve special-purpose neural circuitry for the red-green signal (Chapters 5 and 6; Kaas, 1991). Mollon's (1989) account seems most plausible if early primates were diurnal. If, however, early primates were nocturnal (see below and Chapter 1) there is a problem because a high-resolution neural pathway is unlikely to be useful in dim light.

In considering the evolutionary function of MC and PC pathways, Barlow's (1961) distinction between neurons as filters for coding signals 'of particular significance' as opposed to efficient coding of all information remains relevant. For example, it has been suggested that the red-green mechanism is specifically adapted for finding fruit (Mollon, 1989; Parraga *et al.*, 2001). As fruit are statistically rare, and have characteristic shapes and colors, this specialization might be expected to lead to PC receptive fields, different from those optimal for coding the general information about natural images. There is wide agreement that signals in MC and PC pathways are used for different purposes. For instance, motion perception is mainly based on MC signals, whilst color vision requires PC signals. This functional distinction has been elaborated to suggest that the two retinal pathways feed separate visual streams in the cortex used for seeing motion, color, and form (Livingstone and Hubel, 1988; Zeki, 1990). An implication of this interpretation is that selection has acted separately on MC and PC systems to optimize it for a particular purpose. Theoretically, we might measure signals relevant to the proposed tasks – finding fruit, motion, color, etc. – and relate these separately to PC and MC responses.

An alternative is that retinal ganglion cells constitute a single system that maximizes the overall information capacity of the optic nerve. Parallel pathways could arise because range fractionation allows efficient coding (as with separate ON and OFF pathways). Dan *et al.* (1996) propose such a model for cat retinal coding, while consistent with this view Silveira and De Mello (1998) suggest that the PC and MC signals jointly cover the range of spatial and temporal frequencies represented on the retina (Chapter 5). It would then follow that retinal processing maximizes coding efficiency, and the cortex then uses these inputs according to their usefulness for a given purpose, rather than retinal coding being adapted to the needs of specific cortical pathways (Chapter 6).

Qualitatively the differences between MC and PC cell responses are consistent with predictions of coding theory. For example, the low contrast sensitivity and sluggish responses of PC compared with MC cells are expected because the effects of receptor noise fall with receptive field size, owing to spatial pooling. A PC cell with high contrast sensitivity may simply waste neural bandwidth coding noise (van Hateren, 1992). At the same time optical blurring (including effects of any accommodation errors) reduces the contrast of the retinal image for center-surround receptive fields whose dimensions are close to the size of the Airy disc. Recent work by van Hateren *et al.* (2002) studied these questions quantitatively for primate neurons viewing natural scenes, but it remains unclear whether overall retinal coding optimizes information capacity of the optic nerve.

4.3 Color

'Mammals, even the insectivorous and frugivorous kinds, have very little occasion for a refined color sense, since the great mass of creeping insects are of obscure colors, while the squirrels and allies feed on brown nuts rather than on colored fruits. The evidence seems to show, therefore, that a tolerably perfect color-sense has only been attained, among mammalia, in the monkeys and man, while even in these it is probably very inferior to that of birds. It seems probable, therefore, that the prevalence of color-blindness is really an indication of the color sense in man having been a comparatively recent development, instead of being, as Mr. Allen thinks, a disease of civilisation.' (Wallace, 1879)

4.3.1 Sampling the spectrum: trichromacy and natural spectra

Young (1802) introduced his trichromatic theory as a sampling problem: 'Now, as it is almost impossible to conceive each sensitive point of the retina to contain an infinite number of particles, each capable of vibrating in perfect unison with every possible undulation, it becomes necessary to suppose the number limited, for instance, to the three principal colors, red, yellow, and blue . . .' The insight that many separately tuned receptors cannot be located at each point in the retina is correct; even with three there is a compromise between spatial and color vision. We will return to interactions between spatial and spectral sampling, but start with the simpler question of sampling the spectrum alone.

Barlow (1982) drew attention to the similarities between spectral and spatial sampling. Just as optical blurring limits spatial resolution, the width of receptor spectral sensitivity functions limits spectral resolution (e.g. discrimination of spectra with multiple peaks from those with single peaks). Barlow concluded that for rhodopsin sensitivity functions four or five samples are required for the spectral range from 440 to 650 nm, but three is too few. However, he pointed out that the model did not take account either of the properties of natural spectra or of physiological noise. For reasons similar to those that lead to spatial undersampling (section 4.2), spectral sampling may be optimized with fewer spectral types than predicted.

In fact, the statistics of spatial and spectral signals are different. In spatial vision, zooming-in on a scene nearly always reveals more detail. This means that spatial information rises with resolution, assuming sensitivity is not compromised. By comparison, natural reflectance spectra vary relatively smoothly, with little fine-scale detail – rather like the world viewed through fog (Figure 4.4A). Even if receptors were narrowly tuned

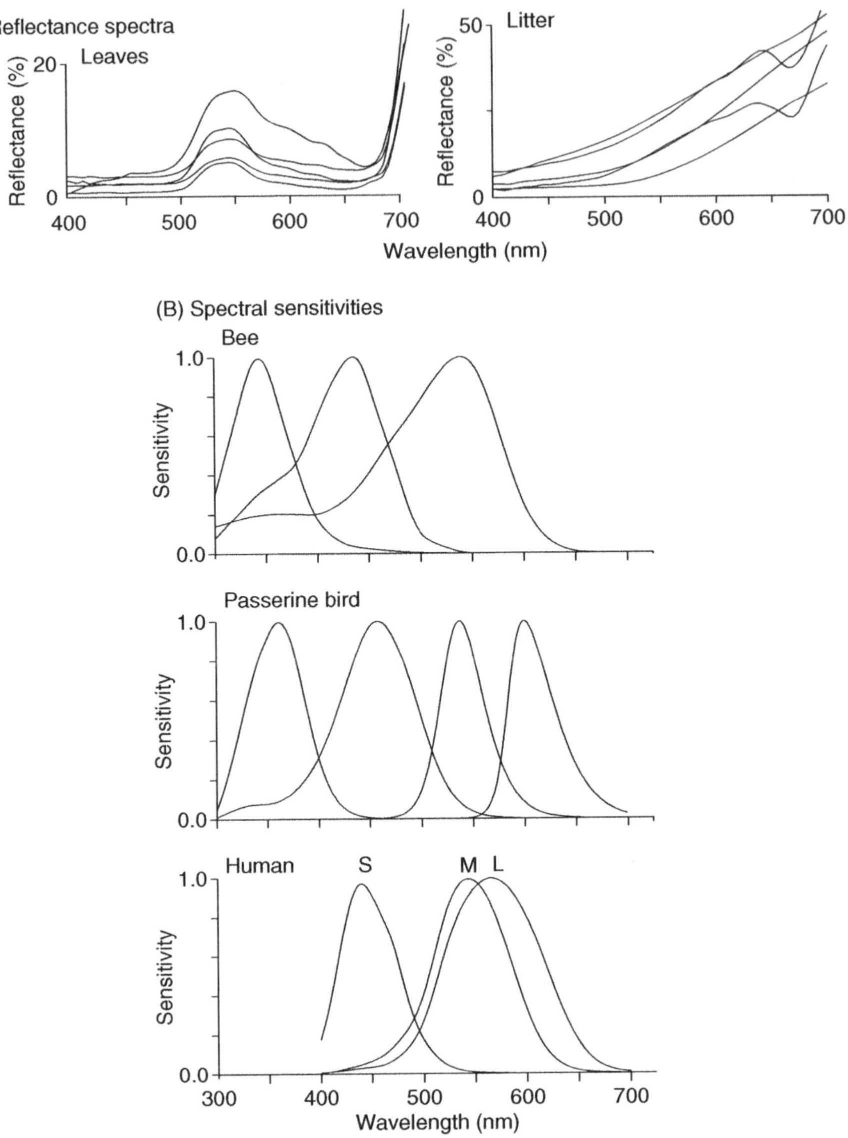

Figure 4.4 Spectral reflectances and spectral sensitivities. (A) Reflectance of leaves and other common surfaces such as bark and litter. Leaves have peak due to chlorophyll near 555 nm, and very low reflectance below 500 nm. The reflectance of most other objects tends to increase monotonically with wavelength. The smooth and predictable reflectance functions of natural objects mean that a small number of receptor types is sufficient to encode spectral information. (B) There is, however, no obvious explanation for the differences between taxonomic groups such as bees, birds, and catarrhine primates, or the relative uniformity of receptor sensitivities within these groups

this lack of detail would limit the useful number of spectral types. Cohen (1964) found that about 99 percent of the variation in the reflectance spectra of 150 randomly selected Munsell chips (the best-known set of commercial color standards) can be represented by a basis set of three parameters, such as principal components. Maloney (1986) used a dataset of natural spectra, and found that (for the human-visible range) two principal components would account for almost 99 percent of the variation and three all but 0.4 percent. This implies that three appropriately chosen spectrally tuned detectors (although not necessarily cone sensitivity functions) will suffice to encode almost all information in spectral signals. van Hateren (1993b) observed that in the presence of noise three detectors with spectral tuning curves close to the natural value for rhodopsin (half-width ca. 100 nm) may maximize coding of information about natural spectra.

Application of principles of sampling and coding efficiency suggests that trichromacy is more or less adequate for coding natural spectra. However, models such as those used by Barlow and van Hateren are adapted from work on spatio-temporal sampling. They assume that the amount variation is roughly uniform across the visible spectrum, and cannot easily account for the specific locations of primate S, M, and L cone sensitivity maxima, or the low sensitivity of the S compared with the L and M mechanisms.

Casual observation shows that the forms of natural reflectance spectra are restricted (Figure 4.4A). Plant pigments are drawn from a limited chemical palette, based on chlorophylls, carotenoids, and anthocyanins, whilst the reflectance of other commonplace materials (such as soil, bark, and dead vegetation) normally increases with wavelength (Osorio and Bossomaier, 1992). If both spectral variation and overall reflectance are indeed greatest at long wavelengths, this should favor placing most sensitivity in this part of the spectrum. Lythgoe (1979) noted that leaves vary more in their reflectance above the chlorophyll reflectance peak at 550 nm than below, and suggested that for discriminating leaf colors it would be beneficial to have receptors with sensitivity maxima at about 600 nm, as are found in birds (Figure 4.4B; Hart, 2001).

4.3.2 Visual pigment tuning

Across the animal kingdom photoreceptor sensitivity maxima range from about 330 to 670 nm, with from one to over a dozen spectral types in a single eye (Kelber *et al.*, 2003). In some groups including birds and reptiles, colored filters substantially modify photopigment spectral sensitivities. What is the evolutionary basis for this variation in receptor spectral sensitivities? Sensitivities of visual photopigments can vary adaptively amongst related species, as is especially obvious amongst fish. Aquatic illumination spectra vary with depth and with the presence of suspended or dissolved material (Lythgoe, 1979). Spectral sensitivities of fish rod pigments tend to match the water color, which maximizes light capture (Bowmaker, 1995). By comparison there is less variation in illumination on land, and most terrestrial species use receptors with sensitivity maxima in the range of 500–560 nm for achromatic (or luminance) vision. These pigment sensitivities roughly match the prevailing wavelengths of illumination and of reflected light (Figure 4.4A). Beyond this there is no obvious relationship between photoreceptor spectral sensitivities and either illumination or reflectance spectra. What seems to count more in some cases is phylogenetic relatedness. Recent surveys of both birds and hymenopteran insects find little evidence that the environment molds receptor spectral sensitivity (Figure 4.4B;

Briscoe and Chittka, 2001; Hart, 2001). Primates have a different set of visual pigments from either insects or birds, but again the 'routine trichromats' (catarrhines and howling monkeys) share the same three cone pigment sensitivities regardless of their lifestyle. The apparent absence of adaptive variation within taxonomic groups is puzzling. It may be that each group is at some type of adaptive peak, which is relatively insensitive to details of lifestyle. An alternative is that molecular constraints limit the spectral positioning of pigments (Goldsmith, 1990).

In diurnal primates S and L cone pigment sensitivity maxima are at about 425 and 560 nm, respectively (Chapter 3). These limits are probably not determined by the demands of visual coding. For short wavelengths one might ask why primates are insensitive to UV, unlike birds and bees (Figure 4.4B). Primate S opsins are derived from the SWS1 gene family (Chapter 3), and yields cone pigments whose λ_{max} values range from about 417 to 430 nm. It is currently believed that the ancestral form of the vertebrate SWS1 pigment had its maximum absorption in the UV (Hunt *et al.*, 2001; Shi and Yokoyama, 2003). Some rodents retain UV SWS1 pigments, but most mammals use longer positions. This shift to the longer wavelengths might allow animals with longer lifespans to exclude UV from their eyes, and hence avoid cumulative UV photodamage.

At long wavelengths, inferences drawn from molecular phylogenies imply that the single long wave-sensitive (LWS) pigment of the ancestral primate had a spectral peak of about 553 nm (Yokoyama and Radlwimmer 1998). However, primates that have more than one version of the M/L gene, and thus two spectrally discrete M/L pigments, *all* share an L pigment with a peak at around 560 nm (Table 4.1). This location is probably the longest wavelength available to a photopigment based on the retinal chromophore; at least none is so far known to be longer than about 565 nm.

Given that the L and S peaks are 'fixed', spectral tuning of primate M/L pigments with sensitivity maxima below 560 nm is of particular interest. The obvious question is whether the 535-nm M pigment peak is adaptive, or alternatively reflects molecular constraints that prevent it from shifting to shorter wavelengths. Primate M/L pigments belong to the LWS opsin gene family (Hisatomi and Tokunaga, 2002), and have sensitivity maxima ranging from about 535 to 560 nm (Table 4.1), but these are discretely stepped, so that there are in total probably six M/L spectral positions. The mechanism of spectral tuning is conservative in that amino acid substitutions in only five positions (out of more than 350) are thought to account for the spectral positioning of all mammalian pigments from the LWS gene family and, among primates, most of the variation is due to changes at only three such locations (Chapter 3; Neitz *et al.*, 1991; Asenjo *et al.*, 1994; Yokoyama and Radlwimmer, 1998).

Although no primates are known to have M pigments with sensitivity maxima below 535 nm, other mammalian LWS family pigments have shorter wavelength peaks, for example, dolphins at 515 nm and mice at 510 nm. As there are no obvious molecular limitations on primate M pigments from shifting to wavelengths below 535 nm, the fact that none have done so implies that natural selection does not favor such 'short-wavelength' M pigments. This conclusion is supported by the distribution of L/M pigment alleles in New World monkeys (section 4.3.5).

What then are selective influences on the tuning of the M pigment, and the concomitant overlap of the M and L cone sensitivities? There are at least three types of adaptive explanation:

(i) That the M/L overlap reflects a trade-off between the demands of luminance and color vision, e.g. where the luminance mechanism would be best with a single type of receptor, and color vision best with more widely separated M and L receptors (section 4.3.4).

(ii) That pigment sensitivities optimize coding of spectral variation (i.e. color discrimination) of objects such as fruit or leaves (section 4.3.3).

(iii) That the spectral tuning of receptors does not necessarily maximize spectral variation, but is adapted to represent signals in a way which simplifies subsequent processing. One suggestion is that the L and M pigments and macular pigmentation may minimize the range of red-green signals presented by leaves, which would facilitate detection of ripe fruit (Mollon and Regan 1999; Regan *et al.*, 2001).

4.3.3 Color vision and food

In his book *The Colour Sense* Allen (1879) proposed that animal color vision co-evolved with fruit and flowers. He noted that it is in the interests of fruit and flowers to be found amongst green leaves, whilst animals need to find them to feed. Also a given species needs to be discriminable from others: pollen has to move within species, and fruiting plants do not want their favored dispersers to be confused, or the fruit to be consumed before they are ripe. Whereas insect color vision seems unlikely to have co-evolved with flowers (Briscoe and Chittka, 2001), the possibility that primate trichromacy, and the red-green mechanism, is a specialization for finding fruit (and conversely that there are fruit adapted to be dispersed by primates) has found support in recent years, especially from Mollon (1989) and his collaborators (Sumner and Mollon, 2000a,b; Regan *et al.*, 2001; also Osorio and Vorobyev, 1996; Parraga *et al.*, 2001; Surridge *et al.*, 2003).

A possible corollary of the 'frugivory' hypothesis for trichromacy is that species that eat little if any ripe fruit should have different color vision from those that do. However, this does not appear to be the case; routinely trichromatic colobine monkeys specialize on a diet of unripe fruit and young leaves (Lambert, 1998); at the same time howler monkeys, the only routinely trichromatic platyrrhine genus, are also the most folivorous of New World species. By comparison, many of those species with a single M/L gene eat few colored leaves at any time, and many are primarily frugivorous. Thus Lucas *et al.* (1998, 2003) draw attention to the importance of the colorful young leaves, which are commonly found on tropical plants, in primate diets. Consistent with this proposal, there is evidence that colorful young leaves are an important part of the diet of routinely trichromatic species, but not the other groups (Lucas *et al.*, 2003). The plants' interests in having colored leaves are, however, obscure (Dominy *et al.*, 2002).

Modeling discrimination and detection of fruit

It is possible to demonstrate experimentally in non-human primates (Caine and Mundy, 2000; Smith *et al.*, 2003) that trichromats have an advantage over dichromats for finding fruit amongst leaves. To complement experiments one can simulate the effects of both real and hypothetical evolutionary shifts in pigment sensitivities on color signals of primate food plants (Osorio and Vorobyev, 1996; Sumner and Mollon, 2000a,b; Regan *et al.*, 2001; Lucas *et al.*, 2003). Unfortunately the effects of varying receptor spectral sensitivities on color discrimination are not straightforward (Kelber *et al.*, 2003), and recently two different models have been used to investigate the question, both of which propose that receptor noise sets thresholds.

One model (Sumner and Mollon, 2000a,b; Regan *et al.*, 2001) assumes that there are separate red-green and yellow-blue mechanisms whose performances are limited by photon noise. It is proposed that behavioral decisions are based on the outputs of either one or other opponent signal. Treating the red-green and yellow-blue signals separately in this way is appropriate if they are used for different behavioral tasks, perhaps with the red-green mechanism being specialized for finding fruit. An alternative model (Osorio and Vorobyev, 1996; Vorobyev and Osorio, 1998) proposes that full use is made of chromatic information encoded by noisy receptors, without further limitations being imposed by opponent mechanisms. This model accurately predicts color thresholds of di- and trichromatic humans, as well as other species (Vorobyev and Osorio, 1998; Vorobyev *et al.*, 2001; Kelber *et al.*, 2003).

Despite their differences, the models which relate color thresholds to receptor sensitivities tend to agree on three main conclusions: (i) for detection of fruit against leaves, given fixed S and L cone sensitivities, the 535-nm M pigment is near or slightly long wave-shifted compared to the optimum; but (ii) discrimination between different species of fruit would favor an M peak nearer to 500 nm; and (iii) that the optima are sensitive to the level of receptor noise, with increasing levels favoring a short-wavelength shift of the M peak.

Overall there is some support for the notion that M pigment tuning is optimal for finding fruit and young leaves, but perhaps more interestingly modeling studies draw attention to the need to precisely specify the behavioral task. Related tasks such as detection of fruit against leaves, identification of fruit species, and judgment of ripeness impose differing demands on color discrimination and color constancy, and could favor different sets of visual pigments (Mollon, 1989; Osorio and Vorobyev, 1996; Sumner and Mollon, 2000a,b; Regan *et al.*, 2001). Likewise the level of noise has a substantial effect on the optimal set of pigments for color discrimination (van Hateren, 1993b; Osorio and Vorobyev, 1996).

4.3.4 Costs of trichromacy

The previous section has outlined possible reasons why routine trichromacy may be optimal for color vision, but as Young (1802) recognized there is likely to be a trade-off between spectral and spatial sampling, so they could not be optimized separately. Primates are not atypical in using achromatic and chromatic signals for different purposes (Livingstone and Hubel, 1988; Zeki, 1990). For instance, motion detection seems to be color-blind in many types of animal (Srinivasan, 1985), and poultry chicks are color-blind for a simple texture recognition task (Jones and Osorio, 2004). However, most animals use one spectral type of receptor for achromatic vision, and primates are exceptional in that they combine M and L cone responses to give a luminance signal. This arrangement probably reflects evolution from dichromatic ancestors, where the luminance system was driven by a single L cone (Mollon, 1989).

Retaining two spectral inputs to the luminance system means there is no loss of sensitivity, but there may be disadvantages. Perhaps the most obvious disadvantage is chromatic aberration (Barlow, 1964; Kröger, 2000); however, we know of no estimates of the magnitude of this problem for natural stimuli. A more likely cost of trichromacy arises when an image is sampled by receptors with differing spectral sensitivities, because it is not possible to separate chromatic from luminance information (Figure 4.2B). This leads to errors that resemble color aliasing in TV pictures, where fine periodic patterns, such as pinstriped clothing, are rendered as shimmering colored Moiré patterns (Osorio *et al.*, 1998; Williams and Hofer, 2004). A comparable effect in human vision where fine high-contrast patterns such as map hachuring appear to be faintly colored is called *Brewster's colors* by Williams and Hofer (2004).

Is the ambiguity between chromatic and achromatic signals sufficient to have influenced the evolution of cone spectral sensitivities, or other aspects of eye design? The answer may depend upon whether aliased signals are comparable to, or exceed, other sources of noise in the visual system. Osorio *et al.* (1998) estimated uncertainty in luminance signals caused by having separate L and M cones. The magnitude of any effect depends upon the spatial distribution of cones and also spatio-chromatic image statistics (Williams *et al.*, 1993; Osorio *et al.*, 1998). For cells with a receptive field of about four cones in diameter, similar to MC neurons, and a random array of L and M cones (Williams and Hofer, 2004) chromatic noise is approximately equal to luminance noise (estimated from contrast sensitivity thresholds). Thus the presence of separate L and M cones does have the potential to corrupt luminance signals. Moreover, the magnitude of chromatic signals increases rapidly with spectral separation of receptors (Nagle and Osorio, 1993), so that it is possible that this effect has indeed limited the spectral separation of L and M cones. There is experimental evidence that human dichromats may do better than trichromats in dim light (Verhulst and Maes, 1998). However, scotopic thresholds of dichromats appear to be similar to those of trichromats (Simunovic *et al.*, 2001).

Another possibility is that trichromacy is a problem for stimuli that exceed conventional psychophysical detection thresholds. Red-green color deficient soldiers are said to be good at breaking military camouflage, presumably because they do not rely on – or are not distracted by – color (Morgan *et al.*, 1992). Unfortunately, whereas this type of problem is easy to demonstrate in the laboratory (e.g. by Ishihara plates), convincing evidence that trichromats do worse in the field than dichromats is lacking, either for humans or for other primates.

In summary, there are currently two main types of adaptive explanation to account for the large spectral overlap between the L and M cones. One is that this arrangement is optimal for color vision, particularly for discriminating fruit and leaves. The alternative is that chromatic coding favors relatively evenly spaced L, M, and S sensitivities, and the uneven distribution is enforced by the cost to luminance vision of having separate L and M receptor signals. At present both types of account remain plausible. Theoretically, they might be distinguished if it was possible to relate differences in eye design to animals' lifestyles, but as all routine trichromats (catarrhines and platyrrhine howler monkeys) share the same set of three pigments this is not feasible.

4.3.5 Maintenance and selection of M/L polymorphism

What does the diversity of visual pigment alleles and color vision amongst New World platyrrhine and strepsirrhine species (Table 4.2; Chapter 3) say about selection on cone pigments? Most platyrhines and some strepsirrhines have a single M/L gene with multiple alleles on their X-chromosomes. All species have a 560-nm allele, but the range of shorter wavelength alleles varies (Table 4.1; Chapter 3). Behavioral work, using Rayleigh matches (Jacobs, 1984), indicates that color discrimination by heterozygote female monkeys resembles that of routinely trichromatic species with similar cone spectral sensitivities. The conclusion that heterozygote New World monkeys have trichromatic color vision is reinforced by the observation that PC pathways are anatomically similar in New and Old World monkeys. Further, physiological properties of PC and MC cells in the heterozygote animals are similar to those in catarrhines (Chapters 5 and 6).

The maintenance of polymorphism is a central problem in population genetics (Futuyma, 1997). If multiple alleles are present in a population at equilibrium, the implication is that their fitnesses are frequency-dependent. Specifically, as an allele becomes rarer its fitness increases. There is convincing evidence for frequency-dependent selection on M/L alleles in New World species because the polymorphism has persisted in separate lineages for at least 15 million years (Surridge *et al.*, 2003). Mollon *et al.* (1984) pointed out that two main types of process could maintain this polymorphism. One is that the

Table 4.2 Spectral sensitivities of primate M/L cone pigments

			Number of genes	λ_{max} of known pigments	Reference
Catarrhini		Human	2	535, 562	Jacobs and Deegan, 1999
Platyrrhini	Atelidae	*Alouatta*	2	535, 562	Jacobs *et al.*, 1996
		Ateles.	1	550, 562	Jacobs and Deegan, 2001
	Cebidae	*Saimiri*	1	535, 550, 562	Jacobs and Neitz, 1987
		Aotus[1]	1	540	Jacobs *et al.*, 1993
		Saguinus	1	540, 555, 562	Jacobs and Deegan, 2003
Strepsirrhini[2]	Lemuridae	*Propithecus*	1	545, 562	Jacobs *et al.*, 2002

[1] Owl monkeys (*Aotus* spp.), have a defective S cone pigment, and so are cone monochromats.
[2] Some tarsiers and strepsirrhines may have only one allele (Tan and Li, 1999).
Note: Summary of the M/L cones or pigment genes recorded from various primate genera (*Alouatta*, howler monkey; *Ateles*, spider monkey; *Saimiri*, squirrel monkey; *Aotus*, owl monkey; *Saguinus*, tamarin; *Propithecus*, sifaka). The pigment is designated by the wavelength of the absorbance maximum, λ_{max}. These λ_{max} values are not necessarily precisely those published in the cited references – as these vary by a few nanometers due to experimental method and error – but typical values, which are used. Note that the apparent absence of an allele from a species may simply mean that it has been overlooked, especially where samples are drawn from a small number of individuals or an inbred group. Further details on the species studied and on other genera are given in the cited literature. At present there is no strong evidence for differences in the alleles present between species of a single genus.

fitness of various color vision *phenotypes* is frequency-dependent – perhaps because they can exploit different types of food. Alternatively, trichromacy may give an advantage to heterozygous females. These two types of explanation are not mutually exclusive.

If fitness of the phenotypes is frequency-dependent, an implication is that individuals with rare types of color vision are at an advantage. One possibility is that members of a troop could specialize in different tasks according to their color vision. Perhaps dichromats are good at spotting camouflaged predators, whilst trichromats lead the troop to colored fruit (Mollon *et al.*, 1984; Regan *et al.*, 2001). Such an arrangement would be of considerable interest in evolutionary biology, as we are not aware of other cases where genetic polymorphism is maintained by specialization within a co-operative social system. There are, however, simpler possibilities, for example, different individuals might specialize on different types of food, thereby reducing competition. Alternatively, if males are fitter as dichromats and females as trichromats, this might stabilize the polymorphism, preventing invasion by individuals with a duplicated M/L gene. Unfortunately all these possibilities remain hypothetical, as we know of no field data to support the hypothesis that there is frequency-dependent selection on color vision phenotypes.

On the other hand, the benefits of trichromatic color vision could well stabilize M/L polymorphism through heterozygote advantage. An additional consideration here is the uniformity of the standard M and L alleles in howler monkeys and catarrhines. Anomalous recombination of the M and L genes gives rise to a range of color vision phenotypes which we recognize as the various types of red-green color deficiency (Chapter 3). Although red-green color deficiencies do occur in humans and (at lower but unknown frequencies) in other primates (Onishi *et al.*, 1999; Saito *et al.*, 2003), there is no evidence that selection favors dichromatic or anomalous trichromatic color vision amongst those species with separate M and L genes.

Selection of M/L alleles in dichromats

The universal occurrence of 535- and 560-nm pigments amongst catarrhines implies that in trichromats these are favored over intermediate variants. Similarly, in those groups with a single M/L gene, one might expect 535- and 560-nm alleles to be favored amongst trichromatic females. However, the overall fitness of the various M/L alleles depends not only upon their fitness in heterozygote females, but also in homozygotes.

Were all M/L homozygote (i.e. dichromatic) phenotypes equally fit, at equilibrium 535- and 560-nm alleles should be equally common, and intermediate 550-nm alleles somewhat rarer (assuming that they give useful but inferior trichromatic color vision). In practice, allele frequencies do not match this pattern (Table 4.1; Osorio *et al.*, 2004). Often the 535-nm allele is absent, and in a study of squirrel monkey populations it is rarer than the two longer (550 and 560 nm) wavelength alleles (Cropp *et al.*, 2002), although it is not possible to say if this effect is statistically significant. The implication is that in dichromats selection favors longer-wavelength M/L alleles. Thus the frequencies of the 535–556-nm alleles in different species with a single M/L gene may reflect a balance between their fitnesses in homozygote and heterozygote phenotypes. Trichromacy probably favors relatively widely separated pigments and equal frequencies of 535- and 560-nm alleles, whilst in dichromats long-wavelength alleles are advantageous. It is unclear whether this advantage arises because of performance in (dichromatic) color vision, or in luminance coding.

4.3.6 Neural coding of color

Psychophysics indicates that S, M, and L cone signals are combined into color channels (mechanisms). An achromatic mechanism can be defined as one sensitive to the changes in intensity of a light stimulus, and based on summation of cone responses. Chromatic channels are formed by computing differences between cone excitations, and hence have an opponent character; ideally they are sensitive to the changes in the shape of the spectrum, but not intensity. Psychophysically chromatic mechanisms of humans and other primates are nearly ideal in that they have minimal sensitivity to stimuli that differ from the achromatic background only in intensity (Sperling and Harwerth, 1971; Loop and Crossman, 2000).

A common consensus is that color is coded by one achromatic (luminance) and by only two chromatic channels. The cone inputs to these mechanisms closely approximate 'cardinal directions' in the color space identified by Krauskopf *et al.* (1982; also Cole *et al.*, 1993; Krauskopf, 1999): the luminance mechanism receives sums of L and M cone signals, the 'blue-yellow' channel compares the response of S cone signal with combined L and M cone signals, $[S-(M+L)]$, and the 'red-green' channel compares L and M cone signals, $[L-M]$. Electrophysiology reveals the presence of opponent and non-opponent ganglion cells and LGN neurons, which broadly correspond to the mechanisms revealed by psychophysics (Chapters 5 and 6), but there are important differences (see below).

As with spatial vision it is interesting to ask how retinal coding of color is adapted to natural signals. Given the separate evolutionary origins of the yellow-blue and red-green systems it would not be surprising if cone inputs to the opponent mechanisms were determined by developmental or phylogenetic constraints (Mollon, 1989; Martin, 1998). An alternative is that chromatic and achromatic mechanisms are adapted for efficient transfer of information via the optic nerve (Barlow, 1961), and so can be predicted from theoretical considerations, such as those applied to spatial coding (section 4.2; Atick, 1992; van Hateren, 1992). A straightforward prediction is that owing to their poor signal-to-noise ratio, chromatic mechanisms should be more sluggish and have lower spatial resolution than luminance mechanisms, as is indeed the case (Atick *et al.*, 1992).

As we have seen (section 4.2) redundancy is minimized (for Gaussian signals) if messages in separate channels are not correlated. In a pioneering paper Buchsbaum and Gottschalk (1983) used principal component analysis (PCA) to find a way of combining L, M, and S cone outputs so as to produce three uncorrelated signals for white noise input spectra. White noise has uncorrelated values for closely separated wavelengths. Although the spectra of natural objects do not resemble white noise, the model predicted luminance, 'yellow-blue' and 'red-green' mechanisms that broadly resemble the cardinal directions in the color space. Buchsbaum and Gottschalk (1983) did not consider the spectra of natural stimuli, and so could not be sure that their code minimizes redundancy. The data needed for such an analysis became available from hyperspectral imaging, where a complete spectrum is measured at each pixel in an image (Burton and Moorehead, 1987). Ruderman *et al.* (1998) did a PCA of spatio-chromatic images and found similar mechanisms to those proposed by Buchsbaum and Gottschalk. However, as Ruderman *et al.* (1998) said, PCA finds an orthogonal transformation, and it is difficult to see why neural mechanisms should use orthogonal codes. If non-orthogonal mechanisms are permitted, decorrelation can probably be achieved by an infinite number of solutions. Indeed, a more careful examination of cardinal directions in the color space shows that the luminance [L+M] mechanism is not orthogonal to the 'yellow-blue' [S−(L+M)], because the luminance mechanism receives negligible S cone input. Thus PCA cannot predict psychophysical mechanisms. Such considerations have led to the application of methods such as independent component analysis (ICA) to color coding. ICA finds a linear non-orthogonal transformation that decorrelates signals, and hence predicts a unique set of mechanisms that are (approximately) statistically independent – for example, because they are generated by separate processes. Recently it has been shown that the luminance and 'yellow-blue' and 'red-green' mechanisms can be predicted from the ICA analysis of natural scenes (Lee *et al.*, 2002). Overall the main conclusion from the analysis of spectra from natural scenes is that signals in achromatic and chromatic mechanisms are uncorrelated, which will go some way to reducing redundancy.

While the 'cardinal directions' in the color space correspond to perceptual attributes of color, and may be represented by cortical neurons, the redundancy reduction hypothesis (Barlow, 1961) referred to retinal coding. Does the prediction that retinal neurons decorrelate cone signals agree with electrophysiological data from retinal ganglion and LGN cells? Cone inputs to neural mechanism can be expressed as a set of relative weights of S, M, and L cone inputs, where the absolute value of components sums to unity. Then the cardinal directions in the color space (Cole *et al.*, 1993) can be represented as: $W_{lum} = (0, 0.5, 0.5)$ for the achromatic system, $W_{blue\text{-}yellow} = (0.5, -0.25, -0.25)$, $W_{red\text{-}green} = (0, -0.5, 0.5)$ for the two chromatic systems. The sum of weights for the chromatic mechanisms is 0, indicating that these mechanisms are not sensitive to an achromatic stimulation.

By comparison, physiological recordings from macaque LGN reveal three groups of neurons (Derrington *et al.*, 1984). There is a lot of variability in cone weights established in physiological recordings from individual cells. Martin (2004) suggests the following cone weights for typical chromatic and achromatic neurons from the LGN of macaque monkey: a typical achromatic neuron sums signals of L and M cones in a ratio close to 2:1 (observed catarrhine L:M cone ratios are close to 1.5:1; Deeb *et al.*, 2000), with little (approx. 1 percent) input from S cones, i.e. $W_{lum} = (S{:}0.01, M{:}0.33, L{:}0.66)$; the responses of a typical 'blue-yellow' opponent cell might be described as $W_{blue\text{-}yellow} = (0.6, -0.1, -0.2)$, and $W_{red\text{-}green} = (0, -0.4, 0.6)$ describes the chromatic properties of a typical 'red-green' cell. Note that sum of weights of chromatic cells differs substantially from 0. Thus the most important difference between the psychophysically derived chromatic mechanisms and those revealed physiologically is that chromatic cells carry a achromatic information. In fact, for natural stimuli the PC pathway probably codes substantially more achromatic than chromatic information (van Hateren *et al.*, 2002). At present then it is unclear to what extent signals in MC and PC neurons are in fact redundant, or, more generally are adapted to optimize coding efficiency (section 4.2). An answer to this question awaits further measurement of the responses of these neurons to natural images.

4.4 Nocturnality and the origins of primate vision

There is good evidence that primates are primitively nocturnal (Chapter 1), and at the same time they have several characters that suggest they are more reliant on vision than most mammals. What can we say about the vision of ancestral primates, and how (or whether) the characteristics that distinguish primates from other mammals might have emerged in a nocturnal animal?

Martin and Ross (Chapter 1) refer to various characters associated with nocturnality in living species, including large eyes and a tapetum.[2] On the basis of comparative and fossil evidence they argue that primates were primitively nocturnal. On the other hand, the presence in all extant primates of midget ganglion cells (Chapters 5 and 6) implies that the common ancestor had high spatial acuity, which is characteristic of diurnality.

Nocturnal animals tend to have larger eyes than diurnal species, and eye size has been used to make inferences about extinct primates (Ross, 2000; Ni *et al.*, 2004; Chapter 1). However, as explained in section 4.1 the relationship between eye size and mean light level is not straightforward. Large pupil apertures (or low f-numbers) are beneficial if photon-noise limits performance because they increase image brightness, but large focal length eyes have the disadvantage that the image is spread over many receptors, which is undesirable if dark noise sets thresholds. Moreover, large eyes are useful even in diurnal animals, partly because photon catch may determine thresholds at relatively high intensities (i.e. when cones are active). Also focal length determines image size, and pupil diameter (ultimately) determines the minimum size of the blur circle, so that increasing eye size benefits spatial resolution as well as sensitivity. A bigger eye is nearly always better, and eye size may then mainly reflect the importance of vision to a species. This point is nicely illustrated by the observation that birds of prey have consistently larger eyes than other birds of the same size (Laughlin, 2001b).

One might also ask whether the evolution of cone pigments is consistent with nocturnal ancestry. Nocturnality is associated with loss of color vision, and this is thought to be the reason why placental mammals lost two of the four classes of vertebrate cone opsin genes (Chapter 3). The process is taken further by many strictly nocturnal mammals (Ahnelt and Kolb, 2000), which have only a single functional cone opsin gene. This arrangement was first detected in two primates, the anthropoid owl monkey, *Aotus*, and a strepsirrhine, the bush baby, *Otolemur* (Jacobs *et al.*, 1996b), which have a single X-chromosome opsin gene yielding an L pigment. The S cone opsin genes are highly homologous to those of other primates, but the gene sequences harbor mutations that render them incapable of supporting normal protein expression. This arrangement predicts monochromacy, a limitation that has been established by behavioral measurements (Jacobs *et al.*, 1993). Recent sequence data point to a similar loss of S cone function in all species of the strepsirrhine groups, Loridae and Galagonidae (Kawamura and Kubotera, 2004).

The loss of the S cones probably occurs because in nearly all mammals they are substantially less sensitive than the L cones (Wyszecki and Stiles, 1982). Moreover, photon flux in reflected light falls substantially at short wavelengths. As a result there is a significant intensity range over which L (and M) but not S cones can operate. This leads to the paradoxical prediction that having separate M and L cones, and red-green color mechanism may be especially advantageous for primate color vision when photon noise limits performance (i.e. at intermediate light levels, see above; Osorio *et al.*, 2004).

That ancestral primates retained their S cones seems to suggest that they were less strictly nocturnal than some contemporary species. In fact more detailed consideration of

[2] There is some question about whether *Aotus* monkeys have eyeshine and a tapetum. Some publications suggest that they do not (Chapter 1), but Walls (1942) said that the owl monkey's eyeshine is 'brighter than a cat's', and this is the case for wild Panamanian owl monkeys; E.J. Warrant, personal communication.

primate lifestyles and the evolutionary genetics of S pigments (and their pattern of loss) presents a confusing picture. On the one hand many animals described as nocturnal are on occasions active at times when color vision may well be both possible and useful. A case in point is the owl monkeys, where at least one species (*Aotus azarae*) is classified as cathemeral, because it is as active during daylight as during the night (Fernandez-Duque, 2003). Yet *A. azarae* has non-functional S cone opsin genes (unpublished observations). On the other hand, a number of stringently nocturnal strepsirrhines have retained functional S cones (e.g. gray mouse lemur, *Microcebus murinus*). There is even evidence that the S cone opsin genes in these nocturnal strepsirrhines have been under purifying selection (Kawamura and Kubotera, 2004; Chapter 3). There is presently no simple functional explanation for why some primates are monochromatic. Adding to the mystery the gene sequence changes that have led to S pigment loss are homogenous across species (Jacobs *et al.*, 1996a), which indicates that this loss of function is in some way adaptive.

References

Ahnelt, P.K. and Kolb, H., 2000. The mammalian photoreceptor mosaic: Adaptive design. *Prog. Retin. Eye Res.* **19**, 711–777.

Ala-Laurila, P., Donner, K., and Koskelainen, A., 2004. Thermal activation and photoactivation of visual pigments. *Biophys. J.* **86**, 3653–3662.

Allen, G., 1879. *The Colour Sense: Its Origin and Development*, Trubner and Co., London.

Andrade da Costa, B.L. and Hokoc, J.N., 2000. Photoreceptor topography of the retina in articles, links the New World monkey *Cebus apella. Vision Res.* **40**, 2395–2409.

Asenjo, A.B., Rim, J., and Oprian, D.D., 1994. Molecular determinants of human red/green color discrimination. *Neuron* **12**, 1131–1138.

Atick, J.J., 1992. Could information-theory provide an ecological theory of sensory processing. *Netw. Comput. Neural Syst.* **3**, 213–251.

Atick, J.J., Li, Z.P., and Redlich, A.N., 1992. Understanding retinal color coding from first principles. *Neural Comput.* **4**, 559–572.

Baddeley, R., 1997. The correlational structure of natural images and the calibration of spatial representations. *Cogn. Sci.* **21**, 351–372.

Barlow, H.B., 1961. Possible principles underlying the transforms of sensory messages, in *Sensory Communication* (ed W.A. Rosenblith), MIT Press, Cambridge, Massachusetts, pp. 217–234.

Barlow, H.B., 1964. The physical limits of visual discrimination, in *Photophysiology*, Chapter 16 (ed A.C. Giese), Academic Press, New York, pp. 163–203.

Barlow, H.B., 1977. Retinal and central factors in human vision limited by noise, in *Vertebrate Photoreception* (ed H.B. Barlow and P. Fatt), Academic Press, London, pp. 337–358.

Barlow, H.B., 1982. What causes trichromacy? A theoretical analysis using comb-filtered spectra. *Vision Res.* **22**, 635–643.

Bowmaker, J.K., 1995. The visual pigments of fish. *Prog. Retin. Eye Res.* **15**, 1–31.

Briscoe, A.D. and Chittka, L., 2001. The evolution of color vision in insects. *Annu. Rev. Entomol.* **46**, 471–510.

Buchsbaum, G. and Gottschalk, A., 1983. Trichromacy, opponent colours coding and optimum colour information transmission in the retina. *Proc. R. Soc. Lond. B* **220**, 89–113.

Burton, G.J. and Moorehead, I.R., 1987. Color and spatial structure in natural scenes. *Appl. Opt.* **26**, 157–170.

Caine, N.G. and Mundy, N.I., 2000. Demonstration of a foraging advantage for trichromatic marmosets (*Callithrix geoffroyi*) dependent on food color. *Proc. R. Soc. Lond. B* **267**, 439–444.

Cohen, J., 1964. Dependency of the spectral reflectance curves of the Munsell color chips. *Psychon. Sci.* **1**, 369–370.

Cohn, T.E., 2004. Detection and sampling, in *The Visual Neurosciences* (eds L.M. Chalupa and J.S. Werner), MIT Press, Cambridge, Massachusetts, pp. 811–824.

Cole, G.R., Hine, T., and McIlhagga, W., 1993. Detection mechanisms in L-, M-, and S-cone contrast space. *J. Opt. Soc. Am. A* **10**, 38–51.

Cronin, H., 1993. *The Ant and the Peacock: Altruism and Sexual Selection from Darwin to Today*, Cambridge University Press, Cambridge.

Cropp, S., Boinski, S., and Li, W.H., 2002. Allelic variation in the squirrel monkey x-linked color vision gene: Biogeographical and behavioral correlates. *J. Mol. Evol.* **54**, 734–745.

Curcio, C.A., Sloan, K.R., Kalina, R.E. *et al.*, 1990. Human photoreceptor topography. *J. Comp. Neurol.* **292**, 497–523.

Curcio, C.A., Sloan, K.R., Packer, O. *et al.*, 1987. Distribution of cones in human and monkey retina: Individual variability and radial asymmetry. *Science* **236**, 579–582.

Dan, Y., Atick, J.J., and Reid, R.C., 1996. Efficient coding of natural scenes in the lateral geniculate nucleus: Experimental test of a computational theory. *J. Neurosci.* **16**, 3351–3362.

Deeb, S.S., Diller, L.C., Williams, D.R. *et al.*, 2000. Interindividual and topographical variation of L:M cone ratios in monkey retinas. *J. Opt. Soc. Am. A* **17**, 538–544.

Derrington, A.M., Krauskopf, J., and Lennie, P., 1984. Chromatic mechanisms in lateral geniculate nucleus of macaque. *J. Physiol.* **357**, 241–265.

Diller, L., Packer, O.S., Verweij, J. *et al.*, 2004. L and M cone contributions to the midget and parasol ganglion cell receptive fields of macaque monkey retina. *J. Neurosci.* **24**, 1079–1088.

Dkhissi-Benyahya, O., Szel, A., Degrip, W.J. *et al.*, 2001. Short and mid-wavelength cone distribution in a nocturnal Strepsirhine primate (*Microcebus murinus*). *J. Comp. Neurol.* **438**, 490–504.

Dominy, N.J., Lucas, P.W., Ramsden, L.W. *et al.*, 2002. Why are young leaves red? *Oikos* **98**, 163–176.

Donner, K., 1992. Noise and the absolute thresholds of cone and rod vision. *Vision Res.* **32**, 853–866.

Ewert, J.P., 1997. Neural correlates of key stimulus and releasing mechanism: A case study and two concepts. *Trends Neurosci.* **20**, 332–339.

Fernandez-Duque, E., 2003. Influences of moonlight, ambient temperature, and food availability on the diurnal and nocturnal activity of owl monkeys (*Aotus azarae*). *Behav. Ecol. Sociobiol.* **54**, 431–440.

Field, D.J., 1987. Relations between statistics of natural images and the response properties of cortical cells. *J. Opt. Soc. Am. A* **4**, 2379–2394.

Franco, E.C., Finlay, B.L., Silveira, L.C. *et al.*, 2000. Conservation of absolute foveal area in New World monkeys: A constraint on eye size and conformation. *Brain Behav. Evol.* **56**, 276–286.

Futuyma, D.J., 1997. *Evolutionary Biology*, 3rd edn, Sinauer, San Francisco.

Goldsmith, T.H., 1990. Optimization, constraint, and history in the evolution of eyes. *Q. Rev. Biol.* **65**, 281–322.

Hart, N.S., 2001. The visual ecology of avian photoreceptors. *Prog. Retin. Eye Res.* **20**, 675–703.

Hecht, S., Schlaer, S., and Pirenne, H., 1942. Energy, quanta and vision. *J. Opt. Soc. Am.* **38**, 196–208.

Hendrickson, A., Djajadi, H.R., Nakamura, L. *et al.*, 2000. Nocturnal tarsier retina has both short and long/mediumwavelength cones in an unusual topography. *J. Comp. Neurol.* **424**, 718–730.

Hirsch, J. and Hylton, R., 1984. Quality of the primate photoreceptor lattice and limits of spatial vision. *Vision Res.* **24**, 347–355.

Hisatomi, O. and Tokunaga, F., 2002. Molecular evolution of proteins involved in vertebrate phototransduction. *Comp. Biochem. Physiol. B* **133**, 509–522.

Hughes, A., 1977. The topography of vision in mammals of contrasting lifestyle: Comparative optics and retinal organization, in *Handbook of Sensory Physiology*, Vol. VII/5 (ed F. Crescitelli), Springer, Berlin, pp. 613–756.

Hughes, A., 1996. Seeing cones in living eyes. *Nature* **380**, 393–394.

Hunt, D.M., Wilkie, S.E., Bowmaker, J.K. *et al.*, 2001. Vision in the ultraviolet. *Cell. Mol. Life Sci.* **58**, 1583–1598.

Jacobs, G.H., 1984. Within-species variations in visual capacity among squirrel monkeys (*Saimiri sciureus*): Color vision. *Vision Res.* **24**, 1267–1277.

Jacobs, G.H. and Deegan, J.F., 1999. Uniformity of colour vision in Old World monkeys. *Proc. R. Soc. Lond. B* **266**, 2023–2028.

Jacobs, G.H. and Deegan, J.F., 2001. Photopigments and colour vision in New World monkeys from the family Atelidae. *Proc. R. Soc. Lond. B* **268**, 695–702.

Jacobs, G.H. and Deegan, J.F., 2003. Cone pigment variations in four genera of New World monkeys. *Vision Res.* **43**, 227–236.

Jacobs, G.H. and Neitz, J., 1987. Inheritance of color vision in a New World monkey *Saimiri sciureus*. *Proc. Natl. Acad. Sci. USA* **848**, 2545–2549.

Jacobs, G.H., Deegan, J.F., Neitz, J. *et al.*, 1993. Photopigments and color vision in the nocturnal monkey, *Aotus*. *Vision Res.* **33**, 1773–1783.

Jacobs, G.H., Neitz, M., Deegan, J.F. *et al.*, 1996a. Trichromatic colour vision in New World monkeys. *Nature* **382**, 156–158.

Jacobs, G.H., Neitz, M., and Neitz, J., 1996b. Mutations in S-cone pigment genes and the absence of colour vision in two species of nocturnal primate. *Proc. R. Soc. Lond. B* **263**, 705–710.

Jacobs, G.H., Deegan, J.F., Tan, Y. *et al.*, 2002. Opsin gene and photopigment polymorphism in a prosimian primate. *Vision Res.* **42**, 11–18.

Jones, C.D. and Osorio, D., 2004. Discrimination of oriented visual textures by poultry chicks. *Vision Res.* **44**, 83–89.

Kaas, J.H., 1991. Plasticity of sensory and motor maps in adult mammals. *Annu. Rev. Neurosci.* **14**, 137–167.

Kawamura, S. and Kubotera, N., 2004. Ancestral loss of short wave-sensitive cone visual pigment in lorsiform prosimians, contrasting with its strict conservation in other prosimians. *J. Mol. Evol.* **58**, 314–321.

Kefalov, V., Fu, Y., Marsh-Armstrong, N. *et al.*, 2003. Role of visual pigment properties in rod and cone phototransduction. *Nature* **425**, 526–531.

Kelber, A., Vorobyev, M., and Osorio, D., 2003. Animal colour vision–behavioural tests and physiological concepts. *Biol. Rev. Camb. Philos. Soc.* **78**, 81–118.

Krauskopf, J., 1999. Higher order color mechanisms, in *Color Vision: From Genes to Perception* (eds K.R. Gegenfurtner and L.T. Sharpe), Cambridge University Press, New York, pp. 303–317.

Krauskopf, J., Williams, D.R., and Heeley, D.W., 1982. The cardinal directions of color space. *Vision Res.* **22**, 1123–1131.

Kremers, J., 1998. Spatial and temporal response properties of the major retino-geniculate pathways of Old and New World monkeys. *Doc. Ophthalmol.* **95**, 229–245.

Kröger, R.H.H., 2000. Optical and developmental constraints on colour vision with lens eyes. *J. Opt. A* **2**, R39–R43.

Lambert, J.E., 1998. Primate digestion: Interactions among anatomy, physiology, and feeding ecology. *Evol. Anthropol.* **7**, 8–20.

Land, M.F. and Nilsson, D.E., 2002. *Animal Eyes*, Oxford University Press, Oxford.

Lapuerta, P. and Schein, S.J., 1995. A four-surface schematic eye of macaque monkey obtained by an optical method. *Vision Res.* **35**, 2245–2254.

Laughlin, S.B., 1992. Retinal information capacity and the function of the pupil. *Ophthalmic Physiol. Opt.* **12**, 161–164.

Laughlin, S.B., 2001a. Energy as a constraint on the coding and processing of sensory information. *Curr. Opin. Neurobiol.* **11**, 475–80.

Laughlin, S., 2001b. The metabolic cost of information – a fundamental factor in visual ecology, in *Ecology of Sensing* (eds F.G. Barth and A. Schmid), Springer, Berlin, pp. 169–185.

Lee, T.W., Wachtler, T., and Sejnowski, T.J., 2002. Color opponency is an efficient representation of spectral properties in natural scenes. *Vision Res.* **42**, 2095–2103.

Lettvin, J.Y., Maturana, H.R., McCulloch, W.S. *et al.*, 1959. What the frog's eye tells the frog's brain. *Proc. Inst. Radio Eng.* **47**, 1940–1951.

Livingstone, M. and Hubel, D., 1988. Segregation of form, color, movement, and depth: Anatomy, physiology, and perception. *Science* **240**, 740–749.

Loop, M.S. and Crossman, D.K., 2000. High color-vision sensitivity in macaque and humans. *Vis. Neurosci.* **17**, 119–125.

Lucas, P.W., Darvell, B.W., Lee, P.K. *et al.*, 1998. Colour cues for leaf food selection by long-tailed macaques (*Macaca fascicularis*) with a new suggestion for the evolution of trichromatic colour vision. *Folia Primatol.* **69**, 139–152.

Lucas, P.W., Dominy, N.J., Riba-Hernandez, P. *et al.*, 2003. Evolution and function of routine trichromatic vision in primates. *Evolution* **57**, 2636–2643.

Lythgoe, J.N., 1979. *The Ecology of Vision*, Oxford University Press, Oxford.

Makous, W.L., 1997. Fourier models and the loci of adaptation. *J. Opt. Soc. Am. A* **14**, 2323–2345.

Maloney, L.T., 1986. Evaluation of linear models of surface spectral reflectance with small numbers of parameters. *J. Opt. Soc. Am. A* **3**, 1673–1683.

Martin, P.R., 1998. Colour processing in the primate retina: Recent progress. *J. Physiol.* **513**, 631–638.

Martin, P.R., 2004. Colour through the thalamus. *Clin. Exp. Optom.* **87**, 249–257.

Mollon, J.D., 1989. "Tho' she kneel'd in that place where they grew . . ." The uses and origins of primate colour vision. *J. Exp. Biol.* **146**, 21–38.

Mollon, J.D. and Regan, B.C., 1999. The spectral distribution of primate cones and of the macular pigment: Matched to properties of the world? *J. Opt. Technol.* **66**, 847–852.

Mollon, J.D., Bowmaker, J.K., and Jacobs, G.H., 1984. Variations of colour vision in a New World primate can be explained by polymorphism of retinal photopigments. *Proc. R. Soc. Lond. B* **222**, 373–399.

Morgan, M.J., Adam, A., and Mollon, J.D., 1992. Dichromats detect colour-camouflaged objects that are not detected by trichromats. *Proc. R. Soc. Lond. B* **248**, 291–295.

Nagle, M.G. and Osorio, D., 1993. The tuning of human photopigments may minimize red-green chromatic signals in natural conditions. *Proc. R. Soc. Lond. B* **252**, 209–213.

Neitz, M., Neitz, J., and Jacobs, G.H., 1991. Spectral tuning of pigments underlying red-green color vision. *Science* **252**, 971–974.

Ni, X., Wang, Y., Hu, Y. *et al.*, 2004. A euprimate skull from the early Eocene of China. *Nature* **427**, 65–68.

Onishi, A., Koike, S., Ida, M. *et al.*, 1999. Dichromatism in macaque monkeys. *Nature* **402**, 139–140.

Osorio, D. and Bossomaier, T.R., 1992. Human cone-pigment spectral sensitivities and the reflectances of natural surfaces. *Biol. Cybern.* **67**, 217–222.

Osorio, D. and Vorobyev, M., 1996. Colour vision as an adaptation to frugivory in primates. *Proc. R. Soc. Lond. B* **263**, 593–599.

Osorio, D., Ruderman, D.L., and Cronin, T.W., 1998. Estimation of errors in luminance signals encoded by primate retina resulting from sampling of natural images with red and green cones. *J. Opt. Soc. Am. A* **15**, 16–22.

Osorio, D., Smith, A.C., Vorobyev, M. *et al.*, 2004. Detection of fruit and the selection of primate visual pigments for color vision. *Am. Nat.* **164**, 696–708.

Parraga, C.A., Troscianko, T., and Tolhurst, D.J., 2001. Spatiochromatic properties of natural images and human vision. *Curr. Biol.* **12**, 483–487.

Peichl, L., Rakotondraparany, F., and Kappeler, P., 2001. Photoreceptor types and distributions in nocturnal and diurnal Malagasy primates. *Invest. Ophthalmol. Vis. Sci.* **42**, 48.

Rao, R.P.N., Olshausen, B.A., and Lewicki, M.S., 2002. *Probabilistic Models of the Brain: Perception and Neural Function*, MIT Press, Cambridge, Massachusetts.

Regan, B.C., Julliot, C., Simmen, B. *et al.*, 2001. Fruits, foliage and the evolution of primate colour vision. *Philos. Trans. R. Soc. Lond. B* **356**, 229–283.

Roorda, A. and Williams, D.R., 1999. The arrangement of the three cone classes in the living human eye. *Nature* **397**, 520–522.

Ross, C.F., 2000. Into the light: The origin of Anthropoidea. *Annu. Rev. Anthropol.* **29**, 147–194.

Rovamo, J.M., Kankaanpaa, M.I., and Hallikainen, J., 2001. Spatial neural modulation transfer function of human foveal visual system for equiluminous chromatic gratings. *Vision Res.* **41**, 1659–1667.

Ruderman, D.L. and Bialek, W., 1994. Statistics of natural images: Scaling in the woods. *Phys. Rev. Lett.* **73**, 814–817.

Ruderman, D.R., Cronin, T.W., and Chiao, C.C., 1998. Statistics of cone responses to natural images: Implications for visual coding. *J. Opt. Soc. Am. A* **15**, 2036–2045.

Saito, A., Mikami, A., Hasegawa, T. *et al.*, 2003. Behavioral evidence of color vision deficiency in a protanomalia chimpanzee (*Pan troglodytes*). *Primates* **44**, 171–176.

Samy, C.N. and Hirsch, J., 1989. Comparison of human and monkey retinal photoreceptor sampling mosaics. *Vis. Neurosci.* **3**, 281–285.
Schneeweis, D.M. and Schnapf, J.L., 1999. The photovoltage of macaque cone photoreceptors: Adaptation, noise, and kinetics. *J. Neurosci.* **19**, 1203–1216.
Shi, Y., and Yokoyama, S., 2003. Molecular analysis of the evolutionary significance of ultraviolet vision in vertebrates. *Proc. Natl. Acad. Sci. USA* **100**, 8308–8333.
Silveira, L.C.L. and De Mello, H.D., 1998. Parallel pathways of primate vision: Sampling of information in the Fourier space by M and P cells, in *Development and Organization of the Retina*, Chapter 10 (eds L.M. Chalupa and B.L. Finlay), Plenum Press, New York, pp. 173–199.
Simoncelli, E. and Olshausen, B., 2001. Natural image statistics and neural representation. *Annu. Rev. Neurosci.* **24**, 1193–1216.
Simunovic, M.P., Regan, B.C., and Mollon, J.D., 2001. Is color vision deficiency an advantage under scotopic conditions? *Invest. Ophthalmol. Vis. Sci.* **42**, 3357–3336.
Smith, A.C., Buchanan-Smith, H.M., Surridge, A.K. *et al.*, 2003. The effect of colour vision status on the detection and selection of fruits by tamarins (*Saguinus* spp.). *J. Exp. Biol.* **206**, 3159–3165.
Snyder, A.W. and Miller, W.H., 1977. Photoreceptor diameter and spacing for highest resolving power. *J. Opt. Soc. Am.* **67**, 696–698.
Snyder, A.W., Laughlin, S.B., and Stavenga, D.G., 1977. Information capacity of eyes. *Vision Res.* **17**, 1163–1175.
Snyder, A.W., Bossomaier, T.R., and Hughes, A., 1986. Optical image quality and the cone mosaic. *Science* **231**, 499–501.
Sperling, H.G. and Harwerth, R.S., 1971. Red-green cone interactions in the increment-threshold spectral sensitivity of primates. *Science* **172**, 180–184.
Srinivasan, M.V., 1985. Shouldn't directional movement detection necessarily be "colour-blind"? *Vision Res.* **25**, 997–1000.
Srinivasan, M.V., Laughlin, S.B., and Dubs, A., 1982. Predictive coding: A fresh view of inhibition in the retina. *Proc. R. Soc. Lond. B* **216**, 427–459.
Stone, J. and Johnston, E., 1981. The topography of primate retina: A study of the human, bushbaby, and new- and old-world monkeys. *J. Comp. Neurol.* **196**, 205–223.
Sumner, P. and Mollon, J.D., 2000a. Catarrhine photopigments are optimized for detecting targets against a foliage background. *J. Exp. Biol.* **203**, 1963–1986.
Sumner, P. and Mollon, J.D., 2000b. Chromaticity as a signal of ripeness in fruits taken by primates. *J. Exp. Biol.* **203**, 1987–2000.
Surridge, A.K., Osorio, D., and Mundy, N.I., 2003. Evolution and selection of trichromatic vision in primates. *Trends Ecol. Evol.* **18**, 198–204.
Tan, Y. and Li, W.H., 1999. Trichromatic vision in prosimians. *Nature* **402**, 36.
Troilo, D., Howland, H.C., and Judge, S.J., 1993. Visual optics and retinal cone topography in the common marmoset (*Callithrix jacchus*). *Vision Res.* **33**, 1301–1310.
van Hateren, J.H., 1992. A theory of maximizing sensory information. *Biol. Cybern.* **68**, 23–29.
van Hateren, J.H., 1993a. Spatiotemporal contrast sensitivity of early vision. *Vision Res.* **33**, 257–267.
van Hateren, J.H., 1993b. Spatial, temporal and spectral pre-processing for colour vision. *Proc. R. Soc. Lond. B* **251**, 61–68.
van Hateren, J.H., Ruttiger, L., Sun, H. *et al.*, 2002. Processing of natural temporal stimuli by macaque retinal ganglion cells. *J. Neurosci.* **22**, 9945–9960.
Verhulst, S. and Maes, F.W., 1998. Scotopic vision in colour-blinds. *Vision Res.* **38**, 3387–3390.
Vorobyev, M., Brandt, R., Peithsch, D., Laughlin, S.B., and Menzel, R. 2001. Colour thresholds and receptor noise: Behaviour and physiology compared. *Vision Res.* **41**, 639–653.
Vorobyev, M. and Osorio, D., 1998. Receptor noise as a determinant of colour thresholds. *Proc. R. Soc. Lond. B* **265**, 351–358.
Wallace, A.R., 1879. Colour in nature [review of The colour sense, its origin and development; an essay in comparative psychology by Grant Allen, 1879]. *Nature* **19**, 501–505.
Walls, G.L., 1942. *The Vertebrate Eye and its Adaptive Radiation*, The Cranbrook Press, Bloomfield Hills, Michigan.

Williams, D.R. and Hofer, H., 2004. Formation and acquisition of the retinal image, in *The Visual Neurosciences* (eds L.M. Chalupa and J.S. Werner), MIT Press, Cambridge, Massachusetts, pp. 795–810.

Williams, D., Sekiguchi, N., and Brainard, D., 1993. Color, contrast sensitivity and the cone mosaic. *Proc. Natl. Acad. Sci. USA* **90**, 9770–9777.

Williams, D.R., Artal, P., Navarro, R. *et al.*, 1996. Off-axis optical quality and retinal sampling in the human eye. *Vision Res.* **36**, 1103–1114.

Wyszecki, G. and Stiles, W.S., 1982. *Color Science: Concepts and Methods, Quantitative Data and Formulae*, 2nd edn, Wiley, New York.

Yamada, E.S., Marshak, D.W., Silveira, L.C. *et al.*, 1998. Morphology of P and M retinal ganglion cells of the bush baby. *Vision Res.* **38**, 3345–3352.

Yokoyama, S. and Radlwimmer, F.B., 1998. The "five-sites" rule and the evolution of red and green color vision in mammals. *Mol. Biol. Evol.* **15**, 560–567.

Young, T., 1802. On the theory of light and colours. *Philos. Trans. R. Soc. Lond.* **92**, 12–48.

Zeki, S., 1990. Parallelism and functional specialization in human visual cortex. *Cold Spring Harb. Symp. Quant. Biol.* **55**, 651–661.

5

Comparative Anatomy and Physiology of the Primate Retina

Luiz Carlos L. Silveira, Ulrike Grünert, Jan Kremers, Barry B. Lee, and *Paul R. Martin*

5.1 Introduction

In this chapter we review the current state of knowledge about retinal anatomy and physiology in primates. We deal mostly with the vertical wiring of the retina, the neurons and direct pathways that take visual information from the photoreceptors to the brain. We do not discuss photoreceptors, as these are discussed in Chapters 1–4. Also, we have kept to a minimum the sections dedicated to lateral retinal elements, horizontal and amacrine cells, as, specially in inner retina, we only partially understand their functional role in primate vision.

5.2 Outer retina

5.2.1 Photoreceptors

Cell recording techniques have been employed to study photoreceptor physiology in macaques at a single cell level. Photocurrent recording from single outer segments with a suction electrode has been used to measure cone and rod spectral sensitivities (Baylor *et al.*, 1984, 1987) and quantitative aspects of phototransduction (Schnapf *et al.*, 1990), whilst whole-cell voltage and current recordings were used to study several aspects of cone and rod physiology, including the influence of nearby photoreceptors by gap-junction electrical coupling, on center-surround antagonism, response noise, and light

The Primate Visual System: A Comparative Approach Edited by Jan Kremers

adaptation (Schneeweis and Schnapf, 1995, 1999, 2000; Verweij *et al.*, 2003). Center-surround antagonism, which is a hallmark feature of the receptive fields of the majority of retinal neurons, has been investigated at the level of primate photoreceptors using such whole-cell patch recording. It has been shown that light-evoked responses of cones in macaques were antagonized when surrounding cones were illuminated (Verweij *et al.*, 2003). The spatial and spectral properties of this antagonism indicate that it results from inhibition by horizontal cells and its pharmacological sensitivity suggested that it is GABA-independent.

The discovery that platyrrhines (Mollon *et al.*, 1984) and prosimians (Tan and Li, 1999) exhibit a variety of color vision phenotypes, which do not exhibit the 'routine' trichromacy found in catarrhines, instigated a renewed interest in the physiology, morphology, and distribution of photoreceptors in primates. Small-eyed primates (e.g. *Callithrix*) exhibit a low rod-to-cone ratio, whereas those with medium to large eyes (e.g. *Cebus*, *Macaca*) have a retina with high rod-to-cone ratio. Nocturnal primates, such as *Aotus*, have a very high rod-to-cone ratio and lack a well-developed fovea.

5.2.2 Horizontal cells

Most authors recognize only two classes of horizontal cells in the primate retina, H1 (Polyak, 1941) and H2 (Kolb *et al.*, 1980), which form separate mosaics and make specific cone contacts in their dendritic fields (Dacey *et al.*, 1996). H1 horizontal cells contact almost exclusively M and L cones, whilst H2 horizontal cells contact preferentially S cones, but also M and L cones (Dacey *et al.*, 1996; Goodchild *et al.*, 1996a). The H1 horizontal cells contact rods with their axon terminals, whilst H2 axons seem to contact only cones (Kolb *et al.*, 1980).

Svaetichin (1953) was the first to perform intracellular recordings of light-evoked responses from the vertebrate retina. Later studies, combining intracellular recording and labeling, showed that S-potentials originated from horizontal cells (Werblin and Dowling, 1969; Kaneko, 1970). In cold-blooded vertebrates, S-potentials depend on stimulus wavelength: some horizontal cells respond with the same polarity to all wavelengths (luminosity, L horizontal cells), whilst others exhibit color opponency (chromatic, C horizontal cells) (Svaetichin and MacNichol, 1958). It has been proposed that their color opponency in non-mammals is due to negative feedback from horizontal cells to cones (Kamermans and Spekreijse, 1995).

Recent data obtained from the macaque retina indicate a different role in color vision for horizontal cells in primates (Dacey *et al.*, 1996). Both types of horizontal cells have ionotropic glutamate receptors and show hyperpolarizing responses to excitation of all cones providing input. Thus, both horizontal cells, H1 and H2, lack color opponency, and the reason for cone specificity of their contacts is unknown (Dacey *et al.*, 1996).

5.3 Bipolar cell circuitry

Visual information is processed in parallel pathways that originate at the first synapse made by rod and cone photoreceptors in the retina. In the outer plexiform layer, bipolar cells make invaginating or flat (basal) synapses with photoreceptors. Invaginating synapses were previously thought to be indicative for ON bipolar types and flat synapses for OFF bipolar types. There are, however, some exceptions; more recently it has been established that some flat synapses are made by ON bipolars; it is the type of glutamate receptor expressed at bipolar dendrites that determines whether they are ON or OFF types (Boycott and Wässle, 1999). The ON bipolar cells express the metabotropic glutamate receptor mGluR6 (Vardi *et al.*, 2000) which is sign-inverting. The OFF bipolar cells express ionotropic AMPA (α-amino-3-hydroxy-5-methyl-4-isoxazolepropionate) and/or kainate glutamate receptors (Haverkamp *et al.*, 2001a,b) which are sign-conserving. Bipolar cells transfer the input from photoreceptors to the inner plexiform layer (IPL) where they make ribbon synapses (dyads) with two postsynaptic processes, usually consisting of an amacrine cell process and a ganglion cell process (Dowling and Boycott, 1966; Boycott and Dowling, 1969). The IPL can be subdivided into an ON and an OFF sublamina. The axons of OFF bipolar cells contact OFF ganglion cells in the outer half of the IPL; the dendrites of ON bipolar cells contact ON ganglion cells in the inner half of the IPL.

As recognized by Ramón y Cajal (1893, 1911) there is one class of rod bipolar cell and several classes of cone bipolar cell in the mammalian retina. Rod bipolar cells contact rods exclusively. Cone bipolar cell classes differ with respect to the number of cones they contact and the stratification level of their axon terminals in the IPL. Boycott and Wässle (1991) distinguished 10 morphological classes of bipolar cell in Golgi-impregnated retina of a catarrhine, the macaque monkey (Figure 5.1), but it is possible that more classes exist, e.g. a giant bistratified bipolar cell described by Mariani (1983) and Rodieck (1988). Comparable bipolar cell classes were identified in the retina of a platyrrhine, the common marmoset *Callithrix jacchus* (Chan *et al.*, 2001a). Many of the bipolar cell classes can be labeled using immunohistochemical markers enabling studies of their density, distribution, and synaptic connectivity (Grünert *et al.*, 1994; Chan *et al.*, 2001a).

Midget bipolar (MB) cells

Polyak (1941) first described midget bipolar cells (MB cells) in the retina of Old World anthropoids. The MB cells have small dendritic and axonal trees, which in central retina

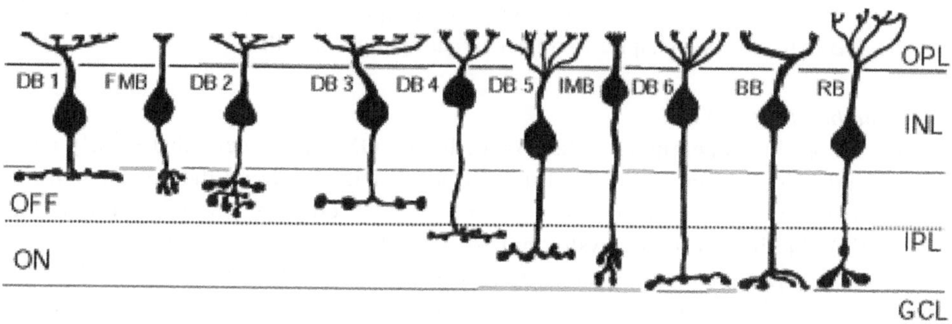

Figure 5.1 Schematic drawing of bipolar cell classes in primate retina (modified from Boycott and Wässle, 1991). OPL, outer plexiform layer; INL, inner nuclear layer; IPL, inner plexiform layer; GCL, ganglion cell layer; DB, diffuse bipolar cell; FMB, flat midget bipolar cell; IMB, invaginating midget bipolar cell; BB, S cone bipolar cell; RB, rod bipolar cell

appear to contact a single cone and a single midget ganglion cell, respectively. They are the most numerous class of bipolar cell. The MB cells can be subdivided into flat MB (FMB) and invaginating MB (IMB) subclasses. The dendrites of FMB cells make flat contacts while those of IMB cells make invaginating synapses with cone pedicles (Kolb, 1970).

In central retina MB cells carry the chromatic signal of the medium wavelength-sensitive (M) or long wavelength-sensitive (L) cone they contact. In peripheral retina MB cells contact up to five cones (Wässle *et al.*, 1994) and thus probably receive mixed cone input. Thus, it has been suggested that the midget system has evolved to meet the requirement for high spatial resolution and not specifically for the processing of color signals (Lennie *et al.*, 1990b; Wässle and Boycott, 1991). This idea is further supported by the demonstration of 'single-headed' (single cone contacting) MB cells in the retinas of mono-, di-, and trichromatic platyrrhines (Silveira *et al.*, 1998, 2001; Chan *et al.*, 2001a).

The connectivity of short wavelength-sensitive (S) cones with MB cells is different from that of M/L cones: the blue-ON signal is not transmitted by MB cells but via a special class of bipolar cell, the blue-cone bipolar (BB) cell (see below). Electron microscopic reconstruction of a macaque fovea has given evidence that the blue-OFF signal in central retina is transmitted via FMB cells (Klug *et al.*, 2003).

Diffuse bipolar (DB) cells

Diffuse bipolar cells (DB cells) have wide dendritic and axonal trees (Figure 5.1). Their dendrites make non-selective contact with all cones (between 5 and 10) in their dendritic field. It is assumed that all classes of DB cells contact all cone types. However, since S cones make up less than 10 percent of the cone population, specialized connectivity with S cones would be difficult to detect. The DB1, DB2, and DB3 classes are OFF cells, whereas DB4, DB5, and DB6 classes are ON cells. Electron microscopy showed that OFF-type cells make flat synapses with cones, whereas the ON-type cells make both invaginating and flat synapses (Hopkins and Boycott, 1997). This is consistent with the idea that the type of glutamate receptor expressed at bipolar dendrites (rather than the flat/invaginating distinction) determines whether a bipolar cell is an ON or OFF cell. In the IPL, DB cells are presynaptic to amacrine and ganglion cells.

Blue-cone bipolar (BB) cells

The S cone or blue-cone bipolar cells (BB cells) are ON bipolar cells that stratify close to the ganglion cell layer. Their dendrites are long, and run across the outer plexiform layer to make invaginating connections with S cones exclusively (Mariani, 1984; Herr *et al.*, 2003). Antibodies to colecystokinin were used to stain BB cells in macaque and marmoset (Kouyama and Marshak, 1992; Wässle *et al.*, 1994; Luo *et al.*, 1999). The connectivity and distribution were similar in both species. In the IPL they make synapses with amacrine and ganglion cells.

Rod bipolar cells

Rod bipolar cells (RB cells) are ON bipolar cells. Their dendrites extend deep into the outer plexiform layer where they form invaginating synapses with rods. The axon terminals of RB cells terminate close to the ganglion cell layer, where they synapse onto amacrine cells (see below).

5.3.1 Density and distribution of bipolar cell classes

Like other cell types in the retina, each class of bipolar cell forms a semi-regular mosaic covering the entire retina (Wässle and Riemann, 1978). An example of a bipolar cell mosaic is shown in Figure 5.2. The images were taken from a macaque retina that was processed with an antibody to label the diffuse bipolar cell class DB6 (Chan *et al.*, 2001b). In macaque and marmoset retinas the spatial densities of some bipolar cell classes were analyzed using immunohistochemical markers and compared to the densities of ganglion cells and photoreceptors (Kouyama and Marshak, 1992; Martin and Grünert, 1992; Milam *et al.*, 1993; Grünert *et al.*, 1994; Wässle *et al.*, 1994; Chan *et al.*, 2001b) (see also Table 5.1). Like the cone density, the density of each cone bipolar class decreases with the distance from the fovea. It is thought that the number of cones contacted by the dendrites of an individual DB cell does not change much with eccentricity and thus the cone-to-cone bipolar ratio does not change across the retina. For RB cells the situation is different. The rod-to-rod bipolar ratio is low in central retina and higher in peripheral retina. The change in relative density of rod bipolar cells is outweighed by an increase in the number of rods. For all bipolar cell classes described so far, except for MB cells, the density ratio of photoreceptors to bipolar cells is lower than the actual convergence (the number of photoreceptors contacted by each bipolar cell). This shows that the signal from the receptor array is shared, not only among different classes of bipolar cell, but also among bipolar cells of the same class.

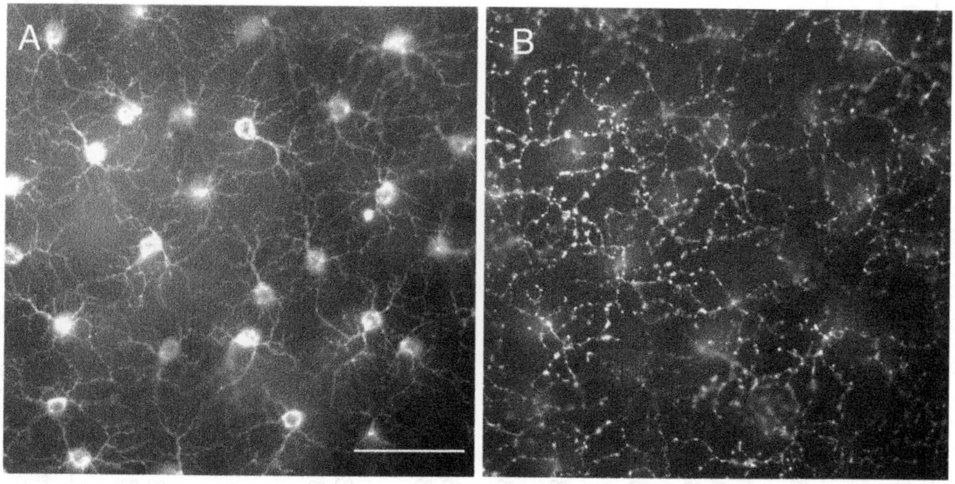

Figure 5.2 Light micrographs of a whole-mounted macaque retina that was processed with an antibody against the cell adhesion molecule CD15. A single population of bipolar cell is labeled. The cells have the morphology of DB6 bipolar cells. (A) The focus is on the somata and dendrites of the cells. (B) The focus is on the axon terminals of the cells. Scale bar = 100 μm

Table 5.1 Density and convergence of bipolar cell types in primate retina

Genus	Cell class	Cell density (cells/mm²)		Numerical convergence*		Actual convergence**		Reference
		Central	Peripheral	Central	Peripheral	Central	Peripheral	
Macaca	MB	22 300	3890	1.0	–	1	2–4	Boycott and Wässle (1991), Wässle *et al.* (1994)
Callithrix	FMB	15 400	1821	0.9	9.4	1–3	5–9	Chan *et al.* (2001a)
Macaca	DB1	n.d.	1382	n.d.	3.7	n.d.	5–8	Boycott and Wässle (1991)
Macaca	DB2	n.d.	2935	n.d.	2	n.d.	5–8	Boycott and Wässle (1991), Grünert *et al.* (1994)
Macaca	DB3	6000	968	n.d.	6.7	n.d.	8–10	Boycott and Wässle (1991), Grünert *et al.* (1994), Martin and Grünert (1992)
Callithrix	DB3	n.d.	1000	n.d.	9–13	n.d.	7–19	Chan *et al.* (2001a), Luo *et al.* (1999)
Macaca	DB4	6600	1000	n.d.	2.7	n.d.	7–8	Boycott and Wässle (1991), Grünert *et al.* (1994)
Macaca	DB5	n.d.	2950	n.d.	1.8	n.d.	5–7	Boycott and Wässle (1991)
Macaca	DB6	1900	270	n.d.	8.1	n.d.	8–10	Boycott and Wässle (1991), Chan *et al.* (2001b), Lee *et al.* (2004)
Callithrix	DB6	1290	440	n.d.	39	n.d.	14–15	Chan *et al.* (2001a)
Macaca	BB	n.d.	500–1300	n.d.	1.4–5	n.d.	1–5	Boycott and Wässle (1991), Wässle *et al.* (1994), Kouyama and Marshak (1992), Kouyama and Marshak (1997)
Callithrix	BB	n.d.	400	n.d.	1	n.d.	1–3	Luo *et al.* (1999)
Macaca	RB	15 000	5000	8	20	15	60	Grünert *et al.* (1994), Grünert and Martin (1991)

*Density ratio: number of photoreceptors per number of bipolar cells.
**Number of photoreceptors presynaptic to each bipolar cell. Abbreviations: DB, diffuse bipolar cell; FMB, flat midget bipolar cell; IMB, invaginating midget bipolar cell; MB, midget bipolar cell (FMB + IMB subclasses); BB, S cone or blue-cone bipolar cell; RB, rod bipolar cell. Bipolar cell density rounded to 10 ganglion cells/mm^2.

5.4 Parallel pathways

5.4.1 The parvocellular (PC) pathway

The PC ganglion cells make up approximately 80 percent of the ganglion cells and project to the PC layers of the lateral geniculate nucleus (LGN; Chapter 6). The stratification of PC ganglion cell dendrites corresponds to that of MB axon terminals, which provide about 50 percent of the input to PC ganglion cells, the other 50 percent coming from amacrine cells. As discussed above, in central and mid-peripheral retina ($<50°$ of eccentricity) MB cells receive input from a single cone and in turn contact a single PC ganglion cell (Figure 5.3). In far peripheral retina ($\geq 50°$) MB cells receive input from two to five cones and several MB cells provide input to an individual PC ganglion cell (Wässle *et al.*, 1994). Thus, in far peripheral retina PC ganglion cells probably receive mixed input from L and M cones. This is consistent with data from *in vitro* recordings of macaque ganglion cells from far peripheral retina (Diller *et al.*, 2004). *In vivo* recordings from midget ganglion cells in mid-peripheral retina have shown that red-green color-opponent responses are present suggesting cone selective input (Martin *et al.*, 2001). The anatomical substrate for this selectivity, however, is still unclear.

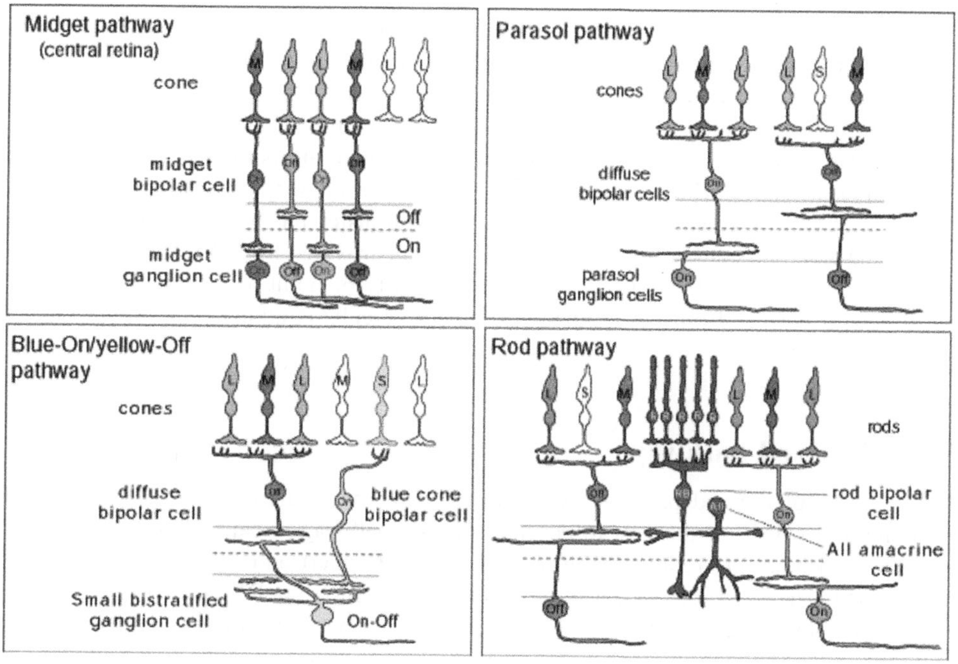

Figure 5.3 Schematic drawing of the major pathways in primate retina (modified from Martin, 1998). Please note that in central retina each cone type connects to one ON and one OFF bipolar cell. For simplicity reasons the two bipolar cell types are drawn separately here

5.4.2 The magnocellular (MC) pathway

The MC ganglion cells make up less than 10 percent of the ganglion cell population. They project to the MC layers of the LGN. The large dendritic fields of MC ganglion cells are narrowly stratified close to the center of the IPL where they receive input from DB2 and DB3 cells (OFF cells), and DB4 and DB5 cells (ON cells, Figures 5.1 and 5.3) (Jacoby *et al.*, 2000; Marshak *et al.*, 2002). Thus the input to MC ganglion cells is dominated by mixed input from M and L cones, via DB cells. The question whether MC ganglion cells receive significant input from S cones is still unresolved (Dacey and Lee, 1994; Chatterjee and Callaway, 2002), although the most recent physiological evidence suggests no S cone input (Sun *et al.*, 2004). Even less well understood is the fact that the vast majority (about 80 percent) of synaptic input to MC ganglion cells is inhibitory and originates from still unidentified amacrine (GABA or glycinergic) cell types (Jacoby *et al.*, 1996; Lin *et al.*, 2000; Macri *et al.*, 2000; Marshak *et al.*, 2002).

5.4.3 The blue-ON/yellow-OFF pathway

Intracellular recordings from small bistratified ganglion cells in the macaque retina showed that they have a blue-ON/yellow-OFF receptive field (Dacey and Lee, 1994). Since similar responses have been recorded from the koniocellular (KC) layers of the LGN, it is assumed that small bistratified ganglion cells project to the KC layers (Martin, 1998; Hendry and Reid, 2000). Ganglion cells with small bistratified morphology have also been found in two diurnal platyrrhines (Ghosh *et al.*, 1997; Silveira *et al.*, 1999) suggesting that the blue-ON pathway is a common feature of all diurnal anthropoids, but they are absent in the owl monkey, a nocturnal anthropoid that lacks S cones (Yamada *et al.*, 2001). The blue-ON pathway receives excitatory input from the S cones (Figure 5.3). The S cones contact BB cells which transfer the blue-ON input to small bistratified ganglion cells whose yellow-OFF input is provided by DB2 and/or DB3 cells (Ghosh *et al.*, 1997; Calkins *et al.*, 1998). The majority of input to the small bistratified ganglion cell derives from yet unidentified amacrine cells (Calkins *et al.*, 1998; Ghosh and Grünert, 1999).

5.4.4 The rod pathway

The anatomy of the rod pathway is common to all mammals (Figure 5.3). Kolb and Famiglietti (1974) were the first to describe the rod pathway in cat, followed by studies in rabbit and macaque retinas (Strettoi *et al.*, 1990; Grünert and Martin, 1991; Wässle *et al.*, 1995). The rod signal is transferred to the inner retina via RB cells. The RB axons contact two amacrine cell types, a GABAergic amacrine cell and the glycinergic AII amacrine cell. The AII amacrine cell is a crucial interneuron in the rod pathway. It transfers the ON signal into the ON ganglion cells via electrical synapses (gap junctions) with ON cone bipolar cells and the OFF signal into the OFF pathway via a sign-inverting glycinergic synapse with OFF cone bipolar cells. Anatomical studies of macaque retina showed that the AII cell makes glycinergic synapses with different OFF bipolar types, including DB3 and FMB cells, providing a pathway for rods to reach the MC and PC pathways (Grünert and Wässle, 1996; Grünert, 1997). This anatomical finding is consistent with physiological studies, although the rod input to the PC pathway is weak (Lee *et al.*, 1997; Weiss *et al.*, 1997). The relationship between scotopic acuity and the density of different elements of the rod pathway have been described elsewhere (Lennie and Fairchild, 1994; Wässle *et al.*, 1995; Mills and Massey, 1999).

5.5 Ganglion cell morphology

Primate ganglion cells comprise several classes which were first identified in human (Dogiel, 1891), macaques, and other catarrhines (Polyak, 1941; Boycott and Dowling, 1969). Data collected in the last three decades showed that all primates have similar sets of ganglion cell classes, which can be well characterized by their morphology, connections, and physiological responses to visual stimuli (Silveira *et al.*, 2004). Those linking the retina to the thalamus and the visual cortex are the origin of conscious perception of color, movement, and recognition of spatial patterns (Lee, 1996, 2004; Silveira and de Mello, 1998; Silveira *et al.*, 2004).

The distribution of ganglion cells is strongly dominated by specialization for central vision in an *area* or *fovea centralis*, but there are important quantitative differences across species (Silveira, 2004; see also Chapter 2). For instance, it seems that an important adaptation to a nocturnal life style consists in the sacrifice of detailed sampling of the central visual field for the benefit of a smaller number of cells with large receptive fields, which are more efficient light collectors. Measurements of optic foramen size and orbit size in both living and extinct primates support the view that the haplorrhine ancestor common to primitive anthropoids and tarsiers was diurnal and had a retina with a highly developed central specialization (Kirk and Kay, 2003). The major ganglion cell classes of the primate retina are distributed according to this general pattern (Silveira and Perry, 1991), but little is known about the distribution of less well studied ganglion cells, such as the several classes of wide-field ganglion cells.

5.5.1 The morphology of MC and PC ganglion cells

Classical stains, retrograde tracers, and dye intracellular injection to reveal retinal ganglion cell morphology

Retinal ganglion cells are readily stained and some classical stains, such as the *in vivo* use of methylene blue (Dogiel, 1891) as well as the Golgi method (Polyak, 1941; Boycott and Dowling, 1969; Rodieck *et al.*, 1985; Kolb *et al.*, 1992), reveal enough cell morphology to allow extensive quantitative analysis of cell bodies and dendritic trees. More recently, fine details of dendritic morphology have been revealed by the use of retrograde transport of neurotracers deposited in the optic nerve (Perry and Cowey, 1981) or injected in the mesencephalic and diencephalic retinal targets (Leventhal *et al.*, 1981), as well as by iontophoretical intracellular injections of fluorescent dyes or other neurotracers (Watanabe and Rodieck, 1989). The combination of neurotracer retrograde transport with dye intracellular injection (Rodieck and Watanabe, 1993), or more recently using the technique of photodynamic staining (Dacey *et al.*, 2003), has become a powerful tool to identify the projection site of specific ganglion cell classes. Electron microscopy has been used to disentangle the details of ganglion cell connections to other retinal neurons (Kolb and DeKorver, 1991; Calkins *et al.*, 1998). Finally, immunocytochemistry has allowed neuropharmacological identification of ganglion cells and their afferents, associating specific neurotransmitter pharmacology to every microcircuit of the primate retina (e.g. Grünert and Ghosh, 1999; Grünert *et al.*, 2002, 2003).

Side and plain views of retinal ganglion cells

Once labeled, retinal sections and flat mounts of the whole retina can be observed using several techniques, including classical light microscopy, Nomarski interference contrast, fluorescence microscopy, and confocal microscopy. The basic wiring plan of the retina can be grasped from transverse sections, in which one can assess how information in the image plan is passed through the retinal layers. Most classical studies on the primate retina were performed on Golgi-stained transverse sections (Polyak, 1941; Boycott and Dowling, 1969). Retinal whole mounts were first employed to study ganglion cells of the human retina more than 100 years ago (Dogiel, 1891) and have recently become popular once more (Leventhal *et al.*, 1981; Perry and Cowey, 1981). The observer can study the retina perpendicularly to the view in transverse sections, to provide ganglion cell morphology, including soma size, dendritic field size, cell density, and dendritic coverage factor, in relation to retinal eccentricity.

MC and PC ganglion cells have been identified in all primates so far studied

The MC and PC ganglion cells have been identified in all diurnal primates so far studied, including human (Rodieck *et al.*, 1985; Dacey and Petersen, 1992; Kolb *et al.*, 1992; Dacey, 1993a); other catarrhines such as chimpanzee (*Pan*: Polyak, 1941), baboon (*Papio*: Watanabe and Rodieck, 1989), and macaques (*Macaca*: Polyak, 1941; Boycott and Dowling, 1969; Leventhal *et al.*, 1981; Perry and Cowey, 1981; Perry *et al.*, 1984; Watanabe and Rodieck, 1989; Silveira and Perry, 1991; Grünert *et al.*, 1993); and also in platyrrhines such as squirrel monkey (*Saimiri*: Leventhal *et al.*, 1989), capuchin monkey (*Cebus*: Lima *et al.*, 1993, 1996; Silveira *et al.*, 1994; Yamada *et al.*, 1996a,b), and common marmosets (*Callithrix*: Ghosh *et al.*, 1996; Goodchild *et al.*, 1996b; Yamada *et al.*, 1996b). In addition, MC and PC ganglion cells have also been identified in nocturnal primates, including the single genus of nocturnal anthropoid, the platyrrhine owl monkey (*Aotus*: Lima *et al.*, 1993, 1996; Silveira *et al.*, 1994; Yamada *et al.*, 1996b, 2001), as well as the prosimian greater bush baby (*Otolemur*: Yamada *et al.*, 1998).

In all primates, MC ganglion cells have large somata, thick axons, and large dendritic trees with a radial branching pattern, whereas PC ganglion cells have small cell bodies, thin axons, and small dendritic trees with a more bushy and dense branching pattern (Figure 5.4). Both cell classes comprise two distinct subpopulations, one having dendrites branching in the outer half and another branching in the inner half of IPL. This branching pattern generates four ganglion cell subclasses, each one in a position to contact a particular set of bipolar cells as described above, and thus originating parallel streams of visual processing that extend from the retina to the LGN and beyond (Perry *et al.*, 1984; Watanabe and Rodieck, 1989; see above).

The MC and PC ganglion cells display substantial morphological and physiological differences at every retinal location. On the other hand, outer and inner branching subclasses of each show only small morphological differences, apart from the stratification level of their dendrites in the IPL. However, outer ganglion cells of these two classes are excited by light decrements and are OFF-center, whilst inner ganglion cells are excited by light increments and are ON-center (Dacey and Lee, 1994). At photopic levels, these properties are determined by bipolar connections, but at scotopic levels, the OFF and ON

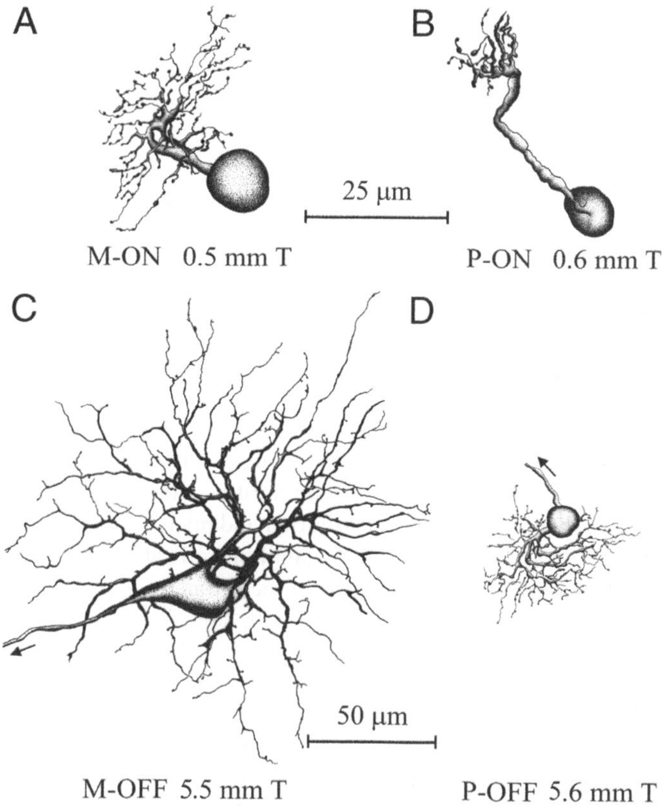

Figure 5.4 Ganglion cells from a diurnal anthropoid, the capuchin monkey *Cebus apella*. (A, B) MC and PC ganglion cells from the central retinal region. (C, D) MC and PC ganglion cells from the peripheral retinal region. (E, F) Small bistratified ganglion cell from the peripheral retinal region, illustrating the inner (E) and outer (F) dendritic tiers, respectively. Arrows indicate cell axons. All the cells in the figure were found in dichromatic male *Cebus*. Cells from dichromatic and trichromatic diurnal anthropoids are very similar in both morphology and physiology, except that PC ganglion cells are color-blind (Yamada *et al.*, 1996a; Silveira *et al.*, 1999; Lee *et al.*, 2000). Ganglion cells were labeled after retrograde transport of Biocytin deposits in the optic nerve. The labeling was revealed using ABC Vectastain (Vector Laboratories, Burlingmane, CA) and HRP histochemistry. The retina was then mounted flat and coverslipped. Drawings were made using a binocular microscope and a drawing apparatus. See Yamada *et al.* (1996a) for additional details

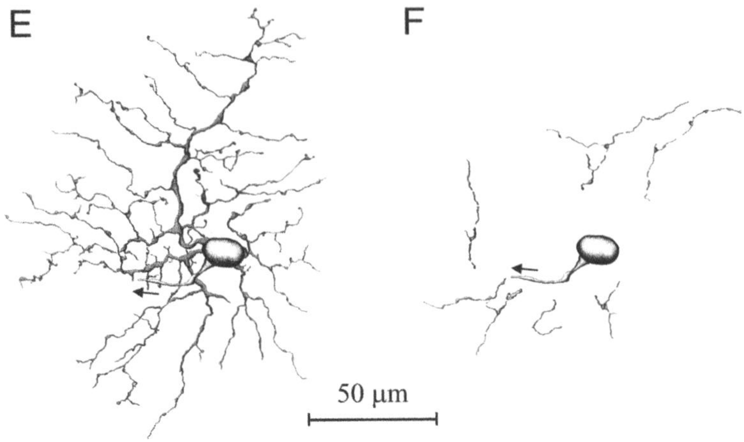

Figure 5.4 *(continued)*

responses of MC and PC ganglion cells are the consequence of a special microcircuitry in the IPL (see above).

Dendritic field size of MC and PC ganglion cells

The prevalence of MC and PC ganglion cells among primates deserves a few additional comments. First, in spite of major qualitative similarities, there are some quantitative differences between the dendritic field sizes of diurnal and nocturnal anthropoids. Figure 5.5 shows the dendritic field sizes of MC and PC ganglion cells at several distances from the central specialization in the human retina and in the retina of five other primate species. The comparison between species is easier when both dendritic field size and eccentricity are expressed in angular metrics, eliminating the large differences in eye size and optical design between different primates.

Both MC and PC ganglion cells increase in size from the fovea toward the retinal periphery. But MC dendritic fields are generally about two to three times larger than PC dendritic fields in all primate species (Perry *et al.*, 1984; Watanabe and Rodieck, 1989; Dacey and Petersen, 1992; Ghosh *et al.*, 1996; Yamada *et al.*, 1996a,b, 2001). In all primates studied so far, the dendritic field sizes of MC and PC ganglion cells located in the nasal retinal quadrant are smaller than those located in the other quadrants, at similar eccentricities (Perry *et al.*, 1984; Watanabe and Rodieck, 1989; Dacey and Petersen, 1992; Ghosh *et al.*, 1996; Yamada *et al.*, 1996a,b, 2001).

There is a large scatter in dendritic tree size of a factor between 2 and 4 for both ganglion cell classes and all primate species, but diurnal MC and PC ganglion cells are generally smaller than those of the nocturnal greater bush baby and owl monkey (Figure 5.5) (Yamada *et al.*, 1998, 2001). Cone convergence – obtained by multiplying cone density by ganglion cell dendritic field area and representing the number of cones that could potentially send their output to ganglion cells of a given class – is, however, similar in diurnal and nocturnal species owing to the lower cone density in nocturnal animals (Figure 5.6) (Yamada *et al.*, 2001). Possibly, dendritic trees of ganglion cells of

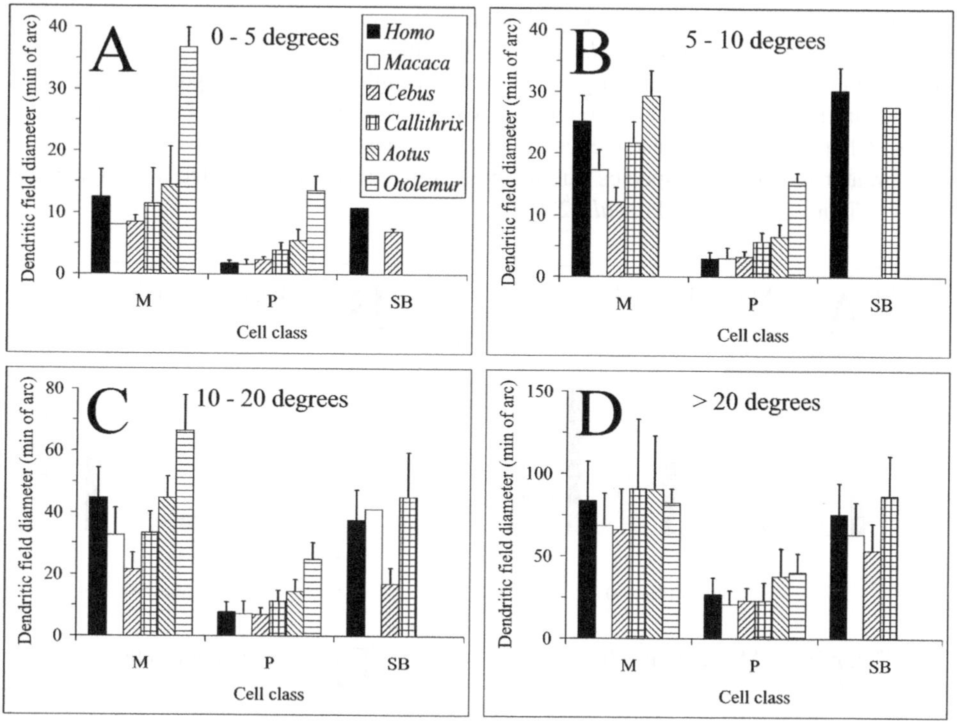

Figure 5.5 Dendritic field sizes of MC, PC, and small bistratified (SB) ganglion cells at different distances from fovea. Angular distance measurements were obtained for both dendritic field size (diameter of a circle of equivalent area) and eccentricity (distance from fovea). For SB ganglion cells only the sizes of inner dendritic tier are represented. MC and PC ganglion cells from both diurnal and nocturnal primates are represented. Only SB ganglion cells from diurnal species are represented because these cells have not been found in nocturnal primates yet. Cells located in the temporal, dorsal, and ventral quadrants were pooled together as there are no significant differences for dendritic field size in those quadrants for the species considered. Nasal cells are significantly smaller than cells located in other quadrants (not illustrated). In the central retina, ganglion cells from diurnal primates are generally smaller than their nocturnal counterparts, but this difference tends to disappear toward the retinal periphery. To convert linear in angular measurements at different eccentricities, the following equations were used, which express retinal magnification factor, y in μm/degree, as a function of eccentricity, x in degrees: *Homo*, $y = -1.0636x + 285.6$; *Macaca*, $y = -0.6454x + 232$; *Cebus*, $y = -0.4738x + 204.2$; *Callithrix*, $y = -0.012x^2 + 0.4x + 126.3$; *Aotus*, $y = -0.4536x + 192.4$; *Otolemur*, $y = 140$. Dendritic field size means and standard deviations, as well as the equations for retinal magnification factor, were obtained from published results (*Homo*: Dacey and Petersen, 1992; Dacey, 1993b; *Macaca*: Watanabe and Rodieck, 1989; Dacey, 1993b; *Cebus*: Yamada *et al.*, 1996a; Silveira *et al.*, 1999; *Callithrix*: Ghosh *et al.*, 1996, 1997; Goodchild *et al.*, 1996b; Wilder *et al.*, 1996; Yamada *et al.*, 1996b; Silveira *et al.*, 1999; *Aotus*: Yamada *et al.*, 2001; *Otolemur*: Yamada *et al.*, 1998)

different primate species adjust during development to receive the signals from a constant number of cones (Goodchild *et al.*, 1996b; Yamada *et al.*, 2001). The match between dendritic field size and cone density might arise through generation of ganglion cells and cones at similar stages during neurogenesis, while rods develop later (LaVail *et al.*, 1991; see also Chapter 2). As a consequence, and unlike cone convergence, rod convergence is much greater in nocturnal species due to their larger dendritic field size and higher rod density (Figure 5.6) (Yamada *et al.*, 2001).

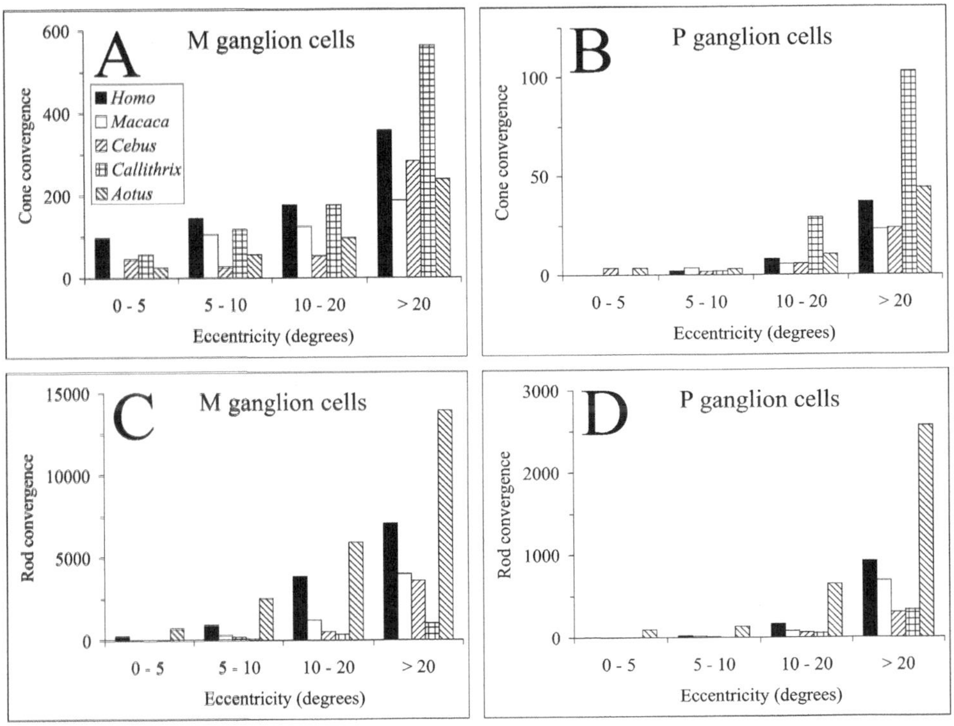

Figure 5.6 Photoreceptor 'potential' convergence for several anthropoid species. The histograms illustrate the number of cones and rods encompassed by the dendritic fields of MC and PC ganglion cells as a function of distance from fovea along the temporal horizontal meridian. The convergence values were estimated by multiplication cone and rod densities with MC and PC dendritic field areas. Conversion of linear to angular measurements at different eccentricities was performed using the equations mentioned in Figure 5.5. The diurnal *Macaca* and *Cebus*, as well as the nocturnal *Aotus*, have similar cone convergences, whilst *Aotus* has a higher rod convergence than the two diurnal anthropoids. The similarities between *Macaca* and *Cebus* are not unexpected because these two primates have similar ganglion cell sizes and photoreceptor densities. The human retina has relatively high cone and rod convergences, especially for MC ganglion cells, a consequence of the relatively large dendritic field size of such cells in comparison with other diurnal anthropoids. The *Callithrix* retina has a higher cone and lower rod density than all the other anthropoids depicted in the figure and, consequently, exhibits a high cone convergence and a low rod convergence for both MC and PC ganglion cells

The private lines of the central region of the primate retina

A second important point concerns the presence of the midget system at the central retina, first described by Polyak (1941) for catarrhines, and now considered to be a common feature of all diurnal anthropoids, both catarrhines and platyrrhines (Boycott and Dowling, 1969; Perry *et al.*, 1984; Rodieck *et al.*, 1985; Kolb and DeKorver, 1991; Kolb *et al.*, 1992; Dacey, 1993a; Calkins *et al.*, 1994; Ghosh *et al.*, 1996; Yamada *et al.*, 1996a; Goodchild *et al.*, 1996b; Silveira *et al.*, 1998). The dendritic field size of PC ganglion cells is constant at low eccentricities (Shapley and Perry, 1986). This is because PC cells located at less than 2 mm from the fovea have a dendritic field small enough to contact a single MB cell axon terminal in the IPL (Polyak, 1941; Boycott and Dowling, 1969; Kolb and DeKorver, 1991; Silveira *et al.*, 1998). Furthermore, MB cells at these retinal locations receive signals from a single M or L cone (Polyak, 1941; Boycott and Dowling, 1969; Calkins *et al.*, 1994; Silveira *et al.*, 1998). Thus, in the central retinal region, the PC circuit comprises connections of single M or L cones to single MB cells and to single PC ganglion cells (the so-called private line). Electron microscopy reconstructions of synaptic connections between cones, MB cells, and PC ganglion cells in the central retinal region of diurnal anthropoids have confirmed the presence of one-to-one neural circuits (Kolb and DeKorver, 1991; Calkins *et al.*, 1994).

One hypothesis to explain the functional roles of these unique private lines is that the small dendritic fields of foveal PC ganglion cells were first a specialization for high spatial acuity which is present in all diurnal primates (Polyak, 1941). According to this idea, PC ganglion cells became the substrate for red-green color opponency (Mollon and Jordan, 1988; Kremers *et al.*, 1999). An alternative hypothesis proposes that the one-to-one connections between M or L cones, MB cells, and PC ganglion cells evolved specifically to subserve red-green color opponency and primate trichromatic vision. They are discussed later and by Kremers *et al.* (1999).

Among living anthropoids, the platyrrhines differ from catarrhines in that they comprise diurnal and nocturnal species and a variety of color vision phenotypes (Mollon *et al.*, 1984; Jacobs, 1998). The PC ganglion cells have been studied in three platyrrhines: the diurnal capuchin monkey, the common marmoset, and the nocturnal owl monkey. In the capuchin monkey and the marmoset, all males and homozygotic females are dichromats having S cones and an M/L cone, whilst the heterozygotic females are trichromats (*Cebus*: Jacobs and Neitz, 1987; *Callithrix*: Travis *et al.*, 1988; Tovée *et al.*, 1992; see Chapters 3 and 4). In the owl monkey both males and females are monochromats because the gene for the S opsin is non-functional and there is no polymorphism for the X-chromosome opsin gene (Jacobs *et al.*, 1993, 1996). Thus, the PC pathway in the mono- and dichromatic animals does not play a role in color vision. However, as in diurnal trichromatic catarrhines, in the central retina of dichromatic capuchin monkeys and monochromatic owl monkeys, the 'private line' is present (Silveira *et al.*, 1998, 2001). This may be due to its role in visual acuity; alternatively, it may be the cost these individuals pay for their siblings to be trichromats.

It remains open what ecological pressures have driven the primates to develop trichromacy. Mollon and colleagues have proposed that primate trichromacy has coevolved with trees with yellow and orange fruits, given the importance of primates in disseminating seeds of such tree species (Mollon and Jordan, 1988; Mollon, 1991; Regan *et al.*, 1998, 2001). Ripe fruits can most effectively be detected by a system with two pigments in the long-wavelength range. However, primates exhibiting full trichromacy (the catarrhines and the platyrrhine genus *Alouatta*) are to some extent folivorous and folivory may provide additional evolutionary pressure toward trichromacy to distinguish withered and yellow or brown leaves (Kremers *et al.*, 1999). A similar argument has been developed by Lucas and colleagues showing that some primates select reddish leaves to complement their diet due to their higher proteic value (Lucas *et al.*, 1998; Dominy and Lucas, 2001).

The number of MC and PC ganglion cells

The MC ganglion cells can be selectively stained by using neurofibrillar stains (Silveira and Perry, 1990, 1991; Lima *et al.*, 1993, 1996) or immunocytochemistry (Grünert *et al.*, 1993). In addition, after retrograde labeling or intracellular injection of large numbers of such cells, direct counting or statistical analysis of their mosaics can be used to estimate MC and PC cell densities (Perry *et al.*, 1984; Dacey and Petersen, 1992; Dacey, 1993a; Yamada *et al.*, 1996a). Using these methods, it was established that MC and PC ganglion cells comprise the majority of ganglion cells of the primate retina. In macaques, it was estimated that MC and PC ganglion cells correspond, roughly, to 10 percent and 80 percent of the total ganglion cell population (Perry *et al.*, 1984; Silveira and Perry, 1991). Whole-mount preparations of platyrrhine retinas showed that about 10 percent and 15 percent of the total retinal ganglion cell population are MC ganglion cells in these primates (Lima *et al.*, 1996).

5.5.2 Koniocellular (KC) ganglion cells

The primate retina has several other ganglion cell classes

In addition to MC and PC ganglion cells, the primate retina has several other ganglion cell classes which were recognized in early anatomical (Dogiel, 1891; Polyak, 1941; Boycott and Dowling, 1969; Leventhal *et al.*, 1981; Perry and Cowey, 1984) and physiological studies (de Monasterio and Gouras, 1975; de Monasterio *et al.*, 1975a,b; de Monasterio, 1978a,b,c). These ganglion cells project to many sites, including the superior colliculus, the pretectal nuclei, the nuclei of the accessory optic system, the LGN, the pregeniculate nucleus, the pulvinar, and the suprachiasmatic nucleus. They can be regarded as part of a heterogeneous group of less well studied visual information channels running in parallel with the better understood MC and PC pathways.

KC ganglion cells and LGN koniocellular layers

Interspaced between the MC and PC layers of the primate LGN is a series of layers noted in the first cytoarchitectonic studies (Le Gros Clark, 1941a,b), called the koniocellular layers (Kaas *et al.*, 1978). More recently, several studies have characterized the properties of KC cells, including protein-expression specificity, receptive fields, and cortical projection sites (Hendry and Yoshioka, 1994; Solomon *et al.*, 1999; White *et al.*, 2001; reviewed by Hendry and Reid, 2000; see also Chapter 6).

Small bistratified ganglion cells: a blue-ON/yellow-OFF ganglion cell

At least three main groups of KC ganglion cells can be distinguished on the basis of their dendritic field morphology (Dacey and Packer, 2003). One group comprises the small-field bistratified ganglion cells, which have dendritic field sizes in the same range as MC ganglion cells. All the other ganglion cell classes described so far have dendritic field sizes larger than MC, PC, and small bistratified ganglion cells (Leventhal *et al.*,

1981; Perry and Cowey, 1984; Rodieck and Watanabe, 1993; Peterson and Dacey, 1999, 2000; Dacey and Packer, 2003).

Small-field bistratified ganglion cells were first described by Rodieck *et al.* (1987) who made large retrograde tracer injections in the dorsal layers of the LGN *in vivo* and one week later, in an *in vitro* retinal preparation, performed intracellular injections to reveal the fine dendritic morphology of cell bodies larger than those of PC ganglion cells. A substantial proportion of the middle- to large-sized cell bodies were small-field bistratified cells. Dacey and Lee (1994), using *in vitro* intracellular recording and labeling, were able to demonstrate that these were indeed, blue-ON/yellow-OFF color-opponent ganglion cells.

Small bistratified ganglion cells seem to constitute a neuronal population shared by all diurnal anthropoids, both catarrhines and platyrrhines, but they seem to be absent in the nocturnal owl monkey (Yamada *et al.*, 2001) which lacks S cones (Wikler and Rakic, 1990; Jacobs *et al.*, 1993). The evidence indicates that small-field bistratified ganglion cells are part of a neural pathway that conveys blue-yellow color-opponent information. These cells are part of a microcircuitry comprising the S cones, H2 horizontal cells, and BB cells (see preceding text). In macaque and marmoset LGN, blue-ON/yellow-OFF responses have been recorded from neurons in KC layers (Martin *et al.*, 1997; Reid *et al.*, 1997).

Several morphological aspects of the small bistratified ganglion cells have been quantified in human (Dacey, 1993b), macaque (Rodieck and Watanabe, 1993; Dacey, 1993b), capuchin monkey (Silveira *et al.*, 1999), and marmoset retinas (Ghosh *et al.*, 1997; Silveira *et al.*, 1999); there are many common features (Figure 5.4). Their somata have sizes between those of MC and PC ganglion cells. In the central retinal region, the whole dendritic tree originates from one stout primary dendrite, whilst in the retinal periphery up to five primary dendrites are found. There are two dendritic tiers, which branch in separate regions of the IPL and are morphologically distinct; the outer dendritic tier originates from several locations in the inner tier and is generally smaller, less dense, and less regular. The outer tier/inner tier diameter ratio ranges from 0.35 to 0.75. The inner tier sizes are in the same range as those of MC ganglion cells (Figure 5.5).

Two small bistratified ganglion cells from the central region of the macaque retina were studied in detail by Calkins *et al.* (1998) using electron microscopy of serial sections. They received the majority of their S cone (blue-ON) input from a single BB cell axon terminal branching on the inner layers of the IPL and received their yellow-OFF input at the outer layers of the IPL from three to four diffuse (DB2 and DB3) bipolar cells. In addition, the cells received input from amacrine cells. Calkins and coworkers estimated that in this retinal region the small-field bistratified ganglion cells receive input from about 20 M and L cones for the yellow-OFF signal.

It has been reported that the inner dendritic tier of the small-field bistratified ganglion cells has a diameter of about 50 μm in the central region of the human retina (Dacey, 1993b), with smaller sizes for macaque (13–23 μm; Calkins *et al.*, 1998) and capuchin monkey retinas (23–25 μm; Silveira *et al.*, 1999). The inner dendritic tier diameter increases from about 130 μm at 2.5 mm to between 200 and 450 μm in the far temporal periphery (Dacey, 1993b). A similar increase is also observed in other diurnal anthropoids, but the human cells are generally larger than those from other primates (Figure 5.5) (Dacey, 1993b; Ghosh *et al.*, 1997; Silveira *et al.*, 1999). However, the quantitative differences in dendritic field size of small bistratified cells are substantially reduced when both eccentricity and dendritic field diameter are transformed from linear to angular space (Figure 5.5).

So far, no specific stains have been developed for small-field bistratified ganglion cells. Thus, one approach to estimate their cell density is to perform a statistical analysis of their mosaic. Using this method, Dacey (1993b) calculated that in the human and macaque retinas the small-field bistratified ganglion cells comprise about 1 percent of the total population in the central retinal region and between 6 percent and 10 percent in the temporal periphery (Dacey, 1993b).

Primate wide-field ganglion cells: a heterogeneous group

A number of ganglion cell classes have been described in the primate retina that have larger dendritic fields than MC, PC, and small bistratified ganglion cells. This very heterogeneous group comprises cells projecting to all retinal brain targets mentioned earlier (Leventhal *et al.*, 1981; Perry *et al.*, 1984; Rodieck and Watanabe, 1993; Dacey and Packer, 2003; Dacey *et al.*, 2003). As in other mammals, axons of some wide-field ganglion cells may bifurcate to more than one site. They represent less than 10 percent of all primate ganglion cells. There are no less than 11 identified classes, plus several subclasses. Thus each cell class makes up only a small proportion of the whole ganglion cell population, making quantitative studies difficult (Peterson and Dacey, 1999, 2000; Dacey and Packer, 2003; Dacey *et al.*, 2003). Morphologically, these cells can be distinguished according to their dendritic level of IPL stratification (as monostratified with outer and inner stratifying varieties, bistratified, or diffuse), dendritic field size (giant, large, and medium), and number of dendritic branch points (very sparse, sparse, moderate, and dense).

LGN projecting wide-field ganglion cells

Using their photodynamic staining technique, Dacey and colleagues characterized several wide-field ganglion cell classes that project to the primate LGN, most likely to the KC layers (Dacey and Packer, 2003; Dacey *et al.*, 2003): three monostratified ganglion cells, one large bistratified ganglion cell, and one diffuse ganglion cell. Of particular interest is the discovery that some of these cell classes receive input from S cones and are likely candidates to convey blue-yellow information to the LGN KC layers and the visual cortex: the large bistratified cell is a blue-ON/yellow-OFF cell and the inner variety of the large sparse monostratified cell is a blue-OFF/yellow-ON cell.

5.6 Ganglion cell physiology – information processing and transfer

The physiological properties of retinal ganglion cells and LGN neurons, that relay the information provided by ganglion cells to the visual cortex, are very similar. This is established for cells belonging to the MC and PC pathways, but probably is also true for the several classes of KC cells. Chapter 6 contains a detailed discussion of properties shared by retinal ganglion cells and their LGN relay neurons. We here discuss the relationships between anatomical and physiological properties of retinal ganglion cells and some physiological properties that relate to the locus of retinal ganglion cells at the junction between retinal circuits and higher visual centers.

5.6.1 Correlation between anatomy and physiology

The classification of ganglion cells is based on the study of their response as a function of achromatic or chromatic contrast in space and time. Most response variability along these stimulus dimensions can be approximated in the first instance by linear models (Kremers *et al.*, 1993; Lee *et al.*, 1994, 1998; Bernadete and Kaplan, 1997a), but full descriptions must include nonlinearities, especially involving gain controls (Lee *et al.*, 1989a, 1993b; Bernadete and Kaplan, 1997b).

Receptive field size and dendritic field size
Dendritic tree geometry and size set constraints on how a neuron handles incoming information. Dendritic field and receptive field sizes generally correlate, but the relationship between other aspects of the dendritic tree geometry and receptive field properties is less clear (e.g. Martin *et al.*, 2001).

Receptive fields from retinal ganglion cells and other visual neurons can be estimated in two different ways, using either discrete spatial stimuli (e.g. spots) (de Monasterio and Gouras, 1975; Lee *et al.*, 1998; McMahon *et al.*, 2004) or periodic spatial stimuli (gratings) (e.g. Derrington and Lennie, 1984; Blakemore and Vital-Durand, 1986; Crook *et al.*, 1988; Croner and Kaplan, 1995). We combine here ganglion cell and LGN data due to the similarity of neuronal responses at these loci.

The first measurements of primate ganglion cell receptive fields (de Monasterio and Gouras, 1975) determined the energy of a small spot (0.01°) required to elicit a threshold response across a cell's receptive field. They reported a two- to threefold difference in receptive field center size between putative MC ganglion cells and PC ganglion cells across a range of eccentricities from the fovea up to 20°–40° (Figures 5.7 and 5.8). However, more recent studies have shown that for low contrasts the spatial capacity of PC and MC cells is quite similar (Derrington and Lennie, 1984; Blakemore and Vital-Durand, 1986; Crook *et al.*, 1988; Croner and Kaplan, 1995; Lee *et al.*, 1998). One reason for this is likely to have been the small spot technique used by de Monasterio and Gouras, which may have confounded a difference in contrast sensitivity (see below) with a difference in center size.

Figures 5.7 and 5.8 summarize the results for MC and PC cells from both the retina and the LGN obtained in different studies. Mean values for receptive field sizes expressed as Gaussian diameters are generally larger for MC cells in comparison with PC cells at all eccentricities.

In the cat, the receptive field centers of α and β retinal ganglion cells have about the same size as their dendritic fields (Peichl and Wässle, 1979). For primates this relationship is less clear for some ganglion cell classes. Anatomical data suggest that the receptive field centers of foveal PC ganglion cells receive their input from a single cone (Kolb and DeKorver, 1991; Calkins *et al.*, 1994; see previous sections). This suggests that the receptive field centers of PC ganglion cells should be smaller than can be expected on the basis of their dendritic field sizes, i.e. the size of a single cone (~0.01° in the fovea). However, direct measurements reveal that the receptive field centers of PC ganglion cells are often larger than can be expected on the basis of single cone (Lee *et al.*, 1998; McMahon *et al.*, 2000). Anatomy and physiology may be partially reconciled considering the effect of optical blur on the retinal image, and then the smallest PC cell receptive field centers have sizes roughly in the range of cone size plus blurring by eye

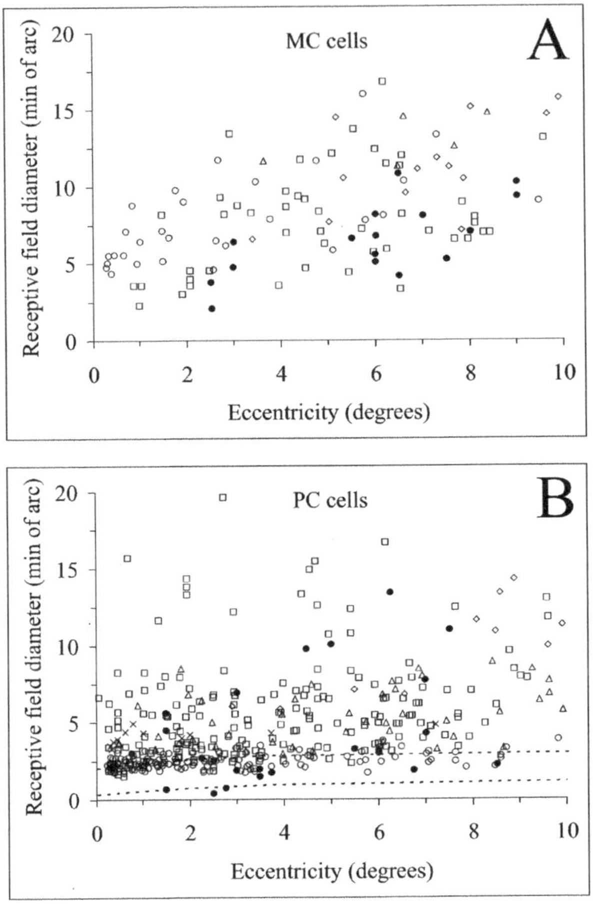

Figure 5.7 Receptive field sizes of macaque MC and PC retinal or LGN cells as a function of eccentricity for the central 10° of visual field. Data source: empty circles and crosses, de Monasterio and Gouras (1975) who described two cell types that now are both considered to be PC cells. They are represented by two different symbols in (B); empty squares, Derrington and Lennie (1984); empty diamonds, Crook *et al.* (1988); empty triangles, Croner and Kaplan (1995); filled circles and dotted lines, Lee *et al.* (1988). Data from de Monasterio and Gouras comprise cells from both nasal and temporal retinal regions. Data from the other studies were from the temporal retina or were from nasal retina and then converted to temporal equivalent eccentricity (Watanabe and Rodieck, 1989). All data were converted to give receptive field size in Gaussian diameter (Peichl and Wässle, 1979). (A) Diameter of receptive field center for MC cells. (B) Diameter of receptive center for PC cells. Data from de Monasterio and Gouras (1975) comprise two sets: concentric color-opponent ganglion cells receiving input from one-cone mechanism to the receptive field center (PC ganglion cells; empty circles) and concentric color-opponent ganglion cells receiving input from two-cone mechanism to the receptive field center (crosses). Also plotted is cone diameter with and without the effect of blurring by the eye optics (top and bottom dotted lines, respectively)

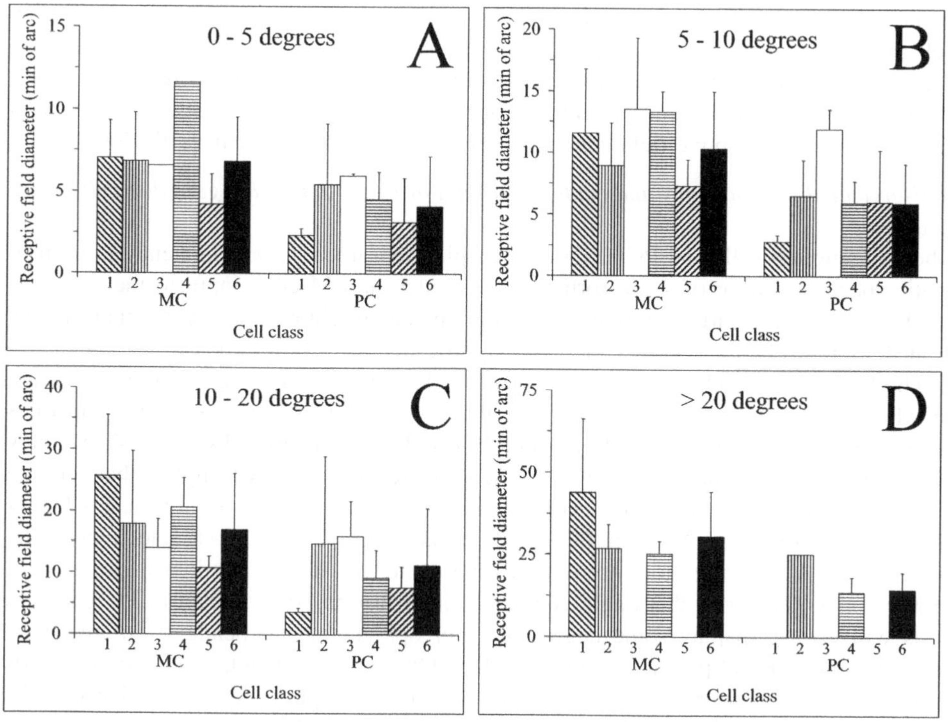

Figure 5.8 Receptive field sizes of macaque MC and PC retinal or LGN cells at different distances from fovea. Data source: (1) de Monasterio and Gouras (1975); (2) Derrington and Lennie (1984); (3) Crook *et al.* (1988); (4) Croner and Kaplan (1995); (5) Lee *et al.* (1998); (6) mean values for all measured cells from these five studies. PC ganglion cell data from de Monasterio and Gouras (1975) comprise only concentric color-opponent ganglion cells receiving input from one-cone mechanism to the receptive field center. Data from de Monasterio and Gouras comprise cells from both nasal and temporal retinal regions, whilst those from the other studies were from the temporal retina or were from nasal retina and then converted to temporal equivalent eccentricity (Watanabe and Rodieck, 1989). All data were converted to give receptive field size in Gaussian diameter (Peichl and Wässle, 1979)

optics (Lee *et al.*, 1998; Lee, 2003) (Figure 5.7B). This would account for the finding of de Monasterio and Gouras that center size does not change much across central retina; the point spread function changes little in this region. However, investigation of the receptive fields of LGN PC cells using laser interference fringes imaged directly on the retina, thus bypassing the eye's optics, revealed an additional complication. Some cell receptive field centers receive a major input from a single cone together with weak input from nearby cones (McMahon *et al.*, 2000). This implies that lateral connections may play a role in receptive field centers even for central PC ganglion cells through cone coupling by gap junctions (Tsukamoto *et al.*, 1992) or minor bipolar cell contributions besides the main synaptic input (Kolb and DeKorver, 1991; but see Calkins *et al.*, 1994). Nevertheless, there is a general agreement that the center responses of foveal PC ganglion cells will be strongly dominated by the responses of the cone that is connected to that retinal ganglion cell through its 'private line'. In summary, PC cell receptive center size is larger than expected in central retina, primarily due to optical blur. In peripheral retina, dendritic field size becomes larger and optical blur plays less of a role; the expected difference in MC and PC cell receptive field center size becomes apparent (Martin *et al.*, 2001).

Red-green color opponency and center-surround organization of PC ganglion cell receptive fields

The PC ganglion cells and PC LGN cells display red-green color opponency in trichromatic species because of some form of opponent L and M cone input to the receptive field center and the surround. Some PC ganglion cells might receive a very small S cone contribution, probably of little functional significance (Sun *et al.*, 2004).

The PC ganglion cell receptive fields can be described by a center-surround structure in which both center and surround have approximately Gaussian responsivity profiles (Derrington and Lennie, 1984; Lee *et al.*, 1998). The 'hit-or-miss hypothesis' (Shapley and Perry, 1986; Lennie *et al.*, 1991; Mullen and Kingdom, 1996) implies that 'private lines' cause PC cell centers to become color-coded. Presently, it is not clear whether the receptive field surrounds of PC ganglion cells receive their input selectively from the cone class that does not contribute to the center, or from a random mixture of M and L cones. Owing to the single cone-class input to the centers, central PC ganglion cells will always display some degree of color opponency independently of receptive surround built from specific or mixed cones. On the other hand, selective surrounds would improve the chromatic signal, and there is some physiological evidence in its favor (Reid and Shapley, 1992; Lee *et al.*, 1998; Martin *et al.*, 2001). This implies some sort of developmental mechanism that directs the establishment of specific connections for receptive field surrounds at most retinal eccentricities.

The ideas of random or selective cone inputs can be tested at the peripheral retina ($>20°$), where the center of PC cell receptive fields becomes large enough to receive input from many MB cells, which according to random models should lead to a loss of chromatic selectivity. Recently, Martin *et al.* (2001) specifically targeted peripheral ganglion cells and recorded from 54 PC ganglion cells at eccentricities between 20° and 50°. The majority of the peripheral PC ganglion cells were overtly red-green color-opponent, supporting the 'selective connection hypothesis' (Figure 5.9). *In vitro* studies recently performed in the macaque retina showed less opponency (Diller *et al.*, 2004), but the high temporal frequency used in testing and the greater eccentricities tested make comparison difficult. In the periphery a substantial proportion of MB cells receive their input from more than one cone and there is no indication of cone selectivity in their connections (Wässle *et al.*, 1994; see above). In summary, it seems likely that some degree of chromatic opponency is lost in the far periphery, but for most retinal regions the PC pathway can provide at least some chromatic signal.

The PC ganglion cells and PC LGN neurons in dichromatic New World primates receive their input from just one cone class (M or L) and therefore do not display color opponency. The temporal properties, including temporal contrast sensitivity, of the PC ganglion cells are very similar in dichromats and trichromats (Lee *et al.*, 2000). Similar results were obtained in the PC LGN neurons of dichromatic and trichromatic marmosets (Chapter 6). These results further suggest that the PC pathway is not a specific adaptation to trichromatic color vision and became color-coded due to a 'hit-and-miss'

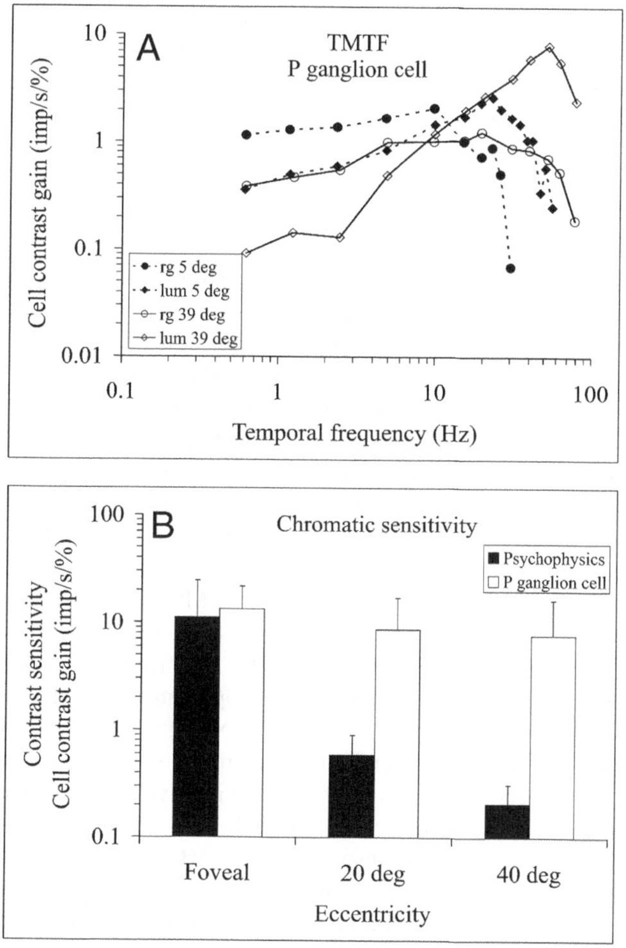

Figure 5.9 (A) Luminance (lum) and red-green chromatic (rg) temporal modulation transfer function (TMTF) for a pool of central and peripheral macaque PC ganglion cells (Martin *et al.*, 2001). A large proportion of peripheral PC ganglion cells are red-green color-opponent. Peripheral ganglion cells have a TMTF displaced to high temporal frequencies for both achromatic and chromatic stimuli. (B) Comparison of chromatic contrast sensitivity of PC ganglion cells with data from human psychophysics. Peripheral PC ganglion cells are very sensitive to red-green chromatic contrast, whilst human red-green color vision quickly degrades with eccentricity. Reproduced by the permission of Macmillan

mechanism. Thus the PC pathway must have arisen in evolution to serve other functions either in the temporal or in the spatial domain, or both (Silveira, 1996; Silveira and de Mello, 1998; Kremers, 1999; Kremers *et al.*, 1999, 2004; Blessing *et al.*, 2004).

Does the characteristic midget morphology of PC cells show that they evolved for spatial vision? In the centralmost fovea, a single cone center would seem of little advantage because the point spread function covers several cones across (Lee *et al.*, 1998;

Lee, 2003). But if an ancestral PC system had centers with few cones, the presence of different opsins would automatically confer spectral sensitivity by the 'random wiring' principle outlined earlier. Thus it is plausible that during evolution, chromatic specificity in the PC pathway arose by chance through random connectivity. Then, improvement of the chromatic signal was achieved by development of selective connectivity in the receptive field.

Structure of MC ganglion cell receptive fields

Although MC ganglion cells do not show a cone opponent response, they are spatially opponent because their receptive fields consist of a center region and an antagonistic surround. Measurements obtained by different groups indicate that the sizes of receptive field centers of MC ganglion cells match the sizes of their dendritic trees (de Monasterio and Gouras, 1975; Perry *et al.*, 1984; Watanabe and Rodieck, 1989; Croner and Kaplan, 1995; Lee *et al.*, 1998). Dendritic field sizes are shown in Figure 5.5 and can be compared with the receptive field center sizes of Figures 5.7 and 5.8.

The MC ganglion cells are more sensitive to luminance contrast than PC ganglion cells. Similar to PC ganglion cells, MC ganglion cells receive input from M and L cones but probably not from S cones (Sun *et al.*, 2004). To red-green isoluminant stimuli they exhibit a frequency-doubled response suggesting that MC ganglion cells can encode the appearance of a color change but not its direction (from red to green or from green to red) (Lee *et al.*, 1989a,b, 1990; Kaiser *et al.*, 1990). The frequency-doubled component may be the result of nonlinear processing of cone-opponent signals and it is possible that signals from PC cells are nonlinearly 'fed' into the MC ganglion cell responses.

In heterochromatic flicker photometry, two differently colored stimuli are modulated in counterphase. The observer's task is to change the relative modulation depth so that the percept of flicker is minimal. Then, by definition, the two stimuli are isoluminant. When the percept of flicker is minimal the response in the MC pathway is also minimal (Lee *et al.*, 1989c; Kaiser *et al.*, 1990). At isoluminance still some residual flicker is present, which may be due to the frequency-doubled response of MC cells or to small differences in the M/L cone balance of individual MC ganglion cells. This is strong evidence that the MC pathway is the physiological basis of the luminance channel of human vision. There might be a difference in the spectral sensitivity of the luminance channels of humans and other primates such as the macaque. The ratio of L to M cone input to the luminance channel of humans is on average about 2:1 but can be highly variable between individuals. This L/M ratio probably reflects the ratio of L to M cone density in the retina (Brainard *et al.*, 2000; Kremers *et al.*, 2000). However, electroretinographic data suggest that the L/M ratio in monkeys is about 1:1 (Jacobs and Deegan, 1987).

5.6.2 Division of labor between different ganglion cell classes

Early evidence indicated that psychophysical detection of luminance and chromatic modulation appears to be mediated by separate mechanisms (Kelly and van Norren, 1977). Lee and colleagues have systematically compared the human performance in several psychophysical tasks with the behavior of MC and PC ganglion cells. They have found that MC ganglion cell responses are the physiological substrate for flicker photometry, as discussed above (Lee *et al.*, 1988), the minimally distinct border (Kaiser *et al.*, 1990; Valberg *et al.*, 1992), detection of luminance modulation (Lee *et al.*, 1989a), as well as displacement and vernier hyperacuity with achromatic patterns (Lee *et al.*, 1993a, 1995). On the other hand, there is a correspondence between PC ganglion cell responsivity and psychophysically determined red-green sensitivity when expressed in terms of cone contrast (Cole *et al.*, 1993; Lee *et al.*, 1993b).

The MC and PC ganglion cells may together extend the range of primate achromatic contrast sensitivity

The MC and PC pathways show overlapping but complementary response properties in several domains when probed with achromatic stimuli. When tested with luminance steps or pulses, MC cell responses are more phasic (transient) than PC cell responses, which are more tonic (sustained) (Gouras, 1968; Purpura *et al.*, 1990; Lee *et al.*, 1994; Kremers, 1999; Kremers *et al.*, 1999). Similarly, when tested with rotating hemicircles the temporal properties of MC and PC responses differ (Kremers *et al.*, 2004), showing that MC neurons encode the changes in the visual scene whereas PC neurons respond proportionally to the level of illumination as can be expected from a sustained mechanism (Kremers, 1999; Kremers *et al.*, 1999).

When tested with sinusoidal luminance modulations, MC neurons display higher temporal cut-off frequencies than PC cells (Derrington and Lennie, 1984; Lee *et al.*, 1989a, 1990; Kremers *et al.*, 1992, 1993). However, PC cells respond to higher frequencies of chromatic modulation than can be perceived psychophysically and it is necessary to postulate some low-pass filter of their signals at a central site. Since such a filter would not be able to distinguish if a response were derived from luminance or chromatic modulation, or a mixture of the two, it must act on the achromatic response of the PC cell as well. This would be in keeping with this pathway being primarily concerned with sustained signals.

The MC neurons are very sensitive to achromatic stimuli at low contrasts, but their responses saturate at moderate contrast levels, whilst PC neurons are relatively contrast-insensitive, but their response increases linearly until high contrast levels (Kaplan and Shapley, 1986; Lee *et al.*, 1990, 1994; Kremers *et al.*, 1993) and both pathways may participate in the perception of achromatic contrast (e.g. Schiller and Logothetis, 1990; Merigan and Maunsell, 1993). The high gain and high sensitivity of MC cells are probably important for spatial vision at low contrast, whilst the PC ganglion cells may convey information about high contrasts, where MC ganglion cell responses saturate. There is psychophysical evidence that detection of achromatic contrast may be mediated by the MC pathway, but discrimination between suprathreshold contrasts may be mediated by the PC pathway (Pokorny and Smith, 1997; see Chapter 11). Assessment of achromatic brightness may also derive from the PC pathway (Valberg and Lee, 1992). Finally, most of the information in the PC pathway's response to natural scenes derives from achromatic rather than chromatic sources, because the range of achromatic contrast in the natural environment is much greater than the chromatic contrasts.

It has been proposed, using computational arguments, that for the performance of many tasks, higher levels of the visual system need access to information from both MC and PC pathways (Van Essen *et al.*, 1992). Similar arguments can be considered when attempting to map the MC and PC pathways in the dorsal and ventral cortical streams (Ungerleider and Mishkin, 1982) or in the cortical pathways for action and perception (Milner and Goodale, 1995), although the dorsal, motion pathway appears to be MC-dominated.

One hypothesis stresses the fact that the visual system needs two or more 'devices' with properties that largely overlap in the spatio-temporal domain, to extract reliable high-resolution information (Silveira, 1996; Silveira and De Mello, 1998). The region of overlap corresponds to the intermediate range of spatio-temporal frequencies, a region of occurrence of relevant features of the visual environment such as faces at conversational distance (Watt, 1991). This suggests that what is important in the MC and PC performance is not their ability to respond in isolation to high contrast stimuli close to the spatial and temporal cut-off frequencies, but on the contrary, their synergic action at intermediate frequencies, where vision must integrate their activities over different contrasts and chromaticities.

5.7 Conclusion

In this chapter, we have described the anatomy and physiology of primate retina, with emphasis on the main classes of retinal ganglion cells and their afferents. There is compelling evidence that the anatomical and physiological properties of MC ganglion cells and small bistratified cells are very similar in all diurnal Anthropoidea so far studied, both catarrhines and platyrrhines. The PC ganglion cells also have similar properties in these animals with the exception that they are color-blind in the monochromatic and dichromatic platyrrhines.

Whatever the original role of the MC and PC pathways, they are likely to have evolved prior to the divergence of catarrhines and platyrrhines. This suggests that they should also be present in prosimians. Very little is known about retinal ganglion cells of prosimians, but the few studies that have been done in these primates indicate that indeed they follow a general primate scheme, having MC and PC ganglion cells similar to those of Anthropoidea. MC, PC, and small bistratified ganglion cells appear to be strongly conserved in the various diurnal primate species. The reasons for this may lie in the roles of these systems for both achromatic and chromatic vision.

References

Baylor, D.A., Nunn, B.J., and Schnapf, J.L., 1984. The photocurrent, noise and spectral sensitivity of rods of the monkey *Macaca fascicularis*. *J. Physiol. (Lond.)* **357**, 575–607.

Baylor, D.A., Nunn, B.J., and Schnapf J.L., 1987. Spectral sensitivity of cones of the monkey *Macaca fascicularis*. *J. Physiol. (Lond.)* **390**, 145–160.

Bernadete, E.A. and Kaplan, E., 1997a. The receptive field of the primate P retinal ganglion cell. I. Linear dynamics. *Vis. Neurosci.* **14**, 169–185.

Bernadete, E.A. and Kaplan, E., 1997b. The receptive field of the primate P retinal ganglion cell. II. Nonlinear dynamics. *Vis. Neurosci.* **14**, 187–205.

Blakemore, C. and Vital-Durand, F., 1986. Organisation and post-natal development of the monkey's lateral geniculate nucleus. *J. Physiol. (Lond.)* **380**, 453–492.

Blessing, E.M., Solomon, S.G., Hashemi-Nezhad, M. *et al.*, 2004. Chromatic and spatial properties of parvocellular cells in the lateral geniculate nucleus of the marmoset (*Callithrix jacchus*). *J. Physiol. (Lond.)* **557**, 229–245.

Boycott, B.B. and Dowling, J.E., 1969. Organization of the primate retina: Light microscopy. *Philos. Trans. R. Soc. Lond. B* **255**, 109–176.

Boycott, B.B. and Wässle, H., 1991. Morphological classification of bipolar cells of the primate retina. *Eur. J. Neurosci.* **3**, 1069–1088.

Boycott, B. and Wässle, H., 1999. Parallel processing in the mammalian retina. *Invest. Ophthalmol. Vis. Sci.* **40**, 1313–1327.

Brainard, D.H., Roorda, A., Yamauchi, Y. *et al.*, 2000. Functional consequences of the relative numbers of L and M cones. *J. Opt. Soc. Am. A* **17**, 607–614.

Calkins, D.J., Schein, S.J., Tsukamoto, Y. *et al.*, 1994. M and L cones in macaque fovea connect to midget ganglion cells by different numbers of excitatory synapses. *Nature (Lond.)* **371**, 70–72.

Calkins, D.J., Tsukamoto, Y., and Sterling, P., 1998. Microcircuitry and mosaic of a blue-yellow ganglion cell in the primate retina. *J. Neurosci.* **18**, 3373–3385.

Chan, T.L., Martin, P.R., Clunas, N. *et al.*, 2001a. Bipolar cell diversity in the primate retina: Morphologic and immunocytochemical analysis of a New World monkey, the marmoset *Callithrix jacchus*. *J. Comp. Neurol.* **437**, 219–239.

Chan, T.L., Martin, P.R., and Grünert, U., 2001b. Immunocytochemical identification and analysis of the diffuse bipolar cell type DB6 in macaque monkey retina. *Eur. J. Neurosci.* **13**, 829–832.

Chatterjee, S. and Callaway, E.M., 2002. S cone contributions to the magnocellular visual pathway in macaque monkey. *Neuron* **35**, 1135–1146.

Cole, G.R., Hine, T., and McIlhagga, W., 1993. Detection mechanisms in L-, M- and S-cone contrast space. *J. Opt. Soc. Am. A* **10**, 38–51.

Crook, J.M., Lange-Malecki, B., Lee, B.B. *et al.*, 1988. Visual resolution of macaque retinal ganglion cells. *J. Physiol. (Lond.)* **396**, 205–224.

Croner, L.J. and Kaplan, E., 1995. Receptive fields of P and M ganglion cells across the primate retina. *Vision Res.* **35**, 7–24.

Dacey, D.M., 1993a. The mosaic of midget ganglion cells in the human retina. *J. Neurosci.* **13**, 5334–5355.

Dacey, D.M., 1993b. Morphology of a small-field bistratified ganglion cell type in the macaque and human retina. *Vis. Neurosci.* **10**, 1081–1098.

Dacey, D.M. and Lee, B.B., 1994. The "blue-on" opponent pathway in primate retina originates from a distinct bistratified ganglion cell type. *Nature (Lond.)* **367**, 731–735.

Dacey, D.M. and Packer, O.S., 2003. Colour coding in the primate retina: Diverse cell types and cone-specific circuitry. *Curr. Opin. Neurobiol.* **13**, 421–427.

Dacey, D.M. and Petersen, M.R., 1992. Dendritic field size and morphology of midget and parasol ganglion cells of the human retina. *Proc. Natl. Acad. Sci. USA* **89**, 9666–9670.

Dacey, D.M., Lee, B.B., Stafford, D.K. *et al.*, 1996. Horizontal cells of the primate retina: Cone specificity without spectral opponency. *Science (New York)* **271**, 656–659.

Dacey, D.M., Peterson, B.B., Robinson F.R. *et al.*, 2003. Fireworks in the primate retina: In vitro photodynamics reveals diverse LGN-projecting ganglion cell types. *Neuron* **37**, 15–27.

de Monasterio, F.M., 1978a. Properties of concentrically organized X and Y ganglion cells of macaque retina. *J. Neurophysiol.* **41**, 1394–1417.

de Monasterio, F.M., 1978b. Center and surround mechanisms of opponent-color X and Y ganglion cells of retina of macaques. *J. Neurophysiol.* **41**, 1418–1434.

de Monasterio, F.M., 1978c. Properties of ganglion cells with atypical receptive-field organization in retina of macaques. *J. Neurophysiol.* **41**, 1435–1449.

de Monasterio, F.M. and Gouras, P., 1975. Functional properties of ganglion cells of the rhesus monkey retina. *J. Physiol. (Lond.)* **251**, 167–195.

de Monasterio, F.M., Gouras, P., and Tolhurst, D.J., 1975a. Trichromatic colour opponency in ganglion cells of the rhesus monkey retina. *J. Physiol. (Lond.)* **251**, 197–216.

de Monasterio, F.M., Gouras, P., and Tolhurst, D.J., 1975b. Concealed colour opponency in ganglion cells of the rhesus monkey retina. *J. Physiol. (Lond.)* **251**, 217–229.

Derrington, A.M. and Lennie, P., 1984. Spatial and temporal contrast sensitivities of neurones in lateral geniculate nucleus of macaque. *J. Physiol. (Lond.)* **357**, 219–240.

Diller, L., Packer, O.S., Verweij, J. *et al.*, 2004. L and M cone contributions to the midget and parasol ganglion cell receptive fields of macaque monkey retina. *J. Neurosci.* **24**, 1079–1088.

Dogiel, A.S., 1891. Ueber die nervösen Element in der Retina des Menschen. *Archiv für Mikroskopische Anatomie und Entwicklungsmechanik* **38**, 317–344.

Dominy, N.J. and Lucas, P.W., 2001. Ecological importance of trichromatic vision to primates. *Nature (Lond.)* **410**, 363–366.

Dowling, J.E. and Boycott, B.B., 1966. Organization of the primate retina: Electron microscopy. *Proc. R. Soc. Lond. B* **166**, 80–111.

Ghosh, K.K. and Grünert, U., 1999. Synaptic input to small bistratified (blue-on) ganglion cells in the retina of a New World monkey, the marmoset *Callithrix jacchus. J. Comp. Neurol.* **413**, 417–428.

Ghosh, K.K., Goodchild, A.K., Sefton, A.E. *et al.*, 1996. The morphology of retinal ganglion cells in the New World marmoset monkey *Callithrix jacchus. J. Comp. Neurol.* **366**, 76–92.

Ghosh, K.K., Martin, P.R., and Grünert, U., 1997. Morphological analysis of the blue cone pathway in the retina of a New World monkey, the marmoset *Callithrix jacchus. J. Comp. Neurol.* **379**, 211–225.

Goodchild, A.K., Chan, T.L., and Grünert, U., 1996a. Horizontal cell connections with short-wavelength-sensitive cones in macaque monkey retina. *Vis. Neurosci.* **13**, 833–845.

Goodchild, A.K, Ghosh, K.K., and Martin, P.R., 1996b. A comparison of photoreceptor spatial density and ganglion cell morphology in the retina of human, macaque monkey, cat, and the marmoset *Callithrix jacchus. J. Comp. Neurol.* **366**, 55–75.

Gouras, P., 1968. Identification of cone mechanisms in monkey ganglion cells. *J. Physiol. (Lond.)* **199**, 533–547.

Grünert, U., 1997. Anatomical evidence for rod input to the parvocellular pathway in the visual system of the primate. *Eur. J. Neurosci.* **9**, 617–621.

Grünert, U. and Ghosh K.K., 1999. Midget and parasol ganglion cells of the primate retina express the $\alpha 1$ subunit of the glycine receptor. *Vis. Neurosci.* **16**, 957–966.

Grünert, U. and Martin, P.R., 1991. Rod bipolar cells in the macaque monkey retina: Immunoreactivity and connectivity. *J. Neurosci.* **11**, 2742–2758.

Grünert, U. and Wässle, H., 1996. Glycine receptors in the rod pathway of the macaque monkey retina. *Vis. Neurosci.* **13**, 101–115.

Grünert, U., Greferath, U., Boycott, B.B. *et al.*, 1993. Parasol (Pα) ganglion-cells of the primate fovea: Immunocytochemical staining with antibodies against $GABA_A$-receptors. *Vision Res.* **33**, 1–14.

Grünert, U., Martin, P.R., and Wässle, H., 1994. Immunocytochemical analysis of bipolar cells in the macaque monkey retina. *J. Comp. Neurol.* **348**, 607–627.

Grünert, U., Haverkamp, S., Fletcher, E.L. *et al.*, 2002. Synaptic distribution of ionotropic glutamate receptors in the inner plexiform layer of the primate retina. *J. Comp. Neurol.* **447**, 138–151.

Grünert, U., Lin, B., and Martin, P.R., 2003. Glutamate receptors at bipolar synapses in the inner plexiform layer of primate retina: Light microscopic analysis. *J. Comp. Neurol.* **466**, 136–147.

Haverkamp, S., Grünert, U., and Wässle, H., 2001a. Localization of kainate receptors at the cone pedicles of the primate retina. *J. Comp. Neurol.* **436**, 471–486.

Haverkamp, S., Grünert, U., and Wässle, H., 2001b. The synaptic architecture of AMPA receptors at the cone pedicle of the primate retina. *J. Neurosci.* **21**, 2488–2500.

Hendry, S.H.C. and Reid, R.C., 2000. The koniocellular pathway in primate vision. *Annu. Rev. Neurosci.* **23**, 127–153.

Hendry, S.H.C. and Yoshioka, T., 1994. A neurochemically distinct third channel in macaque dorsal lateral geniculate nucleus. *Science (New York)* **264**, 575–577.

Herr, S., Klug, K.J., Sterling, P. *et al.*, 2003. Inner S-cone bipolar cells provide all of the central elements for S cones in macaque retina. *J. Comp. Neurol.* **457**, 185–201.

Hopkins, J.M. and Boycott, B.B., 1997. The cone synapses of cone bipolar cells of primate retina. *J. Neurocytol.* **26**, 313–325.

Jacobs, G.H., 1998. Photopigments and seeing – lessons from natural experiments. *Invest. Ophthalmol. Vis. Sci.* **39**, 2205–2216.

Jacobs, G.H. and Deegan II, J.F., 1987. Spectral sensitivity of macaque monkeys measured with ERG flicker photometry. *Vis. Neurosci.* **14**, 921–928.

Jacobs, G.H. and Neitz, J., 1987. Polymorphism of the middle wavelength cone in two species of South American monkey: *Cebus apella* and *Callicebus molloch. Vision Res.* **27**, 1263–1268.

Jacobs, G.H., Deegan II, J.F., Neitz, J. *et al.*, 1993. Photopigments and colour vision in the nocturnal monkey, *Aotus. Vision Res.* **33**, 1773–1783.

Jacobs, G.H., Neitz, M., and Neitz, J., 1996. Mutations in S-cone pigment genes and the absence of colour vision in two species of nocturnal primate. *Proc. R. Soc. Lond. B* **263**, 705–710.

Jacoby, R., Stafford, D., Kouyama, N. *et al.*, 1996. Synaptic inputs to ON parasol ganglion cells in the primate retina. *J. Neurosci.* **16**, 8041–8056.

Jacoby, R.A., Wiechmann, A.F., Amara, S.G. *et al.*, 2000. Diffuse bipolar cells provide input to OFF parasol ganglion cells in the macaque retina. *J. Comp. Neurol.* **416**, 6–18.

Kaas, J.H., Huerta, M.F., Weber, J.T. *et al.*, 1978. Pattern of retinal terminations and laminar organization of the lateral geniculate nucleus of primates. *J. Comp. Neurol.* **182**, 517–554.

Kaiser, P.K., Lee, B.B., Martin, P.R. *et al.*, 1990. The physiological basis of the minimally distinct border demonstrated in the ganglion cells of the macaque retina. *J. Physiol. (Lond.)* **422**, 153–183.

Kamermans, M. and Spekreijse, H., 1995. Spectral behavior of cone-driven horizontal cell in teleost retina. *Prog. Retin. Eye Res.* **14**, 319–360.

Kaneko, A., 1970. Physiological and morphological identification of horizontal, bipolar and amacrine cells in goldfish retina. *J. Physiol. (Lond.)* **207**, 623–633.

Kaplan, E. and Shapley, R.M., 1986. The primate retina contains two types of ganglion cells, with high and low contrast sensitivity. *Proc. Natl. Acad. Sci. USA* **83**, 2755–2757.

Kelly, D.H. and van Norren, D., 1977. Two-band model of heterochromatic flicker. *J. Opt. Soc. Am. USA* **67**, 1081–1091.

Kirk, E.C. and Kay, R.F., 2003. The evolution of high visual acuity in the Anthropoidea, in *Anthropoid Origins: New Visions* (eds C.F. Ross and R.F. Kay), Kluwer Academic/Plenum Publishers, New York, pp. 539–602.

Klug, K., Herr, S., Tran Ngo, I. *et al.*, 2003. Macaque retina contains an S-cone OFF midget pathway. *J. Neurosci.* **23**, 9881–9887.

Kolb, H., 1970. Organization of the outer plexiform layer of the primate retina: Electron microscopy of Golgi-impregnated cells. *Philos. Trans. R. Soc. Lond. B* **258**, 261–283.

Kolb, H. and DeKorver, L., 1991. Midget ganglion cells of the parafovea of the human retina: A study by electron microscopy and serial section reconstructions. *J. Comp. Neurol.* **303**, 617–636.

Kolb, H. and Famiglietti, E.V., 1974. Rod and cone pathways in the inner plexiform layer of cat retina. *Science (New York)* **186**, 47–49.

Kolb, H., Mariani, A., and Gallego, A., 1980. A second type of horizontal cell in the monkey retina. *J. Comp. Neurol.* **189**, 31–44.

Kolb, H., Linberg, K., and Fisher, S.K., 1992. Neurons of the human retina: A Golgi study. *J. Comp. Neurol.* **318**, 147–187.

Kouyama, N. and Marshak, D.W., 1992. Bipolar cells specific for blue cones in the macaque retina. *J. Neurosci.* **12**, 1233–1252.

Kouyama, N. and Marshak, D.W., 1997. The topographical relationship between two neuronal mosaics in the short wavelength-sensitive system of the primate retina. *Vis. Neurosci.* **14**, 159–167.

Kremers, J., 1999. Spatial and temporal response properties of the major retino-geniculate pathways of Old and New World monkeys. *Doc. Ophthalmol.* **95**, 229–245.

Kremers, J., Lee, B.B., and Kaiser, P.K., 1992. Sensitivity of macaque retinal ganglion cells and human observers to combined luminance and chromatic temporal modulation. *J. Opt. Soc Am. A* **9**, 1477–1485.

Kremers, J., Lee, B.B., Pokorny, J. *et al.*, 1993. Response of macaque ganglion cells and human observers to compound periodic waveforms. *Vision Res.* **33**, 1997–2011.

Kremers, J., Silveira, L.C.L., Yamada, E.S. *et al.*, 1999. The ecology and evolution of primate colour vision, in *Color Vision: From Molecular Genetics to Perception* (eds K.R. Gegenfurtner and L.T. Sharpe), Cambridge University Press, Cambridge, pp. 123–142.

Kremers, J., Scholl, H.P.N., Knau, H. *et al.*, 2000. L- and M-cone ratios in human trichromats assessed by psychophysics, electroretinography and retinal densitometry. *J. Opt. Soc. Am. A* **17**, 517–526.

Kremers, J., Weiss, S., and Silveira, L.C.L., 2004. Spatiotemporal properties of the magno and parvocellular neurones of the lateral geniculate nucleus of the marmoset (*Callithrix jacchus*). *Neurociências*, **1**, 138–149.

LaVail, M.M., Rapaport, D.H., and Rakic, P., 1991. Cytogenesis in the monkey retina. *J. Comp. Neurol.* **309**, 86–114.

Lee, B.B., 1996. Receptive field structure in the primate retina. *Vision Res.* **36**, 631–644.

Lee, B.B., 2003. Structure of receptive field centers of midget retinal ganglion cells, in *Normal and Defective Colour Vision* (eds J.D. Mollon, J. Pokorny, and K. Knoblauch), Oxford University Press, Oxford, pp. 63–70.

Lee, B.B., 2004. Paths to colour in the retina. *Clin. Exp. Optom.* **87**, 239–248

Lee, B.B., Martin, P.R., and Valberg, A., 1988. The physiological basis of heterochromatic flicker photometry demonstrated in the ganglion cells of the macaque retina. *J. Physiol. (Lond.)* **404**, 323–347.

Lee, B.B., Martin, P.R., and Valberg, A., 1989a. Sensitivity of macaque retinal ganglion cells to chromatic and luminance flicker. *J. Physiol. (Lond.)* **414**, 223–243.

Lee, B.B., Martin, P.R., and Valberg, A., 1989b. Amplitude and phase of responses of macaque retinal ganglion cells to flickering stimuli. *J. Physiol. (Lond.)* **414**, 245–263.

Lee, B.B., Martin, P.R., and Valberg, A., 1989c. Nonlinear summation of M- and L-cone inputs to phasic retinal ganglion cells of the macaque. *J. Neurosci.* **9**, 1433–1442.

Lee, B.B., Pokorny, J., Smith, V.C. *et al.*, 1990. Luminance and chromatic modulation sensitivity of macaque ganglion cells and human observers. *J. Opt. Soc. Am. A* **7**, 2223–2236.

Lee, B.B., Wehrhahn, C., Westheimer, G. *et al.*, 1993a. Macaque ganglion cell responses to stimuli that elicit hyperacuity in man: Detection of small displacements. *J. Neurosci.* **13**, 1001–1009.

Lee, B.B., Martin, P.R., Valberg, A. *et al.*, 1993b. Physiological mechanisms underlying psychophysical sensitivity to combined luminance and chromatic modulation. *J. Opt. Soc. Am. USA A* **10**, 1403–1412.

Lee, B.B., Pokorny, J., Smith, V.C. *et al.*, 1994. Responses to pulses and sinusoids in macaque ganglion cells. *Vision Res.* **34**, 3081–3096.

Lee, B.B., Wehrhahn, C., Westheimer, G. *et al.*, 1995. The spatial precision of macaque ganglion cell responses in relation to vernier acuity of human observers. *Vision Res.* **35**, 2743–2758.

Lee, B.B., Smith, V.S., Pokorny, J. *et al.*, 1997. Rod inputs to macaque ganglion cells. *Vision Res.* **37**, 2813–2828.

Lee, B.B., Kremers, J., and Yeh, T., 1998. Receptive fields of primate retinal ganglion cells studied with a novel technique. *Vis. Neurosci.* **15**, 161–175.

Lee, B.B., Silveira, L.C.L., Yamada, E.S. *et al.*, 2000. Visual responses of ganglion cells of a New World primate, the capuchin monkey, *Cebus apella. J. Physiol. (Lond.)* **528**(3), 573–590.

Lee, S.C.S., Jusuf, P.R., and Grünert, U., 2004. S-cone connections of the diffuse bipolar cell type DB6 in macaque monkey retina. *J. Comp. Neurol.* **474**, 353–363.

Le Gros Clark, W.E., 1941a. The laminar organization and cell content of the lateral geniculate body in the monkey. *J. Anat.* **75**, 419–433.

Le Gros Clark, W.E., 1941b. The lateral geniculate body in the platyrrhine monkeys. *J. Anat.* **76**, 131–140.

Lennie, P. and Fairchild, M.D., 1994. Ganglion cell pathways for rod vision. *Vision Res.* **34**, 477–482.

Lennie, P., Trevarthen, C., Van Essen, D. *et al.*, 1990b. Parallel processing of visual information, in *Visual Perception: The Neurophysiological Foundations* (eds J. Spillman and J. Werner), Academic Press, San Diego, pp. 103–128.

Lennie, P., Haake, P.W., and Williams, D.R., 1991. The design of chromatically opponent receptive fields, in *Computational Models of Visual Processing* (eds M.S. Landy and J.A. Movshon), MIT Press, Cambridge, Massachusetts, pp. 71–82.

Leventhal, A.G., Rodieck, R.W., and Dreher, B., 1981. Retinal ganglion cell classes in the Old World monkey: Morphology and central projections. *Science (New York)* **213**, 1139–1142.

Leventhal, A.G., Ault, S.J., Vitek, D.J. *et al.*, 1989. Extrinsic determinants of retinal ganglion cell development in primates. *J. Comp. Neurol.* **286**, 170–189.

Lima, S.M.A., Silveira, L.C.L., and Perry, V.H., 1993. The M-ganglion cell density gradient in New World monkeys. *Braz. J. Med. Biol. Res.* **26**, 961–964.

Lima, S.M.A., Silveira, L.C.L., and Perry, V.H., 1996. Distribution of M retinal ganglion cells in diurnal and nocturnal New World monkeys. *J. Comp. Neurol.* **368**, 538–552.

Lin, B., Martin, P.R., Solomon, S.G. *et al.*, 2000. Distribution of glycine receptor subunits on primate retinal ganglion cells: A quantitative analysis. *Eur. J. Neurosci.* **12**, 4155–4170.

Lucas, P.W., Darvell, B.W., Lee, P.K.D. *et al.*, 1998. Colour cues for leaf food selection by long-tailed macaques (*Macaca fascicularis*) with a new suggestion for the evolution of trichromatic colour vision. *Folia Primatol.* **69**, 139–152.

Luo, X., Ghosh, K.K., Martin, P.R. *et al.*, 1999. Analysis of two types of cone bipolar cells in the retina of a New World monkey, the marmoset, *Callithrix jacchus*. *Vis. Neurosci.* **16**, 709–719.

Macri, J., Martin, P.R., and Grünert, U., 2000. Distribution of the $\alpha 1$ subunit of the $GABA_A$ receptor on midget and parasol ganglion cells in the retina of the common marmoset *Callithrix jacchus*. *Vis. Neurosci.* **17**, 437–448.

Mariani, A.P., 1983. Giant bistratified bipolar cells in monkey retina. *Anat. Rec.* **206**, 215–220.

Mariani, A.P., 1984. Bipolar cells in monkey retina selective for the cones likely to be blue-sensitive. *Nature (Lond.)* **308**, 184–186.

Marshak, D.W., Yamada, E.S., Bordt, A.S. *et al.*, 2002. Synaptic input to an ON parasol ganglion cell in the macaque retina: A serial section analysis. *Vis. Neurosci.* **19**, 299–305.

Martin, P.R., 1998. Colour processing in the primate retina: Recent progress. *J. Physiol. (Lond.)* **513**, 631–638.

Martin, P.R. and Grünert, U., 1992. Spatial density and immunoreactivity of bipolar cells in the macaque monkey retina. *J. Comp. Neurol.* **323**, 269–287.

Martin, P.R., White, A.J.R., Goodchild, A.K. *et al.*, 1997. Evidence that blue-on cells are part of the third geniculocortical pathway in primates. *Eur. J. Neurosci.* **9**, 1536–1541.

Martin, P.R., Lee, B.B., White, A.J.R. *et al.*, 2001. Chromatic sensitivity of ganglion cells in the peripheral primate retina. *Nature (Lond.)* **410**, 933–936.

McMahon, M.J., Lankheet, M.J.M., Lennie, P. *et al.*, 2000. Fine structure of parvocellular receptive fields in the primate fovea revealed by laser interferometry. *J. Neurosci.* **20**, 2043–2053.

McMahon, M.J., Packer, O.S., and Dacey, D.M., 2004. The classical receptive field surround of primate parasol ganglion cells is mediated primarily by a non-GABAergic pathway. *J. Neurosci.* **24**, 3736–3745.

Merigan, W.H. and Maunsell, J.H.R., 1993. How parallel are the primate visual pathways? *Annu. Rev. Neurosci.* **16**, 369–402.

Milam, A.H., Dacey, D.M., and Dizhoor, A.M., 1993. Recover in immunoreactivity in mammalian cone bipolar cells. *Vis. Neurosci.* **10**, 1–12.

Mills, S.L. and Massey, S.C., 1999. AII amacrine cells limit scotopic acuity in central macaque retina: A confocal analysis of calretinin labeling. *J. Comp. Neurol.* **411**, 19–34.

Milner, A.D. and Goodale, M.A., 1995. *The Visual Brain in Action*, Oxford University Press, Oxford.

Mollon, J.D., 1991. Uses and evolutionary origins of primate colour vision, in *Vision and Visual Dysfunction, Vol. 2, Evolution of the Eye and Visual System* (eds J.R. Cronly-Dillon and R.L. Gregory), MacMillan Press, Houndmills, pp. 306–319.

Mollon, J.D. and Jordan G., 1988. Eine evolutionäre Interpretation des menschlichen Farbensehens. *Die Farbe* **35**, 139–170.

Mollon, J.D., Bowmaker, J.K., and Jacobs, G.H., 1984. Variations of colour vision in a New World primate can be explained by polymorphism of retinal photopigments. *Proc. R. Soc. Lond. B* **222**, 373–399.

Mullen, K.T. and Kingdom, F.A.A., 1996. Losses in peripheral colour sensitivity predicted from "hit and miss" post-receptoral cone connection. *Vision Res.* **36**, 1995–2000.

Peichl, L. and Wässle, H., 1979. Size, scatter and coverage of ganglion cell receptive field centres in the cat retina. *J. Physiol. (Lond.)* **291**, 117–141.

Perry, V.H. and Cowey, A., 1981. The morphological correlates of X- and Y-like retinal ganglion cells in the retina of monkeys. *Exp. Brain Res.* **43**, 226–228.

Perry, V.H. and Cowey, A., 1984. Retinal ganglion cells that project to the superior colliculus and pretectum in the macaque monkey. *Neuroscience* **12**, 1125–1137.

Perry, V.H., Oehler, R., and Cowey, A., 1984. Retinal ganglion cells that project to the dorsal lateral geniculate nucleus in the macaque monkey. *Neuroscience* **12**, 1101–1123.

Peterson, B.B. and Dacey, D.M., 1999. Morphology of wide-field, monostratified ganglion cells of the human retina. *Vis. Neurosci.* **16**, 107–120.

Peterson, B.B. and Dacey, D.M., 2000. Morphology of wide-field bistratified and diffuse human retinal ganglion cells. *Vis. Neurosci.* **17**, 567–578.

Pokorny, J. and Smith, V.C., 1997. Psychophysical signatures associated with magnocellular and parvocellular pathway contrast gain. *J. Opt. Soc. Am. A* **14**, 2477–2486.

Polyak, S.L., 1941. *The Vertebrate Retina*, University of Chicago Press, Chicago.

Purpura, K., Kaplan, E., Tranchina, D. *et al.*, 1990. Light adaptation in the primate retina: Analysis of changes in gain and dynamics of monkey retinal ganglion cells. *Vis. Neurosci.* **4**, 75–93.

Ramón y Cajal, S., 1893. La Rétine des Vertébrés. *La Cellule* **9**, 121–255.

Ramón y Cajal, S., 1911. *Histologie du Système Nerveux de l'Homme et des Vertebrés* (Paris: A. Maloine. English translation: Swanson, N. and Swanson, L.W.), 1995. *Histology of the Nervous System of Man and Vertebrates*, Oxford University Press, New York.

Reid, R.C. and Shapley, R.M., 1992. Spatial structure of cone inputs to receptive fields in primate lateral geniculate nucleus. *Nature (Lond.)* **356**, 716–718.

Reid, R.C., Alonso, J.M., and Hendry, S.H.C., 1997. S-cone input is relayed to visual cortex from two koniocellular layers of macaque LGN. *Soc. Neurosci. Abs.* **23**, 13.

Regan, B.C., Julliot, C., Simmen, B. *et al.*, 1998. Frugivory and colour vision in *Alouatta seniculus*, a trichromatic platyrrhine monkey. *Vision Res.* **38**, 3321–3327.

Regan, B.C., Julliot, C., Simmen, B. *et al.*, 2001. Fruits, foliage and the evolution of primate colour vision. *Philos. Trans. R. Soc. Lond. B* **356**, 229–283.

Rodieck, R.W., 1988. The primate retina in *Comparative Primate Biology.* (eds H.D. Steklis and J. Erwin), Alan R. Liss, New York, pp. 203–278.

Rodieck, R.W. and Watanabe, M., 1993. Survey of the morphology of macaque retinal ganglion cells that project to the pretectum, superior colliculus, and parvicellular laminae of the lateral geniculate nucleus. *J. Comp. Neurol.* **338**, 289–303.

Rodieck, R.W., Binmoeller, K.F., and Dineen, J., 1985. Parasol and midget ganglion cells of the human retina. *J. Comp. Neurol.* **233**, 115–132.

Rodieck, R.W., Dacey, D.M., and Watanabe, M., 1987. Some other ganglion cell types of the primate retina. *ARVO Abstr., Invest. Ophthalmol. Vis. Sci.* **28**(Suppl), 261.

Schiller, P.H. and Logothetis, N.K., 1990. The color-opponent and broad-band channels of the primate visual system. *Trends Neurosci.* **13**, 392–398.

Schnapf, J.L., Nunn, B.J., Meister, M. *et al.*, 1990. Visual transduction in cones of the monkey *Macaca fascicularis. J. Physiol. (Lond.)* **427**, 681–713.

Schneeweis, D.M. and Schnapf, J.L., 1995. Photovoltage of rods and cones in the macaque retina. *Science (New York)* **268**, 1053–1056.

Schneeweis, D.M. and Schnapf, J.L., 1999. The photovoltage of macaque cone photoreceptors: Adaptation, noise, and kinetics. *J. Neurosci.* **19**, 1203–1216.

Schneeweis, D.M. and Schnapf, J.L., 2000. Noise and light adaptation in rods of the macaque monkey. *Vis. Neurosci.* **17**, 659–666.

Shapley, R.M. and Perry, V.H., 1986. Cat and monkey ganglion cells and their visual functions roles. *Trends Neurosci.* **9**, 229–235.

Silveira, L.C.L., 1996. Joint entropy loci of M and P cells: A hypothesis for parallel processing in the primate visual system. *Braz. J. Biol.* **56**(Suppl 1), 345–367.

Silveira, L.C.L., 2004. Comparative study of the primate retina, in *The Primate Visual System* (eds J.H. Kaas and C.E. Collins), CRC Press, Boca Raton, Florida, pp. 29–51.

Silveira, L.C.L. and Perry, V.H., 1990. A neurofibrillar staining method for retina and skin: A simple modification for improved staining and reliability. *J. Neurosci. Methods* **33**, 11–21.

Silveira, L.C.L. and Perry, V.H., 1991. The topography of magnocellular projecting ganglion cells (M-ganglion cells) in the primate retina. *Neuroscience* **40**, 217–237.

Silveira, L.C.L. and De Mello Jr, H.D., 1998. Parallel pathways of the primate vision: Sampling of the information in the Fourier space by M and P cells, in *Development and Organization of the Retina – From Molecules to Function* (eds L.M. Chalupa and B.L. Finlay), Plenum Press, New York, pp. 173–199.

Silveira, L.C.L., Yamada, E.S., Perry, V.H. *et al.* 1994. M and P retinal ganglion cells of diurnal and nocturnal New World monkeys. *Neuroreport* **5**, 2077–2081.

Silveira, L.C.L., Lee, B.B., Yamada, E.S. *et al.*, 1998. Post-receptoral mechanisms of colour vision in New World primates. *Vision Res.* **38**, 3329–3337.

Silveira, L.C.L., Lee, B.B., Yamada, E.S. *et al.*, 1999. Ganglion cells of a short wavelength sensitive cone pathway in New World monkeys: Morphology and physiology. *Vis. Neurosci.* **16**, 333–343.

Silveira, L.C.L., dos Santos, S.N., and Kremers, J., 2001. The presence of single-cone midget bipolar cells in the retina of the owl-monkey. *ARVO Abstr., Invest. Ophthalmol. Vis. Sci.* **42**, S375.

Silveira, L.C.L., Saito, C.A., Lee, B.B. *et al.*, 2004. Morphology and physiology of primate M and P cells. *Prog. Brain Res.* **144**, 21–46.

Solomon, S.G., White, A.J.R., and Martin, P.R., 1999. Temporal contrast sensitivity in the lateral geniculate nucleus of a New World monkey, the marmoset *Callithrix jacchus. J. Physiol. (Lond.)* **517**, 907–917.

Strettoi, E., Dacheux, R.F., and Raviola, E., 1990. Synaptic connections of rod bipolar cells in the inner plexiform layer of the rabbit retina. *J. Comp. Neurol.* **295**, 449–466.

Sun, H., Smithson, H., Lee, B.B. *et al.*, 2004. A new technique for measuring cone inputs to visual neurons. *ARVO 2004 Annual Meeting Abstract and Program Planner* 4277.

Svaetichin, G., 1953. The cone action potential. *Acta Physiol. Scand.* **29**, 565–600.

Svaetichin, G.S. and MacNichol, E.F., 1958. Retinal mechanisms for chromatic and achromatic vision. *Ann. N.Y. Acad. Sci.* **74**, 385–404.

Tan, Y. and Li, W.H., 1999. Trichromatic vision in prosimians. *Nature* **402**, 36.

Tovée, M.J., Bowmaker, J.K., and Mollon, J.D., 1992. The relationship between cone pigments and behavioural sensitivity in a New World monkey (*Callithrix jacchus jacchus*). *Vision Res.* **32**, 867–878.

Travis, D.S., Bowmaker, J.K., and Mollon, J.D., 1988. Polymorphism of visual pigments in a callitrichid monkey. *Vision Res.* **28**, 481–490.

Tsukamoto, Y., Masarachia, P., Schein, S.J. *et al.*, 1992. Gap junctions between the pedicles of macaque foveal cones. *Vision Res.* **32**, 1809–1815.

Ungerleider, L.G. and Mishkin, M., 1982. Two cortical visual systems, in *The Analysis of Visual Behavior* (eds D.J. Ingle, R.J.W. Mansfield, and M.S. Goodale), MIT Press, Cambridge, Massachusetts, pp. 549–586.

Valberg, A. and Lee, B.B., 1992. Main cell systems in primate visual pathways. *Curr. Opin. Ophthalmol.* **3**, 813–823.

Valberg, A., Lee, B.B., Kaiser, P.K. *et al.*, 1992. Responses of macaque ganglion cells to movement of chromatic borders. *J. Physiol. (Lond.)* **458**, 579–602.

Van Essen, D.C., Anderson, C.H., and Felleman, D.J., 1992. Information processing in the primate visual system: An integrated systems perspective. *Science (New York)* **255**, 419–423.

Vardi, N., Duvoisin, R., Wu, G. *et al.*, 2000. Localization of mGluR6 to dendrites of ON bipolar cells in primate retina. *J. Comp. Neurol.* **423**, 402–412.

Verweij, J., Hornstein, E.P., and Schnapf, J.L., 2003. Surround antagonism in macaque cone photoreceptors. *J. Neurosci.* **23**, 10249–10257.

Wässle, H. and Boycott, B.B., 1991. Functional architecture of the mammalian retina. *Physiol. Rev.* **71**, 447–480.

Wässle, and H. Riemann, H.J., 1978. The mosaic of nerve cells in the mammalian retina. *Proc. R. Soc. Lond. B* **200**, 441–461.

Wässle, H., Grünert, U., Martin, P.R. *et al.*, 1994. Immunocytochemical characterization and spatial distribution of midget bipolar cells in the macaque monkey retina. *Vision Res.* **34**, 561–579.

Wässle, H., Grünert, U., Chun, M.H. *et al.*, 1995. The rod pathway of the macaque monkey retina: Identification of AII-amacrine cells with antibodies against calretinin. *J. Comp. Neurol.* **361**, 537–551.

Watanabe, M. and Rodieck, R.W., 1989. Parasol and midget ganglion cells. *J. Comp. Neurol.* **289**, 434–454.

Watt, R.J., 1991. *Understanding Vision*, Academic Press, London.

Weiss, S., Kremers, J., and Maurer, J., 1997. Interaction between rod and cone signals in responses of lateral geniculate neurons in dichromatic marmosets (*Callithrix jacchus*). *Vis. Neurosci.* **15**, 931–943.

Werblin, F. and Dowling, J.E., 1969. Organization of the retina of the mudpuppy, *Necturus maculosus*. II. Intracellular recording. *J. Neurophysiol.* **32**, 339–355.

White, A.J.R., Solomon, S.G., and Martin, P.R., 2001. Spatial properties of koniocellular cells in the lateral geniculate nucleus of the marmoset *Callithrix jacchus*. *J. Physiol. (Lond.)* **533**, 519–535.

Wikler, K.C. and Rakic, P., 1990. Distribution of photoreceptor subtypes in the retina of diurnal and nocturnal primates. *J. Neurosci.* **10**, 3390–3401.

Wilder, H.D., Grünert, U., Lee, B.B. *et al.*, 1996. Topography of ganglion cells and photoreceptors in the retina of a New World monkey: The marmoset *Callithrix jacchus*. *Vis. Neurosci.* **13**, 335–352.

Yamada, E.S., Silveira, L.C.L., and Perry, V.H., 1996a. Morphology, dendritic field size, somal size, density and coverage of M and P retinal ganglion cells of dichromatic *Cebus* monkeys. *Vis. Neurosci.* **13**, 1011–1029.

Yamada, E.S., Silveira, L.C.L., and Gomes, F.L., 1996b. The retinal ganglion cell classes of New World primates. *Rev. Bras. Biol.* **56**(Suppl 1), 381–396.

Yamada, E.S., Marshak, D.W., Silveira, L.C.L. *et al.*, 1998. Morphology of P and M retinal ganglion cells of the bush baby. *Vision Res.* **38**, 3345–3352.

Yamada, E.S., Silveira, L.C.L., Perry, V.H. *et al.*, 2001. Morphology and dendritic field size of M and P retinal ganglion cells of the owl monkey. *Vision Res.* **41**, 119–131.

6

The Lateral Geniculate Nucleus

Jan Kremers, Jon H. Kaas, Paul R. Martin,
and *Samuel G. Solomon*

6.1 Introduction

The dorsal lateral geniculate nucleus (LGN) is a part of the thalamus. Most sensory input to the thalamus is relayed to the neocortex. In comparison with non-mammalian vertebrates, mammals have a greatly increased neocortex and in parallel also the thalamus is increased in volume (Creutzfeldt, 1993). In non-mammalian vertebrates, the sensory information flow makes use of other pathways.

In most mammals, the retino-geniculo-cortical pathway is the most important visual pathway not only because most retinal ganglion cells project to the LGN but also because the visual information passing along it is used for visual perception. The mammalian homologs of the major visual pathways in non-mammalian vertebrates are also concerned with visual information processing but their function is more related with non-perceptual visual processes such as eye movements, pupil size, and diurnal rhythm. These subcortical pathways of primates are discussed in Chapter 8.

The afferents to the thalamus and thus to the LGN have been divided into two functional categories (Sherman and Guillery, 1998, 2001): the 'drivers' which determine the qualitative characteristics of the receptive fields of the cells and the 'modulators' which can alter quantitative aspects of the receptive field without changing its basic properties. For the LGN the driving input comes from the retina. This chapter is concerned with this retino-geniculate pathway. The modulating non-retinal input will be discussed in Chapter 7.

In the following sections, we will first discuss the anatomical properties of the LGN. Then we will discuss how the LGN neurons are classified. Next, the linear physiological properties of the LGN cells will be described and these properties will be compared with

The Primate Visual System: A Comparative Approach Edited by Jan Kremers

those of the retinal ganglion cells. The properties of the different LGN cells in different primate species will be compared. Finally, nonlinear cell properties will be described.

6.2 The anatomical organization of the LGN

There are several simple organizing features of the LGN that hold for all or nearly all mammals (Kaas *et al.*, 1972). First, the lateral geniculate nucleus is subdivided into layers. Each layer forms a systematic map or retinotopic representation of part of the visual field. The inputs from the two eyes are kept separate, so each layer is activated by one eye or the other. These inputs arise from sectors of the retina that view the contralateral visual hemifield. However, more input is received from the contralateral eye due to inclusion of the lateral extreme of the hemifield. As a result, the layers that represent the contralateral eye always are larger in order to accommodate the representation of this extra visual space. The ipsilaterally projecting retina sees part or most of this same visual hemifield, and the overlap region is the binocular visual hemifield. Because species differ in having more laterally or more forward directing eyes, emphasizing either panoramic or binocular and forward vision, the sizes of the binocular and monocular portions of the hemifield vary. Tree squirrels, for example, having to be concerned with predators from all directions, have visual hemifields of 150°, resulting in a total panoramic view of 300°, but their binocular hemifield is only 30°. Most primates, in contrast, have forwardly directed eyes that create a large binocular field, with the ipsilateral retina input to the LGN devoted to >70°–80° of hemifield, while the contralateral retinal input representing about 90°.

The second main principle of LGN organization is that the number of layers in the LGN varies according to phylogenetic group (taxon), but the number is always more than two. Thus, there must be some reason for layers in addition to bringing the retinotopic maps from the two eyes into register. In the simplest form of lamination, such as seen in the LGN of the rat, the LGN has three layers, two for the contralateral eye and one for the ipsilateral eye, with the ipsilateral layer in the middle. Often, the two layers with contralateral eye inputs demonstrate no clear difference in the size of cells in the layers, the types of input from the retina, and the layers of termination in cortex, and the two layers appear fused where they represent the monocular visual hemifield, and the middle layer disappears. Thus, there is no obvious explanation of why two layers with contralateral inputs exist. In many other species, however, especially those with more than three layers, obvious morphological differences exist between layers, and they have different types of inputs from the retina and cortical projections. Thus, another reason for layers in the LGN is that they group cells together that have similar functional roles, and separate those with different roles. The types of segregations differ across mammalian taxa, and only the types found in primates are further discussed here.

A third general feature of LGN organization is that the retinotopic maps on the separate layers are precisely aligned. Given that the major target of the LGN is the primary visual cortex (V1 or area 17), and this cortex contains a single (or fused) map of the contralateral visual hemifield, this means that each location in V1 receives inputs from a column or row of LGN neurons that extends without fractures across all LGN layers. These projection lines have been demonstrated by placing a lesion in geniculate-recipient layers of V1, with the result that cells in a projection column, crossing all the LGN layers, degenerate.

There are other general features of the mammalian LGN that are useful to keep in mind. First, the LGN receives many other inputs besides those from the retina, although these inputs have less activating power ('modulators'); feedback inputs from visual cortex outnumber all others. Second, not all neurons in the LGN project to the visual cortex. Perhaps 20–25 percent of the neurons are intrinsic, having only local connections in the LGN, and are inhibitory in function. Third, a small projection of LGN relay cells in nearly all species project to extrastriate areas of visual cortex.

Given this general background, the specifics of LGN organization in primates can be discussed. Our main focus is in lamination. All primates have a lateral geniculate nucleus that contains four basic layers. However, this basic laminar organization is elaborated in several ways in the different lines of primate evolution. Early primates diverged over 60 million years ago (mya) into three main branches (Kaas and Preuss, 2003), leading to present-day prosimians (lemurs, lorises, and galagos), tarsiers, and anthropoid primates (monkeys, apes, and humans). Monkeys diverged into New World and Old World simian lines over 40 mya, and the Old World line later separated into monkeys and apes including humans (Chapter 1). All of these primates have layers of two types: those containing large cells and those containing small cells. The layers of large cells are called the magnocellular (MC) layers, while the layers of small cells are called the parvocellular (PC) layers. In anthropoid primates, there are also considerable numbers of even smaller cells between layers. These cells have been traditionally considered too few and too indistinct to justify distinguishing their locations as layers, and instead these locations between the obvious layers have been considered interlaminar zones of the smallest cells. The size and location of these zones relative to MC and PC layers vary, but in all prosimians these smaller cells form two distinct layers, one for each eye, which when added to the other four, make a total of six layers. These layers of smallest cells were termed the koniocellular (KC) layers (Kaas *et al.*, 1978), as the term koniocellular has been used elsewhere to describe those sensory areas of cortex with the smallest cells in layer 4. The small cells in interlaminar zones, not quite elevated to layers, are now called zones of KC cells, or KC zones. The different sizes of the MC, PC, and KC cells in the different layers, of course, reflect differences in function. In general, larger relay neurons support thicker axons with more rapid conduction times. In the LGN of primates, three types of relay cells respond differently to visual stimuli, in large part because they are activated by different classes of retinal ganglion cells (see below). The LGN MC, PC, and KC relay neurons target different layers and sublayers in visual cortex. Thus, primates are characterized by having three distinct processing streams from retina to cortex, and the segregation of these streams is reflected in the layers of the LGN.

In general, the brains of prosimian primates retain more primitive features than other primate brains, but they also have a few advanced or specialized features not found in other primates. In particular, all prosimians have six basic layers in the LGN of three functional types, with a pair of layers (one for each eye) of each type (Figure 6.1). The pair of MC layers is closest to the optic tract; the external or outer layer of the pair (ME) receives inputs from the contralateral eye and this is slightly longer than the internal MC layer (MI) with inputs from the ipsilateral eye. The rest of the LGN contains an external PC layer, PE, furthest from the optic tract and with inputs from the contralateral eye,

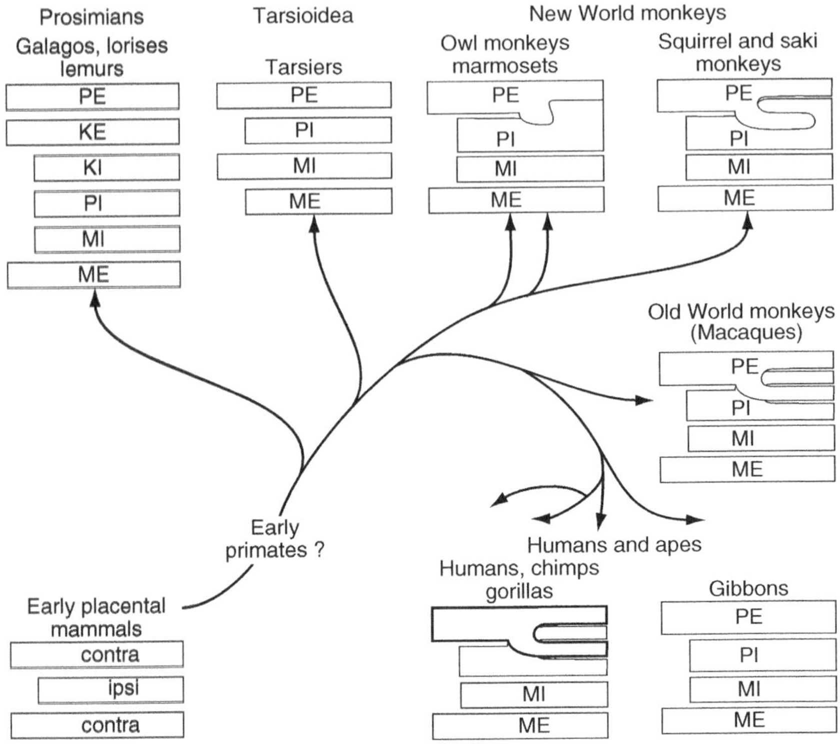

Figure 6.1 Patterns of lamination of the LGN of primates. The first primates probably had pairs of parvocellular (PC) and magnocellular (MC) layers with thin interlaminar zones of koniocellular (KC) neurons. Prosimians evolved distinct pairs of KC layers, while other primates either did not or lost these distinct layers. Most of the anthropoid primates (monkeys, apes, and humans) evolved sublayers of varying extents as PC layers partially subdivided and interdigitated. Pairs of layers are identified here by cell type (MC, PC, or KC) and more external (E) or internal (I) partitions in the nucleus. Layers with inputs from the contralateral eye have a more extensive representation, including the contralateral monocular visual hemifield, and therefore they are longer

and an internal PC layer, PI, with inputs from the ipsilateral eye. However, this pair of layers is separated by a more external KC layer, KE (contralateral eye inputs), and an adjoining more internal KC layer, KI (ipsilateral eye inputs).

All other extant primates, sometimes referred to as haplorrhine primates (Chapter 1), fail to have distinct KC layers between the PE and PI layers, and thus these two layers adjoin each other. However, haplorrhine primates have considerable numbers of koniocells in interlaminar zones, especially in nocturnal tarsiers and nocturnal owl monkeys. However, the PC layers may thicken and partially subdivide and interdigitate to varying extents, depending on species. In some New World monkeys, such as owl monkeys and marmosets, there is little thickening and little evidence of parts of PE and PI protruding into each other. More thickening, with leaflets of PE and PI interdigitating in the caudal part of the LGN (representing more central vision), occurs in squirrel monkeys and the larger New World monkeys. This partial division and interdigitation of PC layers is more pronounced in all Old World monkeys, gorillas, chimps, and humans, but is strangely absent in gibbons. Because of the subdividing of the PC layers, the LGN of 'higher' primates has been traditionally described as having six layers, but there are only four complete layers, corresponding to two major cell types (PC and MC) and two eyes (ipsi and contra). The additional PC 'layers' are incomplete and are better termed sublayers or leaflets.

In all primates, the projections of the PC cells are to the deeper half of layer 4 of primary visual cortex (V1 or area 17). The MC cells project to the more superficial half. The KC cells project predominantly to the supragranular layers (layers 1, 2, 3, and upper layer 4). The projection to layers 2 and 3 is concentrated in cytochrome oxidase blobs.

6.3 The classification of LGN cells

The classification of neurons is not just an intellectual pastime. It is an important exercise because it ideally uses physiological and anatomical characteristics of the cells for the classification and a hierarchy, which is based upon the functional significance of these characteristics. Because neurons can have different physiological and anatomical properties in different animal species, a classification often is limited only to closely related animal species. Neurons in the primate LGN have been classified according to different criteria. The resulting different classifications can be partially congruent with each other, which means that the cells are classified into similar groups despite the use of different criteria. But sometimes the cells may be classified into different cell groups. This can lead to some confusion. For instance, on the basis of anatomical criteria LGN cells are classified into MC, PC, and KC (see above). On the basis of spatial response properties, however, the cells have been classified as X-like, Y-like, and W-like, Y-like cells showing more spatial nonlinearities than X-like cells. On average MC cells are slightly more nonlinear than PC cells (White *et al.*, 2002). Because of this, MC and Y-like cells are often considered to be synonyms. However, the majority of MC cells are spatially linear and therefore cannot be considered to be Y-like.

An ideal classification distinguishes between groups of neurons that have distinct functions in the nervous system. When describing the functions of these cell groups often only a 'typical representative' of each group is used to describe the complete population. However, such a classification may also bear some problems. By looking only at the 'typical representative' it is easily neglected that cells within a group may also show some variability relative to each other. This variability may even be large, so that the distinction between cell groups is not clear. An analogy of this problem can be found, for instance, in the systematics of biological species: all animals are grouped into species. The problem often with this classification is that the distinction between two species is not always clear and only based upon subjective definitions. In summary, while the classification of LGN neurons is helpful for recognizing cell properties, it should be taken into account that not all neurons are identical and that there may be a considerable variability in the cell properties.

As mentioned above, the LGN is layered and the cells in these layers have distinct anatomical properties. These cells have also distinct physiological properties. A similar principle of functional segregation is a feature of the cat LGN. Nevertheless, the classifications in the cat and the primate should not be regarded as homologous. The LGN neurons can be divided into PC, MC, and KC cells. These cells also receive their input from different types of retinal ganglion cells (Rodieck and Watanabe, 1993; Rodieck *et al.*, 1993). The great majority of ganglion cells projecting to the MC layers have large dendritic trees. They have the morphology of so-called parasol cells. The majority of retinal ganglion cells projecting to the PC layers are the so-called midget cells and have much smaller dendritic tree sizes. The retinal ganglion cells projecting to the KC layers are relatively heterogeneous. A subtype of the KC cells shows blue-yellow color opponency (Martin *et al.*, 1997) and most probably receives its input from retinal ganglion cells with a small-field bistratified morphology (Dacey and Lee, 1994). For more detailed description of the anatomy and physiology of retinal cells belonging to the different pathways, see Chapter 5. The cells in the different layers also have different physiological properties. Below, some classifications, based upon physiological properties, are given.

The first classification based upon response properties in trichromatic monkeys was introduced by Wiesel and Hubel (1966). They found that LGN cells can show color and spatial opponency and that most cells have a center-surround organization. Wiesel and Hubel distinguished between four types of LGN cells. Type I cells are color-opponent cells and have a center-surround organization. Most of the cells were red-green opponent. These were predominantly PC cells. Type II cells were rarely found and the great majority showed blue-yellow opponency without spatial opponency. These cells were excited by a blue stimulus. They are probably identical with the so-called blue-ON cells. As described above, these cells may belong to the KC pathway (Martin *et al.*, 1997). Wiesel and Hubel also found cells with a center-surround organization but without color opponency. These type III cells were sometimes found in the PC layers but were predominant in the MC layers. Indeed, MC cells do not show overt color opponency (Smith *et al.*, 1992). Type IV cells were exclusively found in the MC layers. Similar to type III cells they had a center-surround organization and were not color-opponent. But, unlike the type III cells, they displayed a strong suppression by diffuse red light. Type IV cells are probably not very different from other MC cells.

Cat LGN cells and retinal ganglion cells were classified on the basis of the spatial linearity of their responses. X-cells are spatially linear because they show a clear minimum in the response to counterphase modulating gratings when the relative position of these gratings is varied. Moreover, X-cells display sustained responses. On the other hand, Y-cells are spatially more nonlinear because they show a frequency-doubled response for certain positions of the grating. Furthermore, Y-cells display transient responses. W-cells often do not have a clear center-surround organization. Because primate LGN cells can have similar spatial and temporal response properties, they were classified as 'X-, Y-, or W-like', even if this classification is primarily based on an analogy (similar properties) rather than a homology (meaning that the cells in the two species have the same phylogenetic origin). Indeed it was found that the cells with the larger dendritic trees and the larger receptive fields (the MC cells) can have Y-like properties and more transient responses, whereas the PC cells are more linear and respond in a more sustained manner. But the population of MC cells have responses more similar to

X-cells rather than Y-cells in the cat (Kaplan and Shapley, 1982; Derrington and Lennie, 1984; Benardete *et al.*, 1992). There may be differences between primate groups because in the strepsirrhine galago or bush baby (*Galago crassicaudatus*) the PC and MC cells indeed seem to be X- and Y-like, respectively (Norton and Casagrande, 1982; Norton *et al.*, 1988; Irvin *et al.*, 1986), whereas this congruency is absent in anthropoids.

In the following sections, the different physiological properties of LGN cells are described in more detail.

6.4 Basic receptive field properties of LGN cells

6.4.1 The influence of stimulus contrast

The response of LGN cells depends upon the stimulus strength. Stimulus strength can be expressed in different ways.

Michelson contrast is most widely used and is mainly used to quantify the stimulus strength of repetitive stimuli such as sine- or square-waves. It is defined as: $((L_{max} - L_{min})/(L_{max} + L_{min})) \times 100\%$ where L_{min}, L_{max} are the minimal and the maximal luminances in the stimulus. In many repetitive stimuli this quantity equals $((L_{max} - L_{min})/2 \times L_{mean}) \times 100\%$ because $L_{max} + L_{min} = 2 \cdot L_{mean}$ where L_{mean} is the mean luminance. The rationale for using this quantity is that the numerator $(L_{max} - L_{min})$ quantifies the change in the stimulus, whereas the denominator $(L_{max} + L_{min})$ determines the state of adaptation. For flashed stimuli, the state of adaptation is mainly determined by the background light and less by the flash. Therefore the Weber fraction $(\Delta L/L_{min} = (L_{max} - L_{min})/L_{min})$ is a more appropriate quantity to express the strength of flashed stimuli.

Generally, the amplitudes of LGN cell responses increase monotonically but not linearly with increasing stimulus strength. As with the responses of retinal ganglion cells, the response amplitude of LGN cells (R) as a function of Michelson contrast (c) can be described by a general hyperbolic function $(R(c) = R_{max} \cdot c^n/(b^n + c^n))$ in which R_{max} is the maximal response amplitude, b is the contrast for giving half maximal response $(R_{max}/2)$, and n determines the rate of change, or slope, of the function. When n is equal to 1, this general case becomes the often used Naka–Rushton function $R(c) = (R_{max} \cdot c/(b + c))$. A useful quantity that can be derived from the hyperbolic function is the contrast gain: the slope of the function at zero contrast (the gain equals: R_{max}/b^n, and therefore R_{max}/b for a Naka–Rushton function). The gain is often used to quantify the sensitivity (or responsivity) of the cell. Examples of response amplitudes of MC, PC, and KC cells as a function of the stimulus contrast are shown in Figure 6.2 (right panels) together with best-fitting Naka–Rushton functions.

In a linear system the response phase does not change with stimulus strength. However, as in retinal ganglion cells, the response phase of LGN neurons (especially in the MC layers) advances as contrast increases, especially at low and intermediate temporal frequencies. This phenomenon is probably related to a contrast gain control mechanism that was first described by Shapley and Victor (1978, 1979, 1981) in cat retinal ganglion cells (see also below). The contrast gain control mechanism also is manifest as a change in saturation depending on the temporal frequency of the stimulus. As a result, the response amplitude as a function of temporal frequency depends on the stimulus contrast (see also

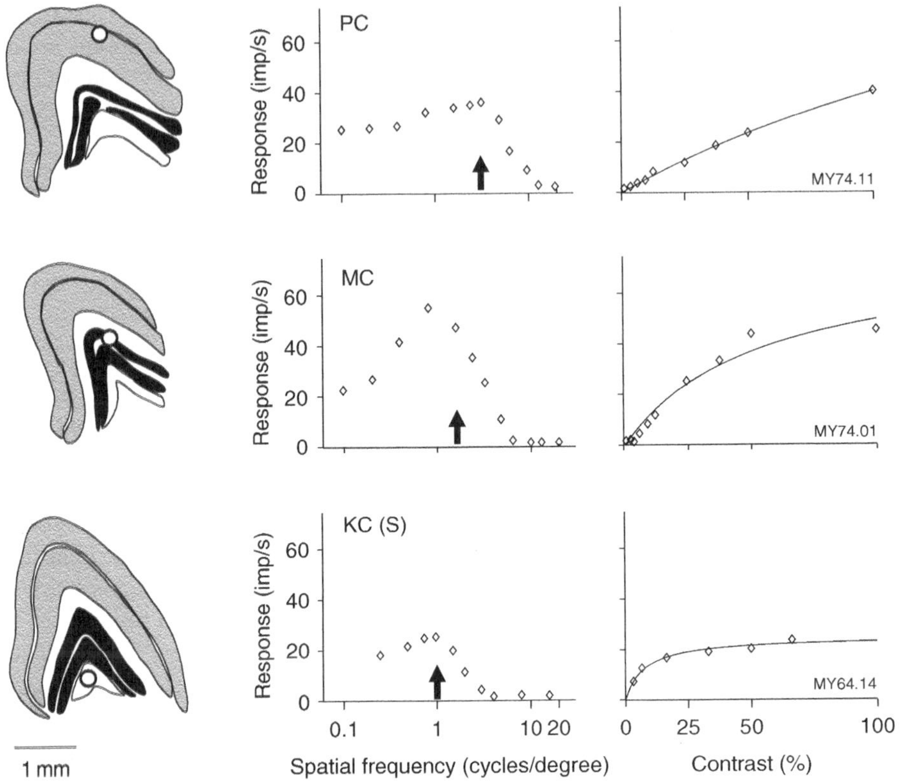

Figure 6.2 Responses to spatial contrast in PC, MC, and KC cells in the marmoset LGN. The recording site within the LGN is shown in the left panels – (grey) PC, (black) MC, (white) KC. The spatial frequency tuning curve for each cell is shown in the center panels. Optimum spatial frequency is indicated by an arrow on each tuning curve. A contrast-response function at the optimal spatial frequency for each cell is shown in the right panels. The lines show best-fitting Naka–Rushton functions, as described in the text

section 6.4.3). This change in saturation was observed in retinal ganglion cells but not in LGN cells, possibly because LGN cells do not respond strongly at the high temporal frequencies at which a change in saturation might be apparent.

Differences between animal species

There are no clear differences between the contrast response behavior in different primate species besides the contrast gain. The responses of LGN cells in all primate species, including the nocturnal owl monkey (Xu *et al.*, 2001; Kilavik, Kremers, and Silveira unpublished data), can be described by a Naka–Rushton function or a hyperbolic function. Thus similar types of saturation seem to be present in the owl monkeys and diurnal monkeys. Other nonlinearities such as contrast-dependent phase changes are also present in owl monkeys as well as in diurnal monkeys. It thus seems that the contrast responses are very similar in different primate species and are relatively independent of anatomical and physiological differences that occur in the owl monkey (see below).

Differences between MC, PC and KC cells
The best-fitting Naka–Rushton function for the PC cell shown in Figure 6.2 resembles a straight line, indicating that there is little saturation. The MC and the KC cells display more saturation. Generally, PC cells indeed show much less saturation to luminance stimuli than MC or KC cells (Kremers *et al.*, 1997; Solomon *et al.*, 1999). Furthermore, MC cells are more sensitive than PC cells. This is also illustrated by the responses in Figure 6.2. The initial slope of the best-fitting Naka–Rushton function is steeper for the MC cell than for the PC cell.

Comparison with retinal ganglion cells
Retinal ganglion cells belonging to the MC pathway are also more sensitive to luminance stimuli than PC cells. However, in the LGN this difference in sensitivity is less clear (Kremers *et al.*, 1997; Solomon *et al.*, 1999). But it seems that the responses of LGN cells in the MC layers still display a better signal-to-noise ratio than those of PC cells (Kremers *et al.*, 2001).

6.4.2 Spatial properties

Most retinal ganglion cells and LGN cells respond better to spatial luminance contrast than they do to diffuse light. This selectivity arises through the center-surround organization of concentric receptive fields (Kuffler, 1953) – the spatial profile of the center is smaller than that of the antagonistic surround. When stimulated with uniform fields, both mechanisms are active and antagonize each other and there is little response. When stimulated with appropriate spatial contrast (e.g. discrete spots of light), the activation of the center is greater than that of the surround, and the response of the cell is thus stronger. For both centers and surrounds, the spatial profile of luminance sensitivity can be approximately described by Gaussian functions (Rodieck and Stone, 1965a,b; Enroth-Cugell and Robson, 1966). The centers and the surrounds respond in an antagonistic manner at low temporal frequencies. This simple description of the receptive field organization is called the Difference of Gaussians (DoG) model. In the DoG model the center and the surround responses have opposite signs to account for the antagonistic responses. However, as also found for retinal ganglion cells (of cat and primates), the responses of the surrounds are delayed by a few milliseconds relative to those of the center (Gouras and Zrenner, 1979; Enroth-Cugell *et al.*, 1983; Frishman *et al.*, 1987; Soodak *et al.*, 1987; Smith *et al.*, 1992; Yeh *et al.*, 1995a,b; Kilavik *et al.*, 2003). As a result, the center and surround responses are not antagonistic (180° phase shifted) when stimulated at high temporal frequencies and the response behavior of a cell can change when the temporal frequency is altered.

The spatial properties of receptive fields can be probed in several ways. One commonly used method is by stimulating with drifting or counterphase modulating gratings of different spatial frequencies (Enroth-Cugell and Robson, 1966). The rationale behind these measurements is that the responses of the centers and the surrounds, if they can be considered to be linear, are mainly driven by overall illuminance changes. When the spatial frequency of the grating is very high and the overall luminance within the center or the surround does not change, then there is no response (i.e. the case when the diameter is larger than about two spatial cycles of the grating). Because the surround is larger than the center, the surrounds stop responding at a lower spatial frequency. Figure 6.2 (left panels)

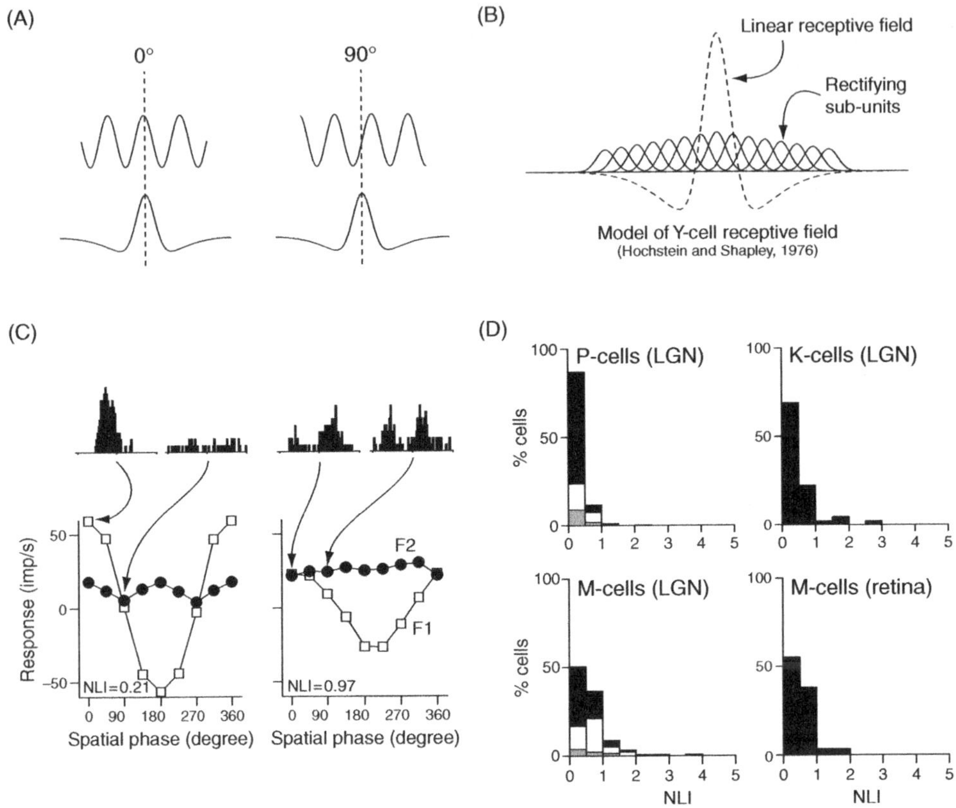

Figure 6.3 Linearity of spatial summation in receptive fields of primate LGN and retina. (A) Probing linear and nonlinear summation using sinusoidal gratings. When gratings are flickered in counterphase, their position over a receptive field is their spatial phase. At 0°, the maximum luminance flux of the grating (top panels) and the peak of the receptive field (lower panels) are aligned: a linear receptive field will respond at the temporal frequency of modulation. At 90° spatial phase, the peak of the receptive field is aligned with the midpoint of the grating: a linearly summating receptive field will not respond since the net flux is 0. (B) Y-cell model of Hochstein and Shapley (1976a). To account for semi-linear responses at low spatial frequency and nonlinear responses at high spatial frequency, in Y-cells of the cat retina, they proposed that the Y-cell consisted of a linear receptive field and a collection of smaller subunits, whose input was rectified before summation. (C) Linear and nonlinear responses in macaque retinal ganglion cells. The responses of two MC-pathway neurons to counterphase flickering gratings of various spatial phases are shown. The response of the linear neuron (left panel) is dominated by the first harmonic (F1), and there is no response at the null spatial phase (90°). The response of the nonlinear neuron (right panel) has some first harmonic component, but the second harmonic component is stronger, and there is no spatial phase where the neuron completely fails to respond. Adapted from White *et al.* (2002). The histograms above the panels show the responses of the cell to one cycle of the stimulus at the peak and null spatial phases. (D) Most primate retinal ganglion cells and LGN cells show linear

shows examples of such measurements. The DoG model predicts a response amplitude of the receptive field center (R_C) that can be described by: $R_C = CK_C\pi r_C^2 \exp(-(\pi r_C v))$ where C and v are the contrast and the spatial frequency of the stimulus. K_C and r_C are the peak sensitivity and size of the receptive field center. Similarly the response of the surround can be calculated. At low temporal frequencies, the center and the surround can be considered to be antagonistic (see above) and thus the cell response (R) can be described by: $R = R_C - R_S = C\left(K_C\pi r_C^2 \exp\left(-(\pi r_C v)^2\right) - K_S\pi r_S^2 \exp\left(-(\pi r_S v)^2\right)\right)$. Because r_S is larger than r_C the efficacy of the cell surround will decrease at lower spatial frequencies. In this measurement the temporal frequency of the stimulus was low enough to ensure an antagonism between center and surround responses and the overall response is obtained by subtracting the surround from the center response. The responses to counterphase modulating gratings not only depend upon spatial frequency but also upon the position of the stimulus relative to the receptive field (Hochstein and Shapley, 1976a,b). Figure 6.3C shows the responses of a linear and a nonlinear cell as a function of the position of the grating. The first harmonic component (F1) is 0 for a certain location of the stimulus. At this minimum, the response phase changes by 180° (in Figure 6.3C this is given by a change of the sign of the response). This stimulus is also used to determine the spatial linearity of the cell because nonlinear cells display frequency-doubled responses (F2>0) when the first harmonic component in the response is 0 (e.g. the cell response on the right of Figure 6.3C).

Bipartite field stimuli are also used to measure receptive field sizes (Enroth-Cugell and Robson, 1966; Kremers and Weiss, 1997; Lee *et al.*, 1998). The stimulus consists of two adjacent fields in each of which a modulation is presented. The modulations in the two fields are identical but have opposite phases. The responses are measured for different locations of stimulus relative to the receptive field. The advantage of this stimulus is that asymmetries in the receptive field structures are easily detected. In Figure 6.4 the response amplitudes (upper panels) and phases (lower panels) of an Off-center PC cell in the marmoset LGN are displayed as a function of the location of the stimulus. Vertical (left panels) and horizontal (right panels) edges were used. Similar data were given by Kremers and Weiss (1997). The curves in the upper panels are the fits of the DoG model to the amplitude data. Similar as with the responses to drifting gratings, the model can describe the data well. It can be seen that the response phase changes by 180° at the minimum of the response amplitude, indicative for antagonism between receptive field center and surround. Indeed the temporal frequency (4 Hz) is low enough to ensure an antagonism. Furthermore, the responses are not always symmetrical (e.g. in Figure 6.4 especially with the horizontal edges) indicating that the receptive field center and surround are not

---

Figure 6.3 *(continued)*
spatial summation. Each of the four panels show distributions of nonlinearity indices (NLI: the ratio of the mean F2 response to the peak F1 response). The distributions of NLI for LGN PC cells ($n = 402$) and MC cells ($n = 194$) are the combined distributions from two studies in macaque (Derrington and Lennie, 1984 – black; Levitt *et al.*, 2001 – white) and one study in marmoset (White *et al.*, 2001 – grey). The NLI distribution for LGN KC cells ($n = 45$) is from marmoset (White *et al.*, 2001), and the distribution for retinal MC cells ($n = 36$) comes from macaque (White *et al.*, 2002)

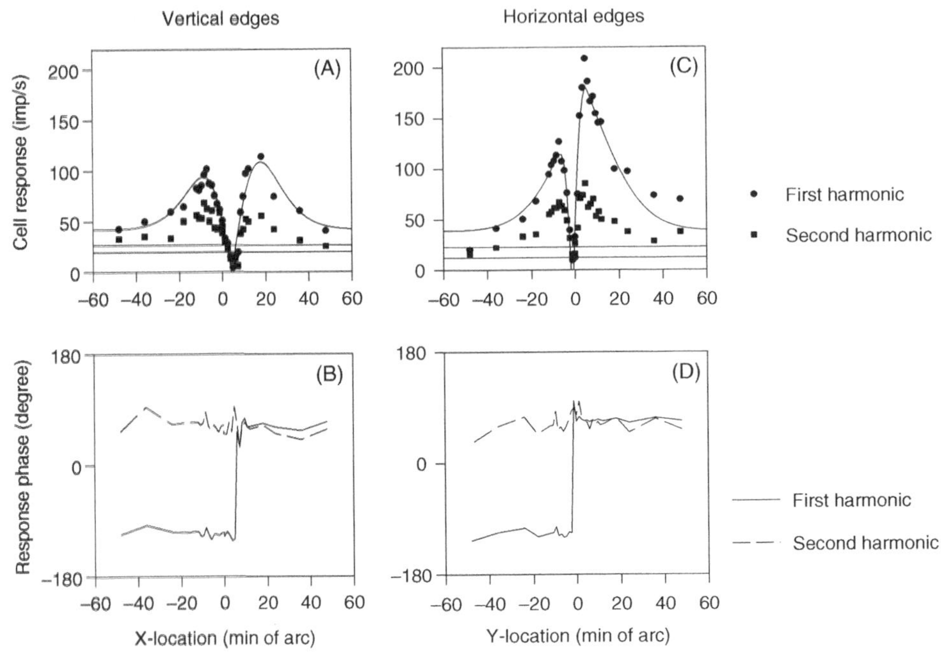

Figure 6.4 The response amplitudes (upper panels) and phases (lower panels) of an OFF-center PC cell in the marmoset LGN to a bipartite field stimulus with the two fields modulating identically (sinusoidal modulation at 4 Hz) but in counterphase. The common edge of this stimulus is positioned at different locations within the receptive field. The responses are displayed as a function of the position of the common edge. The responses are shown when vertical and horizontal common edges are used (left and right plots, respectively). The first harmonic response amplitudes can be described by the DoG model (curves in the upper plots). The receptive field center and surround of this cell are not completely concentric because the response amplitudes to horizontal edges are not symmetrical. The amplitudes and phases of the second harmonic components in the stimulus are also given

completely concentric. The second harmonic components are also displayed. The second harmonic component is approximately proportional to the first harmonic component and is minimal when the first harmonic component is also minimal. This indicates that the cell was spatially linear; the higher harmonics are introduced because the response is not completely sinusoidal. This was found for the vast majority of cells (Kremers and Weiss, 1997) which is in accordance with the hypothesis that in the monkey LGN most cells are X-like and only a few are Y-like. From the measurements to all stimuli described above, the sizes of the centers and the surrounds can be estimated. The estimates of the receptive field center and surround sizes of marmoset and owl monkey LGN cells obtained from measurements using drifting gratings and bipartite field stimuli are shown in Figure 6.5. The two stimulus types yield similar receptive field sizes of LGN cells in different monkey species (Kremers and Weiss, 1997; Usrey and Reid, 2000; Kremers *et al.*, 2001; Kremers, Kilavik, and Silveira, unpublished data).

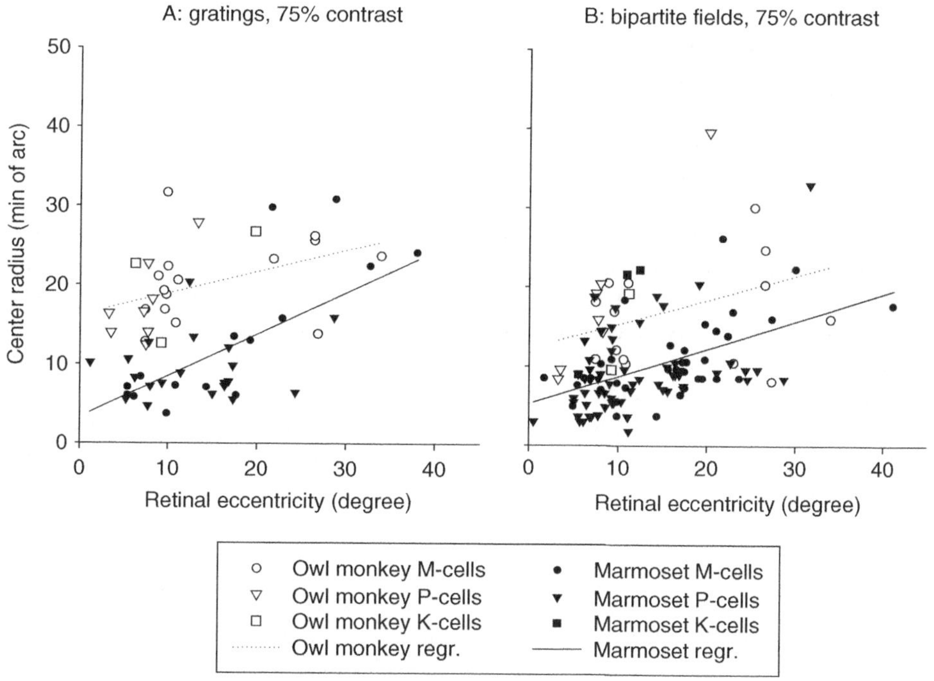

Figure 6.5 Comparison of receptive field center sizes measured in the LGNs of marmosets and owl monkey using drifting gratings (left plots) and bipartite field stimuli (right plots) displayed as a function of retinal eccentricity of the receptive field (Kilavik, Kremers, and Silveira, unpublished data). The contrast in the two stimuli was 75 percent

The receptive field center and surround sizes increase with increasing retinal eccentricity. This is in agreement with anatomical findings that the dendritic tree sizes of retinal ganglion cells increase with retinal eccentricity. In cats, dendritic tree sizes and receptive field center sizes correspond closely (Peichl and Wässle, 1979). In monkeys, however, there is a correlation between the two but the one-to-one relationship is lacking in certain circumstances. This was found for the marmoset (Kremers and Weiss, 1997; White *et al.*, 2001), the owl monkey (Kremers, Kilavik, and Silveira, unpublished data), and the macaque (Wässle and Boycott, 1991). Foveal midget (PC-projecting) cells receive their center input from only a single cone photoreceptor (Chapter 5). However, the receptive field centers of midget cells are larger than predicted by the size of a single cone. It is likely that optical blur is the limiting factor for the receptive field center size rather than dendritic tree sizes or other retinal factors. If that is the case, the question remains what the ecological value might be for a retinal circuit that results in a receptive field center receiving its input from a single cone. A possible answer might be this retinal circuit provides a basis for a chromatic signal in midget cells of trichromatic animals (Lee, 1999). Alternatively, it is possible that even though the spatial acuity of foveal midget cells is decreased by optical blur, the ecological advantage of high-acuity vision makes it imperative to develop a system with maximum resolving capacity, and the 'single cone system' of the midget-PC projecting pathway gives a higher acuity than multiple-cone systems because the array density of PC cells is higher.

In accordance with the anatomy of their retinal inputs, the center sizes of MC cells are larger than those of PC cells. But, the data displayed in Figure 6.5 further show that there is a large variability in the center sizes even at a given eccentricity. As a result, the center sizes of PC, MC, and KC cells overlap substantially. This will be discussed in more detail later.

The receptive fields of a subset of LGN cells lack a surround or cannot be described by a DoG model. This was found for the galago (Irvin *et al.*, 1993), the marmoset (Solomon *et al.*, 1999; Kremers *et al.*, 2001), and the owl monkey (Xu *et al.*, 2001). This group of cells is probably relatively heterogeneous. In some cells, a surround is not present or the surround has the same size as the center. In this case the DoG still can describe the spatial properties of the receptive fields. Others do show a more complex response behavior in that they respond, for example, to flashed stimuli but not to drifting gratings. These cells are difficult to classify and their function in the visual system and for visual perception is not clear yet.

Comparison with retinal ganglion cells

The receptive fields of LGN cells and retinal ganglion cells have comparable sizes, and there is a tight retinotopic organization of retinal afferents to the LGN, suggesting that each LGN cell probably receives dominant input from one or a small number of retinal ganglion cells. This conclusion is in agreement with direct measurements on the input of LGN cells in which it was found that an LGN cell receives the input from only one retinal ganglion cell. Furthermore, an action potential in the LGN cell is nearly always preceded by an action potential in the retinal ganglion cell. However, not every retinal action potential leads to a spike in the LGN cell. The ratio of spikes in the output (the LGN cell) relative to the input (the retinal ganglion cell) is therefore below unity and is generally about 0.5 (Kaplan *et al.*, 1987), although it can vary substantially. It is not expected that this will change the spatial receptive field properties of the LGN cells relative to those of retinal ganglion cells. However, the ratio can depend on the stimulus properties, so that not all properties of the LGN cells are identical with those of the retinal ganglion cells. Direct comparison of S potentials (postsynaptic potentials that reflect the arrival of action potentials from the presynaptic retinal ganglion cells) with the action potentials of the LGN cell shows that, apart from an overall reduction of the number of action potentials, there is little influence of spatial frequency on the transfer ratio (Kaplan *et al.*, 1987), indicating that indeed the spatial properties of retinal ganglion cells and LGN neurons are similar, especially when compared to the transformation that occurs in V1.

Differences between animal species

The center sizes of marmoset LGN neurons are approximately a factor of 1.6 larger than those of macaque retinal ganglion cells and LGN cells (Kremers and Lee, 1998). This difference can be fully explained by differences in eye size. As in the marmoset, a comparison of receptive field sizes of LGN cells in other diurnal primate species (such as the squirrel monkey; Usrey and Reid, 2000) shows that the receptive field sizes are very similar after correction for eye size. In other words, in different species the cells will have similar receptive field sizes when expressed in micrometer on the retina. However, receptive field sizes are expressed in angular units and therefore the receptive field size will be inversely proportional to the eye size. This is consistent with the finding that retinal ganglion cell dendritic tree sizes are fairly similar in different primate species (Chapter 5).

One exception from this scheme is the nocturnal owl monkey (*Aotus*). As can be seen in Figure 6.5, the center sizes of owl monkey LGN cells are larger than those of other monkey species (Usrey and Reid, 2000; Xu *et al.*, 2001; Kremers, Kilavik, and Silveira, unpublished data). This is possibly an adaptation to their nocturnal lifestyle. The larger receptive fields indicate an increased spatial integration of photoreceptor signals in comparison with other monkey species. This might be a mechanism to increase the sensitivity necessary at low-light intensities. The cost would of course be a lower spatial resolution. Apart from differences in the lens, the owl monkey eye is optically similar to those of other monkeys (Detwiler, 1941). Thus, the difference in spatial integration must have a neuronal origin. In agreement with this proposal, center sizes of at least MC cells in the strepsirrhine galago are comparable to those of the owl monkey (Irvin *et al.*, 1993).

Differences between MC, PC, and KC cells

There is some controversy concerning the differences in center sizes between PC and MC cells. Some have reported that retinal ganglion and LGN cells belonging to the PC pathway have smaller centers than the MC cells at corresponding retinal locations and that the two populations were clearly separated (de Monasterio, 1978; Croner and Kaplan, 1995). Others found that in anthropoid and strepsirrhine species, PC cells had on average smaller centers, but there was a considerable overlap in center sizes of the cells belonging to the two cell classes (Norton and Casagrande, 1982; Derrington and Lennie, 1984; Irvin *et al.*, 1993; Kremers and Weiss, 1997; White *et al.*, 2001; Xu *et al.*, 2001). The center sizes of KC cells also overlapped those of PC and MC cells (Norton and Casagrande, 1982; Irvin *et al.*, 1993; White *et al.*, 2001; Xu *et al.*, 2001). The reason for the controversy in the data may in part be caused by a difference in the definition of the cell types. For instance, clear differences in receptive field center sizes were found between X-, Y-, and W-like cells in the LGN of the galago, with Y-like cells having larger centers than the other two cell types (Norton *et al.*, 1988). But as mentioned above, this classification is not equivalent to the classification between PC, MC, and KC cells. Possibly, Y-like MC cells have larger center sizes than X-like MC cells.

Figure 6.2 compares the spatial frequency transfer properties of PC, MC, and KC cells in marmoset. These examples show representative features of each type, but as noted above there is considerable overlap among the PC, MC, and KC populations. The PC cell has the highest peak and cut-off spatial frequency, consistent with a role in high-acuity vision. The MC cell is more responsive than the PC cell over most of the spatial frequency range below the PC cell peak, consistent with a dominant role of this pathway in processing low spatial frequencies. The contrast-response functions also show characteristic differences between these cell pathways. The PC cell response amplitude is roughly proportional to stimulus contrast but the MC cell response shows evidence of saturation for contrasts above 25 percent.

The KC cell has a low-pass spatial transfer characteristic, and a contrast-responsivity function which rapidly saturates at low contrasts. These properties suggest that the majority of the frequency-contrast domain is shared by PC and MC cells, with PC cells specialized for high spatial frequencies and high contrasts, whereas MC cells are specialized for low frequencies and low contrasts. More detail on the contrast-dependent properties of the three pathways is given above.

6.4.3 Temporal properties

Primate LGN cells, as with their retinal afferents, respond well to a wide range of temporal frequencies, though the shape of the tuning curve depends on the spatial structure of the stimulus (Kremers and Lee, 1998). When stimulated with luminance temporal contrast, the receptive field centers and surrounds have temporal response properties that differ from those when the complete receptive field is stimulated (Kilavik *et al.*, 2003). When isolated, the responses of center and surround do not show a clear roll-off at low temporal frequencies (they are largely low-pass). When stimulated together, by using uniform fields, for example, the roll-off seen for full field stimuli is probably caused by a gradual decrease in surround antagonism with increasing temporal frequency, probably due to the time delay between the two mechanisms (Figure 6.6).

Contrast can also modify the response to different temporal frequencies. Nonlinear processes such as response saturation can alter the temporal characteristics. Retinal ganglion cell responses saturate less at high temporal frequencies (Lee *et al.*, 1994). As a result, the temporal frequency for a maximal response increases with increasing stimulus contrast (Figure 6.7). In LGN cells, this effect is less clear because the responses are small

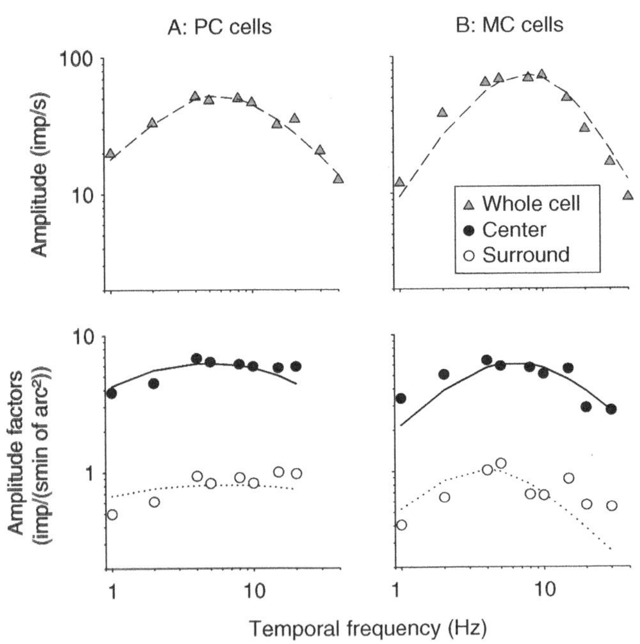

Figure 6.6 Temporal response properties of PC and MC cells after full field stimulation (upper panels) and of the center and surround subfields (lower panels). Data from Kilavik *et al.* (2003) redrawn with kind permission

at those temporal frequencies at which saturation is less strong (Kremers *et al.*, 1997; Solomon *et al.*, 1999).

As temporal frequency increases, the phase of response progressively lags the stimulus, indicative of a fixed time delay in the response of the neuron to changes in the visual stimulus (often called the latency to visual stimulation). Visual latency is obviously longer in the LGN than in the retina (it takes time to pass the information on to the LGN cells) and this progressive phase lag is therefore more pronounced in the LGN (Kremers and Lee, 1998). Other factors also influence the response phase of visual neurons, including the amount of low-pass and high-pass filtering, but the temporal response properties of LGN neurons are generally well captured by nearly linear models (Purpura *et al.*, 1990; Kremers *et al.*, 1997). One exception to this is the observation that LGN neurons operate in two different modes: a tonic mode, which is relatively linear, and a bursty mode, which is much more nonlinear (section 6.5.4). The temporal characteristics strongly depend upon in which mode of operation the cell responds.

Comparison with retinal ganglion cells

The temporal responses of cells in the LGN deviate from those of the retinal ganglion cells. LGN cells have lower optimal temporal frequencies (between 4 and 10 Hz) (Derrington and Lennie, 1984; Kremers *et al.*, 1997; Lankheet *et al.*, 1998; Solomon *et al.*, 1999) than retinal ganglion cells (generally between 20 and 40 Hz) (Lee *et al.*, 1990). It therefore seems that there is a low-pass temporal filter at the synapse between the retinal ganglion cells and the LGN neurons. Direct comparison of the responses of LGN cells with their input confirms this conclusion (Kaplan *et al.*, 1987). These differences between retinal ganglion cells and LGN cells will depend upon the mode of operation (burst or tonic) of the LGN cell.

Differences between animal species

No clear differences in temporal properties have been described with the exception of the owl monkey, where the temporal resolution of LGN and retinal ganglion cells is lower than in other species (Usrey and Reid, 2000; Saito *et al.*, 2001; Xu *et al.*, 2001). This difference is probably related to the stronger rod inputs to owl monkey LGN cells (Saito *et al.*, 2001) which is possibly the result of larger receptive fields and the higher rod density in this species (Chapter 5) and which can be regarded as an adaptation to their nocturnal lifestyle.

Differences between MC, PC, and KC cells

The temporal properties of MC, PC, and KC LGN cells differ. Qualitatively these differences are similar to those found in retinal ganglion cells. MC cell responses are phasic or transient and they are more responsive to high temporal frequencies, whereas PC cells generally respond in a tonic or sustained manner and have lower optimal temporal frequencies. There are indications that the differences between PC and MC LGN cells are less pronounced than in retinal ganglion cells, but they are still substantial (Kremers *et al.*, 1997; Solomon *et al.*, 1999; but see e.g. Usrey and Reid (2000) who found clearer differences).

These differences may be one of the reasons for the division between these pathways. The transient responses of MC cells make them well suited for the detection of temporal changes in the visual scene caused by rapid modulations or motion. When combined with the high sensitivity of the MC pathway to spatial contrast, this suggests that the MC pathway can provide signals for object boundaries, and allow a coarse segmentation of the visual scene. On the other hand, the sustained responses of PC-pathway cells and their sensitivity to fine spatial form could provide a signal for object interiors, and allow the determination of coherent surfaces. Both segmentation and coherence are necessary for object detection and recognition, and thus for early form perception both pathways may be important. Indeed, cortical area V4, which is involved in processing of information on form, receives about equal input from the MC and PC pathways (Ferrera *et al.*, 1992). For other tasks, the brain may extract the necessary information from the subcortical pathway depending on the requirements: those pathways containing more information of the visual aspects of interest may have stronger inputs to the cortical areas processing these aspects. For instance, the transient responses of MC cells contain more information about motion than the sustained responses in the PC pathway. Therefore for motion perception it is more efficient to rely on input from the MC pathway. Consistently, cortical area middle temporal (MT), which is involved in the processing of motion, receives stronger input from the MC pathway than from the PC pathway (Maunsell *et al.*, 1990).

6.4.4 Spectral properties

When more than one class of photoreceptor is present, the retina has the chance to extract information about the wavelength as well as the intensity of the visual image. In primates, at least two classes of ganglion cell exploit this ability. The majority of PC cells of trichromatic animals display color-opponent responses receiving antagonistic input from L and M cones. A small proportion of cells in the LGN of the macaque (Wiesel and Hubel, 1966) and of the trichromatic marmoset (Yeh *et al.*, 1995b; Blessing *et al.*, 2004) lack color-opponent responses. The receptive fields of foveal PC cells receive excitatory input from only one cone (see also Chapter 5), thereby predisposing the receptive field for spectral selectivity. It is not known whether the surround receives its input from exclusively one cone type or a mixture of L and M cone driven signals. No data clearly favor or exclude one of these possibilities, and in both cases where the opponent cone contributes to the surround, substantial color opponency would emerge (Lankheet *et al.*, 1998; Lee *et al.*, 1998). Different primate species can have photoreceptors with different absorption spectra, and many species are polymorphic such that the photopigment complement varies between individuals of the population (as is the case in many New World monkeys; see e.g. Jacobs *et al.*, (1987); Mollon *et al.* (1984); see also Chapters 3 and 4). This variation will confer different spectral sensitivities on cells that nevertheless belong to the same cell class.

Interestingly the L/M cone signal weighting ratio in PC retinal ganglion cells (and probably LGN cells) is close to unity (Smith *et al.*, 1992). This was found also for psychophysical measurements in which detection thresholds, mediated by the chromatic channel, of which the PC pathway is thought be the physiological basis, were determined (Kremers *et al.*, 2000, 2003).

PC cells respond in a temporal band-pass manner to luminance stimuli but show little low-frequency roll-off for isoluminant red-green stimuli (Lankheet *et al.*, 1998). The cause of this change lies in the time delay between the responses of receptive field centers and the surrounds: at low temporal frequencies the center and the surround respond antagonistically. In PC cells the centers and surrounds differ in their spectral sensitivity, and they therefore respond strongly to chromatic stimuli and weakly to luminance stimuli at low temporal frequencies. At high temporal frequencies, the center and surround lose their antagonistic responses and the cells therefore are more responsive to luminance stimuli (Gouras and Zrenner, 1979; Smith *et al.*, 1992).

MC cells also receive input from L and M cones and rods but little if any input from S cones. In general, MC cells sum the signals from the L and M cones, showing no signs of cone opponency. The exception is the sustained inhibitory influence of long-wavelength light that originally gave rise to the 'Type IV' classification of Wiesel and Hubel, and was later shown to be attributable to opponent interactions in the surround of MC cells (Smith *et al.*, 1992). At low illuminance, both MC and PC cells receive rod-driven signals in addition to the cone inputs (Lee *et al.*, 1996; Weiss *et al.*, 1998). KC cells are heterogeneous in their spectral properties. A subpopulation of the KC cells receive strong excitatory S cone input. These cells receive also inhibitory signals from the L and the M cones (Martin *et al.*, 1997; Martin, 1998). The remaining KC cells probably do not display color opponency.

Differences between animal species

We have described the spectral properties of PC cells in trichromatic species with a wide separation between the absorption spectra of L and M cones. In dichromatic animals, the PC and MC cells receive only input from one cone type (Yeh *et al.*, 1995b; Weiss *et al.*, 1998). As a result the PC cells do not display color opponency. The question arises if PC cells of trichromatic animals, in which the absorption spectra of the L and M cones strongly overlap, as can be the case in New World monkey species, also display some degree of color opponency. Recent data (Blessing *et al.*, 2004) suggest that in such animals the strength of red-green opponent signals in PC cells is primarily determined by the spectral separation of the cone pigments.

6.5 Nonlinear response properties of LGN cells

Until now we have discussed the linear response properties of LGN cells and simple nonlinearities such as response saturation. Properties of a linear system are as follows: the responses are additive, which means that the response to the addition of two stimuli equals the addition of the responses to each stimulus; the response to a sinusoidal stimulus is also sinusoidal with the same period (or frequency) as the stimulus; the response phase is constant for all stimulus intensities. As a consequence, the responses of linear systems have also other properties. For example, multiplying stimulus contrast by a certain factor should multiply the response by the same factor. This is the principle of proportionality. A very important consequence of linear systems is that it is possible to predict the response to all types of stimuli if the modulation transfer function (MTF) of the system is known. The MTF describes the amplitude and the phase of the response to sinusoidal stimuli as a function of the spatial or temporal frequency. Any stimulus can be described as a combination of sine waves (called the Fourier series). Because a linear system is additive, the response to the series of sine waves equals the sum of responses to the separate sine wave stimuli which can be derived from the MTF. If the system fails to show any of these properties, it cannot be considered to be linear. However, even if the system is not perfectly linear, there can be stimulus ranges in which the system behaves like a linear system. It is then described as quasi-linear within this stimulus range. Using the above definition of a linear system, saturation is a nonlinearity because the requirement of proportionality is violated: increasing the stimulus intensity leads to a less than proportional increase in the response.

In this section we will describe more complex response properties which involve overt nonlinearities and their effects on the response amplitudes and phases.

6.5.1 Spatial nonlinearities

The response to counterphase modulating sinusoidal gratings depends upon the position of the grating with respect to the receptive field. A linear cell should not respond when luminance increases in one part of the receptive field are completely counterbalanced by luminance decreases in other parts of the receptive field. As the position of the sinusoidal grating is stepped over the receptive field, the balance of luminance increments to luminance decrements changes. The maximum imbalance (and therefore the maximum response) occurs when the grating is aligned with peak sensitivity of the receptive field. Overall the response depends sinusoidally upon the spatial position of the grating (Figure 6.3). At any one spatial position, a linear neuron will respond to either the withdrawal or presentation of the grating, but not to both.

In the cat, many cells display a second harmonic component in the response. That is, they respond to both the withdrawal and presentation of the grating at many or all spatial positions. The strength of the nonlinearity is used for cell classification schemes in the cat (see preceding text). Y-cells show a clear frequency doubling but X-cells do not. This criterion has also been used to classify primate LGN cells, but is little used today. Two factors have led to this classification scheme being discarded. First, few primate LGN cells show strong frequency doubling (between 5 and 20 percent of cells in the MC layers). Second, there is no clear dichotomy between those MC cells that show frequency doubling and those that do (Figure 6.3), rather the responses of MC cells to counterphase gratings seem to form a continuum (Kaplan and Shapley, 1982; Derrington and Lennie, 1984; Benardete *et al.*, 1992).

6.5.2 Contrast gain control

For a linear cell, an increase in contrast results in a proportional increase of the response independent of the temporal frequency. In other words, the MTF (i.e. the curve describing the response as a function of temporal frequency) does not change if contrast is changed. In reality, many cells show a systematic change in the MTF when the contrast of the stimulus is altered, including an increase of response phase (or a decrease in phase lag) with increasing contrasts at intermediate temporal frequencies. This nonlinearity, called the contrast gain control, was first described for cats (Shapley and Victor, 1978, 1979, 1981) and it is also present in the responses of primate retinal ganglion cells belonging to the MC pathway (Yeh *et al.*, 1995a; Benardete and Kaplan, 1999). The presence of the contrast gain control in the responses of retinal ganglion cells means that it should also be apparent in the responses of LGN cells. This seems to be the case (Kremers *et al.*, 1997; Kremers and Lee, 1998; Solomon *et al.*, 1999; Figure 6.7). Contrast-dependent phase

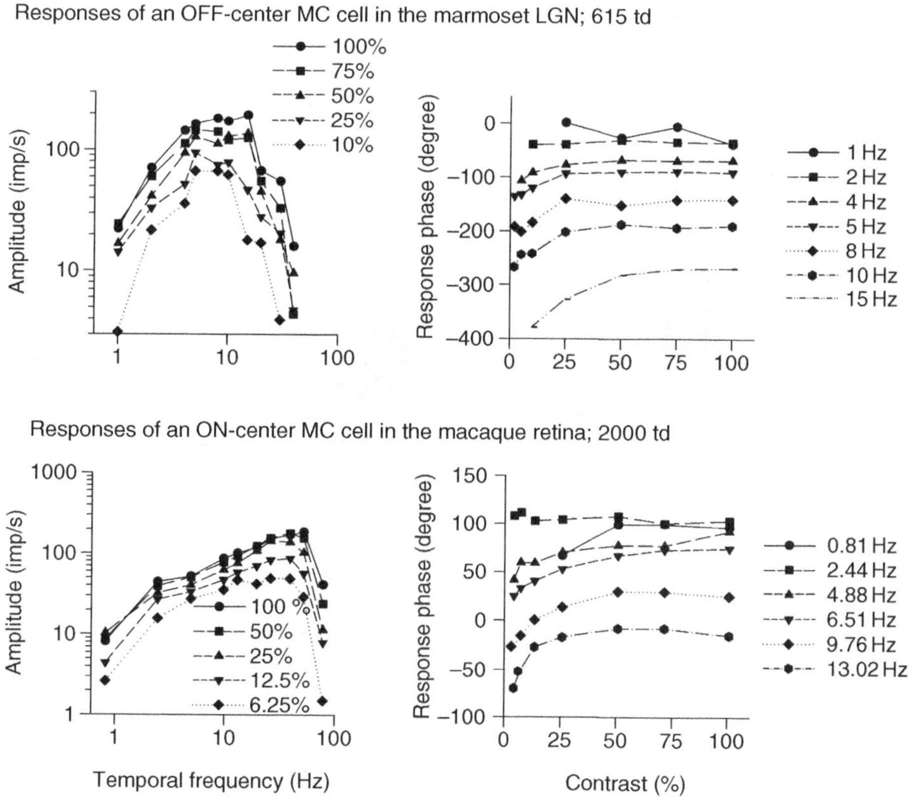

Figure 6.7 Response amplitudes of MC cells in the marmoset LGN (upper panels) and the macaque retina (lower panels) as a function of temporal frequency (left panels). The response phase at different temporal frequencies are displayed as a function of contrast in the right panels. Clearly, the phases increase with increasing contrast. Modified from Kremers and Lee (1998)

changes are even more overt when drifting gratings are used. These response phase changes were found to be present in the responses of MC and PC cells of marmosets (Kremers *et al.*, 2001) and owl monkeys (Kilavik *et al.*, 2001). It is not clear, however, whether this phase alteration is caused by the same contrast gain control mechanism or whether additional nonlinearities are involved.

6.5.3 Nonlinear interactions between receptive field center and surround

The DoG model assumes that the receptive field center and surround are independent so that the responses of the two are subtracted at every instant. The two responses do not interact with each other. Recent data, however, show that this assumption is not completely correct. The presence of a stimulus in the receptive field center can alter the responses to stimuli presented in the receptive field surround (Kremers *et al.*, 2004). When

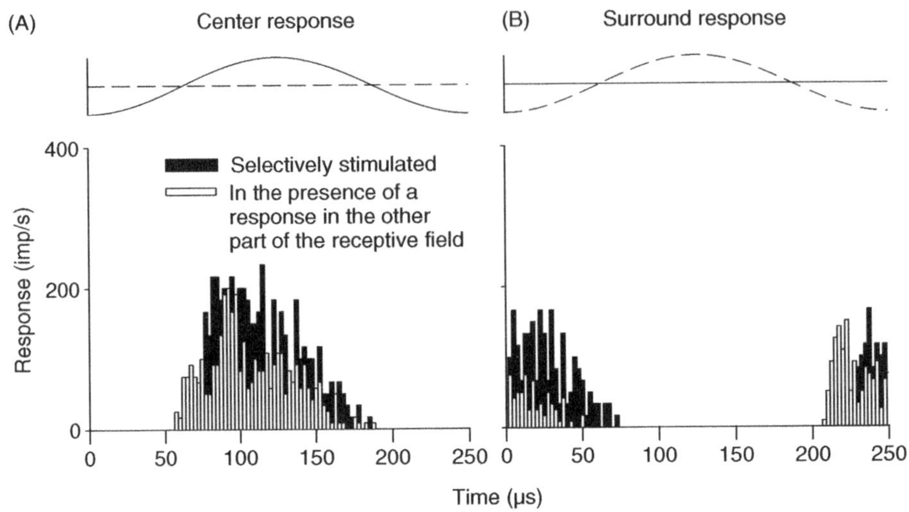

Figure 6.8 (A) Responses of the receptive field center of an ON-center MC cell in the marmoset LGN when selectively stimulated (black histograms) and when the surround is also stimulated (open histograms). Apparently the presence or absence of a response in the receptive field surround does hardly influence the response of the receptive field center. (B) The responses of the receptive field surround in the absence (closed histograms) or presence (open histograms) of a response in the center. Clearly the surround response is more phase-advanced and smaller in the presence of a center response. The change in response phase was consistently found in LGN cells. The change in response amplitude could be different in different cells

the receptive field center is active, the response of the surround is more phase-advanced and can have a different amplitude than in the absence of a center response (Figure 6.8). Thus, centers and surrounds cannot be considered to be completely independent as assumed by the DoG model. The reverse seems not to be the case: the center response does not depend on the presence or absence of a surround response. These nonlinear interactions between center and surround are contrast-dependent. Similar interactions were found in the responses of cat retinal ganglion cells (Shapley and Victor, 1979). Victor and Shapley proposed that this nonlinear interaction is related to the contrast gain mechanism. It is difficult to give a physiological basis for these nonlinearities.

It is also difficult to speculate upon the ecological function of this nonlinearity or whether it has a function at all. But the presence of a center response in general results in an increased antagonism between center and surround, possibly as a result of a synchronization of the center and surround responses. The spatial antagonism is important in the detection of, for example, edges and this might be important for form perception (see preceding text).

6.5.4 Bursty and tonic modes

LGN neurons share some properties with cells in other thalamic nuclei. One of these properties is that they operate in two different modes: tonic and burst. Excellent overviews are given by Sherman and Guillery (1996, 2001). Briefly, the mode in which a cell is operating depends upon the state of polarization of the cell. If the cell is normally polarized, then the cell is in a tonic mode. In this mode a spike is generated in the normal manner, i.e. after a depolarization beyond the threshold for the neuron. In the tonic mode, the response of the LGN cell is more or less proportional to its input faithfully forwarding its retinal afferent. If the cell is in a hyperpolarized state, under normal circumstances no spikes are generated. But then the cell changes its mode of operation (called burst mode) because a particular class of channels in the membrane of the neuron (T channels) are activated. After a small depolarization, these channels may give rise to a calcium spike, depolarizing the cell strongly so that in the wake of the calcium spike a train of normal spikes is generated. The relationship between input and output is less proportional and more nonlinear when the cell is in this burst mode.

The temporal response properties therefore strongly depend upon the mode of operation of the cell. To date there are no data available on the two modes and the temporal response properties of monkey LGN cells. But data from cat LGN cells show that the dependencies of the response amplitudes and phases on temporal frequency are more similar to those of the retinal inputs (recorded with S potentials) when the cells are in tonic mode (Mukherjee and Kaplan, 1995). Cells in the tonic mode have an optimal temporal frequency above 10 Hz and the responses show a response delay relative to their retinal input. On the other hand, cells in the burst mode have optimal temporal frequencies between 4 and 10 Hz and the response delay relative to the retinal input is additionally increased in comparison with the responses in tonic mode. The responses can be delayed relative to their retinal input up to 30 ms when in burst mode, whereas the delay is less than 15 ms when in tonic mode. However, in the burst mode, the LGN cells' response is phase-advanced in comparison with their retinal input at low temporal frequency, suggesting temporal filtering changes in addition to an increased response delay.

Although no direct measurements are available from monkey LGN cells, it has been suggested that monkey LGN cells also operate in two modes (Mukherjee and Kaplan, 1995). Furthermore, the mode of operation will probably have similar influence on the temporal response properties. As mentioned above, LGN cells in different monkey species generally have an optimal temporal frequency of between 4 and 8 Hz, which is lower than the one of retinal ganglion cells. This suggests that in the majority of experimental conditions, the cells are in a bursty mode. However, the response phase characteristics of monkey LGN cells are different from those of bursty cells in the cat LGN: at low temporal frequencies, the responses of the majority of cells do not display a strong phase advance. Furthermore, the response delay relative to the stimulus is about 40–50 ms (recalculated from the data of Kremers *et al.*, 1997). Taking into account a response delay in the retinal ganglion cells of about 25 ms (recalculated from Lee *et al.*, 1990), the delay of the LGN cells relative to the retinal ganglion cells is smaller than the estimates for burst cells and more in line with the cells being in tonic mode.

The ecological value of the two modes of operation could be that the cells transfer the visual information more reliably when they are in a tonic mode (Sherman and Guillery, 1996, 2001). Thus when the retinal input is used for cortical processing then the LGN neurons are in a tonic mode and the response of the LGN cells reliably reflects the response of the retina. However, the responses of many LGN cells are redundant, and may provide too much spiking input to visual cortex if active at all times; it is more efficient if information is transmitted only when it is of interest. If those LGN cells which are not involved in the transmission of 'interesting' information are hyperpolarized (in burst mode), they will not transmit any information. Under certain circumstances stimuli may induce a burst of spikes and the cortex may 'use' these bursts to shift its 'attention' to the part of the visual field in which the stimulus occurred. The cells within this part of the visual field then may change their state into a tonic mode and thus transmit more reliable information. Such a system may protect the cortex from an 'overflow' without a loss of important information.

6.5.5 Extraclassical surrounds

This chapter so far has described the classical receptive field (CRF) properties of LGN cells. However, it has long been recognized that in the visual system the responses to CRF stimulation are subject to influences from surrounding regions of visual space. Stimuli presented in these regions (which we call the extraclassical receptive field or ECRF) generally are not capable of directly eliciting a response from the cell, but can modulate the responses to stimuli presented in the CRF (see Figure 6.9). Thus, although the stimuli in the ECRF are visual, they are not 'drivers' but 'modulators' as defined in the beginning of this chapter. In cortical neurons the ECRF can exert both facilitatory and suppressive influences on the CRF, and these effects are usually dependent on the relative orientation of stimuli in the CRF and ECRF (Allman *et al.*, 1985). By contrast the ECRF effects manifest at subcortical levels are usually suppressive and orientation-independent.

One manifestation of an ECRF, termed the 'suppressive field', has been consistently demonstrated in relay cells within the mammalian LGN (Singer and Creutzfeldt, 1970; Levick *et al.*, 1972; McClurkin and Marrocco, 1984; Marrocco and McClurkin, 1985; Felisberti and Derrington, 2001). The suppressive field was originally attributed to a surround-like mechanism (i.e. with a large summation area), but a more recent study (Solomon *et al.*, 2002) showed that the suppressive ECRF in primate LGN exerts its effects at spatial frequencies above those to which the CRF surround is responsive. It is not clear whether the suppressive ECRF arises from feedforward inhibition at the level of the LGN or feedback action from the visual cortex. Webb *et al.* (2002) provide evidence from the marmoset that ablating visual cortex reduces the effectiveness of the ECRF seen in the LGN. However, they, and studies in the macaque (Przybyszewski *et al.*, 2000), also show that ablating visual cortex reduces the response evoked from the classical receptive field alone. Webb *et al.* (2002) suggest that stimulation of regions around the CRF blocks excitatory feedback from visual cortex to LGN, leading to the appearance of suppressive fields, although the mechanism for this is unclear. It is possible that the LGN ECRF reflects multiple mechanisms – some inherited from the retina, some arising from interactions intrinsic to the LGN, and some arising from feedback to the LGN from visual cortex or the reticular thalamus (a potent inhibitor of neurons in the cat LGN). Regardless of mechanism, the ECRF in primate LGN is relatively insensitive to spatial structure (orientation, spatial frequency), unlike the situation in the cat LGN (Sillito *et al.*, 1993; Cudeiro and Sillito, 1996), which is suggestive of a general role in the gating of retinal signals.

A second manifestation of an ECRF at subcortical levels is elicited by abrupt movement of remote targets, which causes transient inhibition of ongoing activity (the 'shift effect';

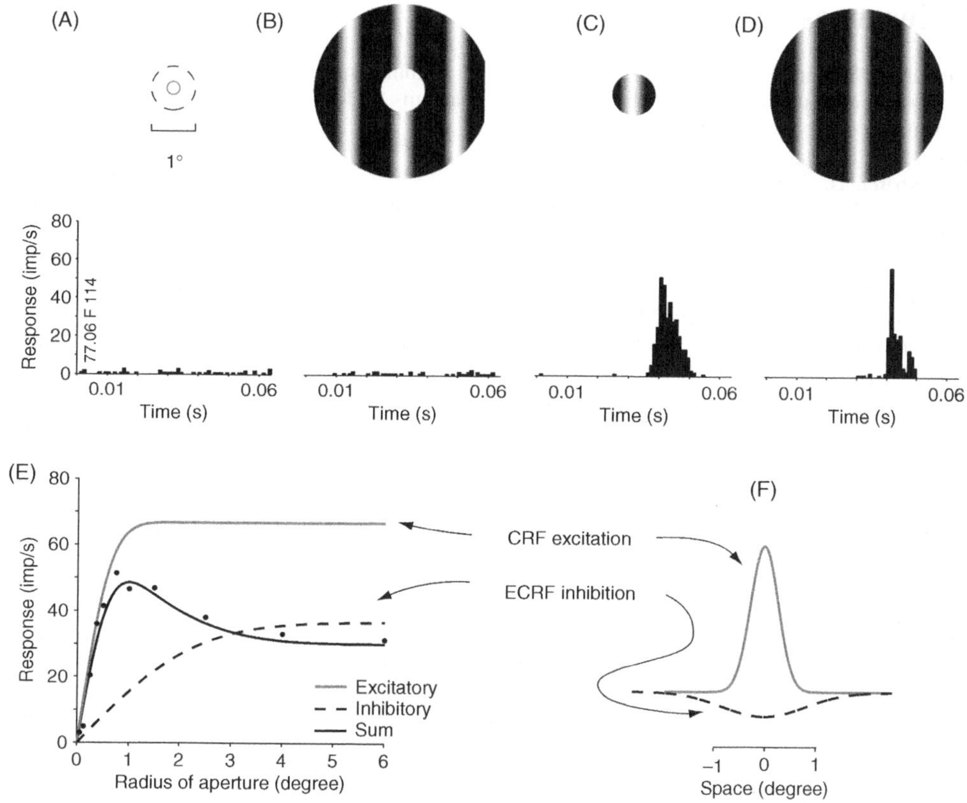

Figure 6.9 (A–D) Responses of a cell in the marmoset LGN to different stimuli: spontaneous activity (A: no stimulation). (B) The response to a drifting grating in an annulus. The central aperture is larger than the classical receptive field (CRF) so that the cell does not respond to the stimulus. (C) Response to the drifting grating in the aperture. (D) Response to the combined stimulation. The response is clearly smaller as in C. Thus, the presence of the stimulus in the annulus inhibits the response to the stimulus within the CRF, although it does not elicit a direct response. (E) The response amplitude (closed circles) to a stimulus with a fixed spatial frequency (similar to the one shown in C) as a function of the size of the stimulus. The response amplitude increases when the stimulus covers larger parts of the CRF. For stimulus sizes beyond the CRF, the response amplitude decreases again, owing to the action of the inhibitory ECRF. The black drawn curve is the best fit of a model that assumes that the CRF and ECRF have Gaussian sensitivity profiles (as sketched in F). Modified from Solomon *et al.* (2002)

McIlwain, 1966; Cleland *et al.*, 1971). The shift effect has been consistently demonstrated in all cell classes studied in marmoset LGN (Felisberti and Derrington, 2001), and in primates is likely to be inherited from the retinal ganglion cells, where the effect was first discovered in recordings from cat retina. It is thought to be involved in saccadic suppression (the suppression of vision during saccades). It is not clear whether the shift effect and the suppressive surround have the same retinal origin.

References

Allman, J., Miezin, F.M., and McGuinness, E., 1985. Stimulus specific responses from beyond the classical receptive field: Neurophysiological mechanisms for local–global comparisons in visual neurons. *Annu. Rev. Neurosci.* **8**, 407–430.

Benardete, E.A. and Kaplan, E., 1999. The dynamics of primate M retinal ganglion cells. *Vis. Neurosci.* **16**, 355–368.

Benardete, E.A., Kaplan, E., and Knight, B.W., 1992. Contrast gain control in the primate retina: P cells are not X-like, some M cells are. *Vis. Neurosci.* **8**, 483–486.

Blessing, E.M., Solomon, S.G., Hashemi-Nezhad, M. *et al.*, 2004. Chromatic and spatial properties of parvocellular cells in the lateral geniculate nucleus of the marmoset (*Callithrix jacchus*). *J. Physiol. (Lond.)* **557**, 229–245.

Cleland, B.G., Dubin, M.W., and Levick, W.R., 1971. Sustained and transient neurons in the cat's retina and lateral geniculate nucleus. *J. Physiol.* **217**, 473–496.

Creutzfeldt, O.D., 1993. *Cortex Cerebri*, Springer-Verlag, Berlin, Heidelberg.

Croner, L.J. and Kaplan, E., 1995. Receptive fields of P and M ganglion cells across the primate retina. *Vision Res.* **35**, 7–24.

Cudeiro, J. and Sillito, A.M., 1996. Spatial frequency tuning of orientation-discontinuity-sensitive cortifugal feedback to the cat lateral geniculate nucleus. *J. Physiol. (Lond.)* **490**, 481–492.

Dacey, D.M. and Lee, B.B., 1994. The 'blue-on' opponent pathway in primate retina originates from a distinct bistratified ganglion cell type. *Nature* **367**, 731–735.

de Monasterio, F.M., 1978. Properties of concentrically organization X and Y ganglion cells of macaque retina. *J. Neurophysiol.* **41**, 1394–1417.

Derrington, A.M. and Lennie, P., 1984. Spatial and temporal contrast sensitivities of neurones in lateral geniculate nucleus of macaque. *J. Physiol.* **357**, 219–240.

Detwiler, S.R., 1941. The eye of the owl monkey (*Nyctipithecus*). *Anat. Rec.* **80**, 233–241.

Enroth-Cugell, C. and Robson, J.G., 1966. The contrast sensitivity of retinal ganglion cells of the cat. *J. Physiol.* **187**, 517–552.

Enroth-Cugell, C., Robson, J.G., Schweitzer-Tong, D.E. *et al.*, 1983. Spatio-temporal interactions in cat retinal ganglion cells showing linear spatial summation. *J. Physiol.* **341**, 279–307.

Felisberti, F. and Derrington, A.M., 2001. Long-range interactions in the lateral geniculate nucleus of the New-World monkey, *Callithrix jacchus*. *Vis. Neurosci.* **18**, 209–218.

Ferrera, V.P., Nealey, T.A., and Maunsell, J.H.R., 1992. Mixed parvocellular and magnocellular geniculate signals in visual area V4. *Nature* **358**, 756–758.

Frishman, L.J., Freeman, A.W., Troy, J.B. *et al.*, 1987. Spatiotemporal frequency responses of cat retinal ganglion cells. *J. Gen. Physiol.* **89**, 599–628.

Gouras, P. and Zrenner, E., 1979. Enhancement of luminance flicker by color-opponent mechanisms. *Science* **205**, 587–589.

Hochstein, S. and Shapley, R.M., 1976a. Linear and nonlinear spatial subunits in Y cat retinal ganglion cells. *J. Physiol. (Lond.)* **262**, 265–284.

Hochstein, S. and Shapley, R.M., 1976b. Quantitative analysis of retinal ganglion cell classifications. *J. Physiol.* **262**, 237–264.

Irvin, G.E., Norton, T.T., Sesma, M.A. *et al.*, 1986. W-like response properties of interlaminar zone cells in the lateral geniculate nucleus of a primate (*Galago crassicaudatus*). *Brain Res.* **362**, 254–270.

Irvin, G.E., Casagrande, V.A., and Norton, T.T., 1993. Center/surround relationships of magnocellular, parvocellular, and koniocellular relay cells in primate lateral geniculate nucleus. *Vis. Neurosci.* **10**, 363–373.

Jacobs, G.H., Neitz, J., and Crognale, M., 1987. Color vision polymorphism and its photopigment basis in a callitrichid monkey (*Saguinus fuscicollis*). *Vision Res.* **27**, 2089–2100.

Kaas, J.H. and Preuss, T.M., 2003. Human brain evolution, Chapter 45, in *Fundamental Neuroscience* (ed L.R. Squire), 2nd edn, Academic Press, San Diego, pp. 1147–1166.

Kaas, J.H., Guillery, R.W., and Allman, J.M., 1972. Some principles of organization in the dorsal lateral geniculate nucleus. *Brain Behav. Evol.* **6**, 253–299.

Kaas, J.H., Huerta, M.F., Weber, J.T. *et al.*, 1978. Patterns of retinal terminations and laminar organization of the lateral geniculate nucleus of primates. *J. Comp. Neurol.* **182**, 517–554.

Kaplan, E. and Shapley, R.M., 1982. X and Y cells in the lateral geniculate nucleus of macaque monkeys. *J. Physiol.* **330**, 125–143.

Kaplan, E., Purpura, K., and Shapley, R.M., 1987. Contrast affects the transmission of visual information through the mammalian lateral geniculate nucleus. *J. Physiol.* **391**, 267–288.

Kilavik, B.E., Kremers, J., and Silveira, L.C.L., 2001. Spatial receptive field properties of lateral geniculate cells in the owl monkey (*Aotus azarae*). *Invest. Ophthalmol. Vis. Sci.* **42** (Suppl), 406 (Abstract).

Kilavik, B.E., Silveira, L.C.L., and Kremers, J., 2003. Center and surround responses of marmoset lateral geniculate neurones at different temporal frequencies. *J. Physiol. (Lond.)* **546.3**, 903–919.

Kremers, J. and Lee, B.B., 1998. Comparative retinal physiology in Anthropoids. *Vision Res.* **38**, 3339–3344.

Kremers, J. and Weiss, S., 1997. Receptive field dimensions of lateral geniculate cells in the common marmoset (*Callithrix jacchus*). *Vision Res.* **37**, 2171–2181.

Kremers, J., Weiss, S., and Zrenner, E., 1997. Temporal properties of marmoset lateral geniculate cells. *Vision Res.* **37**, 2649–2660.

Kremers, J., Scholl, H.P.N., Knau, H. *et al.*, 2000. L- and M-cone ratios in human trichromats assessed by psychophysics, electroretinography and retinal densitometry. *J. Opt. Soc. Am. A* **17**, 517–526.

Kremers, J., Silveira, L.C.L., and Kilavik, B.E., 2001. Influence of contrast on the responses of marmoset lateral geniculate cells to drifting gratings. *J. Neurophysiol.* **85**, 235–246.

Kremers, J., Stepien, M.W., Scholl, H.P.N. *et al.*, 2003. Cone selective adaptation influences L- and M-cone driven signals in electroretinography and psychophysics. *J. Vis.* **3**, 146–160.

Kremers, J., Kozyrev, V., Silveira, L.C.L. *et al.*, 2004. Lateral interactions in the perception of flicker and in the physiology of the lateral geniculate nucleus. *J. Vis.* **4**, 643–663.

Kuffler, S.W., 1953. Discharge patterns and functional organization of mammalian retina. *J. Neurophysiol.* **16**, 37–68.

Lankheet, M.J.M., Lennie, P., and Krauskopf, J., 1998. Temporal-chromatic interactions in LGN P-cells. *Vis. Neurosci.* **15**, 47–54.

Lee, B.B., 1999. Receptor inputs to primate ganglion cells, in *Color Vision: From Genes to Perception* (eds K.R. Gegenfurtner and L.T. Sharpe), Cambridge University Press, Cambridge, New York, Melbourne, Madrid, pp. 203–217.

Lee, B.B., Pokorny, J., Smith, V.C. *et al.*, 1990. Luminance and chromatic modulation sensitivity of macaque ganglion cells and human observers. *J. Opt. Soc. Am. A* **7**, 2223–2236.

Lee, B.B., Pokorny, J., Smith, V.C. *et al.*, 1994. Responses to pulses and sinusoids in macaque ganglion cells. *Vision Res.* **34**, 3081–3095.

Lee, B.B., Smith, V.C., Pokorny, J. *et al.*, 1996. Rod inputs to macaque ganglion cells and their temporal dynamics. *Invest. Ophthalmol. Vis. Sci.* **36** (Suppl), 689 (Abstract).

Lee, B.B., Kremers, J., and Yeh, T., 1998. Receptive fields of primate retinal cells studied with a novel technique. *Vis. Neurosci.* **15**, 161–175.

Levick, W.R., Cleland, B.G., and Dubin, M.W., 1972. Lateral geniculate neurons of cat: Retinal inputs and physiology. *Invest. Ophthalmol.* **11**, 302–311.

Levitt, J.B., Schumer, R.A., Sherman, S.M. *et al.*, 2001. Visual response properties of neurons in the LGN of normally reared and visually deprived macaque monkeys. *J. Neurophysiol.* **85**, 2111–2129.

Marrocco, R.T. and McClurkin, J.W., 1985. Evidence for spatial structure in the cortical input to the monkey lateral geniculate nucleus. *Exp. Brain Res.* **59**, 50–56.

Martin, P.R., 1998. Colour processing in the primate retina: Recent progress. *J. Physiol. (Lond.)* **513.3**, 631–638.

Martin, P.R., White, A.J.R., Goodchild, A.K. *et al.*, 1997. Evidence that blue-on cells are part of the third genicolocortical pathway in primates. *Eur. J. Neurosci.* **9**, 1536–1541.

Maunsell, J.H.R., Nealey, T.A., and DePriest, D.D., 1990. Magnocellular and parvocellular contributions to responses in the middle temporal visual area (MT) of the macaque monkey. *J. Neurosci.* **10**, 3323–3334.

McClurkin, J.W. and Marrocco, R.T., 1984. Visual cortical input alters spatial tuning in monkey lateral geniculate nucleus cells. *J. Physiol. (Lond.)* **348**, 135–152.

McIlwain, J.T., 1966. Some evidence concerning the periphery effect in cat's retina. *Exp. Brain Res.* **1**, 265–271.

Mollon, J.D., Bowmaker, J.K., and Jacobs, G.H., 1984. Variations of colour vision in a new world primate can be explained by polymorphism of retinal photopigments. *Proc. R. Soc. B* **222**, 373–399.

Mukherjee, P. and Kaplan, E., 1995. Dynamics of neurons in the cat lateral geniculate nucleus: In vivo electrophysiology and computational modeling. *J. Neurophysiol.* **74**, 1222–1243.

Norton, T.T. and Casagrande, V.A., 1982. Laminar organization of receptive-field properties in lateral geniculate nucleus of bush baby (*Galago crassicaudatus*). *J. Neurophysiol.* **47**, 715–741.

Norton, T.T., Casagrande, V.A., Irvin, G.E. *et al.*, 1988. Contrast-sensitivity functions of W-, X-, and Y-like relay cells in the lateral geniculate nucleus of bush baby, *Galago crassicaudatus*. *J. Neurophysiol.* **59**, 1639–1655.

Peichl, L. and Wässle, H., 1979. Size, scatter and coverage of ganglion cell receptive field centres in the cat retina. *J. Physiol.* **291**, 117–141.

Przybyszewski, A.W., Gaska, J.P., Foote, W. *et al.*, 2000. Striate cortex increases contrast gain of macaque LGN neurons. *Vis. Neurosci.* **17**, 485–494.

Purpura, K., Tranchina, D., Kaplan, E., and Shapley, R.M., 1990. Light adaptation in the primate retina: Analysis of changes in gain and dynamics of monkey retinal ganglion cells. *Vis. Neurosci.* **4**, 75–93.

Rodieck, R.W. and Stone, J., 1965a. Analysis of receptive fields of cat retinal ganglion cells. *J. Neurophysiol.* **28**, 833–849.

Rodieck, R.W. and Stone, J., 1965b. Analysis of receptive fields of cat retinal ganglion cells to moving visual patterns. *J. Neurophysiol.* **28**, 819–832.

Rodieck, R.W. and Watanabe, M., 1993. Survey of the morphology of macaque retinal ganglion cells that project to the pretectum, superior colliculus, and parvicellular laminae of the lateral geniculate nucleus. *J. Comp. Neurol.* **338**, 289–303.

Rodieck, R.W., Brening, R.K., and Watanabe, M., 1993. The origin of parallel visual pathways, in *Contrast Sensitivity* (eds R. Shapley and D.M.K. Lam), MIT Press, Cambridge, Massachusetts, pp. 117–144.

Saito, C.A., Lee, B.B., Kremers, J. *et al.*, 2001. Rod-cone interactions in the ganglion cell response: Studies using the diurnal capuchin-monkey and the nocturnal owl-monkey. *Invest. Ophthalmol. Vis. Sci.* **42** (Suppl), 676 (Abstract).

Shapley, R.M. and Victor, J.D., 1978. The effect of contrast on the transfer properties of cat retinal ganglion cells. *J. Physiol.* **285**, 299–310.

Shapley, R.M. and Victor, J.D., 1979. Nonlinear spatial summation and the contrast gain control of cat retinal ganglion cells. *J. Physiol.* **290**, 141–161.

Shapley, R.M. and Victor, J.D., 1981. How the contrast gain control modifies the frequency responses of cat retinal ganglion cells. *J. Physiol.* **318**, 162–179.

Sherman, S.M. and Guillery, R.W., 1996. Functional organization of thalamocortical relays. *J. Neurophysiol.* **76**, 1367–1395.

Sherman, S.M. and Guillery, R.W., 1998. On the actions that one nerve cell can have on another: Distinguishing "drivers" from "modulators". *Proc. Natl. Acad. Sci. USA* **95**, 7121–7126.

Sherman, S.M. and Guillery, R.W., 2001. *Exploring the Thalamus*. Academic Press, San Diego.

Sillito, A.M., Cudeiro, J., and Murphy, P.C., 1993. Orientation sensitive elements in the corticofugal influence on centre-surround interactions in the dorsal lateral geniculate nucleus. *Exp. Brain Res.* **93**, 6–16.

Singer, W. and Creutzfeldt, O.D., 1970. Reciprocal lateral inhibition of on- and off-center neurones in the lateral geniculate body of the cat. *Exp. Brain Res.* **10**, 311–330.

Smith, V.C., Lee, B.B., Pokorny, J. *et al.*, 1992. Responses of macaque ganglion cells to the relative phase of heterochromatically modulated lights. *J. Physiol.* **458**, 191–221.

Solomon, S.G., White, A.J.R., and Martin, P.R., 1999. Temporal contrast sensitivity in the lateral geniculate nucleus of a New World monkey, the marmoset *Callithrix jacchus*. *J. Physiol.* **517**, 907–917.

Solomon, S.G., White, A.J.R., and Martin, P.R., 2002. Exraclassical receptive field properties of parvocellular, magnocellular and koniocellular cells in the primate lateral geniculate nucleus. *J. Neurosci.* **22**, 338–349.

Soodak, R.E., Shapley, R.M., and Kaplan, E., 1987. Linear mechanism of orientation tuning in the retina and lateral geniculate nucleus of the cat. *J. Neurophysiol.* **58**, 267–275.

Usrey, W.M. and Reid, R.C., 2000. Visual physiology of the lateral geniculate nucleus in two species of New World monkeys: *Saimiri sciureus* and *Aotus trivirgatis*. *J. Physiol. (Lond.)* **523**, 755–769.

Wässle, H. and Boycott, B.B., 1991. Functional architecture of the mammalian retina. *Physiol. Rev.* **71**, 447–480.

Webb, B.S., Tinsley, C.J., Barraclough, N.E. *et al.*, 2002. Feedback from V1 and inhibition from beyond the classical receptive field modulates the responses of neurons in the primate lateral geniculate nucleus. *Vis. Neurosci.* **19**, 583–592.

Weiss, S., Kremers, J., and Maurer, J., 1998. Interaction between rod and cone signals in responses of lateral geniculate neurons in dichromatic marmosets (*Callithrix jacchus*). *Vis. Neurosci.* **15**, 931–943.

White, A.J.R., Solomon, S.G., and Martin, P.R., 2001. Spatial properties of koniocellular cells in the lateral geniculate of the marmoset *Callithrix jacchus*. *J. Physiol.* **533.2**, 519–535.

White, A.J.R., Sun, H., Swanson, W.H. *et al.*, 2002. An examination of physiological mechanisms underlying the frequency-doubling illusion. *Invest. Ophthalmol. Vis. Sci.* **43**, 3590–3599.

Wiesel, T.N. and Hubel, D.H., 1966. Spatial and chromatic interactions in the lateral geniculate body of the rhesus monkey. *J. Neurophysiol.* **29**, 1115–1156.

Xu, X., Ichida, J.M., Allison, J.D. *et al.*, 2001. A comparison of koniocellular, magnocellular and parvocellular receptive field properties in the lateral geniculate nucleus of the owl monkey (*Aotus trivirgatus*). *J. Physiol. (Lond.)* **531**, 203–218.

Yeh, T., Lee, B.B., and Kremers, J., 1995a. Temporal response of ganglion cells of the macaque retina to cone-specific modulation. *J. Opt. Soc. Am. A* **12**, 456–464.

Yeh, T., Lee, B.B., Kremers, J. *et al.*, 1995b. Visual responses in the lateral geniculate nucleus of dichromatic and trichromatic marmosets (*Callithrix jacchus*). *J. Neurosci.* **15**, 7892–7904.

7

Extraretinal Inputs and Feedback Mechanisms to the Lateral Geniculate Nucleus (LGN)

Vivien A. Casagrande, David W. Royal, and *Gyula Sáry*

7.1 Introduction

Walker argued that the thalamus holds 'the secret of much that goes on in the cerebral cortex' (Walker, 1938). The thalamus is the first point at which most sensory signals arriving from the periphery can be modified by the rest of the brain. Therefore, the essence of what thalamic sensory relays do lies not so much in the quality of the sensory signals that they receive from the periphery but in how those signals are modified on their way to cortex and how these signals contribute to the survival of the organism. Given this perspective, it is surprising how little we actually know about the functional roles of the many modulatory signals that regulate the flow of sensory inputs to cortex. There may be several reasons why the role or roles of thalamic sensory relay nuclei remain unclear. One reason simply could be that most studies of thalamic cell properties have been performed in anesthetized preparations where modulatory inputs are likely not operating or are actively being blocked by the anesthetic. A second reason may be conceptual, namely, the idea that sensory thalamic cells must faithfully represent the periphery in order for percepts to be built up in cortex via a strictly feedforward pathway. The latter view tightly constrains possible roles for modulatory pathways and argues against the value of looking actively for the modifications of sensory signals that may occur at the level of the thalamus as a result of inputs that come via routes other than directly from the periphery especially via feedback from cortex. Finally, the assumption that basic sensory messages sent by thalamic cells can be decoded simply by examining

The Primate Visual System: A Comparative Approach Edited by Jan Kremers

the average firing rate of one cell at a time may have turned attention away from evidence that modulatory pathways can impact the temporal structure of firing of many cells that send convergent and divergent signals to cortex: signals that likely also carry important messages about the significance of sensory information.

In this chapter we focus on the best known of sensory 'relay' nucleus in the thalamus, the lateral geniculate nucleus (LGN). The aim is to review what is known about non-retinal inputs to the LGN in an effort to link these inputs to the function or functions of this nucleus. The goal is not to provide a global overview of LGN structure and function since this has been covered in detail in a number of older as well as recent reviews (Casagrande and Norton, 1991; Sherman and Guillery, 1996, 2001; Hendry and Reid, 2000; Casagrande and Xu, 2004) in addition to contributions to this book (Chapter 6). Instead, the goal is to briefly cover the essential features of LGN structure and circuitry. The bulk of the chapter explores what is known about extraretinal inputs to the LGN and what questions remain about the modulation of visual signals at this level. Given that species differ in the organization of their visual pathways and their LGN organization in particular, this chapter stresses the primate LGN. Since much more research has been done on the LGN of carnivores (especially cats) than on primates, information on other species is also covered where unavailable in primates.

In the past, it has been common to emphasize the feedforward nature of sensory signals. In this model, information goes from retina through the LGN to higher and higher levels of cortex to construct percepts which are then utilized for 'action'. Instead, in this chapter we emphasize the dynamic nature of the system and the importance of feedback pathways. Although the LGN can be considered an early component in the feedforward visual hierarchy, it also can be considered to be at the highest level of the feedback hierarchy. Clearly in awake animals vision is an active process. The visual system is never a one-way street. Each view of the world is the result of a purposeful decision to move the head and eyes to a location to acquire new information. This decision presumably occurs through a combination of pathways including higher cortical areas involved with memory, planning, and decision making, as well as limbic circuits that add emotional tone and motivation. To be efficient for survival the system must acquire essential information quickly and screen out irrelevant material. If an animal is searching for food, then it would make sense to pay attention to the locations where the food is normally found, its size, shape, and other characteristics such as whether it is moving or stationary. If an animal is simultaneously avoiding predators, it is essential that relevant characteristics of such predators also remain available and that information from different sensory modalities be organized to either enhance or inhibit each other depending upon the circumstances such as looking in the direction of the sounds made by prey or predator. Flooding the system with irrelevant sensory detail is wasteful and potentially dangerous in terms of the animal's survival. In this chapter, we argue that the LGN is actively involved in the selection process and receives constant feedback input from cortex directly or indirectly through the midbrain and brainstem, feedback that regulates which retinal signals reach cortex and which are enhanced or suppressed. We further argue that as early as the LGN the visual system is biased and that these biases are never fixed but are dynamically updated moment to moment.

This review is divided into five sections including the introduction. In section 7.2 we provide an overview of LGN cell types and introduce the basic circuitry. It is important to appreciate the basic circuitry to understand the ways extraretinal inputs can modify signals. In section 7.3 we consider the spatial and temporal response properties of individual LGN cells. The purpose here is not to provide a detailed review (Chapter 6) but to introduce the framework against which any modulatory pathway to the LGN must act to have an impact at the next stage of processing in cortex. In section 7.4 we review the circuitry, organization, and possible functions of non-retinal inputs. In the final section we provide a brief summary of key points reviewed in this chapter and list some unanswered questions.

7.2 Cell types and basic circuitry of the LGN

In primates the LGN consists of two principal cell types, relay cells which contain glutamate and send axons mainly to primary visual cortex (V1), and interneurons which contain gamma-aminobutyric acid (GABA) and communicate with other cells in the LGN itself (Casagrande and Ichida, 2002a). Eighty percent of cells in the LGN are relay cells and these cells consist of several classes which are organized into layers (Chapter 6; Conley *et al.*, 1985).

Unlike relay cells, the interneurons of the LGN are scattered relatively evenly throughout the nucleus. These neurons also are morphologically distinct having very thin dendrites which are purported to extend long distances in macaque monkey LGN (Wilson and Hendrickson, 1988). Whether one or several types of LGN interneurons exist in primates remains unclear, but all contain GABA. Given that there exists evidence for two types of LGN interneurons in cat LGN (Bickford *et al.*, 1999) and that there are many classes of interneurons in visual cortex (Hendry *et al.*, 1994) and other areas of the brain, it is likely that LGN interneurons in primates will also eventually be divided into subtypes. The LGN interneurons mainly communicate via dendro-dendritic synapses with relay cells (see also below). The size and extent of the thin branching dendrites of LGN interneurons and the fact that they are presynaptic have led to the proposal that the dendritic compartments may form circuits that are independent of the soma/axonal communication network of the cell (Bloomfield and Sherman, 1989; Erisir *et al.*, 1998). Since there is some evidence in cats (Bloomfield and Sherman, 1989; Erisir *et al.*, 1998) that LGN interneurons also communicate with relay cells via their axons (no axons were identified in reconstructed interneurons in monkeys; Wilson, 1986), this means that one interneuron could communicate different messages to different relay cells simultaneously via different dendrites and via their axons.

LGN relay cells and interneurons receive many inputs as described below. Before considering the complexities of all of these connections, it is worth initially laying out the basic circuitry. Both LGN relay cells and interneurons respond primarily on the basis of the information they receive from the retina. Sherman and Guillery (2001) have made the distinction between *drivers* and *modulators*, with drivers being essential to the response of a cell measured primarily via extracellular single unit recording of action potentials. Basically they argue that 'drivers are the information bearing input . . . to cortex' (Sherman and Guillery, 2001, p. 572). According to this distinction, retinal inputs are the main drivers for most LGN relay cells and interneurons. Given the long latency to respond to chiasm stimulation on the part of some koniocellular (KC) LGN relay cells (maybe also their associated interneurons), it has been speculated that some KC cells could receive their main 'drive' indirectly from the retina via the superior colliculus (Norton and Casagrande, 1982; Casagrande, 1994). This hypothesis remains to be tested empirically. Regardless, if the retina provides the main drive to the vast majority of LGN cells, then all other inputs to the LGN are, by definition, modulators. The problem with this definition in terms of information processing in the LGN, however, is that it assumes that the other pathways are not carrying essential information. LGN cells are never silent, so if a modulatory input causes changes in the spontaneous spike production of LGN cells in the absence of direct retinal input, does this modulator then act as a 'driver' if it is carrying information relative to another modality or when the LGN is active while subjects imagine a visual scene? For example, we know that LGN cells of all classes can be modulated by auditory and somatosensory input and by eye movements in the absence of visual input (Irvin *et al.*, 1986; Royal *et al.*, 2005). Regardless, it is obvious in the case of the LGN that visual receptive fields (both relay cells and interneurons) in awake animals derive their signature receptive field structure (defined by extracellular recording) from their retinal inputs (Chapter 6). Retinogeniculate axons end as large terminals that make multiple synapses on the proximal dendrites of relay cells and on the cell body as well as proximal and distal dendrites of interneurons (Pasik *et al.*, 1986; Wilson, 1986). In addition, relay cells receive *feedforward inhibition* from interneurons via dendro-dendritic synapses that often occur in 'triadic' assemblies where a retinal terminal contacts both an LGN relay cell and a presynaptic dendrite of an interneuron that, in turn, synapses on the same relay cell dendrite (Sherman and Guillery, 2001). In macaque monkeys triads appear to be common in the magnocellular (MC) layers and much rarer in the parvocellular (PC) layers (Wilson, 1993), suggesting a difference in the impact of inhibitory interneurons between these cell classes in primates. Relay cells also project a collateral axon to the thalamic reticular nucleus (TRN) whose cells provide *feedback inhibition* to all LGN cell classes (see following sections for details).

7.3 Response properties: A brief overview

Since the seminal work of Wiesel and Hubel (1966) we have known that LGN cells have visual receptive field properties that are similar to their retinal ganglion cell inputs. It is not the purpose of this section to review, in detail, the spatial and temporal structure of LGN receptive fields (for review, see Casagrande and Norton, 1991; Chapter 6). Instead, we simply summarize information that is basic to understanding the potential impact of extraretinal signals in an effort to unravel their functional messages. LGN cells are never silent even when animals are placed in a dark room or are asleep. These cells are spontaneously active, so even without a visual message these cells will have a differential effect on cortical cells by resetting levels of depolarization in V1 and thereby changing the thresholds of V1 cells. The relative state of LGN cells also affects the visual message sent to cortex since burst and tonic, non-rhythmic and various rhythmic modes can transmit information differently (see below). Different cell classes also respond to stimuli with very different latencies (Irvin *et al.*, 1986; Schmolesky *et al.*, 1998; Royal *et al.*, 2004) and different levels of transience. The timing of these messages to cortex will, of course, be critical in determining which V1 cortical cells reach threshold, which messages are combined, and when and what messages are sent to the next level, or how messages are combined with feedback from higher visual areas in a dynamic network. Within this context the relevant properties of LGN cells are described in the following sections.

7.3.1 Spatial properties

In anesthetized and paralyzed primates the majority of MC, PC, and KC LGN cells have been shown to have center/surround organizations that can be modeled by a Difference of Gaussians (DoG) model (Norton *et al.*, 1988; Irvin *et al.*, 1993; Xu *et al.*, 2002; Chapter 6). Using this model, one can compare also the receptive field structure of ganglion cells that provide input to the receptive field structure of LGN cells. Comparison suggests that either the surround of LGN cells is stronger or the center response weaker presumably because of the addition of feedforward and feedback inhibition in the LGN. The net result also is that the proportion of spikes produced by an LGN cell is generally less than that produced by the ganglion cell input to that cell (i.e. the transfer ratio is less than 1.0) (Casagrande and Norton, 1991). Examining the transfer ratio and modeling the spatial receptive field of LGN cells is useful in understanding the impact of other inputs to these cells (Uhlrich *et al.*, 1995). Viewed from the perspective of the DoG model, extraretinal inputs can impact the structure of the receptive field by either changing the gain or space constant of the inhibitory surround mechanism or changing the gain or space constant of the excitatory center mechanism. Of course, the output of many LGN cells is combined to drive cortical cells and these LGN cells are, in turn, dynamically linked to each other through feedback from cortex and input from other areas. This means that stimuli presented elsewhere in space outside of the classical single LGN receptive field can potentially impact the spatial structure of the receptive field and/or the transfer ratio of signals from retina. The fact that LGN ensembles, not single cells, encode information utilized by cortex is attested to by results showing that natural scenes with recognizable moving objects could be reconstructed from six to eight pairs of ON- and OFF- center LGN cells per point in space using a simple decoding algorithm applied to the population (Stanley *et al.*, 1999).

7.3.2 Temporal properties

It is clear that what is communicated to cortex by the LGN will depend on the postsynaptic impact of spikes produced by these cells in relationship to the state of the recipient cortical cell. Many LGN cells also communicate with each individual V1 cell; therefore, how the spikes are packaged across time within each LGN cell and how these spikes are synchronized across LGN cells will define the postsynaptic response in V1 and beyond. Both LGN relay cells and interneurons have many voltage-dependent channels that control various currents including both high and low threshold Ca^{2+} conductances, K^+ conductances, and Na^+ conductances (Hernandez-Cruz and Pape, 1989). Depending upon the circumstances, relay cells tend to adopt two basic modes of firing referred to as burst and tonic. During tonic firing, action potentials of relay cells are triggered to more faithfully reflect the temporal sequence of retinal inputs. During burst firing, activation of Ca^{2+} spikes in response to retinal input can trigger several action potentials, and the ratio is no longer one-to-one (Sherman and Guillery, 1996, 1998, 2002 [Fig. 3]). These two modes of firing are controlled by a transient (T) type calcium current (I_T). When relay cells are hyperpolarized below −70 mV for approximately 100 ms, I_T is slowly de-inactivated. The following suprathreshold depolarization or excitatory post-synaptic potential (EPSP) above −70 mV then activates I_T which produces an all-or-none Ca^{2+} spike called a 'low threshold Ca^{2+} spike' (Ramcharan *et al.*, 2000). The amplitude of the Ca^{2+} spike depends on the magnitude and the length of the preceding hyper-polarization. Depolarization (inactivation of I_T) and hyperpolarization (de-inactivation of I_T) are the key processes for switching the firing of LGN cells between tonic and burst modes (Hillenbrand and van Hemmen, 2001; Guillery and Sherman, 2002). Inactivation or de-inactivation of I_T depends on the duration of the sustained membrane potentials. Sustained membrane potentials require slow-responding receptors. In the LGN, neurotransmitters act on either ionotropic or metabotropic receptors at postsynaptic terminals. Ionotropic receptors include glutamate-responsive AMPA and NMDA receptors, $GABA_A$ receptors, and nicotinic receptors (see also below). Metabotropic receptors include glutamine receptors (mGLURs 1–8), $GABA_B$ receptors, and acetylcholine M1 and M2 receptors. Ionotropic receptors respond with a fast postsynaptic potential, but the metabotropic receptors act through second messengers and so are much slower (Coutinho and Knopfel, 2002; Salt, 2002). The slow and sustained actions of the metabotropic receptors are necessary for inactivation or de-inactivation of I_T. Interestingly, whereas all retinal inputs to LGN relay cells act on ionotropic receptors, other inputs activate both ionotropic and metabotropic receptors, suggesting that extraretinal inputs play a role in switching between burst and tonic modes of firing. Since burst firing is more effective in causing cortical spikes than tonic firing (Swadlow and Gusev, 2001) and tonic firing more faithfully represents the retinal input message, Sherman has suggested that burst mode in the LGN of awake animals acts as a 'wake-up call' for detection of novel stimuli whereas tonic mode transmits information about stimulus quality. The difficulty with this hypothesis is that bursts occur very rarely in behaving primates that are engaged in routine visual tasks (Royal *et al.*, 2003). On the other hand, bursts are common in sleeping animals and it has been argued that their main function is to disconnect thalamus from cortex when animals are asleep (McCormick and Prince, 1986; Steriade and Llinás, 1988). The timing of oscillatory bursts of activity or the general synchronization of activity between LGN and cortex may also be involved in coordinating the effectiveness of messages within the visual network as suggested by various investigators (Sillito and Jones, 2002; Worgotter *et al.*, 2002).

Regardless, it is clear that different messages may be conveyed to cortex depending upon the temporal structure of the spike train (see also Dan *et al.*, 1998; Usrey and Reid, 1999). The same holds true whether or not LGN cells are conveying visual messages or other non-visual messages concerning the animal's state.

7.4 Organization of extraretinal inputs

The extraretinal inputs to the LGN far outnumber, in terms of synapses, the retinal input (Wilson and Forestner, 1995). From the standpoint of function, it is important to appreciate how these inputs are organized (Figure 7.1). Clearly, if an input is visuotopic and specific to certain layers or cell types, it will be a position to regulate those signals locally. A number of inputs fall within this category although their relative visuotopic specificity varies. These inputs include glutamatergic projections from visual cortical areas and

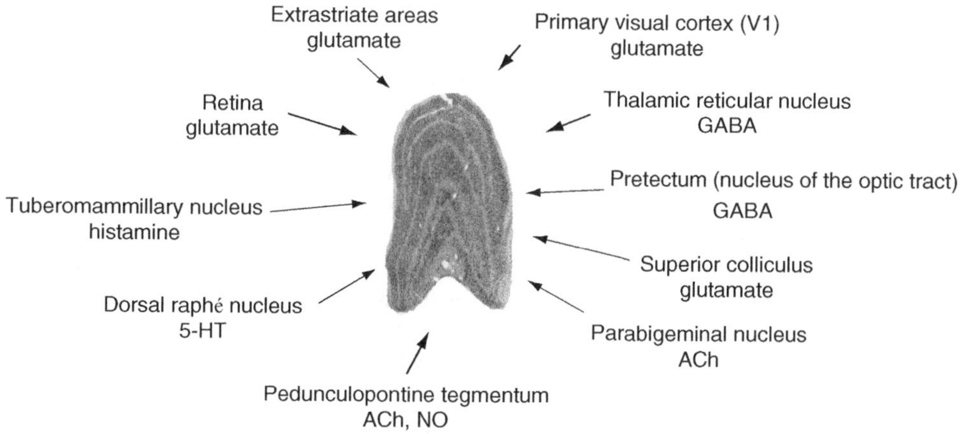

Figure 7.1 Organization of the LGN inputs in primates. The figure shows the known inputs to the LGN in primates. For simplicity, we did not show the input to relay cells and interneurons separately. Arrows indicate the relative weight of the inputs. Transmitters are shown with the name of the input. See text for details. GABA: gamma-aminobutyric acid; ACh: acetylcholine; 5-HT: serotonin; NO: nitric oxide

from the superior colliculus, GABAergic inputs from the pretectum (see also Chapter 8) and TRN, and possibly the cholinergic parabigeminal input. An input could also show a restricted distribution but relate to another type of mapping dimension other than vision. The cholinergic inputs from the pedunculopontine area appear to be of this type. Finally, extraretinal inputs can regulate LGN signals very globally via non-synaptic release of transmitter. The histaminergic and serotonergic inputs to the LGN from the hypothalamus and the brainstem, respectively, fall into this last category. In the following sections we consider the organization and possible functions of each of these inputs in more detail.

7.4.1 Visuotopically organized glutamatergic inputs

Visual cortex (V1 and other cortical areas)

Primary visual cortex (e.g. striate cortex, area 17 or V1) provides the major extraretinal input to the LGN in all species where this input has been examined (see for review Sherman and Guillery, 1996). As in other species, in primates the V1 input to LGN arises in cortical layer 6 (Lund *et al.*, 1975; Conley and Raczkowski, 1990; Fitzpatrick *et al.*, 1994; Casagrande and Ichida, 2002b). Unlike in cats, however, this input to the LGN appears to be more precise both in terms of its visuotopic relationship to the LGN and in terms of the regulation of individual cell classes and layers (Ichida and Casagrande, 2002). For example, in owl monkeys, bush babies, and macaque monkeys, anatomical studies indicate that V1 axons never innervate *both* PC and MC LGN cell layers and that cortical cells that give rise to these axons tend to be segregated to the upper and lower portions of layer VI, respectively (Lund *et al.*, 1975; Conley and Raczkowski, 1990; Fitzpatrick *et al.*, 1994; Ichida and Casagrande, 2002). The situation for KC cells appears different in that KC cells that lie near MC LGN cells receive input from the same cells that innervate MC cells via collateral axons, while KC cells that lie near PC LGN cells share cortical input with neighboring LGN PC cells (Ichida and Casagrande, 2002). Axon reconstructions also suggest that some axons innervate only single eye-specific LGN layers, at least in owl monkeys (Ichida and Casagrande, 2002). At the level of the LGN electron microscopic (EM) immunocytochemical studies indicate that cortico-geniculate axons (which themselves contain glutamate) innervate primarily glutamatergic relay cells and not GABAergic interneurons, suggesting that their primary initial effect is excitatory (Ichida *et al.*, 2004). Taken together, these patterns of cortico-geniculate projections in primates suggest that V1 can modulate activity in the LGN in both a functionally and a retinotopically specific manner in relationship to other input pathways (see also below). It is noteworthy, however, that since V1 receives feedback from a number of higher-order visual cortical areas and sends axons to a number of other subcortical sites that, in turn, send input to the LGN, the functional impact of V1 on LGN activity likely is complex and context-dependent.

Studies in other species, particularly rats, cats, and ferrets, indicate that signals provided to the LGN from visual cortex can be regulated in complex ways depending upon the types of receptors that are activated. When cortico-geniculate projections are active, both fast and slow EPSPs have been identified in the LGN. In cats, slow EPSPs are reduced when glutamate metabotropic receptor (mGluR) antagonists are applied, suggesting that cortical inputs to the LGN activate metabotropic glutamate receptors which, in turn, act more slowly since, as mentioned earlier, second messengers are involved (von Krosigk *et al.*, 1999). Fast EPSPs in cat LGN are mediated by the actions of the ionotropic glutamate receptors (iGluRs) in addition to NMDA and non-NMDA receptors (Godwin *et al.*, 1996a).

Two types of mGluRs have been identified in LGNs of non-primates: mGluR1s that are located in the cortical recipient zone of relay cell distal dendrites and mGluR5s that are found in association with interneuronal dendrites and on proximal dendrites where retinal inputs can terminate (Godwin *et al.*, 1996a). Also, mGluR1s in the LGN are activated in response to cortical inputs (Turner and Salt, 2000) and cortical inputs have been reported to be the sole activators of mGluR1 in cat LGN (Godwin *et al.*, 1996b). Since both cortical and retinal inputs are glutamatergic, unique localization of mGluR1and mGluR5 on dendrites may allow relay cells to respond only to the specific source of inputs (Godwin *et al.*, 1996a). One function of this arrangement may be to regulate the temporal properties of LGN cells. Recently Eyding *et al.* (2003) selectively eliminated cortico-geniculate neurons in cats in order to test the hypothesis that this pathway was important in synchronizing LGN and cortical signals. The neurons in the LGN fire in burst mode when animals exhibit a synchronized electro encephalographic (EEG) state. When the EEG state changes to a desynchronized state indicative of wakefulness, the same LGN neurons respond in tonic mode. Upon elimination of the cortico-geniculate projections, LGN neurons no longer switch to tonic mode to match the change in EEG state. Furthermore, the synchronized EEG state no longer induces higher incidents of burst firing (Eyding *et al.*, 2003). Cooling of the cortex also has been shown to cause LGN relay cells to remain in tonic firing mode (Worgotter *et al.*, 2002). In another experiment, an mGluR1 antagonist was applied to LGN relay cells. As a result low-threshold Ca^{2+} spikes were abolished, and the response mode was shifted from burst to tonic (Godwin *et al.*, 1996b). These findings indicate that cortical inputs influence the firing mode of LGN relay cells. When mGluRs are activated, the potassium leak channels close, leading to depolarization and a shift in firing mode from burst to tonic. On the other hand, when $GABA_B$ receptors are activated, the potassium leak channels open generating hyperpolarization (Hillenbrand and van Hemmen, 2001). It still remains to be demonstrated whether the effects seen in the cat can be translated directly to the primate LGN.

In addition to inputs from V1, it is known that other primate visual areas, including the second (V2), middle temporal (MT), and dorsal-lateral (DL) visual areas (also called V4 and V5), provide a minor input to the LGN as demonstrated in several primate species (Symonds and Kaas, 1978; Graham *et al.*, 1979; Benevento and Yoshida, 1981). Interestingly, the latter appear to target specifically the LGN KC layers for reasons that remain unclear.

Feedback from cortex to the LGN has been suggested to play a variety of roles. In the spatial domain it has been proposed that feedback enhances the contrast gain of PC and MC cells (Przybyszewski *et al.*, 2000), is involved in both global integration (binding) of visual features and segmentation (Sillito and Jones, 2002), and is critical to binocular integration for stereo vision (McIlwain, 1995). In the temporal domain it has been argued that feedback synchronizes the firing of relay cells (Sillito *et al.*, 1994) as well as changing firing from burst to tonic mode (see above). It is clear that given the topographic specificity of V1 to LGN projections this pathway also may be involved in independently regulating different classes of LGN cells as well.

Superior colliculus

The superficial grey layer of superior colliculus sends topographically restricted axons to the LGN in all species that have been studied (Harting *et al.*, 1991a; Feig and Harting, 1994). This collicular input can be found within all KC layers in macaques and bush babies (Harting, 1977; Harting *et al.*, 1991a; Lachica and Casagrande, 1993; Feig and Harting, 1994). In strepsirrhine primates (bush babies) axon reconstruction studies provide evidence for two classes of axons that project from the colliculus to the LGN KC layers. Although both types of collicular axons terminate within restricted zones, the spread of colliculogeniculate arbors is somewhat broader than retinal input to KC cells. At the ultra-structural level, both collicular and retinal inputs to LGN KC cells terminate as asymmetric synapses very close together on distal dendrites, suggesting that visual drive to some KC LGN cells may arise indirectly from the colliculus or that visual drive to these cells requires a combination of retinal and collicular inputs to reach threshold (Feig and Harting, 1994). In cats, data indicate that colliculogeniculate axons contain glutamate which is consistent with ultra-structural evidence in both cats and primates (Feig and Harting, 1994). Since the superior colliculus also is connected reciprocally with the parabigeminal nucleus and both the colliculus and the parabigeminal nucleus favor the KC LGN layers as targets in all species studied, it could be that these inputs provide information about eye movements (Sherk, 1979; see also below and Chapter 8). Additionally, since the superficial layers of the superior colliculus have been implicated as important in visual attention (Wurtz *et al.*, 1982; Newsome, 1996), it also has been suggested that information about attentional shifts might be carried to cortex from the colliculus via LGN KC cells (Casagrande, 1994). Recent evidence, however, indicates that MC, PC, and KC LGN cells can all be modulated by attention in awake behaving macaque monkeys (Royal *et al.*, 2004).

7.4.2 Visuotopically organized gabaergic inputs

Pretectum (nucleus of the optic tract)

In macaque monkeys and bush babies, pretectal input to the LGN arises from the nucleus of the optic tract (NOT; Chapter 8). In cats, pretectal input is GABAergic and terminates primarily in the A layers (e.g. on X- and Y-cells) of the LGN (Cucchiaro *et al.*, 1991; Wahle *et al.*, 1994). In bush babies and likely other primates, this input is also GABAergic (Feig and Harting, 1994). At present, it is unclear whether pretectal input in primates ends preferentially in specific layers although the input appears to show some retinotopic specificity (Bickford *et al.*, 2000). In bush babies, more pretectal input was identified in the PC layers than in the other LGN layers (Harting *et al.*, 1986), whereas in the macaque monkeys some reports have suggested there is more pretectal input to the MC layers (Büttner-Ennever *et al.*, 1996). Bickford *et al.* (2000), however, found that some pretectal cells send input to both MC and PC layers. In bush babies, this input has been found to target principally relay cells where it terminates on distal dendrites (Feig and Harting, 1994). The latter result suggests that in primates pretectogeniculate input may directly inhibit relay cells. Interestingly, the opposite appears in cats where pretectogeniculate input ends primarily on the dendrites of interneurons, indicating that this projection would disinhibit cat relay cells (Schmidt, 1996; Wang *et al.*, 2002). Regardless, it is possible that the pretectal input to the LGN is made up of several pathways to the LGN that have different roles.

In functional terms the NOT can be considered a visuo-motor nucleus. In macaque monkey there is evidence the NOT encodes position, velocity, and acceleration components of retinal error that may be used by the targets of NOT for synthesis of smooth-pursuit eye movements and for image stabilization (Das *et al.*, 2001; see also Chapters 8 and 10). In awake cats and in wallabies, it has also been reported that cells in NOT respond to saccades and to eye blinks. These results could account for the suppression of activity seen in the LGN during saccades and/or the enhancement seen after saccades that we and others have reported in the LGN of awake behaving macaque monkeys (Zuber and Stark, 1966; Montero and Robles, 1971; Riggs *et al.*, 1974; Ross *et al.*, 1996; Zhu and Lo, 1996; Lee and Malpeli, 1998; Ramcharan *et al.*, 2001; Royal *et al.*, 2004, submitted; Thilo *et al.*, 2004). In fact, studies using antidromic activation in cats have demonstrated that pretectogeniculate cells are selectively sensitive to saccadic eye movements (Schmidt, 1996). Taken together with other results that have reported eye movement effects in LGN, it seems likely that the role of the NOT is to inform the LGN about particular aspects of planned ocular movements.

Thalamic reticular nucleus

Perhaps the most important key to understanding how visual information and sensory information, in general, are altered in the thalamus lies in an understanding of the role of the TRN. This interesting nucleus, which forms a shell around the thalamus and contains GABAergic neurons, is subdivided into zones which project to individual thalamic nuclei (Jones, 2002). The portion of this nucleus that sends and receives input from the LGN has been examined most thoroughly. Among primates the relationship between the TRN and the LGN has been best studied in the bush baby where it has been shown that reciprocal connections between all layers of the LGN and the TRN are topographic and specific (Harting *et al.*, 1991a). Similar evidence of a high degree of retinotopic specificity in connections between the TRN and the LGN have been reported also in the macaque monkey (Bickford *et al.*, 2000; Wang *et al.*, 2001). In cats, it has been suggested that cells in the visual portion of the TRN, referred to as the perigeniculate nucleus, project primarily to LGN Y-cells (Fitzgibbon, 2002). This preference for one pathway has not been reported in primates; instead the TRN appears to project to all LGN cell classes in primates (Harting *et al.*, 1991b). The portion of the TRN that projects to the LGN is also innervated by the collateral branches of axons that arise from cells in layer VI of V1 (see preceding text). At present, it is unclear if all corticogeniculate axons provide such collaterals to the TRN or only a subset (Ichida and Casagrande, 2002), but in cats all reconstructed axons from visual cortex were found to send a collateral branch to the perigeniculate nucleus (Murphy and Sillito, 1987). The TRN also receives input from collateral axons of LGN relay cells and sends its output back to these relay cells as well as to LGN interneurons. These circuits allow the TRN not only to provide feedback inhibition to the LGN, but also to regulate LGN cell output in complex ways depending upon other inputs that the TRN receives from both extrastriate visual areas and the brainstem (Sherman and Guillery, 1996; Guillery *et al.*, 1998 for review; Jones, 2002). For example, inhibitory reticular inputs have been shown to affect the temporal correlation between LGN input and output, pushing the neural circuit toward synchronized oscillation (Le Masson *et al.*, 2002). This process could increase the efficiency of signal transmission between LGN and V1 (Sillito and Jones, 2002). Simulation studies of the LGN–V1–TRN pathway show that the TRN activity suppresses the background and improves the signal-to-noise ratio (Bickle *et al.*, 1999). Cortical inputs presumably regulate oscillations in LGN in the following way. When a long enough period of hyperpolarization has occurred, relay cells fire in burst mode when I_T is de-inactivated while the cells recover from an inhibition provided by the TRN. This burst firing excites the TRN cells. The activated TRN cells re-inhibit relay cells, and relay cells fire again in burst mode as they recover from the inhibition produced by the TRN. These events occur repeatedly, generating a low-frequency oscillation. Therefore, the TRN not only induces the burst mode of firing in relay cells, but also generates oscillation by repeatedly inducing the burst firing (Jones, 2002). Activation of the TRN, however, does not always lead to the result one might predict by such a model. Glutamate, which is generally an excitatory neurotransmitter, can also directly inhibit TRN cells. This inhibition occurs when glutamate activates group II mGluRs and the potassium conductance is increased. Activation of group I mGluRs has the opposite effect leading to depolarization; this suggests that the glutamatergic inputs to the TRN can be either excitatory or inhibitory, depending on the group of receptors that is activated (Cox *et al.*, 1998; Cox and Sherman, 1999). Additionally, evidence exists in mice for GABA to act directly on retinal axons via presynaptic $GABA_B$ receptors (Chen and Regehr, 2003). Whether this involves input from interneurons or the TRN remains unclear. Regardless, this sort of receptor-dependent excitation and inhibition may add great flexibility to the modulatory roles of the TRN but also predicts that understanding the role of the TRN requires appreciation of the complexity of the circuit and the fact that the system is dynamic. Although the TRN has been proposed to play specific roles in sleep, arousal, and attention (Crick and Koch, 1990), it seems likely that the TRN is not tied to a specific role relative to LGN activity but is utilized in a variety of ways. Nevertheless, unlike the more global modulatory inputs to the LGN, the visual TRN, like V1 to which it is intimately linked, is in a position to modulate visual activity quite precisely given its retinotopically specific connections with the LGN.

7.4.3 Cholinergic inputs

The largest non-retinal brainstem input to the LGN in primates (as well as other species) is cholinergic (Bickford *et al.*, 2000). This input may account for as much as 25 percent of synapses in the LGN, at least in the cat (Erisir *et al.*, 1997). The cholinergic input comes from two sources, cells in pedunculopontine tegmentum (PPT) and from the parabigeminal nucleus of the midbrain. The PPT source (referred to also as CH5 by Mesulam, 1990) innervates all LGN layers in primates, but appears to show differences in innervation density that correlate with visual lifestyle. Thus, in the nocturnal simian owl monkey and nocturnal prosimian bush baby, acetylcholinesterase (the degradative enzyme for acetylcholine) is densest in the LGN PC layers, whereas it is densest in the MC layers of diurnal simian squirrel monkeys and macaque monkeys (Fitzpatrick and Diamond, 1980; Graybiel and Ragsdale, 1982; Wilson *et al.*, 1999). In contrast to the input from the PPT, the parabigeminal cholinergic input (CH8 of Mesulam, 1990) projects primarily to the LGN KC layers although its primary output is to the superior colliculus (Feig and Harting, 1994); KC layers also receive sparse cholinergic input from the PPT (Bickford *et al.*, 2000). Additionally, the axonal projections from the PPT provide the exclusive source of the neurotransmitter nitric oxide (NO) to the LGN (Bickford *et al.*, 2000) since no bNOS positive cell bodies have been found in primate LGN (Wiencken and Casagrande, 2000; see, however, Bickford *et al.*, 1999 for evidence of bNOS positive interneurons in the cat LGN). PPT projections make asymmetric synaptic contacts onto both proximal and distal relay cell dendrites as well as onto the dendrites of interneurons. Parabigeminal inputs also are found to synapse on both relay and interneuronal cell dendrites in bush babies (Feig and Harting, 1992). Both cholinergic inputs to primate LGN are bilateral, although the ipsilateral input to LGN dominates (Feig and Harting, 1992; Bickford *et al.*, 2000). The fact that projections are bilateral suggests that cholinergic brainstem pathways can potentially send signals that integrate across the two hemifields of visual space. Add to this complexity the fact that cholinergic inputs can act through at least three types of receptors (nicotinic and two muscarinic [M1 and M2] receptors) and it is clear that these cholinergic pathways can potentially influence LGN cell activity in complex ways depending upon the circumstances. Evidence exists that the net effect of activation of the PPT pathway in non-primates (primates have not been studied) is excitation of LGN relay cells. This is accomplished via nicotinic and M1 receptors on relay cells and via M2 receptors on presynaptic dendrites of interneurons (McCormick and Prince, 1986; McCormick and Pape, 1990). The depolarization of relay cells is further enhanced by the co-release of NO (Nucci *et al.*, 2003). Studies of anesthetized cat LGN neuronal responses to visual stimuli (drifting gratings) in the presence of electrical stimulation of the PPT pathway demonstrated that the most common effect of PPT activation was response enhancement. Interestingly, stimulation of the PPT pathway could induce robust responses to visual stimuli even in cases in which LGN cells did not respond at all to the same stimulus (Uhlrich *et al.*, 1995). PPT activation in the study by Uhlrich *et al.* (1995) mainly resulted in an increase in both the center and surround responses of LGN cells, suggesting that the main effect is an increase of the transfer ratio of the retinal signal (see preceding text). They also found a more variable effect on the surround response as well as on the spontaneous activity of LGN cells, presumably because both are affected by the inhibitory circuitry within the LGN itself as well as by other inputs, not just input from the retina. Many functions have been attributed to the PPT. It is beyond the scope of this review to cover all the studies of the PPT but activity in this region has been implicated as important in a variety of behaviors including eye movements, attention, arousal, rapid eye movement, and sleep (Fitzpatrick *et al.*, 1989). Understanding the function of the PPT pathway, or pathways, to the LGN has been difficult given that in primates the cells of origin are scattered and not confined tightly to a nucleus.

So far, we have less information on the impact of the parabigeminal on LGN responses, although given its strong connections with the superior colliculus and given evidence that parabigeminal cells in awake behaving cats reflect retinal position error, it seems likely that this pathway would inform LGN cells about target location (Cui and Malpeli, 2003). It is interesting in this regard that the main cells in cats and primates that receive input from parabigeminal axons are cat LGN W-cells and primate LGN KC cells (Harting *et al.*, 1991c).

7.4.4 Diffuse modulatory inputs (histamine and serotonin)

All of the LGN layers of macaque monkey and squirrel monkey have been shown to receive diffuse input from brainstem and hypothalamic sources that appears capable of globally modulating LGN activity via mainly non-synaptic release of the transmitters serotonin and histamine, respectively (Morrison and Foote, 1986; Pasik *et al.*, 1986; Wilson and Hendrickson, 1988; Uhlrich *et al.*, 1995). There has been some debate about whether serotonergic input is more dense in the KC or MC layers than in the PC layers. Nevertheless, it is generally agreed that all LGN layers receive these modulatory inputs and that these inputs are non-topographic relative to the visual coordinates of the nucleus (Wilson and Hendrickson, 1988; Uhlrich *et al.*, 1995).

Serotonin

Although some reports have suggested that serotonergic input to the monkey LGN is moderately dense (Morrison and Foote, 1986; Pasik *et al.*, 1986; Wilson and Hendrickson, 1988), quantitative estimates suggest that serotonergic input makes up approximately 1 percent of the vesicle-filled profiles (Wilson and Hendrickson, 1988). All studies of the function of serotonin in the LGN have been done in non-primates, typically in slice preparations. In slice preparations of LGN, serotonin has been found to have a complex effect on LGN cell responses which appears to be either excitatory or inhibitory. Given that stimulation of the dorsal raphé or infusion of serotonin *in vivo* causes inhibition of LGN relay cells, suggest that serotonin could act indirectly by exciting GABAergic LGN cells (Funke and Eysel, 1995). In slice preparations of mouse LGN, evidence was found for presynaptic action on retinal axons within the LGN operating via 5HT1A receptors (Nucci *et al.*, 2003). Since no synapses have been reported on retinal axons from labeled serotonin fibers, this transmitter must diffuse hormone-like to reach these receptors. Regardless, the fact that serotonin could block or blunt retinal transmission presynaptically would make it ideal for regulating transmission during sleep, although other pathways have been implicated (Steriade and Deschênes, 1984; Steriade and Llinás, 1988; McCormick and Pape, 1990).

Histamine
Receptor binding studies in LGN indicate that the impact of this pathway could be larger than suggested by the limited population of tuberomammillary hypothalamic cells labeled retrogradely from the LGN in macaque monkeys (Bickford *et al.*, 2000). A high density of dendritic histamine H1 and H2 receptors and presynaptic H3 receptors (Bouthenet *et al.*, 1988; Ruat *et al.*, 1990; Chazot *et al.*, 2001) have been identified in the LGN. Although no direct functional studies of the tuberomammillary to LGN pathway have ever been done in primates, in the thalamus the histaminergic system is thought to play a primary role in general arousal, with tuberomammillary neurons active during the waking state and relatively inactive during sleep (Vanni-Mercier *et al.*, 1984; Lin *et al.*, 1988, 1990; Monti, 1993). Receptors for histamine presumably exist only on LGN relay cells since there is no evidence that GABAergic LGN interneurons respond to this input (Uhlrich *et al.*, 1995). Application of histamine in slice preparations of non-primates changes the firing pattern of LGN neurons from the burst mode of firing to the tonic mode by decreasing K^+ conductance causing depolarization and inactivating I_T (McCormick and Williamson, 1991; McCormick, 1992), mimicking the change in general firing activity recorded in the thalamus as the brain transitions from sleep to waking (Steriade and Deschênes, 1984; Steriade and Llinás, 1988). Uhlrich *et al.* (1995) have shown that stimulation of the tuberomammillary nucleus causes release of histamine within the cat LGN, resulting in an increase in baseline activity as well as an increase in firing activity to a visual stimulus and thus an increase in the transfer ratio of information from the retina. This supports the idea that this pathway is part of a general arousal system.

Since there are several pathways that result in increased transmission through the LGN, it is likely that each pathway brings a different context to bear on the visual signal. Histamine release is often associated with negative stimuli. Thus, one might speculate that this pathway to the LGN functions to increase the transfer ratio of retinal signals in situations where potential danger exists or possibly be more globally tied to enhancing visual signals in relationship to reward and punishment or to general levels of motivation.

7.5 Concluding remarks and remaining questions

The LGN and primary visual cortex (V1) are part of a dynamically linked loop. Unlike in cats, the extrastriate output of LGN cells in primates is relatively small (Casagrande and Norton, 1991). Whether this extra-geniculostriate pathway can function in the absence of primary visual cortex is still hotly debated (Collins *et al.*, 2003) under the heading of 'blind sight'. Regardless, it is generally agreed that area V2 is silenced by the removal of V1 (Schiller and Malpeli, 1977; Merigan *et al.*, 1993); therefore, if the LGN is to communicate visual messages to the rest of the brain, it likely does so via V1 in primates. As we have seen earlier, however, LGN can receive messages that may not directly involve V1 and can arrive via a number of routes. Subcortical sites that send axons to LGN, of course, also receive from additional cortical and subcortical sources, so

LGN cells can be informed about sensory, motor, and limbic activities via these indirect sources. Because the system is dynamically linked, LGN relay cells can potentially carry messages that are non-retinal together with or before or after retinal messages arrive. These non-retinal inputs can be demonstrated by showing modulation of the level and the temporal structure of spontaneous LGN activity in the absence of retinal stimulation (Royal *et al.*, 2004, submitted). In fact, LGN activation has been detected using fMRI in subjects with eyes closed and no direct visual input while these subjects imagined visual scenes (Chen *et al.*, 1998). Given that LGN and V1 are connected dynamically, the latter result also indicates that both areas may be actively involved in processing signals during visual imagery. Furthermore, it has been demonstrated recently in imaging studies using voltage-sensitive dyes that even in the anesthetized state, and in the absence of visual stimulation, visual cortical activity is not random but seems to show intrinsic patterns of activity that evolve over time by switching among different states that resemble the architecture of activity produced in response to visual stimuli (Kenet *et al.*, 2003). This finding indicates that so-called spontaneous activity in visual cortex is not random. Since LGN and visual cortex are so intimately linked, it seems reasonable to propose also that the 'spontaneous' activity of LGN neurons is not random noise in the system but instead reflects different states.

Additionally, since inputs to the LGN act through both fast ionotopic and slow metabotropic receptors, this means that the impact of retinal or other signals to LGN cells could outlast a peripheral stimulus under some circumstances just as easily as they could be truncated by direct or indirect inhibitory inputs. Enhancement of relevant stimuli and suppression of irrelevant stimuli would make sense for species survival. This idea implies, for example, that under conditions where a very negative, potentially painful stimulus (dentist drill, large angry wasp, large knife coming at you) is seen, it may activate LGN cells via direct retinal pathways as well as cortical pathways associated with the learned meaning of the stimulus, pathways that attach emotional tone/arousal (histamine) or allow increased attention (acetylcholine). Temporal coordination of all of these inputs to LGN could lead to a temporary or permanent enhancement of specific types of visual or other signals through the LGN gateway to the cortex. In fact, a variety of functions (reviewed above) have been attributed to each of the non-retinal inputs to LGN in addition to those mentioned earlier. Key related questions for each input pathway to the LGN are as follows: (1) Are some non-retinal messages to the LGN used to communicate non-visual messages to V1 via changes in baseline firing or changes in the temporal structure of LGN cell firing? In other words, does the level or structure of spontaneous activity convey information independent of vision? (2) In addition to simply controlling the transfer ratio between the retina and the cortex (Sillito and Jones, 2002), does the LGN aid in the construction of visual images? Although much more information will be required, the data reviewed above suggest that the answer to both of these key questions is *yes*. Many other more specific questions remain about each of the non-retinal pathways to LGN reviewed above. Several examples follow which are not intended as an exhaustive list. (3) Do the inhibitory pathways from interneurons and the TRN relate differently to different classes of relay cells in primates as suggested for cat relay cells (Sherman and Guillery, 2001)? (4) How many classes of LGN interneurons exist in the LGN of primates? (5) Why are there two or more cholinergic inputs to the LGN? (6) How are LGN indirect retinal inputs (via superior colliculus, pretectum) coordinated with direct retinal messages to the LGN? (7) Why do the superior colliculus, parabigeminal nuclei, and extrastriate visual areas send input primarily to the KC LGN layers? (8) Does the histaminergic input convey information about potential reward or punishment of visual stimuli?

Although we understand a great deal about the anatomy of the LGN, our understanding of the LGN's function, especially with respect to the LGN's extraretinal inputs, is largely a proverbial 'black box'. This is a direct consequence of the fact that for far too long the process of 'vision' has been considered strictly a cortical phenomenon. The bulk of this chapter along with the questions listed above, however, demonstrate that a complete understanding of visual information processing will remain beyond our reach until research shifts subcortically, to the array of non-retinal inputs that continually and selectively modulate the visual stream.

Acknowledgments

We thank Dr Octavio Ruiz, Julia Mavity-Hudson, James L. Enlow, and Maria Couppis for helpful suggestions on the manuscript. Supported by 1F31NS44691 (DWR), EY01778 (VAC), NSF IBN-0234646 (VAC), and core grants EY08126 and HD15052.

References

Benevento, L.A. and Yoshida, K., 1981. The afferent and efferent organization of the lateral geniculo-prestriate pathways in the macaque monkey. *J. Comp. Neurol.* **203**, 455–474.

Bickford, M.E., Carden, W.B. and Patel, N.C., 1999. Two types of interneurons in the cat visual thalamus are distinguished by morphology, synaptic connections, and nitric oxide synthase content. *J. Comp. Neurol.* **413**(1), 83–100.

Bickford, M., Ramcharan, E., Godwin, D. *et al.*, 2000. Neurotransmitters contained in the subcortical extraretinal inputs to the monkey lateral geniculate nucleus. *J. Comp. Neurol.* **424**, 701–717.

Bickle, J., Bernstein, M., Heatley, M. *et al.*, 1999. A functional hypothesis for LGN-V1-TRN connectivities suggested by computer simulation. *J. Comput. Neurosci.* **6**, 251–261.

Bloomfield, S.A. and Sherman, S.M., 1989. Dendritic current flow in relay cells and interneurons of the cat's lateral geniculate nucleus. *PNAS* **86**(10), 3911–3914.

Bouthenet, M.L., Ruat, M., Salès, N. *et al.*, 1988. A detailed mapping of histamine H_1-receptors in guinea pig central nervous system established by autoradiography with [^{125}i] iodobolpyramine. *Neuroscience* **26**, 553–600.

Büttner-Ennever, J.A., Cohen, B., Horn, A.K.E. *et al.*, 1996. Pretectal projections to the oculomotor complex of the monkey and their role in eye movements. *J. Comp. Neurol.* **366**, 348–359.

Casagrande, V.A., 1994. A third visual pathway to primate V1. *Trends Neurosci.* **17**, 305–310.

Casagrande, V.A. and Ichida, J.M., 2002a. The lateral geniculate nucleus, in *Physiology of the Eye* (ed F.H. Adler), C.V. Mosby Company, St Louis, MO, pp. 655–668.

Casagrande, V.A. and Ichida, J.M., 2002b. The visual cortex, in *Physiology of the Eye* (ed F.H. Adler), C.V. Mosby Company, St Louis, MO, pp. 669–685.

Casagrande, V.A. and Norton, T.T., 1991. The lateral geniculate nucleus: A review of its physiology and function, in *The Neural Basis of Visual Function* (ed A.G. Leventhal), *Vision and Visual Disfunction* Vol. 4 (ed J.R. Cronley-Dillon), MacMillan Press, London, pp. 41–84.

Casagrande, V.A. and Xu, X., 2004. Parallel visual pathways: A comparative perspective, in *The Visual Neurosciences* (eds L. Chalupa and J.S. Werner), MIT Press, pp. 494–506.

Chazot, P.L., Hann, V., Wilson, C. *et al.*, 2001. Immunological identification of the mammalian H3 histamine receptor in the mouse brain. *Neuroreport* **12**(2), 259–262.

Chen, C. and Regehr, W.G., 2003. Presynaptic modulation of the retinogeniculate synapse. *J. Neurol.* **23**, 3130–3135.

Chen, W., Kato, T., Zhu, X.H. *et al.*, 1998. Human primary visual cortex and lateral geniculate nucleus activation during visual imagery. *Neuroreport* **9**, 3669–3674.

Collins, C.E., Lyon, D.C., and Kaas, J.H., 2003. Responses of neurons in the middle temporal visual area after long-standing lesions of the primary visual cortex in adult new world monkeys. *J. Neurosci.* **23**(6), 2251–2264.

Conley, M. and Raczkowski, D., 1990. Sublaminar organization within layer VI of the striate cortex in Galago. *J. Comp. Neurol.* **302**, 425–436.

Conley, M., Birecree, E., and Casagrande, V.A., 1985. Neuronal classes and their relation to functional and laminar organization of the lateral geniculate nucleus: A Golgi study of the prosimian primate, *Galago crassicaudatus. J. Comp. Neurol.* **242**, 561–583.

Coutinho, V. and Knopfel, T., 2002. Metabatropic glutamate receptors: Electrical and chemical signaling properties. *Neuroscientist* **8**, 551–561.

Cox, C.L. and Sherman, S.M., 1999. Glutamate inhibits thalamic reticular neurons. *J. Neurosci.* **19**(15), 6694–6699.

Cox, C.L., Zhou, Q., and Sherman, S.M., 1998. Glutamate locally activates dendritic outputs of thalamic interneurons. *Nature* **394**(6692), 478–482.

Crick, F. and Koch, C., 1990. Towards a neurobiological theory of consciousness. *Semin. Neurosci.* **2**, 263–275.

Cucchiaro, J.B., Bickford, M.E., and Sherman, S.M., 1991. A gabaergic projection from the pretectum to the dorsal lateral geniculate-nucleus in the cat. *Neuroscience* **41**(1), 213–226.

Cui, H. and Malpeli, J.G., 2003. Activity in the parabigeminal nucleus during eye movements directed at moving and stationary targets. *J. Neurophysiol.* **89**, 3128–3142.

Dan, Y., Alonso, J.M., Usrey, W.M. *et al.*, 1998. Coding of visual information by precisely correlated spikes in the lateral geniculate nucleus. *Nat. Neurosci.* **1**, 501–507.

Das, V.E., Economides, J.R., and Ono, S., 2001. Information processing by parafoveal cells in the primate nucleus of the optic tract. *Exp. Brain Res.* **140**(3), 301–310.

Erisir, A., Van Horn, S.C., and Sherman, S.M., 1997. Relative numbers of cortical and brainstem inputs to the lateral geniculate nucleus. *PNAS* **94**(4), 1517–1520.

Erisir, A., Van Horn, S.C., and Sherman, S.M., 1998. Distribution of synapses in the lateral geniculate nucleus of the cat: Differences between laminae A and A1 and between relay cells and interneurons. *J. Comp. Neurol.* **390**(2), 247–255.

Eyding, D., Macklis, J.D., Neubacher, U. *et al.*, 2003. Selective elimination of corticogeniculate feedback abolishes the electroencephalogram dependence of primary visual cortical receptive fields and reduces their spatial specificity. *J. Neurosci.* **23**(18), 7021–7033.

Feig, S. and Harting, J.K., 1992. Ultrastructural studies of the primate parabigeminal nucleus – electron-microscopic autoradiographic analysis of the tectoparabigeminal projection in *Galago crassicaudatus. Brain Res.* **595**(2), 334–338.

Feig, S. and Harting, J.K., 1994. Ultrastructural studies of the primate lateral geniculate nucleus: Morphology and spatial relationships of axon terminals arising from the retina, visual cortex (area 17), superior colliculus, parabigeminal nucleus, and pretectum of *Galago crassicaudatus. J. Comp. Neurol.* **343**, 17–34.

Fitzgibbon, T., 2002. Organization of reciprocal connections between the perigeniculate nucleus and dorsal lateral geniculate nucleus in the cat: A transneuronal transport study. *Vis. Neurosci.* **19**(4), 511–520.

Fitzpatrick, D. and Diamond, I.T., 1980. Distribution of acetylcholinesterase in the geniculo striate system of *Galago senegalensis* and *Aotus trivirgatus*: Evidence for the origin of the reaction product in the lateral geniculate body. *J. Comp. Neurol.* **194**(4), 703–719.

Fitzpatrick, D., Diamond, I.T., and Raczkowski, D., 1989. Cholinergic and monoaminergic innervation of the cat's thalamus: Comparison of the lateral geniculate nucleus with other principal sensory nuclei. *J. Comp. Neurol.* **288**, 647–675.

Fitzpatrick, D., Usrey, W.M., Schofield, B.R. *et al.*, 1994. The sublaminar organization of corticogeniculate neurons in layer 6 of macaque striate cortex. *Vis. Neurosci.* **11**(2), 307–315.

Funke, K. and Eysel, U.T., 1995. Possible enhancement of GABAergic input to cat dorsal lateral geniculate relay cells by serotonin. *Neuroreport* **6**, 474–476.

Godwin, D.W., Vaughan, J.W., and Sherman, S.M., 1996a. Metabotropic glutamate receptors switch response mode of LGN neurons from burst to tonic. *J. Neurophysiol.* **76**, 1800–1816.

Godwin, D.W., Van Horn, S.C., Gunluk, A.E. *et al.*, 1996b. Ultrastructural localization suggests that retinal and cortical inputs access different metabotropic glutamate receptors in the cat lateral geniculate nucleus. *J. Neurosci.* **16**, 8181–8192.

Graham, J., Lin, C.S., and Kaas, J.H., 1979. Subcortical projections of six visual cortical areas in the owl monkey, *Aotus trivirgatus. J. Comp. Neurol.* **187**, 557–580.

Graybiel, A.M. and Ragsdale, C.W. Jr, 1982. Pseudocholinesterase staining in the primary visual pathway of the macaque monkey. *Nature* **299**(5882), 439–442.

Guillery, R.W. and Sherman, S.M., 2002. Thalamic relay functions and their role in corticocortical communication: Generalizations from the visual system. *Neuron* **33**, 1–20.

Guillery, R.W., Feig, S.L., and Lozsádi, D.A., 1998. Paying attention to the thalamic reticular nucleus. *Trends Neurosci.* **21**, 28–32.

Harting, J.K., 1977. Descending pathways from the superior collicullus: An autoradiographic analysis in the rhesus monkey (*Macaca mulatta*). *Comp. Neurol.* **173**(3), 583–612.

Harting, J.K., Hashikawa, T., and van Lieshout, D., 1986. Laminar distribution of tectal, parabigeminal and pretectal inputs to the primate dorsal lateral geniculate nucleus: Connectional studies in *Galago crassicaudatus. Brain Res.* **366**(1–2), 358–363.

Harting, J.K., Huerta, M.F., Hashikawa, T. *et al.*, 1991a. Projection of the mammalian superior colliculus upon the dorsal lateral geniculate nucleus: Organization of tectogeniculate pathways in nineteen species. *J. Comp. Neurol.* **304**(2), 275–306.

Harting, J.K., van Lieshout, D.P., and Feig, S., 1991b. Connectional studies of the primate lateral geniculate nucleus: Distribution of axons arising from the thalamic reticular nucleus of *Galago crassicaudatus. J. Comp. Neurol.* **310**(3), 411–427.

Harting, J.K., van Lieshout, D.P., Hashikawa, T. *et al.*, 1991c. The parabigeminogeniculate projection: Connectional studies in 8 mammals. *J. Comp. Neurol.* **305**, 559.

Hendry, S.H.C. and Reid, R.C., 2000. The koniocellular pathway in primate vision. *Annu. Rev. Neurosci.* **23**, 127–153.

Hendry, S.H.C., Huntsman, M.M., Vinuela, A. *et al.*, 1994. Gaba(A) receptor subunit immunoreactivity in primate visual-cortex – distribution in macaques and humans and regulation by visual input in adulthood. *J. Neurosci.* **14**(4), 2383–2401.

Hernandez-Cruz, A. and Pape, H.C., 1989. Identification of two calcium currents in acutely dissociated neurons from the rat lateral geniculate nucleus. *J. Neurophysiol.* **61**(6), 1270–1283.

Hillenbrand, U. and van Hemmen, J.L., 2001. Does corticothalamic feedback control cortical velocity tuning? *Neural Comput.* **13**, 327–355.

Ichida, J.M. and Casagrande, V.A., 2002. Organization of the feedback pathway from striate cortex (V1) to the lateral geniculate nucleus (LGN) in a simian primate (*Aotus trivirgatus*). *J. Comp. Neurol.* **454**, 272–283.

Ichida, J.M., Shostak, Y., Royal, D.W. *et al.*, 2004. Corticogeniculate axons in the prosimian primate, bush baby: Morphology, layer terminations and synaptic contacts. *J. Neurosci.* (submitted).

Irvin, G.E., Norton, T.T., Sesma, M.A. *et al.*, 1986. W-like response properties of interlaminar zone cells in the lateral geniculate nucleus of a primate (*Galago crassicaudatus*). *Brain Res.* **362**(2), 254–270.

Irvin, G.E., Casagrande, V.A., and Norton, T.T., 1993. Center/surround relationships of magnocellular, parvocellular, and koniocellular relay cells in primate lateral geniculate nucleus. *Vis. Neurosci.* **10**, 363–373.

Jones, E.G., 2002. Thalamic circuitry and thalamocortical synchrony. *Philos. Trans. R. Soc. Lond. B Biol. Sci.* **357**(1428), 1659–1673.

Kenet, T., Bibitchkov, D., Tsodyks, M. *et al.*, 2003. Spontaneously emerging cortical representations of visual attributes. *Nature* **425**(6961), 954–956.

Lachica, E.A. and Casagrande, V.A., 1993. The morphology of collicular and retinal axons ending on small relay (W-like) cells of the primate lateral geniculate nucleus. *Vis. Neurosci.* **10**(3), 403–418.

Le Masson, G., Le Masson, S., Debay, D. *et al.*, 2002. Feedback inhibition controls spike transfer in hybrid thalamic circuits. *Nature* **417**(20), 854–858.

Lee, D. and Malpeli, J.G., 1998. Effects of saccades on the activity of neurons in the cat lateral geniculate nucleus. *J. Neurophysiol.* **79**, 922–936.

Lin, J.S., Sakai, K., and Jouvet, M., 1988. Evidence for histaminergic arousal mechanisms in the hypothalamus of cat. *Neuropharmacology* **27**(2), 111–122.

Lin, J.S., Sakai, K., Vanni-Mercier, G. *et al.*, 1990. Involvement of histaminergic neurons in arousal mechanisms demonstrated with H3-receptor ligands in the cat. *Brain Res.* **523**(2), 325–330.

Lund, J.S., Lund, R.D., Bunt, A.H. *et al.*, 1975. The origin of efferent pathways from the primary visual cortex, area 17, of the macaque monkey as shown by retrograde transport of horseradish peroxidase. *J. Comp. Neurol.* **164**, 287–303.

Mesulam, M.M., 1990. Human brain cholinergic pathways. *Prog. Brain Res.* **84**, 231–241.

McIlwain, J.T., 1995. Lateral geniculate lamination and the corticogeniculate projection: A potential role in binocular vision in the quadrants. *J. Theor. Biol.* **172**, 329–333.

McCormick, D.A., 1992. Neurotransmitter actions in the thalamus and cerebral cortex and their role in neuromodulation of thalamocortical activity. *Prog. Neurobiol.* **39**, 337–388.

McCormick, D.A. and Pape, H.-C., 1990. Noradrenergic and serotonergic modulation of a hyperpolarization-activated cation current in thalamic relay neurons. *J. Physiol.* **431**, 319–342.

McCormick, D.A. and Prince, D.A., 1986. Acetylcholine induces burst firing in thalamic reticular neurones by activating a potassium conductance. *Nature* **319**, 402–405.

McCormick, D.A. and Williamson, A., 1991. Modulation of neuronal firing mode in cat and guinea pig LGND by histamine – possible cellular mechanisms of histaminergic control of arousal. *J. Neurosci.* **11**, 3188–3199.

Merigan, W.H., Nealey, T.A., and Maunsell, J.H., 1993. Visual effects of lesions of cortical area V2 in macaques. *J. Neurosci.* **13**(7), 3180–3191.

Montero, V.M. and Robles, L., 1971. Saccadic modulation of cell discharges in the lateral geniculate nucleus. *Vision Res.* **3** (Suppl.), 253–268.

Monti, J.M., 1993. Involvement of histamine in the control of the waking state. *Life Sci.* **53**(17), 1331–1338.

Morrison, J.H. and Foote, S.L., 1986. Noradrenergic and serotonergic innervation of cortical, thalamic, and tectal visual structures in old and new world monkeys. *J. Comp. Neurol.* **243**, 117–138.

Murphy, P.C. and Sillito, A.M., 1987. Corticofugal feedback influences the generation of length tuning in the visual pathway. *Nature* **329**, 727–729.

Newsome, W.T., 1996. Visual attention: Spotlights, highlights and visual awareness. *Curr. Biol.* **6**(4), 357–360.

Norton, T.T. and Casagrande, V.A., 1982. Laminar organization of receptive-field properties in the lateral geniculate nucleus of bush baby (*Galago crassicaudatus*). *J. Neurophysiol.* **47**(4), 715–741.

Norton, T.T., Casagrande, V.A., Irvin, G.E. *et al.*, 1988. Contrast-sensitivity functions of W-, X-, and Y-like relay cells in the lateral geniculate nucleus of bush baby, *Galago crassicaudatus*. *J. Neurophysiol.* **59**, 1639–1656.

Nucci, C., Morrone, L., Rombola, L. *et al.*, 2003. Multifaceted roles of nitric oxide in the lateral geniculate nucleus: From visual signal transduction to neuronal apoptosis. *Toxicol. Lett.* **139**, 163–173.

Pasik, P., Pasik, T., and Hamori, J., 1986. A newly recognized element in the monkey dorsal lateral geniculate nucleus exhibiting both presynaptic and postsynaptic sites. *J. Neurocytol.* **15**(2), 177–186.

Przybyszewski, A.W., Gaska, J.P., Foote, W. *et al.*, 2000. Striate cortex increases contrast gain of macaque LGN neurons. *Vis. Neurosci.* **17**(4), 485–494.

Ramcharan, E.J., Gnadt, J.W., and Sherman, S.M., 2000. Burst and tonic firing in thalamic cells of unanesthetized, behaving monkeys. *Vis. Neurosci.* **17**(1), 55–62.

Ramcharan, E.J., Gnadt, J.W., and Sherman, S.M., 2001. The effects of saccadic eye movements on the activity of geniculate relay neurons in the monkey. *Vis. Neurosci.* **18**, 253–258.

Riggs, L.A., Merton, P.A., and Morton, H.B., 1974. Suppression of visual phosphenes during saccadic eye movements. *Vision Res.* **14**, 997–1011.

Ross, J., Burr, D., and Morrone, C., 1996. Suppression of the magnocellular pathway during saccades. *Behav. Brain Res.* **80**, 1–8.

Royal, D., Sáry, Gy, Schall, J. *et al.*, 2003. Are spike bursts and pseudo-bursts in the lateral geniculate nucleus (LGN) related to behavioral events? *Soc. Neurosci.* (Abstract).

Royal, D., Sáry, Gy, Schall, J. *et al.*, 2004. Does the lateral geniculate nucleus (LGN) pay attention? *Vision Sci. Soc.*, *J. Vis.* **4**(8), 622a (Abstract).

Royal, D., Sáry, Gy, Schall, J. *et al.*, 2005. Evidence of behavioral modulation of macaque monkey lateral geniculate nucleus (LGN) cell activity during saccades. Exp. Brain Res. (submitted).

Ruat, M., Traiffort, E., Bouthenet, M.L. *et al.*, 1990. Reversible and irreversible labeling and autoradiographic localization of the cerebral histamine H_2 receptor using [^{125}i]-iodinated probes. *PNAS* **87**, 1658–1662.

Salt, T.E., 2002. Glutamate receptor functions in sensory relay in the thalamus. *Philos. Trans. R. Soc. Lond. B Biol. Sci.* **357**, 1759–1766.

Schiller, P.H. and Malpeli, J.G., 1977. The effect of striate cortex cooling on area 18 cells in the monkey. *Brain Res.* **126**(2), 366–369.

Schmidt, M., 1996. Neurons in the cat pretectum that project to the dorsal lateral geniculate nucleus are activated during saccades. *J. Physiol.* **76**, 2907–2918.

Schmolesky, M.T., Wang, Y., Hanes, D.P. *et al.*, 1998. Signal timing across the macaque visual system. *J. Neurophysiol.* **79**, 3272–3278.

Sherk, H., 1979. Connections and visual-field mapping in cat's tectoparabigeminal circuit. *J. Neurophysiol.* **42**, 1656–1668.

Sherman, S.M. and Guillery, R.W., 1996. Functional organization of thalamocortical relays. *J. Neurophysiol.* **76**(3), 1367–1395.

Sherman, S.M. and Guillery, R.W., 1998. On the actions that one nerve cell can have on another: Distinguishing "drivers" from "modulators". *PNAS* **95**, 7121–7126.

Sherman, S.M. and Guillery, R.W., 2001. *Exploring the Thalamus*, Academic Press, San Diego.

Sherman, S.M. and Guillery, R.W., 2002. The role of thalamus in the flow of information to cortex. *Philos. Trans. R. Soc. Lond.* **357**, 1695–1708.

Sillito, A.M. and Jones, H.E., 2002. Corticothalamic interactions in the transfer of visual information. *Philos. Trans. R. Soc. Lond.* **357**(1428), 1739–1752.

Sillito, A.M., Jones, H.E., Gerstein, G.L. *et al.*, 1994. Feature-linked synchronization of thalamic relay cell firing induced by feedback from the visual cortex. *Nature* **369**, 479–482.

Stanley, G.B., Li, F.F., and Dan, Y., 1999. Reconstruction of natural scenes from ensemble responses in the lateral geniculate nucleus. *J. Neurosci.* **19**(18), 8036–8042.

Steriade, M. and Deschênes, M., 1984. The thalamus as a neuronal oscillator. *Brain Res.* **32**, 1–63.

Steriade, M. and Llinás, R., 1988. The functional states of the thalamus and the associated neuronal interplay. *Physiol. Rev.* **68**, 649–742.

Swadlow, H.A. and Gusev, A.G., 2001. The impact of 'bursting' thalamic impulses at a neocortical synapse. *Nat. Neurosci.* **4**(4), 402–408.

Symonds, L.L. and Kaas, J.H., 1978. Connections of striate cortex in the prosimian, *Galago senegalensis. J. Comp. Neurol.* **181**, 477–512.

Thilo, K.V., Santoro, L., Walsh, V. *et al.*, 2004. The site of saccadic suppression. *Nat. Neurosci.* **7**, 13–14.

Turner, J.P. and Salt, T.E., 2000. Synaptic activation of the group I metabotropic glutamate receptor mGlu1 on the thalamocortical neurons of the rat dorsal lateral geniculate nucleus in vitro. *Neuroscience* **100**, 493–505.

Uhlrich, D.J., Tamamaki, N., Murphy, P.C. *et al.*, 1995. The effects of brain stem parabrachial activation on receptive field properties of cells in the cat's lateral geniculate nucleus. *J. Neurophysiol.* **73**, 2428–2447.

Usrey, W.M. and Reid, R.C., 1999. Synchronous activity in the visual system. *Annu. Rev. Physiol.* **61**, 435–456.

Vanni-Mercier, G., Sakai, K., and Jouvet, M., 1984. Specific neurons for wakefulness in the posterior hypothalamus in the cat. *C. R. Acad. Sci. III* **298**(7), 195–200.

von Krosigk, M., Monckton, J.E., Reiner, P.B. *et al.*, 1999. Dynamic properties of corticothalamic excitatory postsynaptic potentials and thalamic reticular inhibitory postsynaptic potentials in thalamocortical neurons of the guinea-pig dorsal lateral geniculate nucleus. *Neuroscience* **91**(1), 7–20.

Wahle, P., Stuphorn, V., Schmidt, M. *et al.*, 1994. LGN-projecting neurons of the cat's pretectum express glutamic acid decarboxylase mRNA. *Eur. J. Neurosci.* **6**(3), 454–460.

Walker, A.E., 1938. *The Primate Thalamus*, University Press, Chicago.

Wang, S., Bickford, M.E., Van Horn, S.C. *et al.*, 2001. Synaptic targets of thalamic reticular nucleus terminals in the visual thalamus of the cat. *J. Comp. Neurol.* **440**(4), 321–341.

Wang, S., Eisenback, M., Datskovskaia, A. *et al.*, 2002. GABAergic pretectal terminals contact GABAergic interneurons in the cat dorsal lateral geniculate nucleus. *Neurosci. Lett.* **323**, 141–145.

Wiencken, A.E. and Casagrande, V.A., 2000. The distribution of NADPH diaphorase and nitric oxide synthetase (NOS) in relation to the functional compartments of areas V1 and V2 of primate visual cortex. *Cereb. Cortex* **10**, 490–511.

Wiesel, T.N. and Hubel, D.H., 1966. Spatial and chromatic interactions in the lateral geniculate body of the rhesus monkey. *J. Neurophysiol.* **29**, 1115–1156.

Wilson, J.R., 1986. Synaptic connections of relay and local circuit neurons in the monkey's dorsal lateral geniculate nucleus. *Neurosci. Lett.* **66**, 79–84.

Wilson, J.R., 1993. Circuitry of the dorsal lateral geniculate nucleus in the cat and monkey. *Acta Anat.* **147**, 1–13.

Wilson, J.R. and Forestner, D.M., 1995. Synaptic inputs to single neurons in the lateral geniculate nuclei of normal and monocularly deprived squirrel monkeys. *J. Comp. Neurol.* **362**, 468–488.

Wilson, J.R. and Hendrickson, A.E., 1988. Serotoninergic axons in the monkey's lateral geniculate nucleus. *Vis. Neurosci.* **1**, 125–133.

Wilson, J.R., Manning, K.A., Forestner, D.M. *et al.*, 1999. Comparison of cholinergic and histaminergic axons in the lateral geniculate complex of the macaque monkey. *Anat. Rec.* **255**(3), 295–305.

Worgotter, F., Eyding, D., Macklis, J.D. *et al.*, 2002. The influence of the corticothalamic projection on responses in thalamus and cortex. *Philos. Trans. R. Soc. Lond. B Biol. Sci.* **357**(1428), 1823–1834.

Wurtz, R.H., Goldberg, M.E., and Robinson, D.L., 1982. Brain mechanisms of visual attention. *Sci. Am.* **246**(6), 124–135.

Xu, X., Bonds, A.B., and Casagrande, V.A., 2002. Modeling receptive-field structure of koniocellular, magnocellular, and parvocellular LGN cells in the owl monkey (*Aotus trivirgatus*). *Vis. Neurosci.* **19**(6), 703–710.

Zhu, J.J. and Lo, F.S., 1996. Time course of inhibition induced by a putative saccadic suppression circuit in the dorsal lateral geniculate nucleus of the rabbit. *Brain Res. Bull.* **41**, 281–291.

Zuber, B.L. and Stark, L., 1966. Saccadic suppression: Elevation of visual threshold associated with saccadic eye movements. *Exp. Neurol.* **16**, 65–79.

8

Visual Functions of the Retinorecipient Nuclei in the Midbrain, Pretectum, and Ventral Thalamus of Primates

Michael R. Ibbotson and ***Bogdan Dreher***

We dedicate this chapter to the memory of Robert (Bob) W. Rodieck (1937–2003), a personal friend of one of us (BD), powerful analytical intellect, great and accomplished scholar and one of the main contributors to research on morphology, pattern of projections and physiology of mammalian (especially primate) retinal ganglion cells.

In this chapter we discuss the morphology, connectivity (hodology), and functions of neurons in the visual nuclei of the midbrain, the pretectum, and the pregeniculate complex.

8.1 Superior colliculus

The superior colliculi (SC), which is Latin for upper hills, are the principal retino-recipient nuclei of the mammalian mesencephalon and constitute two symmetrical protuberances on the tectum (Latin for roof) of the midbrain. Morphologically and functionally, primate colliculus, as in non-primate mammals, is usually subdivided into the retino-recipient superficial layers and non-retino-recipient intermediate and deep layers. Overall, the superficial layers play an important role in selecting targets for visually-guided behavior, while intermediate and deep layers have important roles in integrating visual,

The Primate Visual System: A Comparative Approach Edited by Jan Kremers

somatosensory, and auditory inputs and in initiating goal-directed orienting responses toward sensory stimuli.

Early in the previous century, Ramón y Cajal described the neuronal morphology of the vertebrate optic tectum, including that of the SC of some mammals (Cajal, 1911). In visual mammals such as domestic cats and virtually all primates, the most conspicuous manifestation of orienting responses are rapid eye movements called 'saccades', which allow rapid voluntary changes in gaze direction. The retinas of virtually all primates have a small, specialized area of very high visual acuity, the *fovea centralis* (Latin for central pit). The *fovea* is moved around to generate a high resolution map of the visual world across the visual field (Ross *et al.*, 1997). In primates saccades can have large amplitudes (30° or more) and are not necessarily accompanied by head movements (for review, Sparks and Hartwich-Young, 1989; Guitton, 1992). Raymond Dodge (1902) stated the function of what we now call saccades as: 'fundamentally reactions to eccentric retinal stimulation, . . . dependent on the tendency . . . to move the eyes so that the point of interest will be seen with the visual center of the retina'. The term 'saccade' is derived from the French word *saquer*, which refers to jerking a horse's head by pulling on the reins (introduced to describe rapid eye movements in humans, e.g. Landolt, 1891, cited after Westheimer, 1989). There is a strong correlation between saccade amplitude and its mean velocity, larger amplitudes equating to higher mean velocities. Saccades in humans are slower than those in macaque monkeys. Thus, during 10° and 20° saccades in humans the maximal eye velocities are, respectively, 350°/s and 490°/s (Westheimer, 1954), while in macaques maximal eye velocities at saccades of the same displacement are 570°/s and 810°/s (Fuchs, 1967). Saccades require the integration of information about the location of objects in space and the appropriate motor commands to move the eyes. The involvement of the SC in this process had already been suggested as early as the second half of the nineteenth century (e.g. Adamük, 1870; cited by Pasik *et al.*, 1966).

8.1.1 Laminar organization

In all mammals, when the SC is sectioned coronally and stained with cationic dyes for 'Nissl substance' (nucleic acids), at least seven alternating layers are apparent. The layers are arranged concentrically around the dorsolateral periaqueductal grey (PAG), which surrounds the cerebral aqueduct (Figures 8.1B and 8.2; for reviews, Kaas and Huerta, 1988; Rhoades *et al.*, 1991; Sefton *et al.*, 2004).

Superficial layers (I–III): The most superficial layer, the *stratum zonale* (SZ, zonal layer, layer I), is very thin (20–50 μm), relatively cell-poor but rich in myelinated fibers and dendrites arising from neurons in layer II. SZ neurons consist of the horizontal and marginal cells whose somal diameters are about 10 μm and whose dendrites are oriented parallel to the dorsal surface of the layer. The next layer, the *stratum griseum superficiale* (SGS, superficial gray layer, layer II), is thicker (250–400 μm in primates), cell-rich, and usually subdivided into the larger upper and smaller lower sublayers. Both sublayers stain heavily and uniformly for the enzyme acetylcholinesterase (Illing, 1996). The upper sublayer stains heavily for the mitochondrial enzyme cytochrome oxidase but the lower sublayer less so. The third layer, the *stratum opticum* (SO, optic layer, layer III), is cell-poor but thick (200 μm in prosimians and small monkeys, 500 μm in larger

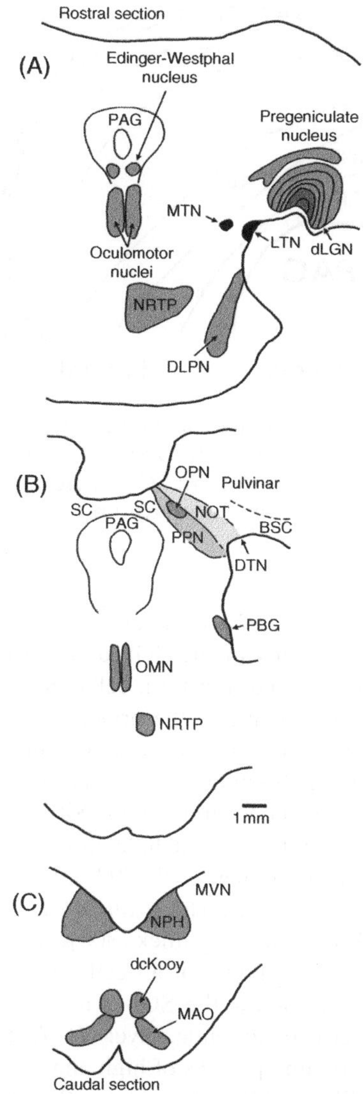

Figure 8.1 Composite coronal sections through the caudal thalamus, midbrain, and medulla of a macaque. PAG (periaqueductal grey), MTN (medial terminal nucleus), LTN (lateral terminal nucleus), DTN (dorsal terminal nucleus – not a well-defined region in macaques), dLGN (dorsal lateral geniculate nucleus), NRTP (*nucleus reticularis tegmentis pontis*), DLPN (dorsolateral pontine nucleus), SC (superior colliculus), BSC (brachium of the superior colliculus), OMN (oculomotor nuclei), OPN (olivary pretectal nucleus), NOT (nucleus of the optic tract), PPN (posterior pretectal nucleus), PBG (parabigeminal nucleus), NPH (*nucleus prepositus hypoglossi*), MVN (medial vestibular nucleus), MAO (medial accessory nucleus of the inferior olive), dcKooy (dorsal cap of Kooy)

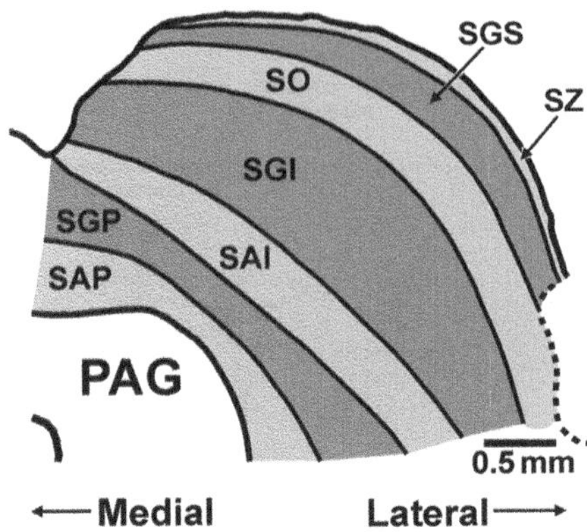

Figure 8.2 Coronal section through the superior colliculus of a squirrel monkey. Dark grey: cell-rich layers; light grey: fiber-rich layers. SZ: *stratum zonale*; SGS: *stratum griseum superficiale*; SO: *stratum opticum*; SGI: *stratum griseum intermediale*; SAI: *stratum album intermediale*; SGP: *stratum griseum profundum*; SAP: *stratum album profundum*

monkeys, apes, and humans). The cells' somata are of medium size and the layer stains lightly for cytochrome oxidase. Cells in the SGS and SO include 'wide-field vertical cells', 'narrow-field vertical cells', and 'pyriform cells' and are relay neurons projecting to the deeper SC. Cells immunoreactive to the principal inhibitory neurotransmitter of the vertebrate central nervous system, GABA (gamma-aminobutyric acid), and/or the GABA synthesizing enzyme GAD (glutamic acid decarboxylase) are present in virtually all collicular layers. However, they concentrate in the SZ and SGS, where they constitute a third of all neurons. GABAergic neurons include all cell types and have small somata (9–15 μm in diameter; Mize, 1992; Behan *et al.*, 2002).

Intermediate and deep layers (IV–VII): The *stratum griseum intermediale* (SGI, intermediate grey, layer IV) is cell-rich and very thick (800 μm in prosimians; up to 1300 μm in macaques). The SGI contains a mixture of all cell sizes, with its dorsal part staining heavily for cytochrome oxidase. Below the SGI is a thin cell-poor layer, the *stratum album intermediale* (SAI, intermediate white, layer V). Acetylcholinesterase activity is distributed in a mosaic of alternating patches of high and low enzyme activities throughout the intermediate layers (Illing, 1996). The next cell-rich layer, the *stratum griseum profundum* (SGP, deep white, layer VI) is separated from the central grey by the cell-poor *stratum album profundum* (SAP, deep white, layer VII). There are fewer GABAergic neurons in the deeper layers but the ratio of GABAergic neurons to the total cell count in all layers is fairly similar (Mize, 1992; Behan *et al.*, 2002).

8.1.2 Superficial layers

Retinotopic organization

In the superficial layers visually responsive neurons are organized into retinotopic maps of visual space. Thus, stimulation of a specific region of the retina generates responses in neurons located in specific 'slabs' of the SC oriented radially to the surface (Figure 8.3).

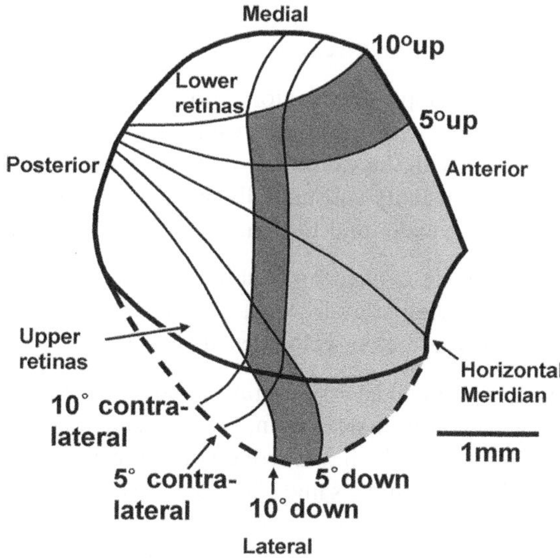

Figure 8.3 Bird's eye view of the retinotopic map in the superficial layers of the macaque superior colliculus (Cynader and Berman, 1972). The light grey area represents regions of the contralateral visual field within 5° of the *fovea*. The light and dark grey areas combined show parts of the contralateral visual field within 10° of the *fovea*. Signals measured in the medial SC (upper segments) arise from the lower retina of both eyes and therefore view the upper visual field. Reproduced by the permission of the American Physiological Society

Visual response properties

The receptive field (RF) properties of neurons in the superficial SC layers of Old World macaque monkeys have been described in numerous studies (anaesthetized: Schiller and Koerner, 1971; Cynader and Berman, 1972; Marrocco and Li, 1977; Moors and Vendrik, 1979a,b; Rizzolatti *et al.*, 1980; non-anaesthetized, alert: Goldberg and Wurtz, 1972a; Schiller *et al.*, 1974). These properties are virtually identical to that of neurons in the superficial layers of anaesthetized New World diurnal *Cebus* monkeys (Updyke, 1974). Response properties can be summarized as follows:

(1) Virtually all neurons respond to visual stimuli presented in their RFs and exhibit moderate levels of 'spontaneous' activity. The spontaneous activity is higher in alert animals.

(2) Most neurons respond to stationary flashing stimuli presented inside an excitatory RF with transient bursts of action potentials at light onset and offset and the ON and OFF discharge regions tend to spatially overlap. This property correlates with the high proportion of SC-projecting retinal ganglion cells that have dendritic trees bistratified into the ON (inner) and OFF (outer) part of the inner plexiform layer where, respectively, the ON- and OFF- center bipolar cells terminate (Wässle and Boycott, 1991).

(3) The excitatory RFs are circular or slightly elliptical in shape, their sizes increasing with retinal eccentricity (distance from fovea centralis). They are rarely orientation selective.

(4) RFs of most cells contain suppressive regions surrounding the excitatory area. Stimulation of suppressive regions *per se* does not result in spiking activity. The suppressive regions partially overlap with the excitatory regions as indicated by the fact that the optimum stimulus size is usually substantially smaller than the size of the excitatory region (Figure 8.4; e.g. Cynader and Berman, 1972).

(5) For most cells the responses evoked by flashing stationary stimuli are much weaker than those evoked by moving stimuli. Most cells respond reliably at low velocities of 0.5–30°/s but the most effective velocities are 30–800°/s.

Neurons in the primate superficial layers generally have RF characteristics very similar to those in the superficial SC of non-primates (Stein, 1981; Chalupa, 1984; Stein and Meredith, 1991). Furthermore, habituation, which is a reduction in the magnitude of responses to repetitive stimulation (0.5–2 s) with the same stimulus, is a common feature of superficial neurons of most mammals (e.g. macaques: Goldberg and Wurtz, 1972a,b; Rizzolatti *et al.*, 1980; cats: Harutunian-Kozak *et al.*, 1971; Binns and Salt, 1995; mice: Dräger and Hubel, 1975; rabbits: Oyster and Takahashi, 1975; hamsters: Chalupa and Rhoades, 1977; rats: Binns and Salt, 1997). The suppressive regions within the RFs of collicular cells appear to be based on the input from inhibitory interneurons containing GABA (Mize, 1992; Behan *et al.*, 2002). Thus, in the rat iontophoretic application of bicuculline, a selective blocker of ionotropic $GABA_A$ receptors, substantially reduces the strength of the suppressive regions in the RFs of superficial collicular cells (Binns and Salt, 1997). Similarly, GABA-mediated inhibition is involved in the mechanism underlying

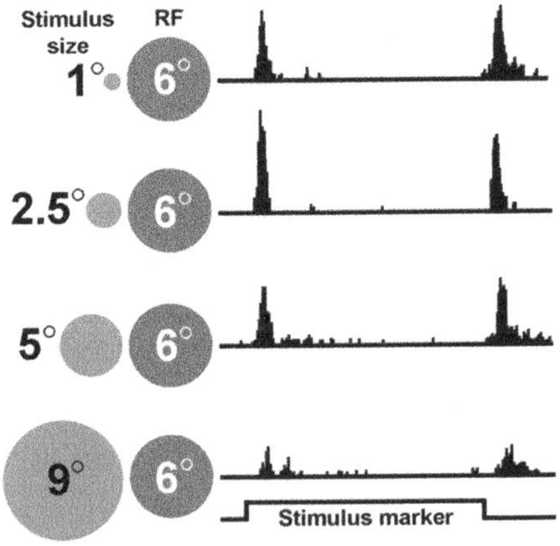

Figure 8.4 Responses of a superficial SC neuron in macaque monkey to spots of light turned ON and OFF (Cynader and Berman, 1972). Note that although the excitatory RF of the cell is 6° in diameter the responses to spot sizes of 1°–2.5° are largest. This suggests that suppressive regions partially overlap with the excitatory RF. Stimulation duration 1s. Reproduced by the permission of the American Physiological Society

habituation of responses in superficial cells. However, response habituation in rats is not reduced by the injection of bicuculline but rather by the iontophoretic application of CGP 35348, a selective blocker of metabotropic $GABA_B$ receptors (Binns and Salt, 1997).

In macaques and *Cebus* monkeys most neurons in the superficial collicular layers are activated by visual stimuli presented via either eye. Consistent with the numerically similar input from the contralateral nasal and ipsilateral temporal retinas, the magnitudes of responses to stimuli presented via each eye are usually very similar. This contrasts with the situation in other mammals with frontally-positioned eyes. For example, in cats the majority of superficial SC cells although binocular respond more strongly to stimuli presented via the contralateral eye (for review, Stein and Meredith, 1991; cf. Waleszczyk *et al.*, 1999; Hashemi-Nezhad *et al.*, 2003). In mammals with laterally-positioned eyes (e.g. nocturnal rodents: Stein, 1981) most cells in the superficial collicular layers are driven exclusively by the contralateral eye. This dominance of the contralateral eye is consistent with the very small input from the ipsilateral retina (Dreher *et al.*, 1985).

In macaque and *Cebus*, only 0–10 percent of superficial SC neurons exhibit direction selectivity (DS), i.e. strong responses to stimuli moving in one direction and weak or no responses to movement in the opposite direction (e.g. Cynader and Berman, 1972; Goldberg and Wurtz, 1972a; Updyke, 1974; Moors and Vendrik, 1979a; Rizzolatti *et al.*, 1980). Even so, in macaques most superficial cells become very sensitive to target direction and velocity when the targets are presented on a textured moving background that is larger than the cell's RF (Bender and Davidson, 1986). The responses of about 80 percent of cells in macaques are strongly suppressed when the target and background stimuli move in the same direction and at the same velocity, but when the directions of the target and background movement differ by 90° or more there is little reduction in response magnitude. The paucity of directional cells in the superficial layers of macaque and *Cebus* contrasts with the prevalence of strongly DS cells in New World squirrel monkeys (Kadoya *et al.*, 1971a) and cats (for reviews, Chalupa, 1984; Stein and Meredith, 1991).

There is a massive input to the superficial SC in macaques from cells in layer 5 of the ipsilateral primary visual cortex (area V1, striate cortex, cytoarchitectonic area 17; Figure 8.5), a third of which are strongly DS (Finlay *et al.*, 1976). Despite this, the removal or transient inactivation of area V1 has little effect on the magnitude of responses of SC neurons to visual stimuli or their RF organization (Schiller *et al.*, 1974). The same is true for those parts of cortical area V2 where the central visual field is represented. Reversible blockade of the parvocellular (Latin for small cells) layers of the dorsal lateral geniculate nucleus (dLGN: Figure 8.1A), which relay information from the midget retinal ganglion cells to area V1 (section on Visual inputs to the superficial SC layers; Rodieck, 1998; Kaplan, 2003), does not affect the magnitude of responses or the RF properties of cells in the superficial and intermediate SC of macaques (Schiller *et al.*, 1979). Similarly, reversible blockade of the magnocellular (Latin for small cells) layers of the dLGN, which relay information from the parasol retinal ganglion cells to area V1 (section on Visual inputs to the superficial SC layers), has little influence on the magnitude of responses to visual stimuli or the RF properties of superficial cells. Interestingly, the blockade of the magnocellular layers strongly reduces response magnitudes to visual stimuli of most cells in the intermediate layers of the ipsilateral SC (section 8.1.2; Schiller *et al.*, 1979). Transient inactivation of ipsilateral area V1 in cats results in strong but reversible attenuations of responses of SC neurons to visual stimuli and changes in

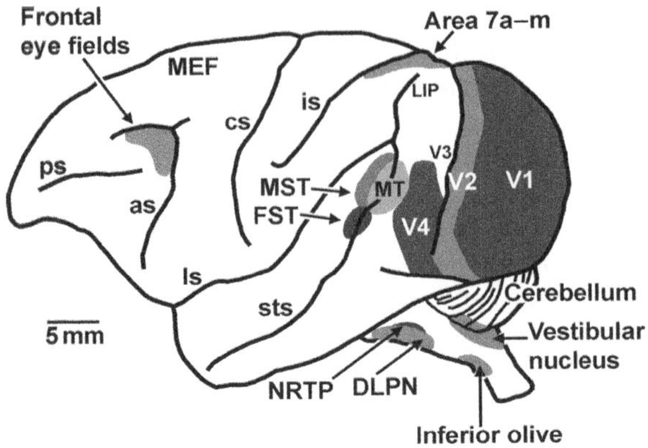

Figure 8.5 Lateral view of the macaque brain. Cortical areas: V1, V2, V3, V4, MT (middle temporal), MST (medial superior temporal), FST (fundus of the superior temporal), LIP (lateral interparietal cortex), MEF (medial eye field). Brainstem regions: NRTP and DLPN (Figure 8.1). Sulci: ps (principal), as (arcuate), ls (lateral), cs (central), is (intraparietal), sts (superior temporal). The inner edge of V2 marks the lunate sulcus (ls). Regions ahead of ls and behind sts are preoccipital cortex (PreOcc) in the text (Leichnetz, 1990)

DS indices (Stein and Meredith, 1991; Hashemi-Nezhad *et al.*, 2003). Furthermore, in cats, chronic removal of V1 usually results in a dramatic reduction in the proportion of DS cells in the superficial layers ipsilateral to the lesioned cortex (Stein and Meredith, 1991).

The effect of selective removal or inactivation of ipsilateral area middle temporal (MT) on the magnitude of responses and RF properties of cells in the primate superficial SC has not been studied. Area MT is a cortical area specializing in motion processing (Figure 8.5). However, in cats reversible inactivation of the posterior suprasylvian areas, which are presumed homologs of primate cortical areas around the superior temporal sulcus (including area MT; Payne, 1993), results in reduced response magnitudes and changes in RF properties in the intermediate collicular layers but not the superficial layers (Stein and Meredith, 1991).

Virtually all of the SC-projecting retinal ganglion cells are characterized by a lack of color opponency, low spontaneous activities, and slow conduction velocities (Marrocco and Li, 1977; Schiller and Malpeli, 1977). In their lack of spectral opponency these cells appear to correspond to some retinal ganglion cells that have atypical RF organizations in macaques (de Monasterio, 1978). SC-projecting retinal ganglion cells have small somata, presumed thin axons, and consequently exhibit slow conduction velocities. In view of the presence of bistratified ganglion cells among the SC-projecting cells and the presence of color-opponent cells in the SC of squirrel monkeys (Kadoya *et al.*, 1971a), it is possible that the primate colliculus receives input from relatively rare bistratified ganglion cells that have input from short-wavelength cone bipolar ON cells and from medium- and long-wavelength cone bipolar OFF cells (Dacey and Lee, 1992). If the blue-ON ganglion cells indeed project to primate SC, the colliculus might play a role in color vision along the blue-yellow axis, which is suggested to be phylogenetically old (Mollon, 1991).

Visual inputs to the superficial SC layers

Retinal input Visual input is carried mainly by axons of the retinal ganglion cells located in the contralateral nasal and ipsilateral temporal retinas and cells in infragranular layer 5 of the ipsilateral visual cortices (corticotectal cells). Together they form the brachium of the superior colliculus and enter the SC via the SO close to its rostral pole. In addition, the superficial SC is reciprocally connected with the parabigeminal nucleus (section on Outputs from the superficial SC layers) and pretectum (section 8.2).

The SC in mammals receives retinal input from both eyes (Figure 8.6). In non-primate mammals with laterally-positioned eyes virtually all ganglion cells project to the SC (rabbits: Vaney *et al.*, 1981; rats: Dreher *et al.*, 1985). In these species optic tract axons projecting to all other retino-recipient nuclei are axonal collaterals of cells projecting to the SC. Similarly, in non-primate species with frontally-positioned eyes approximately 80 percent of ganglion cells in the nasal retina project to the contralateral SC (cat: Wässle and Illing, 1980). The proportions of ganglion cells in the temporal retina projecting to either the ipsilateral or contralateral SC of the cat are small (25 percent in both cases: Wässle and Illing, 1980; Leventhal *et al.*, 1985). In macaque monkeys 6–10 percent of the retinal ganglion cells in the nasal and temporal retinas project, respectively, to the contralateral and ipsilateral SC (Schiller and Malpeli, 1977; Perry and Cowey, 1984; Rodieck and Watanabe, 1993; Williams *et al.*, 1995). The small proportion of cells projecting to the primate SC still represents a large number of neurons since there are 1.5–1.6 million ganglion cells in macaques (Rakic and Riley, 1983a,b; Perry and Cowey, 1985) and 1.1–1.25 million in humans (Provis *et al.*, 1985; Curcio and Allen, 1990). In primates there is virtually no representation of the ipsilateral visual hemifield in the SC (squirrel monkey: Kadoya *et al.*, 1971b; macaque: Cynader and Berman, 1972; *Cebus*: Updyke, 1974; New World owl monkey and prosimian bush baby: Lane *et al.*, 1973). By contrast, in most non-primate mammals the SC receives direct input not only from the contralateral nasal and ipsilateral temporal retinas but also from the contralateral

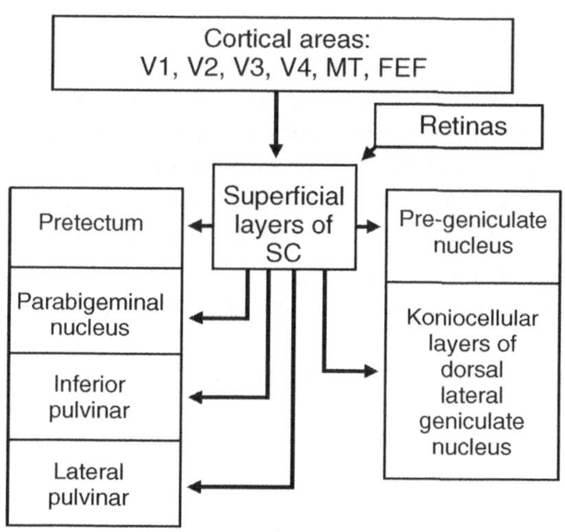

Figure 8.6 Connectivity of the superficial layers of the primate SC

temporal retina (eutherians: Huerta and Harting, 1984; metatherians such as opossum: Volchan *et al.*, 1982 or tammar Wallaby: Mark *et al.*, 1993).

The excitatory RFs of collicular neurons in behaving macaques make small encroachments of approximately 1° into the ipsilateral hemifield (Goldberg and Wurtz, 1972a). Furthermore, in the SC of *Cebus* monkeys some encroachments of suppressive RFs occur into the ipsilateral hemifield (Updyke, 1974). These small encroachments are most likely related to the nasotemporal overlap of retinal ganglion cells with RFs located within 0.5° of the representation of the vertical meridian separating the contralateral and ipsilateral visual hemifields (Stone *et al.*, 1973; Bunt *et al.*, 1977; Leventhal *et al.*, 1988; Fukuda *et al.*, 1989; Chalupa and Lia, 1991). In macaques about 6 percent of retinal ganglion cells on the very edge of the foveal pit project into the SC (Cowey and Perry, 1980; Perry and Cowey, 1984). The SC neurons with RFs in the *fovea* are unaffected when the ipsilateral visual cortex, which provides indirect visual input to the SC, is inactivated (Schiller *et al.*, 1974).

The input from the contralateral eye terminates mainly on neurons in the upper SGS. In prosimians the distribution of contralateral retinal terminals is fairly continuous, while in the New and Old World monkeys and apes the contralateral retinal input terminates in patches (Kaas and Huerta, 1988). In all primates the ipsilateral retinal input terminates in patches in the lower SGS. There is also a small direct retinal input to less numerous neurons in the SZ and SO. In cats there is also a small direct retinal input to the SGI (Berson, 1988).

The region of the retina in the vicinity of the *fovea* is represented in a greater part of the SC than the fraction of the retina occupied by this region (Figure 8.3). In primates, representation of the central 10° occupies about a third of the collicular surface (Kadoya *et al.*, 1971a; Cynader and Berman, 1972). In macaques the somata of perifoveal ganglion cells projecting to the SC are smaller than those of collicular projecting cells in the nasal or temporal periphery (Perry and Cowey, 1984). Nevertheless, the somata of SC-projecting ganglion cells, irrespective of eccentricity, tend to be small (10–16 μm diameter; Rodieck and Watanabe, 1993). However, on the basis of their dendritic morphology, ganglion cells projecting to macaque SC form at least three distinct groups (Rodieck, 1998). One group, termed the T-group, consists of cells with dense and bistratified dendritic fields. The fields reside in the ON and OFF parts of the inner plexiform layer (Wässle and Boycott, 1991). The second, termed the M-group consists of neurons with dense monostratified dendritic trees. The third, termed the S-group consists of cells with sparse and sometimes bistratified dendritic fields. Interestingly, S-cells, unlike T- and M-cells (and most other ganglion cells), do not show increases in dendritic field size with eccentricity (Rodieck, 1998). The three cell groups are morphologically similar to the heterogenous functional group of W-cells, which provide the visual input to the superficial SC in cats (Berson, 1988).

Overall, the diameter of ganglion cell dendritic trees projecting to the macaque SC is 100–150 μm for cells located at 0°–4° eccentricity and 400–700 μm for cells located at 14°–18° eccentricity (Rodieck and Watanabe, 1993). Since the retinal magnification factor in macaques is such that 1° of visual angle equals 223 μm on the retina (Perry and Cowey, 1985), and the diameter of the excitatory RF regions of ganglion cells corresponds closely to the diameter of their dendritic trees, one would expect that the diameters of centrally and peripherally located SC-projecting cells would be 0.5°–0.65°

and 1.8°–3.1°, respectively. At 0°–4° eccentricity, the RFs of superficial collicular cells in macaques vary from 0.75° to 2.0° in diameter, while those at 14°–20° eccentricity vary from 2.0° to 6.0° (Cynader and Berman, 1972). However, RFs as small as 0.25° have been reported around the *fovea* (Updyke, 1974; Marrocco and Li, 1977; Rizzolatti *et al.*, 1980). In alert macaques the excitatory RFs of SC cells located in the superficial layers at small eccentricities are often less than 0.5° in diameter, while at large eccentricities of 15°–20° they are up to 20° in the longest axis (Schiller and Koerner, 1971; Goldberg and Wurtz, 1972a).

Thus, there is a similarity in the sizes of the excitatory RFs of the perifoveal collicular neurons and sizes of RF centers of *perifoveal* ganglion cells that project to the SC. By contrast, the RFs of peripheral collicular neurons are much larger than the RF centers of the peripheral SC-projecting cells. The difference is probably related to a greater scatter of RF positions for peripheral versus perifoveal ganglion cells. Furthermore, it is clear that the spatial resolutions of perifoveal and peripheral parts of the retino-collicular system are substantially lower than those of the parvocellular and magnocellular retino-geniculo-cortical pathways, which relay signals from the midget and parasol ganglion cells, respectively (Rodieck, 1998; Kaplan, 2003). The estimated maximal spatial resolutions of the parafoveal collicular cells of macaques are 2–4 cycles/degree. This resolution is thus far lower than the acuity of 46 cycles/degree determined in behavioral tests (De Valois *et al.*, 1974).

There is substantial evidence indicating that in carnivores there is much parallel processing of visual information in the superficial SC, e.g. most cells in the retino-recipient layers of the SC do not receive excitatory retinal input from more than one functional class of retinal ganglion cells (Hoffmann, 1973; Berson, 1988; Waleszczyk *et al.*, 1999). However, in primates we do not know to what extent the morphological groups of SC-projecting ganglion cells constitute distinct functional classes.

The most common (70 percent) and second most common (10 percent) primate retinal ganglion cells are, respectively, the midget and parasol cells (Polyak, 1941; Rodieck, 1998). Midget ganglion cells are also referred to as B cells, Pβ cells, or P cells, while parasols cells are referred to as A cells, Pα cells, or M cells (Leventhal *et al.*, 1981; Perry *et al.*, 1984; Dacey and Petersen, 1992; Silveira *et al.*, 1994; Ghosh *et al.*, 1996). Despite their large numbers, virtually none of these cells project to the SC (Perry and Cowey, 1984; Rodieck and Watanabe, 1993; Williams *et al.*, 1995). Similarly, in cats very few of the β- or X-type retinal ganglion cells, which are suggested homologs of midget cells (Dreher *et al.*, 1976; Leventhal *et al.*, 1981; Rowe, 1991; Rodieck, 1998), project to the SC (for review, Berson, 1988). However, the α- or Y-type ganglion cells, which are suggested homologs of parasol cells (Dreher *et al.*, 1976; Leventhal *et al.*, 1981; Rowe, 1991; Rodieck, 1998), provide a strong input to the cat SC (for reviews, Berson, 1988; Tamamaki *et al.*, 1995).

The diameters of the dendritic trees of perifoveal (0°–4° eccentricity) and peripheral (14°–18°) midget cells in macaques are, respectively, in the range of 5–15 μm and 65–130 μm (Rodieck and Watanabe, 1993). The diameters of the dendritic trees of perifoveal (35–85 μm) and peripheral (175–370 μm) parasol cells are substantially larger than those of midget ganglion cells but still smaller than those of the SC-projecting ganglion cells. The RF diameters in degrees of visual angle for both midget (0.3°–0.6°)

and parasol (0.8°–1.65°) cells are clearly far smaller than the RF sizes of ganglion cells projecting to the SC.

Cortical inputs In macaques, injections of retrograde neuronal tracers into the superficial SC label pyramidal cells in the infragranular layer 5 of three cortical regions (Figure 8.6): (1) many cells in ipsilateral V1 of the occipital lobe; (2) few cells in ipsilateral area V4; and (3) many cells in ipsilateral area MT (Wurtz and Albano, 1980; Graham, 1982; Fries, 1984; Huerta and Harting, 1984; Ungerleider *et al.*, 1984; Distler and Hoffmann, 2001). In macaque, superficial collicular layers receive direct but numerically small projections from ipsilateral cortical areas V2 and V3 (Distler and Hoffmann, 2001). Similarly, in both squirrel and nocturnal owl monkeys, the principal cortical input to the superficial SC arises from ipsilateral areas V1, MT, and the dorsomedial area (Cusick, 1988; Kaas and Huerta, 1988). Input from ipsilateral area MT to the SC has also been reported in the prosimian bush baby (Wall *et al.*, 1982). In macaques, numerically small inputs to the superficial SC arise from the frontal eye field (FEF) in the frontal lobe (Figure 8.5; Distler and Hoffmann, 2001).

In both Old World and New World monkeys the input from area V1 terminates in the upper part of the SGS, while that from 'higher-order' areas, such as area V2, terminates in both the upper and part of the deeper SGS. Finally, the corticotectal fibers from ipsilateral area MT terminate throughout the entire dorso-ventral extent of the SGS (Cusick, 1988).

Outputs from the superficial SC layers

Ventral thalamus and epithalamus: The superficial layers project to the ipsilateral pregeniculate complex (PrGC: section 8.4). There are also direct and reciprocal connections with the ipsilateral pretectum (section 8.2). The input from the pretectum to the superficial SC is numerically larger than the reverse projection from the SC.

Midbrain: *The parabigeminal nucleus* is located in the lateral midbrain tegmentum (Figure 8.1B) and is considered a satellite system of the SC (Graybiel, 1978; Sherk, 1978, 1979). It is a presumed mammalian homolog of the isthmo-optic nucleus of other vertebrates (Le Gros Clark, 1933). In primates, the parabigeminal nucleus receives almost its entire visual input from the ipsilateral superficial SC and the bulk of the axons originating in the parabigeminal nucleus terminate in those layers (Baizer *et al.*, 1991; Feig *et al.*, 1992; Feig and Harting, 1994). In cats, the RF properties of parabigeminal neurons are virtually identical to those of their collicular inputs (Graybiel, 1978). In macaques collicular neurons projecting to the parabigeminal nucleus are (1) in the SGS, the SO, and occasionally the SGI; (2) are often direction-selective; and (3) have relatively thick axons, as indicated by the short (0.5–3.0 ms) latencies of their antidromic responses to electrical stimulation of the ipsilateral parabigeminal nucleus (Marrocco *et al.*, 1981). In addition to the reciprocal connections with the ipsilateral SC, the parabigeminal nucleus projects to the superficial layers of the contralateral SC. Furthermore, the parabigeminal nucleus of primates projects to the very small-cell, the so-called koniocellular (from Greek Κονισ – dust) layers in the ipsilateral and contralateral dLGN (Hashikawa *et al.*, 1986). All projection cells of the parabigeminal nucleus are cholinergic (Fitzpatrick and Raczkowski, 1991). Activity of cholinergic cells in the brainstem increases during arousal and ACh acts as an excitatory neurotransmitter. Since cholinergic fibers in the superficial SC terminate not only on the projection neurons but also on the inhibitory GABAergic interneurons (Feig and Harting, 1994), it is likely that the main function of the colliculo-parabigeminal connection is the arousal-dependent modulation of responsiveness of visual SC neurons.

Dorsal lateral geniculate nucleus: In primates SC cells located mainly in the upper part of the SGS project to the K layers of the ipsilateral dLGN. In turn, the cells in the K layers project to area V1 and, albeit to a lesser extent, the extrastriate visual cortices (Kaas and Huerta, 1988; Casagrande, 1994; Hendry and Reid, 2000; Kaplan, 2003). Very recently (Sincich *et al.*, 2004), a substantial direct projection from the K layers of dLGN to the extrastriate middle-temporal (MT) motion area in the parietal lobe was discovered in macaque monkeys. Interestingly, koniocellular geniculate neurons projecting to area MT do not send collateral branches to area V1 (Sincich *et al.*, 2004). In the rat, the activity of projection neurons located in the SGS is enhanced by a specific antagonist of ionotropic $GABA_C$ receptors but is not influenced by bicuculline, which is a specific blocker of $GABA_A$ receptors (Schmidt *et al.*, 2001).

Pulvinar: In primates the cells located in the lower part of the SGS project to the pulvinar (Figure 8.1B). The main targets of the colliculo-pulvinar projections are several subdivisions of the inferior pulvinar and one subdivision of the lateral pulvinar (Kaas and Huerta, 1988; Stepniewska, 2004). In macaques only non-DS cells in the SGS or SO can be activated antidromically from the inferior pulvinar (Marrocco *et al.*, 1981). The inferior pulvinar in turn projects to several visual cortical areas (including areas V1–V4, MT, medial superior temporal (MST), and fundus of the superior temporal (FST): Figure 8.5), as well as temporal occipital (TEO) and temporal (TE) areas in the inferotemporal cortex (Stepniewska, 2004). However, few of these projections originate from the parts of the inferior pulvinar that receive input from the superficial layers of the SC. In *Cebus* monkeys, injections of the retrogradely transported herpes-simplex virus into the lateral intraparietal (LIP) area in the inferior parietal lobule (Figure 8.5) label many cells in the ipsilateral inferior pulvinar. Such injections also lead to transynaptic labeling of many cells in the ipsilateral, superficial SC (Clower *et al.*, 2001).

Functional considerations

What are the functional roles of the projections from the superficial collicular layers that relay information via the dLGN to the visual cortex? In the cat there is emerging evidence that the superficial SC has a role in visual attention and pattern discrimination. Thus, unilateral reversible inactivation of the superficial and intermediate layers of the cat SC results in reversible neglect of stimuli presented in the contralateral visual hemifield (for reviews, Wallace *et al.*, 1989; Payne *et al.*, 1996; Sprague, 1996). Furthermore, reversible inactivation of the superficial SC results in significant retardation of learning of global, but not local, elements of visual patterns (Lomber, 2002). We do not know the perceptual effects of selective inactivation or ablation of the superficial SC in primates. However, in macaques radiofrequency lesions (which spare fibers of passage) of the SC and some pretectal nuclei result in long-lasting increases in the search time and/or percentage of errors in a visual search task that involves identifying a small target displayed amongst distracters (Bender and Butter, 1987). Furthermore, unilateral ablations of the SC (along with the pretectal complex and parts of the dorsal thalamus) result in a deficit in stimulus detection of visual stimuli in the contralateral hemifield lasting two weeks (Albano *et al.*, 1982). Similarly, bilateral lesions of the SC and pretectum result in an initial impairment of movement detection, which is followed by a long-lasting elevation in motion detection thresholds in the entire visual field (Cowey *et al.*, 1984). These 'perceptual' deficits are, however, always accompanied by a visuomotor deficit that persists after sensory function returns. In the case of unilateral ablations, the visuomotor deficit consists of a decrease in the frequency of exploratory saccadic eye movements toward visual targets that are contralateral to the lesion (Albano *et al.*, 1982). Thus, it is often thought that in primates the transient perceptual deficits following ablations of the SC (and pretectum) are side effects of more severe and permanent visuomotor deficits (Cowey *et al.*, 1984).

Radiofrequency lesions of the macaque SC result in (1) small reductions in the percentage of neurons responsive to visual, somatosensory, and auditory stimuli in the polysensory area in the superior temporal sulcus (STS); (2) a reduction in the magnitude of responses to visual stimuli and RF sizes of STS cells (Bruce *et al.*, 1986); and (3) a small *increase* in the magnitude of responses of neurons in area MT, which is not accompanied by changes in the proportion of visually responsive MT cells or their response properties (Rodman *et al.*, 1990). More significantly, the projections from the superficial SC to the dorsal thalamus appear to underlie the retention of visual responsiveness of some cortical cells in the parietal motion processing cortical stream after ablation of area V1 (Bullier *et al.*, 1994). Indeed, when ablation of area V1 is followed by radiofrequency ablations of the SC, the residual responses of STS and MT cells to visual stimuli are totally abolished (Bruce *et al.*, 1986; Rodman *et al.*, 1990).

In both humans and macaques a complete removal of area V1, with the retention of the extrastriate cortical areas, results in 'blindsight' (Humphrey, 1974; e.g. Weiskrantz *et al.*, 1974; Cowey and Stoerig, 1991, 1995; Stoerig and Cowey, 1997; Weiskrantz, 2002). This phenomenon is defined as a condition in which the sufferer responds to visual stimuli without consciously perceiving them. Thus, untrained human blindsight subjects respond to visual stimuli by making saccades directed correctly toward visual stimuli presented at different eccentricities in the blind part of the visual field. Nearly 40 years ago, Trevarthen (1968) proposed that in primates there are two types of visual perception: (1) 'focal' vision based on the retino-geniculo-cortical pathway and (2) 'ambient' vision based on the retinal projections to the midbrain and pretectum. Focal vision allows the recognition of visual stimuli, while ambient vision allows the detection of visual stimuli without necessarily identifying them. It appears that the blindsight might be based on the ambient vision when there in no accompanying focal vision. What are the neuronal circuitries underlying the focal vision and those underlying ambient/blindsight vision? Focal vision is dependent on the integrity of the striate cortex and both the midget (via parvocellular dLGN layers) and parasol ganglion cells (via magnocellular dLGN layers) project almost exclusively to the striate cortex, while a small proportion of koniocellular geniculate cells projects to the extrastriate cortices (Hendry and Reid, 2000). Recently discovered substantial direct projection from the K layers of the dLGN to the MT area is likely to contribute to blindsight (Sincich *et al.*, 2004). Indeed, since koniocellular geniculate neurons projecting to area MT do not send collateral branches to area V1, destruction of area V1 should not affect the integrity of this projection. However, as mentioned earlier, the unilateral ablation of the striate cortex followed by ablations of the ipsilateral SC result in abolition of the responses of neurons in two extrastriate areas (STS and MT) to visual stimuli (Bruce *et al.*, 1986; Rodman *et al.*, 1990). Therefore, it appears that the information relayed from the retina to area MT via K layers of the dLGN is not sufficient to maintain the blindsight. On the other hand, the pathways relaying visual information from the superficial SC via the dorsal thalamus to the extrastriate cortices are likely to contribute to blindsight. Indeed, consistent with the low spatial resolution of retinal ganglion cells projecting to the SC, blindsight is also characterized by very poor spatial resolution (Stoerig and Cowey, 1997). Furthermore, macaques in which part of area V1 on one side has been removed could detect and make usually inaccurate saccades to 0.25° targets presented in the 'striate-blind' region only when the brightness of the spot was dramatically increased (Mohler and Wurtz, 1977). If both area V1 and the SC are electrolitically removed, monkeys are unable to detect and make saccades to very bright small visual targets presented in the 'striate/SC blind' part of the visual hemifield (Mohler and Wurtz, 1977).

Several colliculo-dorsal thalamic pathways might be involved in blindsight. Most important of them appears to be the projection from the superficial layers of the SC to the koniocellular layers of the dLGN (Hartin *et al.*, 1991) and hence to area MT (Sincich *et al.*, 2004). By contrast, the colliculo-pulvinar-cortical pathway, which until very recently was favored as the main pathway contributing to blindsight, is numerically small and very few cells in the colliculo-recipient part of the inferior pulvinar appear to project to area MT (Stepniewska, 2004). However, one cannot exclude the contribution to blindsight of direct retinal projections to that part of the inferior pulvinar that projects to area MT (Cowey *et al.*, 1994). Interestingly, a small proportion of retinal inputs to the macaque pulvinar are constituted by midget and parasol ganglion cells (Cowey *et al.*, 1994).

In conclusion, there is much evidence indicating that the SC contributes to identifying the spatial location of visual and other sensory stimuli. Indeed, Schneider (1969) proposed that the colliculi are involved mainly in determining 'where' stimuli are rather than 'what' they are. In primates, the function of the colliculi is most apparent in their selection of visual targets for saccades.

Saccade-related response enhancement

In about half of the cells recorded from the superficial SC layers of awake monkeys that are fixating a very small target (0.05° in diameter), responses to stationary visual stimuli become more regular and vigorous when the monkey is required to make a saccade from a fixation point toward a target located in the recorded cell's RF (Figure 8.7: Goldberg and Wurtz, 1972b; Wurtz and Mohler, 1976). This 'response enhancement' has three partially overlapping manifestations: (1) an increase in the magnitude of the initial ON response to stationary stimuli, with response prolongation; (2) an enlargement of the excitatory RF; and/or (3) a lack of habituation to repeated stimulus presentations (Goldberg and Wurtz, 1972b).

Visual response enhancement occurs in the period before the animal makes a saccade and does not represent a generalized alerting effect since it is spatially specific. That is, it occurs only before saccades directed at the recorded cell's RF and not before saccades directed elsewhere in the visual field or saccades in the dark (Figure 8.7; Goldberg and Wurtz, 1972b). The enhancement is also temporally specific, occurring 200 ms before saccade onset but not when the monkey fails to make a saccade toward the target, even if the target is located in the cell's RF (Wurtz and Mohler, 1976). Thus, superficial cells appear to have a role in selecting visual targets for saccades (Wurtz and Albano, 1980). The visual enhancement is probably related to the strong input to the superficial layers

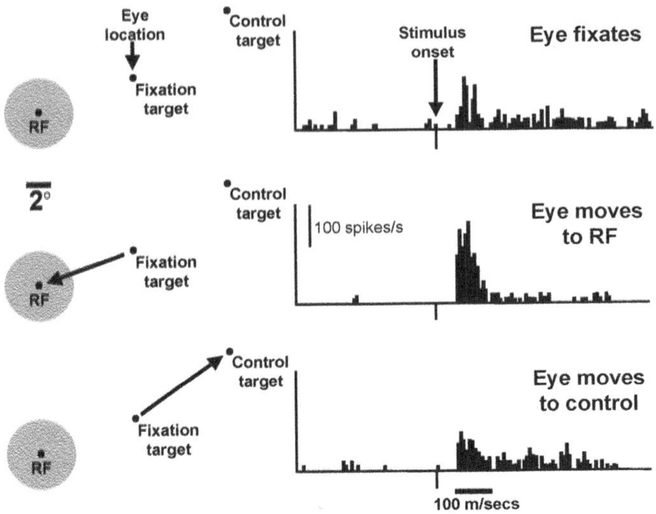

Figure 8.7 Saccadic enhancement in the superficial SC (Wurtz and Mohler, 1976). Diagrams to the left: upper scheme shows a stimulus presentation in the RF; middle scheme shows the response when a stimulus is presented in the RF prior to a saccade to the RF center; lower scheme shows the response when a stimulus is presented in the RF before a saccade to a control target. Reproduced by the permission of the American Physiological Society

from the ipsilateral cortical areas around the STS (Figure 8.5). This is especially the case for area MT, which primarily contains DS cells whose responses are strongly enhanced when a monkey is attending to the region of visual space represented in a cell's RF (Treue and Maunsell, 1996).

8.1.3 Deep SC layers

Integration of inputs from different sensory modalities

In primates, as in other mammals, the majority of cells in the deeper layers, which include both the intermediate and deep layers, receive visual, auditory, and/or somatosensory input. Furthermore, their auditory and/or somatosensory RFs are in most cases largely spatially overlapping with visual RFs (for reviews, Stein and Meredith, 1991; Stein *et al.*, 2001; Gandhi and Sparks, 2003). As a result of integration of cross-modal sensory information the responses of multisensory SC neurons to visual stimuli can be strongly enhanced or depressed by cues from other sensory modalities. The reduction in response magnitudes to visual stimuli occurs mainly (but not exclusively) in cells in which the visual and non-visual RFs are in different spatial locations. The multisensory integration in the SC is established only after a relatively long period of postnatal experience and is largely dependent on the cortico-tectal input from a number of 'polysensory' cortical areas. Detailed analysis of the mechanisms underlying the multisensory integration in the SC and its role in general orienting behavior is however beyond the scope of a book devoted to primate vision.

Role of the deep layers in initiating saccades

In awake macaques, electrical stimulation of the SC results in conjugate saccades of both eyes in the direction of the visual hemifield contralateral to the stimulated SC (Sparks and Hartwich-Young, 1989). The direction and magnitude of saccades evoked by SC stimulation depend on the position of the electrode tips in relation to the topographic map of the contralateral visual hemifield. For stimulation of a particular region of the superficial or intermediate SC layers, saccades are driven to the location of the visual RFs of the stimulated cells. For stimulation of the deep SC layers, in which cells do not have visual RFs, the saccades move the eyes to the location of the visual RFs of cells located above them in the superficial layers (Robinson, 1972; Schiller and Stryker, 1972).

When the tips of stimulating electrodes are located in the superficial, intermediate, and deep layers the current thresholds for evoking saccades are, respectively, 400–800 μA, 200 μA, and below 20 μA (Robinson, 1972; Schiller and Stryker, 1972). Clearly, the deepest layers are critical for the final motor commands that initiate saccades. The stimulation experiments are consistent with the pattern of SC connections (section on Visual inputs to the superficial SC layers), in which only cells in the deepest layers project directly to the brainstem 'saccade generators' (Harting, 1977; Edwards and Henkel, 1983). Repeated stimulation of a point in the SC results in saccades of the same direction, amplitude, and velocity, regardless of the frequency or intensity of super-threshold electrical stimulation (Robinson, 1972; Schiller and Stryker, 1972). Long-lasting (150–200 ms) electrical stimulation of SC results in a series of identical saccades interrupted by short-lasting fixations.

Most deep cells discharge 20–150 ms before saccades to visual targets as well as to spontaneous saccades in the light and dark (Schiller and Stryker, 1972). They also discharge prior to the rapid phases of vestibular ocular responses (VOR; Schiller and Koerner, 1971) and during the fast phases of optokinetic nystagmus (OKN; Schiller and Stryker, 1972). For more detail on VOR and OKN, see section on NOT output regions and functions. Each cell only discharges if the subsequent saccade has the appropriate range of directions and amplitudes. The region in space where a saccade must land for the pre-saccadic activity to occur is the *movement field*. Some cells in the intermediate layers have a *movement field* but no visual RF (*saccade-related burst neurons*), while others have both a visual RF and a *movement field* (*visuomotor cells*) (for reviews, Wurtz and Albano, 1980; Sparks and Hartwich-Young, 1989; Wurtz, 1996). In visuomotor cells the visual RFs and *movement fields* overlap but are not necessarily the same shapes and sizes. Their responses might consist of a burst of firing to target appearance in their visual RFs and then a second before the saccade. In behaving macaques with one surgically immobilized eye, cells in the intermediate layers respond well to stationary flashing stimuli and rapid square-wave displacements of stimuli located within their visual RFs. However, they do not respond to smooth back and forth movements (Schiller and Koerner, 1971; Schiller and Stryker, 1972).

So-called visually triggered cells do not respond during spontaneous saccades in the dark or light but do discharge when saccades are directed at a visual target (Figure 8.8). These cells always have a *movement field* but not necessarily a visual RF. Therefore, visual dependence of the response does not require a visual RF that generates spikes (Mohler and Wurtz, 1976). Cells located in the rostral pole of the SC are often referred to as fixation cells because they generate low-level activity during fixation but are silent during ipsiversive saccades (Munoz and Wurtz, 1993a,b). Ipsiversive saccades are

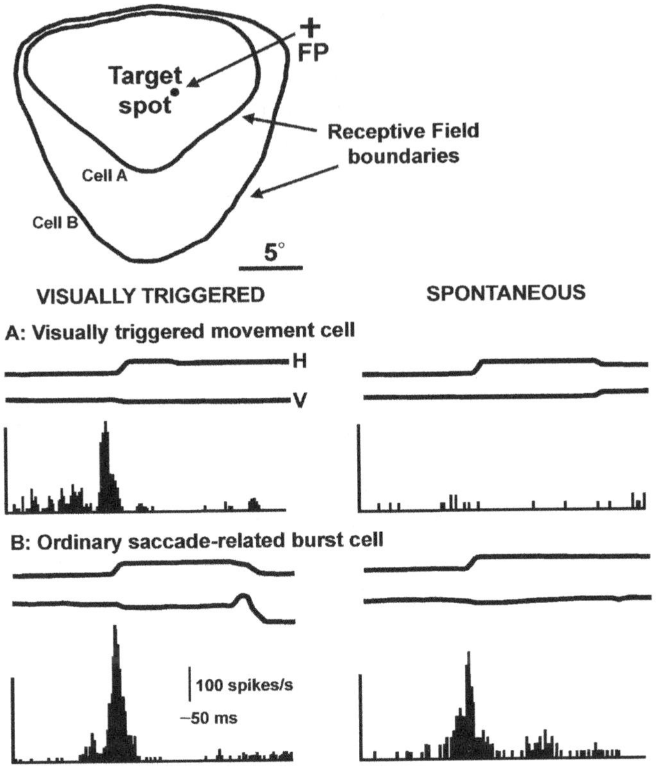

Figure 8.8 (Top left) Stimulus and RF structure of two cells when a macaque makes saccades from the fixation point (FP) to a target (Wurtz and Mohler, 1976). (A) The appearance of the target prior to a target-directed saccade generates spikes. Spontaneous saccades of the same direction and size with no target do not elicit discharges. (B) A cell responding to a target-directed saccade and to a spontaneous saccade. Reproduced by the permission of the American Physiological Society

directed toward the side of the brain from which the recording is being made. While they do not respond to ipsiversive saccades, some can generate pre-saccadic bursts during contraversive saccades produced to fixate parafoveal targets.

An important cell type (quasi-visual cells) may signal the difference between current and desired eye positions (Mays and Sparks, 1980; for review, Gandhi and Sparks, 2003). Imagine a cell that has an RF 5° to the left of the *fovea*. If the monkey fixates an object and a stimulus is flashed 5° to the left of the fixation target the cell will respond. Now the monkey is required to move its eyes to a new fixation location 10° to the right of the first. Prior to the appearance of the new fixation target a stimulus is briefly flashed 5° to the left of the new fixation target but disappears before the monkey changes gaze direction. Despite the second flash never falling inside the cell's visual RF, a quasi-visual cell will respond to the flash. This demonstrates that the cells are provided with a signal 'informing' them that a target was in the location in space where its RF will be. Such response properties suggest that the system controlling saccades has a capacity for neural re-mapping of visual space. This capacity matches that of some cells in the LIP area (Duhamel *et al.*, 1992), which provides input to the deeper SC (Paré and Wurtz, 1997).

Connectivity of the deep layers

In primates there is no direct retinal input to the deep SC. However, there is clear evidence of visual input to the intermediate layers from cells in LIP (Figure 8.9: Paré and Wurtz, 1997) and the frontal eye field (FEF) (Figure 8.9: Segraves and Goldberg, 1987). In macaques, blockade of the magnocellular layers of the dLGN strongly affects the magnitude of responses to visual stimuli of most cells in the intermediate layers of the ipsilateral SC (Schiller *et al.*, 1979). Therefore, it appears that the input from the parasol retinal ganglion cells, which do not project directly to the SC (section on Visual inputs to the superficial SC layers), is relayed via the dLGN to the cortex and then to the intermediate SC layers. In several non-primate species signals from the superficial collicular layers are sent directly to the intermediate layers (Mooney *et al.*, 1992; Lee *et al.*, 1997; Isa *et al.*, 1998).

In the awake macaque, electrical stimulation of several visual cortical areas results in saccades directed toward the contralateral visual hemifields (Schiller and Tehovnik, 2001). These cortical areas include areas V1 and V2, area LIP (Andersen *et al.*, 1990), and the medial and frontal eye fields (Figure 8.5: Bruce *et al.*, 1985). After unilateral ablation of the SC, electrical stimulation of the occipital or parietal lobes (e.g. areas V1, V2, and LIP) ipsilateral to the removed SC no longer evokes saccades. However, saccades can be generated by stimulation of the FEF or medial eye field (MEF) in the frontal lobe (Schiller and Tehovnik, 2001). Thus, for the occipital and parietal cortices the SC constitutes the only access to the brainstem saccade generators, while the frontal cortices can access them without the SC.

In behaving macaques saccade latencies toward visual targets have a bimodal distribution (Figure 8.10; Schiller and Tehovnik, 2001). Irrespective of the target location in the visual field, one group of saccades (express saccades) are characterized by latencies

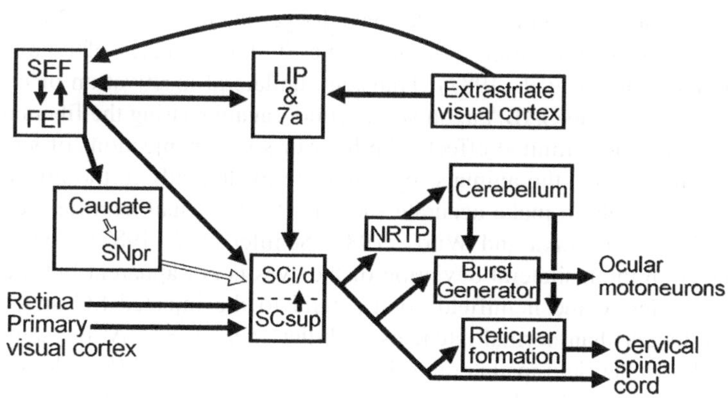

Figure 8.9 Connectivity of the intermediate and deep layers of the primate SC (Scudder *et al.*, 2002). SEF (supplementary or medial eye field), sNpr (*substantia nigra pars reticulata*), sup (superficial SC layers), i/d (intermediate and deep layers). Reproduced by the permission of Elsevier B.V

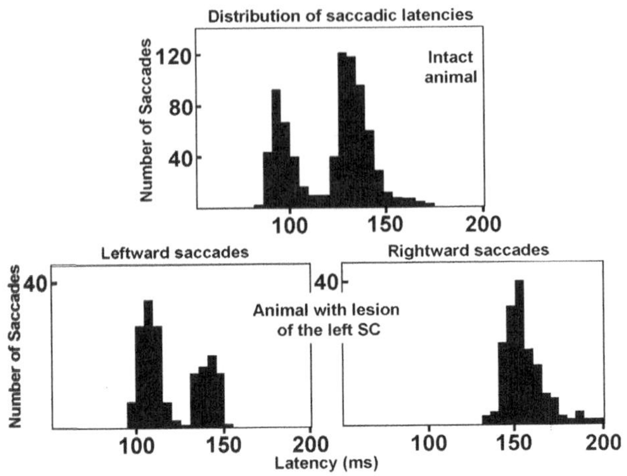

Figure 8.10 Distribution of saccadic latencies for an intact animal (upper) and one with a left SC lesion (Schiller and Tehovnik, 2001). In the latter case there are no rightward express saccades and regular saccades are delayed. Reproduced by permission of Elsevier B.V

of 80–110 ms, while a second group, the regular saccades, are characterized by latencies of 125–170 ms (Fischer and Boch, 1983). In both macaques and humans (Fischer and Ramsperger, 1984) the frequency of express saccades progressively increases with increases in the time interval between the extinguishing of the fixation spot and onset of a target (Fischer and Weber, 1993; Schiller *et al.*, 2004). In macaques, after unilateral radiofrequency ablation of the SC (thus sparing fibers of passage), only saccades toward the stimuli presented in the visual hemifield ipsilateral to the ablated SC exhibit a bimodal latency distribution (Figure 8.10; Schiller and Tehovnik, 2001). For stimuli presented in the hemifield contralateral to the ablated SC there are never express saccades and 3–20 days after the ablation the regular saccades have longer latencies of 135–200 ms (Schiller *et al.*, 1987). About 70 days after ablation of the SC the latencies of regular saccades return to normal (Schiller *et al.*, 1987). Transient, unilateral inactivation of macaque SC neurons by injection of the GABA agonist muscimol (again sparing the fibers of passage) has very similar but time-limited effects. An hour or so after injections of less than 1 μg of muscimol into one SC, the animals' eyes deviate ipsilaterally to the injected SC and the animals are not able to make accurate saccades to the visual hemifield contralateral to the injected SC (Hikosaka and Wurtz, 1985; Schiller *et al.*, 1987). Four hours after muscimol injection the ipsilateral deviation of the eyes is not apparent but there are still no express saccades to the hemifield contralateral to the injected SC and the regular saccades have longer than normal latencies (Schiller *et al.*, 1987). At the same time, the express saccades to the field ipsilateral to injected SC are plentiful (Schiller *et al.*, 1987). Thus, the SC is essential for driving express saccades and contributes to 'speeding-up' the onset of regular saccades.

The deeper layers of the SC provide outputs to the dorsal medullary reticular formation, the rostral paramedian pontine reticular formation, and *nucleus reticularis tegmentis pontis* (NRTP) (Figure 8.1; Harting, 1977; Edwards and Henkel, 1983). These areas contain the circuits that provide the necessary motor signals to control saccades (Figure 8.9: Scudder *et al.*, 2002). An important component of the saccade-generating circuit is the *substantia nigra pars reticulata* (SNPR), which resides in the cerebral pedunculi of the midbrain but is functionally considered a part of the basal ganglia. It sends strong GABAergic (inhibitory) projections to the SC (Scudder *et al.*, 2002). When saccades are not occurring, cells in the *substantia nigra* are spontaneously active and generate ongoing suppression of the SC. Before voluntary saccades to the contralateral visual hemifield the firing of cells in the *substantia nigra* is suppressed by signals from the caudate nucleus. As a result of suppression of ongoing activity in the SNPR the SC cells innervated by the *substantia nigra* are released from inhibition and can initiate a chain of events leading to a saccade.

The FEF (Figure 8.5) contains many movement-related cells that fire before and during all saccades and do not respond to visual stimuli that are not the targets for saccades (Bruce and Goldberg, 1985). These cells project primarily to the intermediate SC and excite cells there (Stanton *et al.*, 1988b). Simultaneously, neighboring cells in the FEF excite cells in the caudate nucleus, which inhibit the *substantia nigra* (Stanton *et al.*, 1988a; Hikosaka *et al.*, 1989). Consequently, the movement-related cells in the FEF excite cells in the SC and release it from inhibition (Figure 8.9).

Signals are also sent through an ascending pathway from the SC to the FEF via the mediodorsal (MD) nucleus of the dorsal thalamus (Sommer and Wurtz, 2004a). Both visual and saccade-related activity is conveyed via this pathway. Sommer and Wurtz (2004b) suggest that the SC–MD–FEF pathway conveys a corollary discharge used for coordinating sequential saccades into the contralateral visual field and for stabilizing vision across saccades. Corollary discharges provide a mechanism for tracking eye movements because they provide internal copies of movement commands.

Cells in the deeper reaches of the intermediate layers of the rostral pole of the SC (fixation cells: Munoz and Wurtz, 1993a) project to the dorsal raphé nucleus. They prevent saccades from occurring until required by exciting omnipause cells in the raphe nucleus, which have the role of inhibiting saccades (e.g. Scudder *et al.*, 2002). An electrically induced increase in the activity of fixation cells delays the initiation of saccades. With bilateral stimulation, monkeys make saccades only after stimulation stops. Also stimulation delivered during saccades interrupts them in midflight. Injection of bicuculline, an antagonist of GABA, into the SC's rostral pole increases cell activity, thus increasing saccade latencies. Lesions of the rostral pole make it difficult for monkeys to maintain visual fixation and suppress unwanted saccades. The fixation cells in the rostral SC are critical for controlling the frequency of express saccades (Munoz and Wurtz, 1993b).

Role of the deep layers in eye–hand coordination

Recently, evidence has suggested that apart from their important role in controlling saccades, the deep layers of the primate SC might have a role in eye–hand coordination (for reviews, Lünenburger *et al.*, 2001; Gandhi and Sparks, 2003). Thus, the deep layers of macaque SC, as well as the underlying mesencephalic reticular formation, contain the 'reach' neurons that are activated during arm movements (usually contralateral) to a particular spatial location. Furthermore, the activity of a subclass of such neurons (gaze-related reach neurons) is strongly dependent on the spatial location of the visual objects fixated by the animal in a given moment. See section 8.2.1 for a discussion of visual reaching in relation to the pretectum and SC.

8.2 Pretectum

The mammalian pretectal complex is located at the junction of the diencephalon and mesencephalon. From an embryological viewpoint it should be considered part of the epithalamus (Rose, 1942). Hutchins and Weber (1985) used standardized terminology established for studying the pretectum of other mammals (Rose, 1942) and on anatomical grounds distinguished five nuclei in primate pretectum. The largest neuronal assembly constitutes the nucleus of the optic tract (NOT), which is embedded amongst the fibers of the brachium of the superior colliculus. At its most lateral edge the NOT is juxtaposed with the dorsal terminal nucleus of the accessory optic system (section 8.3, Figure 8.1B). The olivary pretectal nucleus (OPN) is located in the center of the pretectum and the remaining three nuclei are, respectively, the anterior (APN), medial (MPN), and posterior (PPN) pretectal nuclei (Figure 8.11).

8.2.1 Anterior, medial, and posterior pretectal nuclei

Presumed functions

The APN is the most rostral nuclear group in the pretectum located between the nucleus limitans of the thalamus and the OPN (Hutchins and Weber, 1985). The MPN is a small nuclear group situated lateral to the base of the pineal body while the PPN is a large structure situated ventro-medially to the OPN (Figures 8.1 and 8.11). The lack of data concerning receptive field properties and the effects of selective damage of the APN,

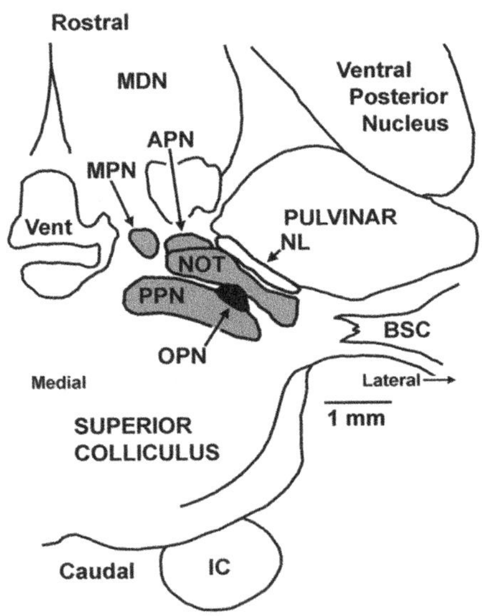

Figure 8.11 Horizontal section through the squirrel monkey pretectum (Hutchins and Weber, 1985). APN (anterior pretectal nucleus), MPN (medial pretectal nucleus), MDN (medial dorsal nucleus), NL (nucleus limitans), IC (inferior colliculus), Vent (third ventricle). PPN, OPN, NOT, BSC (Figure 8.1). Reproduced by permission of John Wiley & Sons Ltd

MPN, and PPN make it difficult to be certain about their functions. However, it is generally suggested on hodological grounds that these nuclei play a role in accommodation, which involves changing the lens shape to alter focal length, and vergence eye movements, which act to point the *foveas* of both eyes at a target in a given depth plane.

Connectivity of the pretectal nuclei
The connectivity of the pretectal nuclei is illustrated schematically in Figure 8.12. In all primates studied, including the New and Old World monkeys and chimpanzee, the retinal input to the pretectal nuclei is sparse and originates predominantly in the contralateral retinas (Tigges and Tigges, 1970; Tigges *et al.*, 1977; Benevento and Standage, 1983). In the squirrel monkey both the contralateral and ipsilateral retinal fibers terminate in single loci within the MPN and PPN (Weber and Hutchins, 1982). In macaques, the retinal input originates from several morphological types, many indistinguishable from the SC-projecting cells (Rodieck, 1998). Their somata at any eccentricity are small (10–15 μm), while their dendritic trees are fairly large (210–840 μm). The pretectum does not receive direct input from the midget ganglion cells but unlike the SC receives some input from the parasol cells (Rodieck, 1998).

In addition to their retinal input, the PPN and the APN of macaque and squirrel monkeys receive direct visual input from the superficial SC (e.g. Harting *et al.*, 1980; Huerta and Harting, 1983). By contrast, MPN does not receive input from the superficial SC but it does get moderate input from the deep SC. The existence of direct projections to the APN and PPN from ipsilateral V1 (Figure 8.5) remains controversial (Lui *et al.*, 1995). However, in macaques the PPN receives a direct input from ipsilateral cortical area V4 (Dineen and Hendrickson, 1983). Both the APN and PPN of macaques receive direct input from the inferior parietal cortex, including areas 7a–7b (Figure 8.5; Asanuma *et al.*, 1985; Lui *et al.*, 1995). In macaque and *Cebus*, both the APN and PPN receive direct input from cortical area 7 m (Leichnetz, 2001). Although this area is generally considered a supplementary somatosensory rather than visual associational cortex (Pandya and Barnes, 1987), it appears to be also a multisensory area that may act as part of a system involved in controlling 'visual reaching' (section on Role of the deep layers in eye–hand coordination; Leichnetz, 2001). Indeed, area 7 m is reciprocally connected

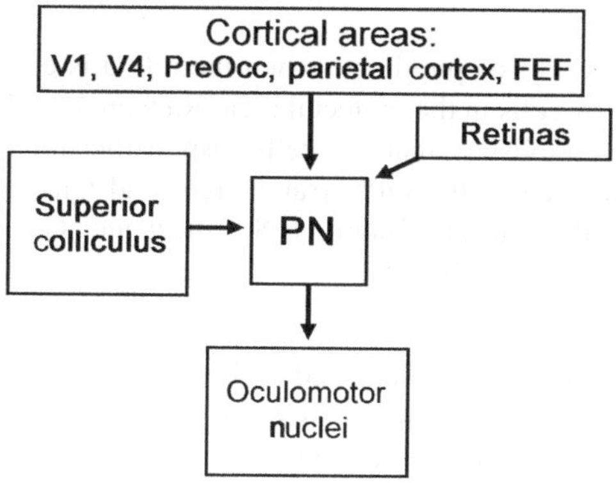

Figure 8.12 Connectivity of the anterior, medial, and posterior pretectal nuclei (PN)

with motion processing areas of the visual cortex (e.g. MST) as well as with the eye movement specialized areas (FEF and MEF; Figure 8.5) in the frontal lobe, which in turn project to brain stem nuclei controlling saccadic and smooth pursuit eye movements (Leichnetz, 2001). The connections between area 7 m and the APN and PPN suggest that the latter might be components of the neural network relating to eye movement control during visual reaching.

The middle preoccipital cortex receives input from the FEF. These interconnections underlie the notion that the APN and PPN, along with the neighboring OPN, play a role in 'reaching' responses that require the coordination of changes in focal length through accommodation and vergence eye movements. Support for this hypothesis comes from the fact that neurons in the FEF of squirrel, owl, and macaque project directly to the pretectal complex including terminals in the APN, MPN, and PPN (Leichnetz, 1982a; Huerta *et al.*, 1986; Stanton *et al.*, 1988a,b).

Büttner-Ennever *et al.* (1996b) placed neuronal tracers into the oculomotor complex (Figure 8.1), which contains not only the oculomotor neurons innervating the extrinsic striated oculomotor muscles but also preganglionic neurons in the accessory oculomotor nucleus (Edinger–Westphal nucleus: EW), which in turn innervate the cilliary ganglion neurons innervating the smooth, intrinsic oculomotor muscles. They retrogradely filled cells in the OPN and the NOT but also filled cells which Leichnetz (2001) claims are located in APN and PPN. Furthermore, when Büttner-Ennever *et al.* (1996b) placed neural tracer into the NOT, OPN, and surrounding nuclei, they labeled terminals mainly in the contralateral side just outside the oculomotor nucleus (nIII), in the 'C-group' of motor neurons. There were also projections to the lateral visceral cell (LVC) column of the EW complex. No terminals were found in the EW itself, which is known to have a role in controlling pupil constriction (Figure 8.1). Other fibers terminated in the ipsilateral medial accessory nucleus of Bechterew, which from its connectivity appears to be involved in internuclear connections (Leichnetz, 1982b). The 'C-group' motor neurons are involved in vergence eye movements, while the LVC cells-play a role in pupillary constriction and changing depth of focus through accommodation. The results suggest that APN and PPN might have the correct motor connectivity to drive vergence and accommodation.

8.2.2 Olivary pretectal nucleus

Summary of functions

The pupillary light reflex adjusts pupil size when the retina is exposed to different light intensities and is driven by cells in the pretectal area (Ranson and Magoun, 1933; Magoun *et al.*, 1935). It is well established that the cells responsible for the pupillary reflex in several mammals are located in the OPN (rats: Trejo and Cicerone, 1984; Clarke and Ikeda, 1985a,b; cat: Distler and Hoffmann, 1989; macaque: Gamlin *et al.*, 1995; Pong and Fuchs, 2000b; Clarke *et al.*, 2003b).

OPN physiology

Three neuronal cell types have been identified in the OPN of behaving macaques (Clarke *et al.*, 2003b), all of which respond to an increase in brightness with a transient burst of action potentials followed by a sustained lower frequency response lasting for the duration of the stimulus (Gamlin *et al.*, 1995; Pong and Fuchs, 2000b). Consistent with this, OPN-projecting wide-field retinal ganglion cells show sustained ON responses to maintained levels of illumination and have poor temporal resolution (Gamlin *et al.*, 2001; Dacey *et al.*, 2005). Unlike most other types of retinal ganglion cells, those projecting to the OPN are sensitive to the level of sustained light intensity (Gamlin *et al.*, 2001; Dacey *et al.*, 2005). About 40 percent of OPN cells respond when visual stimuli are presented in either the ipsilateral or contralateral hemifields (bilateral cells). The RFs of bilateral cells extend for at least 30° into the contralateral and ipsilateral hemifields. Almost 30 percent of neurons respond only when visual stimuli are close to the central fixation spot (macular cells) and the others respond only when stimuli are presented in the contralateral hemifield (contralateral cells). The mean maximal firing rate of macular cells to stimuli presented near the *fovea* is significantly greater than that to stimuli presented more peripherally. However, they are not significantly greater than the maximal firing rates of bilateral or contralateral cells to stimuli presented in the peripheral contralateral visual hemifield (Clarke *et al.*, 2003b). For all classes of OPN cells the dynamic range of responses is 3 log units of light intensity (Clarke *et al.*, 2003b).

Almost 85 percent of macaque OPN cells (of all types) are driven by inputs from both the contralateral and ipsilateral retinas. While about 44 percent of cells can be driven equally strongly by inputs from either eye, 16 percent can be driven exclusively from the contralateral eye, and none are driven exclusively by input from the ipsilateral eye (Clarke *et al.*, 2003b). The high proportion of binocular neurons in primate OPN contrasts with the relatively low proportion (22 percent) of binocular cells in cats (Distler and Hoffmann, 1989). The high proportion of binocular OPN neurons in primates might account for the similar amplitude of direct and consensual (illumination of one eye results in changes in pupil size of the other eye) pupillary light reflexes. Although in non-primate mammals (rat: Trejo *et al.*, 1989) there are also direct and consensual pupillary light reflexes, the latter are far weaker. The pupillary light reflex of alert macaques shows characteristics similar to those in alert humans (Pong and Fuchs, 2000a; Clarke *et al.*, 2003c).

In macaques bilateral and contralateral OPN cells have unusually large RFs (bilateral: $4800^{\circ 2}$; contralateral: $2500^{\circ 2}$). The magnitudes of their responses increase with luminance and stimulus size. The spatial extent of the RFs of bilateral and contralateral OPN neurons in macaque is substantially greater than that of any OPN neurons recorded in rats and cats (Trejo and Cicerone, 1984; Distler and Hoffmann, 1989). In the case of macular cells in macaque the spatial summation extends only for approximately 20° (RFs do not exceed $100–900^{\circ 2}$. In all three cell types further extension of stimuli beyond the summation area does not produce changes in firing rate (Clarke *et al.*, 2003b). OPN cells do not have suppressive surrounds.

OPN connectivity and function

The connectivity of the OPN is shown in Figure 8.13. A small proportion (see further) of the retinal ganglion cells project to the OPN and synapse with neurons that project bilaterally to preganglionic parasympathetic neurons in the EW nucleus. The retinal ganglion cells projecting to the OPN of macaques appear to constitute a single morphological and functional type (Gamlin *et al.*, 2001; Dacey *et al.*, 2005). They also send collaternal branches to the dLGN (Dacey *et al.*, 2005). The peak density (20–25 cells/mm^2) of the OPN-projecting ganglion cells is in the *parafoveal* retina and the density drops to $\sim$5 cells/mm^2 at about 8 mm from fovea (Dacey *et al.*, 2005). OPN-projecting ganglion cells are characterized by relatively large somata (mean 17 μm) located mainly in the ganglion cell layer (Gamlin *et al.*, 2001; Dacey *et al.*, 2005) but occasionally displaced to the inner

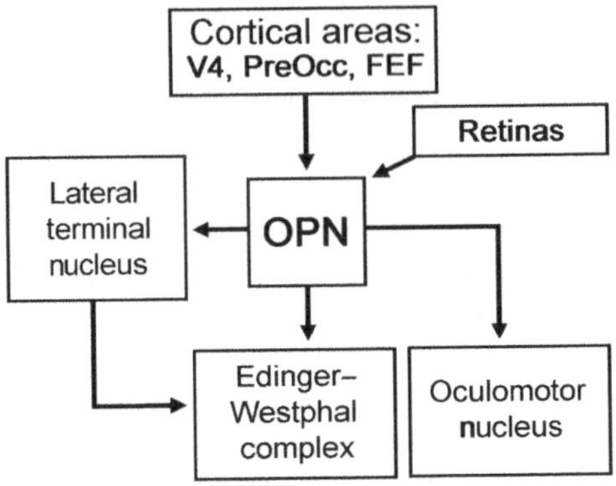

Figure 8.13 Connectivity of the olivary pretectal nucleus (OPN)

nuclear layer (Dacey *et al.*, 2005). Their thick, sparsely branching, smooth dendrites are narrowly stratified either in the extreme inner (ON portion; 40 percent of cells) or extreme outer (OFF portion; 60 percent of cells) portions of the inner plexiform layer (Dacey *et al.*, 2005). Surprisingly, not only ganglion cells with dendrites stratified in the ON portion of the inner plexiform layer but also those with dendrites stratified in the OFF portion of the inner plexiform layer show sustained ON responses to maintained levels of illumination (Dacey *et al.*, 2005). Overall, the dendritic fields of the OPN-projecting retinal ganglion cells are large and the size of the dendritic trees increases with eccentricity (Dacey *et al.*, 2005). Thus, the range of dendritic tree sizes of OPN-projecting ganglion cells receiving their input from the region cantered at the *fovea* is 260–520 μm, while that of OPN-projecting ganglion cells receiving their input from the region cantered at 10° is 590–1050 μm (Dacey *et al.*, 2005). Since in macaques 1° of visual angle equals 223 μm on the retina (Perry and Cowey, 1985), even the smallest dendritic trees of the OPN-projecting ganglion cells exceed 1° of visual angle, while the largest exceed 4.5° of the visual angle. On the basis of the RF sizes of OPN-projecting ganglion cells, it is estimated that each bilateral or contralateral OPN cell receives excitatory input from hundreds of retinal ganglion cells, while macular cells receive excitatory input only from 20–100 cells (Clarke *et al.*, 2003b). By contrast, in rats typical OPN neurons receive input from 10 ganglion cells (Trejo and Cicerone, 1984).

In macaques, OPN-projecting retinal ganglion cells receive both rod and cone inputs (Dacey *et al.*, 2005). Cone-mediated receptive fields approximate the sizes of dendritic trees (Dacey *et al.*, 2005). Surprisingly, the OPN-projecting ganglion cells appear to be 'colour-opponent', that is, the short wavelength S cone-mediated OFF response is antagonistic to the ON response mediated by the combined input from the long (L) and medium (M) wavelength cones (Dacey *et al.*, 2005). Apart from the rod and cone inputs both somata and dendrites of the OPN-projecting retinal ganglion cells contain putative photopigment melanopsin presumably underlying their intrinsic photosensitivity when rod and cone transmission to the inner retina is blocked pharmacologically (Dacey *et al.*, 2005). In humans and macaques the total number of the melanopsin-expressing and presumably OPN-projecting retinal ganglion cells does not exceed ~3000, that is, in these species they constitute ~0.2 percent of all retinal ganglion cells (Dacey *et al.*,

2005). In nocturnal rodents such as rats and mice, despite a far lower total number of retinal ganglion cells, the number of melanopsin-expressing ganglion cells, is very similar to that in humans and macaques (Hattar *et al.*, 2002). In nocturnal rodents the principal target of melanopsin-expressing retinal ganglion cells appears to be the suprachiasmatic nucleus of the hypothalamus (SCN; Berson *et al.*, 2002), the circadian pacemaker of the brain which determines daily cycles in activity and hormonal levels. It is likely therefore that in primates the melanopsin-expressing retinal ganglion cells project not only to the OPN and dLGN (Dacey *et al.*, 2005) but also to the SCN.

In primates the OPN on each side receives direct input from the contralateral and ipsilateral retinas (Hutchins and Weber, 1985; Gamlin and Clarke, 1995). The retinal ganglion cells in the nasal retina project to the contralateral OPN, while the ganglion cells in the temporal retina project to the ipsilateral OPN (Perry and Cowey, 1984; Kondo *et al.*, 1992; Rodieck and Watanabe, 1993). In view of the pattern of the nasotemporal division in primates such a pattern of retinal projection implies that OPN cells should respond exclusively to visual stimuli presented in the contralateral hemifield. However, the existence of bilateral cells undermines such a conclusion (Clarke *et al.*, 2003b). Anatomical investigations reveal no connection between the OPNs (Baleydier *et al.*, 1990; Mustari *et al.*, 1994), thus leaving the cortex as the likely source of the visual input from the ipsilateral hemifield. An input to macaque OPN from the dorsal and ventral parts of the ipsilateral preoccipital cortex has been reported (Dineen and Hendrickson, 1983; Asanuma *et al.*, 1985; Leichnetz, 1990; Lui *et al.*, 1995). In that part of the visual cortex both the contralateral and much of the ipsilateral visual hemifield is represented (Maunsell and Newsome, 1987). The ipsilateral representation is likely to arise from the callosal input from the contralateral visual cortices to the ipsilateral cortex, which in turn projects to the OPN (Hoffmann *et al.*, 1992). Also, strong inputs to the OPN arise from the FEF (Leichnetz, 1982a).

In primates, there is a direct projection from the OPN to the ipsilateral and contralateral EW nuclei and the contralateral projection crosses in the posterior commissure (Gamlin and Clarke, 1995; Clarke *et al.*, 2003a). Neurons in the EW project to the ipsilateral ciliary ganglion through the preganglionic parasympathetic fiber bundle of the oculomotor nerve (N.III). In turn, neurons from the ciliary ganglion innervate the smooth muscle intrinsic to the eye (pupillary sphincter), which constricts the pupil (Büttner-Ennever *et al.*, 1996a; Carpenter and Pierson, 1973; Pierson and Carpenter, 1974; Benevento *et al.*, 1977). Apart from the direct projections from OPN to EW in both macaques (Carpenter and Pierson, 1973; Pierson and Carpenter, 1974, Baleydier *et al.*, 1990; Cooper *et al.*, 1990) and New World marmoset monkeys (Blanks *et al.*, 1995; Clarke *et al.*, 2003a), there are indirect projections to the contralateral EW, which relay through the contralateral lateral terminal nucleus (LTN). Electrical stimulation of the OPN generates pupil constriction, revealing the functional importance of the area in the pupillary light reflex (Gamlin *et al.*, 1995; Pong and Fuchs, 2000b). All three of the neuronal types in OPN contribute to the behaviorally observed characteristics in which stimulation of the macular region generates stronger pupillary responses than stimulation of the periphery (Clarke *et al.*, 2003c).

Outputs from the OPN innervate the C-group motoneurons in the oculomotor nucleus (N.III), the LVC column of the EW complex, and the ipsilateral medial accessory nucleus of Bechterew (Büttner-Ennever *et al.*, 1996b). The C-group motor neurons are involved in vergence eye movements, while the LVC have a role in pupillary constriction and changing depth of focus through accommodation. The role of the OPN might therefore include controlling extraocular muscles in addition to the intrinsic ocular muscles.

8.2.3 Nucleus of the optic tract

Summary of functions

The principal functions of the NOT can be summarized as follows: (1) NOT drives the slow phases of horizontal optokinetic nystagmus (hOKN), which is characterized by alternating horizontal slow and fast eye movements in which the eyes first track wide-field moving images then saccade back to the center of the orbit. The eye movements that follow the background movement reduce the relative motion over the retina (retinal slip), which is critical for high-resolution vision. (2) It provides drive for smooth pursuit eye movements where the eyes track small moving objects. (3) It plays some role in gating information passing between the retina and the visual cortex. (4) It assists in controlling pupil size. (5) It influences saccade generation and visual fixation. There is a degree of uncertainty concerning the last three points as they are postulated on hodological grounds only.

NOT physiology

Three physiological cell types have been identified in the macaque NOT.

Wide-field direction-selective cells: In these cells wide-field image motion in one direction (preferred) increases spiking activity, while motion in the opposite (antipreferred) direction suppresses the ongoing activity (Figure 8.14A; Hoffmann and Distler, 1989; Mustari and Fuchs, 1990). As the axis of motion moves away from the preferred–antipreferred axis the responses decrease in magnitude in an approximately cosine-like fashion until motion is perpendicular. Perpendicular motion evokes no change in firing rate. Concurrently, as the axis of motion moves away from the preferred axis the suppression for antipreferred directions follows a similar cosine-like pattern. The mean half-width, defined as the directions generating >50 percent of the maximum response, is $127° \pm 25°$ (Hoffmann and Distler, 1989). Most NOT cells are tuned such that their preferred directions are for horizontal fronto-parallel motion toward the recorded nucleus (ipsiversive motion). For most cells, large visual stimuli provide the strongest drive (RFs $>10° \times 10°$), many covering large parts of the binocular visual fields of both eyes (Mustari and Fuchs, 1990). RFs typically include the *fovea* and all cells receive input from both eyes (Hoffmann and Distler, 1989). These cells rarely respond when the monkeys track a small moving spot with their eyes in an otherwise dark room (Figure 8.14B; Mustari and Fuchs, 1990). These neurons respond strongly to induced motion of the background when a monkey tracks a small target across a textured pattern, thus being referred to as *background velocity cells* (Ilg and Hoffmann, 1996; Ilg, 1997).

Approximately 18 percent of NOT cells that show strong motion responses cease firing for 50 ms after a saccade (even for saccades in darkness). The suppression actively inhibits their spontaneous activities, such that the spiking rate is virtually 0. The reduction in firing rate is not simply due to a loss of visual stimulation because this would bring the cells back to the spontaneous rate, not to a lower level. Rotating the monkeys from side to side in the dark does not generate responses in these cells, thus revealing a lack of vestibular input (Mustari and Fuchs, 1990). Optimal speeds of stimulus movement for wide-field cells are 4–200°/s (Hoffmann and Distler, 1989; Mustari and Fuchs, 1990).

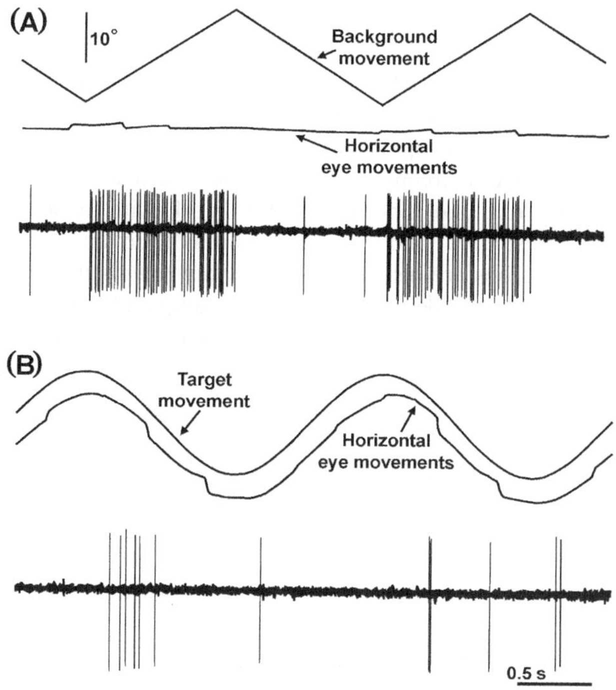

Figure 8.14 Spiking activity of a NOT neuron in the right-hand brain of a macaque (Mustari and Fuchs, 1990). (A) The position of the background at top (upwards is rightwards). Below it is the horizontal eye position. This cell spikes during rightward motion. (B) The monkey follows the movement of a small oscillating target in the dark. When the spot moves leftward, spikes are sometimes generated due to rightward target slippage. Reproduced by permission of the American Physiological Society

Small-field direction-selective cells: A smaller population of cells in NOT respond more vigorously during smooth pursuit eye movements in the dark than during wide-field stimulation in the light. These cells typically have small RFs ($<10° \times 10°$) that include the *fovea*. For this cell type, briefly removing the spot during smooth pursuit in darkness reduces the firing rate to the spontaneous level, thus showing that retinal slip and not the eye movements themselves generates the responses (Mustari and Fuchs, 1990). The preferred speeds of the small-field cells are low, while those for wide-field cells cover the whole speed range (Mustari and Fuchs, 1990). This cell type is peculiar to the primate NOT. Ilg and Hoffmann (1996) found that these cells responded when a pursuit target was moved across a textured background but not when the monkey fixated a stationary target and the background moved. They referred to these neurons as *target velocity cells* (Ilg and Hoffmann 1991, 1996; Ilg, 1997).

Wide-field non-directional cells: A non-directional cell type can be recorded in regions of the NOT ventral to the DS neurons (Hoffmann and Distler, 1989). These cells respond to the movement in any direction of rapidly moving images (500–1000°/s). These speeds usually occur only during saccades (section 8.1), suggesting a role in detecting the visual motion associated with saccades (section on NOT outputs to midbrain, diencephalon, and cortex). More is known about these neurons in non-primate species, notably the cat (e.g. Schmidt, 1996; Schmidt and Hoffmann, 1992) and wallaby (Ibbotson and Mark, 1994; Price and Ibbotson, 2001). In the wallaby the cells exhibit short response latencies, have large RFs, and respond to rapid image motion, with a preference for low spatial frequencies and high temporal frequencies. Tests indicate that they receive Y-like cell input from the retina (Price and Ibbotson, 2001). Work with awake cats reveals two populations of these cells: 'jerk' and 'saccade' neurons (Sudkamp and Schmidt, 1995; Schmidt, 1996). Jerk neurons are insensitive to eye movements in the dark but respond to rapid image motion and saccades in the light (Sudkamp and Schmidt, 1995). Saccade cells are responsive to rapid image displacements and to saccades in the light and dark (Schmidt, 1996).

NOT inputs

The NOT is a terminal area for the axons of some retinal ganglion cells that have traveled along the optic tract (Figure 8.15). In macaques retinal ganglion cells terminate in both the contralateral and ipsilateral NOTs, with a ratio of approximately 2:1 in favor of the former (Telkes *et al.*, 2000). In other mammals with frontally-positioned eyes (e.g. cats) the proportion of ganglion cells projecting to the ipsilateral NOT is much lower (about 10 percent; Ballas *et al.*, 1981; Distler and Hoffmann, 1993). Telkes *et al.* (2000) have shown that in macaques 1.2–2.4 percent of the ganglion cells in the contralateral retina

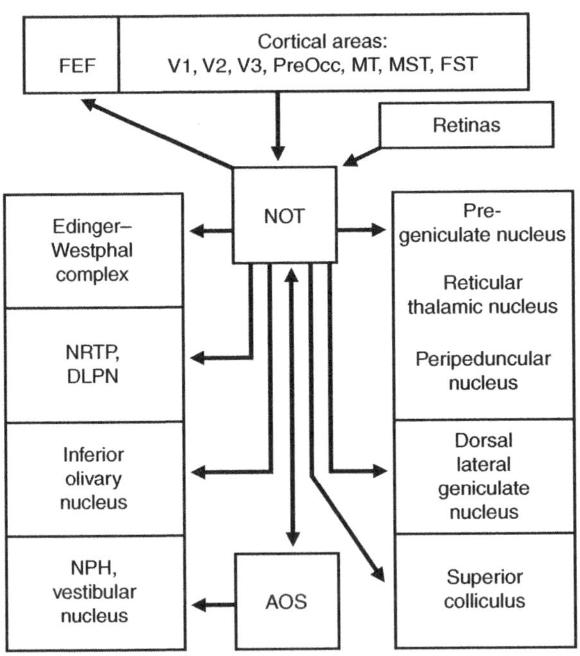

Figure 8.15 NOT connectivity

and 0.7–2.8 percent of the ganglion cells in the ipsilateral retina project to the NOT and dorsal terminal nucleus (DTN), thus making the NOT a major recipient area compared with other pretectal sites (Perry and Cowey, 1984; Rodieck and Watanabe, 1993). Inputs from the contralateral and ipsilateral eyes originate in the nasal and temporal retinas, respectively. Therefore, the input to the right-hand NOT arises primarily from cells viewing the left-hand visual field and vice versa. The largest density of retinal ganglion cells projecting to the NOT–DTN is along the horizontal strip of retina where the zero horizontal meridian is represented (Telkes *et al.*, 2000). In general, the retinal cells that terminate in the macaque NOT have slowly conducting axons (Hoffmann *et al.*, 1988). They have small cell bodies (50 μm^2) in areas within 1 mm of the *fovea*, around 150 μm^2 in areas 3–4 mm from the *fovea*, and over 250 μm^2 in areas 4–5 mm from the *fovea* (Telkes *et al.*, 2000). No midget and very few parasol ganglion cells project to the pretectum (Perry and Cowey, 1984; Rodieck, 1998; Telkes *et al.*, 2000). In macaques the pretectal-projecting ganglion cells, including the NOT–DTN (Rodieck and Watanabe, 1993; Telkes *et al.*, 2000), typically have large, sparse, and unistratified dendritic fields (200–800 μm in diameter). These features identify these ganglion cells as the specialized cell types that are rarely encountered (<3 percent; de Monasterio and Gouras, 1975). The responses of cells providing the NOT input have not been recorded, so we do not know if primate NOT-projecting retinal cells are DS, as is the case in other species (rabbit: Oyster *et al.*, 1972; cat: Hoffmann and Stone, 1985).

Numerically, the input which NOT receives from the visual cortex is about 10-fold larger than that from the retina (Figure 8.15). There is also a large cortical input to the cat NOT (Schoppmann, 1981). By contrast, in marsupials there is very little evidence of cortical input to the NOT (e.g. Ibbotson *et al.*, 2002). Injection of neuronal tracers into the NOT leads to the retrograde labeling of cell bodies in the ipsilateral areas V1, V2, and V3 of owl monkeys (Graham *et al.*, 1979), and area V1 of squirrel (Spatz *et al.*, 1970) and macaque monkeys (Hoffmann *et al.*, 1991; Lui *et al.*, 1995; Distler and Hoffmann, 2001; Distler *et al.*, 2002). However, the major input to the NOT arises from cortical areas in the vicinity of the STS (Figure 8.5), which contain cells specialized for motion processing and eye movement control (Distler and Hoffmann, 2001; Distler *et al.*, 2002). The major areas of interest are areas MT and MST and the fundus of the superior temporal area (area FST). Area MT in the primate specializes in motion processing and has a large proportion of DS cells (owl monkey: Allman and Kaas, 1971; macaques: Zeki, 1974; Maunsell and Van Essen, 1983a,b,c; Albright, 1984). In macaques, the RFs of MT cells are fairly large (3°–10°: Maunsell and Van Essen, 1983c). Furthermore, RFs of many MT cells extend well into the ipsilateral hemifield (Hoffmann *et al.*, 1992; Raiguel *et al.*, 1995). There is a strong projection from area MT to the ipsilateral NOT. All MT cells projecting to NOT are pyramidal neurons located in the infragranular layer 5 of the hemisphere ipsilateral to the NOT. Sectioning the corpus callosum, thus depriving MT of the input from the other hemisphere, results in the loss of the representation of the ipsilateral hemifield (Hoffmann *et al.*, 1992).

Most cells in area MST and a third of cells in area FST are direction selective (Desimone and Ungerleider, 1986; Saito *et al.*, 1986; Tanaka *et al.*, 1986; Komatsu and Wurtz, 1988). RF sizes of MST and FST cells are larger than those of MT cells (Duffy and Wurtz, 1991). Some MST neurons are selective for the direction of pursuit eye movements (Komatsu and Wurtz, 1988), and lesions in this area impair smooth pursuit

(Dürsteller *et al.*, 1987). Both areas MST and FST provide significant non-reciprocal input to the NOT (Boussaoud *et al.*, 1992; Distler *et al.*, 2002). MST neurons projecting to the ipsilateral NOT–DTN are invariably pyramidal cells of layer 5 (Distler and Hoffmann, 2001; Distler *et al.*, 2002). There is a bias in directional tuning amongst MT and MST cells that matches the tuning of NOT cells (Ilg and Hoffmann, 1993; Hoffmann *et al.*, 1992; Hoffmann *et al.*, 2002). Although few parasol ganglion cells project directly to the pretectum, the retinal input to visual cortical areas MT and MST originates predominantly (Merigan and Maunsell, 1993) although not excusively (Sawatari and Callaway, 1996; Yabuta *et al.*, 2001; Sincich *et al.*, 2004) from parasol cells. Since the cortical input to the pretectum is excitatory, in primate pretectum there must be an excitatory convergence of different functional information channels, relayed through different cell types originating in the retina (Merigan and Maunsell, 1993; Sawatari and Callaway, 1996; Vidyasagar *et al.*, 2001; Yabuta *et al.*, 2001; Sincich *et al.*, 2004).

NOT output regions and functions

The traditional view of the NOT is that it is involved in generating hOKN. The functionally important phase of hOKN is a slow tracking period where the eyes follow a moving scene at a speed close to that of the image. When the head turns about its vertical axis the vestibular apparatus in the inner ear also detects the acceleration of the head. If primates are rotated about their vertical axis in the dark, a pattern of eye movements similar to OKN occurs: the VOR. In the light, VOR and OKN work together to generate eye movements that stabilize retinal images during head turning (Waespe and Henn, 1987). The VOR is largely controlled by a neural network including the brain and vestibular nuclei (Figure 8.5). The vestibular nuclei receive input from the vestibular apparatus and also respond to corresponding movements of the visual surround (Waespe and Henn, 1987). These brain stem nuclei and associated areas are major targets for the NOT, showing that visual and vestibular information is integrated in the brain stem to produce optimal retinal stabilization through combined VOR and OKN responses.

In alert macaques, electrical stimulation of the NOT in the dark generates horizontal nystagmus of both eyes with ipsiversive slow phases (Schiff *et al.*, 1988). Ipsiversive motion in the NOT is the optimal stimulus, so there is a good correspondence between the directional tuning of the cells and the behavioral output of the system. Lesions of the NOT greatly attenuate the slow-phase speeds of hOKN (Schiff *et al.*, 1990). While the role of the NOT in hOKN is clearly important, the NOT also provides output to other brain areas not involved in hOKN.

NOT descending brainstem connections

Inferior Olive (IO): Hoffmann *et al.* (1988) showed that the NOT projects to the ipsilateral inferior olivary nucleus in the *medulla oblongata* of macaques. Electrical stimulation of the inferior olive generates responses with short latencies in NOT cells, showing that a direct connection is present. Cells that can be activated from the inferior olive are all DS and respond to both small- and wide-field patterns. The NOT sends its heaviest projection to the ipsilateral dorsal cap of Kooy (dcKooy) in the inferior olive (Mustari *et al.*, 1994). Only weak projections are evident to the contralateral inferior olive. Neurons in the dcKooy discharge during motion stimulation (Fuchs *et al.*, 1992), perhaps as a result of receiving NOT input. Evidence suggests that the visual modification of the VOR relies on the NOT-inferior olive pathway (Lisberger *et al.*, 1984; Miyashita, 1986; Yakushin *et al.*, 2000a,b). The inferior olive provides climbing fiber input to Purkinje cells in the contralateral cerebellar cortex (Eccles *et al.*, 1966; Thach, 1969). Signals in the ventral paraflocculus of the cerebellum might carry motor commands relating to the control of ocular following (Shidara *et al.*, 1993; Gomi *et al.*, 1998; Kobayashi *et al.*, 1998).

Nucleus Prepositus Hypoglossi (NPH) and *Medial Vestibular Nucleus* (MVN): In macaques the NOT provides weak projections (5 percent of cells) to the NPH and MVN (Mustari *et al.*, 1994). This sparse output suggests that the NOT–NPH projection could be involved in controlling hOKN (Büttner-Ennever *et al.*, 1996a). The MVN is thought to play a role in driving hOKN and storing eye velocity signals (Waespe and Henn, 1987; Fukishida *et al.*, 1992).

Dorsolateral Pontine Nucleus (DLPN) and *Nucleus Reticularis Tegmentis Pontis* (NRTP): The NOT projects to the DLPN and NRTP, which in turn project to the flocculus and vermis of the cerebellar cortex, the former connecting with oculomotor regions of the vestibular nuclei (Figure 8.5). It is probable that the NOT provides input to those neurons in the DLPN and NRTP that respond to wide-field motion stimulation (Keller and Crandall, 1983; Mustari *et al.*, 1988a). These circuits control smooth pursuit eye movements and probably assist in driving the slow phases of ocular following (Kawano *et al.*, 1992). Connections with the NRTP probably also provide input to the smooth pursuit system (Ilg and Hoffmann, 1991).

Oculomotor complex: NOT cells innervate directly the contralateral oculomotor complex, including the EW nucleus (Büttner-Ennever *et al.*, 1996b). Motoneurons of the oculomotor complex innervate several extraocular striated muscles, namely the inferior oblique and the inferior, superior, and medial recti. However, the interpretation of results from dye injections into the pretectal complex is fraught with difficulty because it is impossible to prevent dye dispersion into adjacent areas. As outlined in section 8.2.1, projections to the oculomotor complex may partially arise from other pretectal areas (Leichnetz, 2001).

NOT ouputs to midbrain, diencephalon, and cortex

Accessory optic system: The NOT projects strongly to the four nuclei of the ipsilateral accessory optic system (DTN, medial terminal nucleus (MTN), LTN, interstitial terminal nucleus (ITN)) and to the NOT and DTN in the contralateral brain (Büttner-Ennever *et al.*, 1996a). These connections are reciprocal and many are GABAergic, which suggests an inhibitory role. The NOT and accessory optic system (AOS) may modulate each other. For example, as NOT is horizontally tuned, while LTN is vertically tuned (Mustari and Fuchs, 1989), reciprocal inhibition might prevent responses to orthogonal motion (Mustari *et al.*, 1994). The nuclei of the AOS have a role in driving OKN, so the NOT probably influences OKN via the accessory optic nuclei.

Superior colliculus: NOT neurons terminate throughout the upper layers of the SC, with strong projections to the rostral SC where the *fovea* is represented (Büttner-Ennever *et al.*, 1996a). Many of the cells are GABAergic, thus inhibitory, but other transmitters have also been identified. The SC is generally involved in driving saccades (Sparks and Hartwich-Young, 1989), while rostrally located cells are involved in maintaining fixation (Büttner-Ennever and Horn, 1994). It is likely that the SC-projecting fibers from the NOT are involved in suppressing saccades and fixation during wide-field visual stimulation.

Ventral thalamus: One of the strongest outputs from the NOT is to the ventral thalamus, which in macaque includes the pregeniculate complex (PrGC: Büttner-Ennever *et al.*, 1996a). The PrGC is covered in detail in section 8.4. Its connections with the NOT may relate to a range of oculomotor activity and/or the processing of sensory information during and after saccades (Livingston and Fedder, 2003).

dLGN: Pretectal cells of macaque monkeys terminate throughout the ipsilateral dLGN but far more densely in the magnocellular layers (Büttner-Ennever *et al.*, 1996a). A weak projection also travels to the contralateral dLGN. In the prosimian Galago, the NOT provides synapses with inhibitory morphologies to many cells in the dLGN (Feig and Harting, 1994). The physiological responses of the non-directional cells in the primate NOT (Hoffmann and Distler, 1989) suggest similarities with those of non-primate species (Ibbotson and Mark, 1994; Schmidt, 1996). In all species they are tuned to detect the visual disturbances associated with saccades and blinks. To prevent conscious awareness of image motion during those ocular disturbances visual perception is suppressed (Ridder and Tomlinson, 1997). In cats the non-directional cells project to the dLGN and possibly have GABAergic synapses onto inhibitory interneurons, which in turn synapse onto thalamic relay cells (Cucchiaro *et al.*, 1991; Wahle *et al.*, 1994). Given that relay cells are part of the pathway from the retina to the cortex, it is possible that the non-directional NOT cells have a role in controlling cortical processing during saccades. Certainly, inactivation of the pretectum disinhibits dorsal thalamic relay cells during saccades in awake cats (Fischer *et al.*, 1996, 1998).

Pulvinar: Büttner-Ennever *et al.* (1996a) describe projections from the NOT to the pulvinar of macaques. The pulvinar is thought to be involved in visual spatial attention such as the early attention-shifting decisions that precede saccades (Robinson and McClurkin, 1989; Robinson and Peterson, 1992). The role of the pretectal NOT in this function is unclear but it could relate to the need for information relating to ongoing eye movements during attentional shifting. The fact that many of the visually responsive cells in the pulvinar become unresponsive to visual stimuli during eye movements might be important perceptually (sections on Outputs from the superficial SC layers and on Functional considerations). Perhaps pretectal cells inhibit visual activity during eye movements?

Frontal eye field: The FEF is involved in controlling saccades and smooth pursuit eye movements. The FEF receives a substantial direct input from the NOT (Leichnetz, 1982a). Thus, NOT appears to provide more direct visual input to the FEF than that relayed via the visual cortex.

8.2.4 Following omnidirectional pause neurons (FOPNs)

Mustari *et al.* (1997) have identified a region in the pretectum that constitutes a thin sheet of cells in the posterior pretectum on the dorsal border of the NOT. Cells in this area cease firing immediately after saccades and are referred to as FOPNs. The pause in firing occurs approximately 50 ms after the onset of saccades to targets and spontaneously in the dark. The pause in firing is not linked to visual input but rather to the motor command relating to saccade generation. About half of FOPNs generate spikes at or within 20 ms of saccade onset. This presaccadic burst is not related to the properties of the subsequent saccade but could signal the occurrence of an impending saccade. The specific connectivity of the FOPNs is difficult to assess because injections of tracer into the area with the FOPNs will inevitably spread to adjacent areas. As outlined in sections 8.2.3 and 8.4, fibers from this area project to the macaque PrGC. It is quite possible that the so-called NOT projections are actually projections from FOPNs. Mustari *et al.* (1997) have proposed that PrGC neurons are released from ongoing inhibition by FOPNs when the latter have a saccade-related pause in firing. As PrGC cells are connected with the dLGN via inhibitory terminals, so PrGC activity could gate information flow to the visual cortex.

8.3 Accessory optic system

In primates, the AOS is less prominent than in mammals with laterally-positioned eyes (e.g. rabbits and rats). Nevertheless, in primates the AOS consists of at least four retinorecipient nuclei located in the midbrain that are supplied by a subdivision of the optic tract – the accessory optic tract (AOT). The AOT diverges from the main optic tract at the brachium of the superior colliculus. The three principal accessory optic nuclei (AON) found in mammals are the DTN, LTN, and MTN. An additional collection of AON reside within the afferent limbs of the AOT – the interstitial terminal nucleus (ITN; Cooper *et al.*, 1990; Büttner-Ennever *et al.*, 1996a).

8.3.1 Summary of functions

The AOS has two principal functions: a role in driving OKN during horizontal (DTN) and vertical image motion (LTN and MTN) and a secondary role in providing input useful for the control of pupil size.

8.3.2 AOS physiology

Owing to its dorsal location in the midbrain and its juxtaposition with the NOT, the DTN is the most thoroughly studied region of the primate AOS. The DTN contains wide-field DS neurons that are sensitive to horizontal ipsiversive motion over both eyes. All DTN neurons receive input from both eyes, the input from the ipsilateral eye being indirect, and have large RFs (Hoffmann and Distler, 1989). Given the similarities between the physiology of the NOT and the DTN, readers are directed to section on NOT physiology, which describes the physiology of DS neurons in the NOT–DTN complex.

The LTN of primates, unlike rabbits or cats, is unequivocally larger than the MTN. The direct retinal input to the LTN of macaques originates exclusively from the contralateral eye (Mustari *et al.*, 1988b). The LTN of primates is located under the medial LGN, which is the principal auditory dorsal thalamic relay. Recordings from behaving macaques trained to fixate stationary targets and to track targets moving across a patterned background or in the dark have revealed several functional types of LTN neurons (Mustari *et al.*, 1988b; Mustari and Fuchs, 1989). Around 50 percent of cells are DS and respond best when stimulated by large stimuli (5°–70° diameter) and show only weak and variable activity when animals perform smooth pursuit in the dark. In the few cells that discharge during smooth pursuit in the dark, removal of the target causes complete cessation of spiking activity, showing that the cells are responding to the retinal slip generated by the spot during tracking. A further 36 percent of LTN cells (visuo-oculomotor cells) have similar DS properties but also respond to smooth pursuit in the dark (Figure 8.16). During repeated, predictable sinusoidal movements of the target the monkey will continue to move its eyes as if tracking the dot even when the dot is briefly turned off. The visuo-oculomotor cells continue to respond to the eye movement during this phase even without a visual target (Figure 8.16). The remaining 14 percent of LTN cells do not respond to visual stimulation but become active during smooth pursuit, suggesting that their input is entirely related to motor commands. Most LTN cells are DS, responding during vertical background movement and/or eye movements (Figure 8.16). Eighty-eight percent of DS cells prefer upward movement. On the basis of their speed tuning the LTN cells can be divided into two categories. The first group responds equally strongly over a wide range of velocities (4–80°/s). The second group exhibits narrow speed tuning with

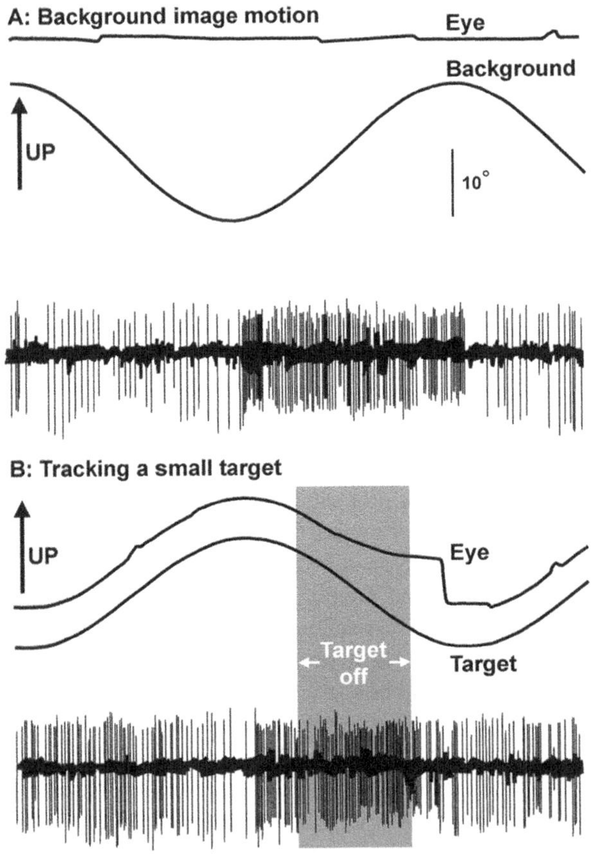

Figure 8.16 Responses of a visuo-oculomotor neuron in the macaque LTN (Mustari *et al.*, 1988b). (A) Upward motion of a background pattern increases spiking. (B) Tracking a small downward moving target in the dark generates increased spiking activity. Even with the target temporarily switched off, the neuron continues to respond. Reproduced by permission of Elsevier B.V

peaks averaging 13°/s. Two percent of LTN neurons respond to visual stimuli but are not DS (luminance cells).

We know little about the physiology of the MTN or ITN in primates. Westheimer and Blair (1974) recorded from neurons in the AOS of awake behaving macaques, referring to the recording site as the transpeduncular tract. This may be the region referred to in other studies as the MTN. The cells recorded in the transpeduncular tract respond to either horizontal or vertical motion of large visual patterns, respond during smooth pursuit, and exhibit high background activities. Consistent with colocalization of the transpeduncular tract and the MTN, in cats 75 percent of neurons recorded from the MTN exhibit high background activities and are DS (Grasse and Cynader, 1982). The remaining 25 percent of cat cells are responsive to changes in background illumination but are not directional. Most directional cells in cat MTN are excited by off-axis downward motion but 18 percent prefer upward motion. Cells are maximally excited by slow speeds (<1°/s). RFs of cat MTN cells are large.

8.3.3 AOS connectivity and functions

In all mammals, including frontal-eyed species (cats and primates), the direct retinal input to the AON originates predominantly from the contralateral eye (Figure 8.17; Cooper and Magnin, 1986). While in non-primates both the nasal and temporal retinas of the contralateral eye provide the AON with input, in primates the AON input from both temporal retinas is virtually absent. The morphology of the retinal ganglion cells projecting

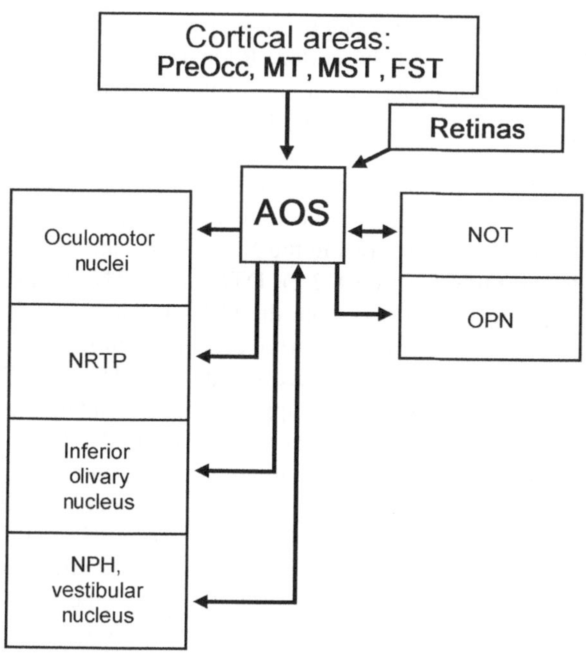

Figure 8.17 AOS connectivity

to the AON of primates is not known. However, in several mammals, including cats, AON-projecting retinal cells exhibit very similar morphology. The cells have widespread, sparse dendritic arbors which lie in the inner portion of the inner plexiform layer. In humans, using a marker of degenerated myelinated axons Fredericks *et al.* (1988) identified the retinal inputs to the DTN, LTN, and ITN, 10–15 years following either monocular enucleation or onset of bitemporal hemianopsia resulting from a pituitary tumor. The density of degenerated fibers was highest in the LTN. The retinal input to the MTN has not been studied in humans.

The DTN is located lateral to the NOT, below the brachium of the superior colliculus, and its precise borders are difficult to delineate (Figure 8.1). Moreover, the response properties of DTN and NOT neurons are largely indistinguishable, except that the non-directional neurons may be absent in the DTN. The similarity between NOT and DTN has led many to refer to the NOT–DTN as a single entity. However, there are differences between the two nuclei (Büttner-Ennever *et al.*, 1996a). For example, the NOT receives retinal input via the fibers traveling in the brachium of the SC, while the DTN receives its from the AOT.

The DTN receives direct input from the motion processing cortical areas in the ipsilateral and contralateral MT and MST (Hoffmann *et al.*, 1992; Lui *et al.*, 1995). The DTN projects to the inferior olive (Hoffmann and Distler, 1989), so it is assumed that it is able to influence visual adaptation of the VOR and drive ocular following (section on NOT descending brainstem connections). Cells in the DTN project to the NPH in the medulla, thus providing a direct pathway for control of hOKN (Büttner-Ennever *et al.*, 1996a). The DTN is fully interconnected with other AOS nuclei and with the NOT (Baleydier *et al.*, 1990).

The LTN is the most conspicuous nucleus in the primate AOS. It receives input from both the retina (Weber, 1985) and the cortex. Retinal ganglion cells projecting to the LTN are mainly distributed near the fovea and in the nasal region of the contralateral eye (Baleydier *et al.*, 1990). In macaques the LTN receives ipsilateral input from areas MT and MST (Maioli *et al.*, 1989; Lui *et al.*, 1995) and from the middle preoccipital cortex (Leichnetz, 1990). Importantly, no inputs have been identified arising from V1. In the common marmoset, the largest projection from the LTN is to the pretectal and AOS nuclei. Ipsilateral projections travel to the NOT, DTN, and ITN, while contralateral projections travel to the OPN (section 8.2.2), DTN, and LTN (Blanks *et al.*, 1995). In macaques, a main input to the LTN is from the OPN and it is entirely of contralateral origins. This uniquely primate projection originates from large multipolar cells in the OPN (Baleydier *et al.*, 1990). A direct connection between LTN and OPN suggests a role for the former in the pupillary light reflex, supported by the finding that some LTN cells respond only to luminance changes (Mustari and Fuchs, 1989). Other projections from primate LTN have been traced into the oculomotor nuclei and the ipsilateral reticular formation, NRTP, and inferior olive (Figure 8.1). Finally, fibers are supplied bilaterally to the dorsal cap of Kooy, the vestibular nuclei, and the NPH (Büttner-Ennever *et al.*, 1996a). These connections presumably provide pathways to control vertical OKN.

According to Cooper and Magnin (1986) in non-primates the MTN is the largest of the AOS nuclei and is usually subdivided into a ventral (MTNv) and dorsal division (MTNd). The MTNd is present in a very similar position in all mammals. While the MTNd in most primates is rather small, in the gibbon (an ape), it extends rostrocaudally for over 2 mm (Cooper and Magnin, 1986). The AOT fibers reach the MTNd through the cerebral peduncle and form obliquely oriented fiber bundles within the *substantia nigra*. The apparent absence of the MTNv in most primates might reflect the regression of the MTN in primate evolution (Weber and Giolli, 1986) or be a result of dorsal displacement of many MTN cells by development of a large pyramidal tract (Cooper and Magnin, 1986). Büttner-Ennever *et al.* (1996a) report the existence of a small MTNv in macaques. Anatomically, MTNd contains relatively large pale neurons ovoid or fusiform in shape with reticulum-like dendrites (Lenn, 1972; Cooper and Magnin, 1986). The MTN neurons, which are partially embedded in the *substantia nigra*, can be easily distinguished from the large polygonal-shaped cells of the *substantia nigra* (Cooper and Magnin, 1986). The MTN receives input from cortical area MT (Figure 8.5; Lui *et al.*, 1995) and projects to the NPH through which it can drive OKN.

The retino-recipient ITN reside amongst the input fibers of the AOT between the cerebral peduncle and the *substantia nigra* (Cooper *et al.*, 1990). In addition to the retina, the ITN receives input from cortical areas MT and MST (Lui *et al.*, 1995).

8.4 Pregeniculate complex

The PrGC is the only retino-recipient region of the ventral thalamus and is a presumed homolog of the ventral geniculate of other mammals (Kaas and Huerta, 1988; Harrington, 1997). The PrGC is a multi-layered region that extends over the medial and lateral aspects of the dLGN (Figure 8.1; Polyak, 1957; Niimi *et al.*, 1963).

8.4.1 Summary of functions

The PrGC appears to act as a relay between the occipital and parietal cortices and the brain stem oculomotor pathways. Consistent with this pattern of connections, it is involved in the integration of oculomotor signals with visual inputs (Büttner and Fuchs, 1973; Putkonen *et al.*, 1973; Magnin *et al.*, 1974; Livingston and Fedder, 2003).

8.4.2 PrGC physiology

In behaving macaques responses of around 85 percent of PrGC cells are modulated by saccades (Livingston and Fedder, 2003). Saccade-related cells discharge during saccades in the dark and the discharges are often modulated by visual input from cells with large RFs (Büttner and Fuchs, 1973). Approximately 33 percent of saccade-related cells show either post-saccadic bursts of spikes or inhibition of tonic spiking activity for saccades of all directions and amplitudes (Figure 8.18; Livingston and Fedder, 2003). The responses of approximately 45 percent of these cells coincide with or precede the saccade and 43 percent are DS. Sixteen percent of PrGC cells respond during smooth pursuit eye movements and half of these respond to visual motion during fixation (Livingston and Fedder, 2003). The latter responses are similar to those of NOT cells. Around 18 percent of cells have responses that are modulated by eye position (Livingston and Fedder, 2003).

Two cell types in the PrGC, probably both in the retino-recipient sublayer, have firing rates that are either tonically increased or decreased by prolonged exposure to diffuse

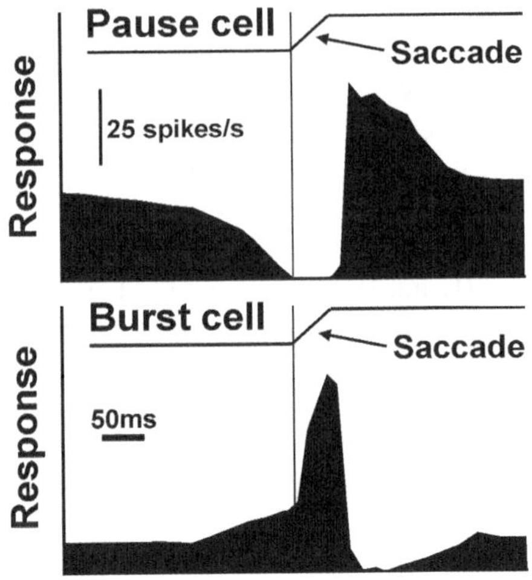

Figure 8.18 Spike density functions from two types of omnidirectional neurons in the macaque pregeniculate complex: pause cells and burst cells (Livingston and Fedder, 2003). Reproduced by permission of the American Physiological Society

light. These visual cells have response properties similar to cells found in the ventral LGN of other mammals (for review, Harrington, 1997).

8.4.3 Connectivity and possible functions of the PrGC

In macaque monkeys the PrGC can be divided into four subnuclei: a weakly developed non-retino-recipient dorsal lamina, the retinorecipient sublayer, the superior sublayer, and the medial division (Livingston and Mustari, 2000). A so-called lateral division is also located lateral to the dLGN (Babb, 1980). This is not strictly part of the PrGC but its proximity and similarity in connectivity merits inclusion here. Retinal afferents terminate exclusively in the retinorecipient sublayer. Injections into one eye label equal numbers of terminals in both PrGCs. The retinorecipient layer and the lateral division project to the SC and NOT, while the medial division and superior layer have reciprocal connections with the SC and NOT.

The PrGC receives input from many areas of the cortex (Figure 8.5), including areas V1 and V2 and several regions of the parietal cortex (e.g. 7a, LIP, MT, MST, and the polysensory area of the superior temporal sulcus or STS) (Spatz and Tigges, 1973; Norden *et al.*, 1978; Graham *et al.*, 1979; Maunsell and Van Essen, 1983b; Maioli *et al.*, 1984; Ungerleider *et al.*, 1984; Leichnetz, 1990). These regions can be divided into two subsets on the basis of the similarities of responses of cells in the PrGC with cells in a given cortical region. Areas MT, MST, and STS have cells with responses that are similar to the PrGC cells that respond during visual stimulation and smooth eye movements. Neurons in LIP and area 7a carry information related to eye position and these compare to the eye position signals observed in some PrGC cells. The PrGC projects to the NPH (Sherk, 1978; 1979; Belknap and McCrea, 1988; McCrea, 1988). Collectively, these outputs provide pathways by which signals can be relayed from the PrGC to oculomotor regions. Moreover, similarities between the physiology of PrGC cells and those in the pretectum suggest a close association. For example, the response properties of FOPNs around the pretectal NOT are identical to post-saccadic pause neurons in the PrGC in terms of latency and duration (section 8.2.4). Some of the pretectal FOPNs have a distinct excitatory burst prior to the pause in firing, which equates well with the post-saccadic burst neurons in the PrGC. The velocity tuning of PrGC cells to visual stimulation (peak at 4–80°/s) is also comparable to that in the pretectal NOT.

References

Adamük, E., 1870. Über die Innervation der Augenbewegungen. *Zentr. Med. Wiss.* **8**, 65–67.

Albano, J.E., Mishkin, M., Westbrook, L.E. *et al.*, 1982. Visuomotor deficits following ablation of monkey superior colliculus. *J. Neurophysiol.* **48**, 338–351.

Albright, T.D., 1984. Direction and orientation selectivity of neurons in visual area MT of the macaque. *J. Neurophysiol.* **52**, 1106–1130.

Allman, J.M. and Kaas, J.H., 1971. A representation of the visual field in the caudal third of the middle temporal gyrus of the owl monkey (*Aotus trivirgatus*). *Brain Res.* **31**, 85–105.

Andersen, R.A., Asanuma, C., Essick, G. *et al.*, 1990. Corticocortical connections of anatomically and physiologically defined subdivisions within the inferior parietal lobule. *J. Comp. Neurol.* **296**, 65–113.

Asanuma, C., Andersen, R.A., and Cowan, W.M., 1985. The thalamic relations of the caudal inferior parietal lobule and the lateral prefrontal cortex in monkeys: Divergent cortical projections from cell clusters in the medial pulvinar nucleus. *J. Comp. Neurol.* **241**, 357–381.

Babb, R.S., 1980. The pregeniculate nucleus of monkey (*Macacca mullatta*). I. A study at the light microscopy level. *J. Comp. Neurol.* **190**, 651–672.

Baizer, J.S., Whitney, J.F., and Bender, D.B., 1991. Bilateral projections from the parabigeminal nucleus to the superior colliculus of the monkey. *Exp. Brain Res.* **86**, 467–470.

Baleydier, C., Magnin, M., and Cooper, H.M., 1990. Macaque accessory optic system. II. Connections with the pretectum. *J. Comp. Neurol.* **302**, 405–416.

Ballas, I., Hoffmann, K.-P., and Wagner, H.J., 1981. Retinal projection to the nucleus of optic tract in the cat as revealed by retrograde transport of horseradish peroxidase. *Neurosci. Lett.* **26**, 197–202.

Behan, M., Steinhacker, K., Jeffrey-Borger, S. *et al.*, 2002. Chemoarchitecture of GABAergic neurons in the ferret superior colliculus. *J. Comp. Neurol.* **452**, 334–359.

Belknap, D.B. and McCrea, R.A., 1988. Anatomical connections of the prepositus and abducens nuclei in the squirrel monkey. *J. Comp. Neurol.* **268**, 13–28.

Bender, D.B. and Butter, C.M., 1987. Comparison of the effects of superior colliculus and pulvinar lesions on the visual search and tachistoscopic pattern discrimination in monkeys. *Exp. Brain Res.* **69**, 140–154.

Bender, D.B. and Davidson, R.M., 1986. Global visual processing in the monkey superior colliculus. *Brain Res.* **381**, 372–375.

Benevento, L.A. and Standage, G.P., 1983. The organization of projections of the retinorecipient and non-retinorecipient nuclei of the pretectal complex and layers of the superior colliculus to the lateral pulvinar and medial pulvinar in the macaque monkey. *J. Comp. Neurol.* **217**, 307–336.

Benevento, L.A., Rezak, M., and Santos-Anderson, 1977. An autoradiographic study of projections of the pretectum in the rhesus monkey (*Macaca mulatta*): Evidence for sensorimotor links with the thalamus and oculomotor nuclei. *Brain Res.* **127**, 197–218.

Berson, D.M., 1988. Retinal and cortical inputs to cat superior colliculus, convergence and laminar specificity. *Prog. Brain Res.* **75**, 17–26.

Berson, D.M., Dunn, F.A., and Takao, M., 2002. Phototranduction by retinal ganglion cells that set the circadian clock. *Science* **295**, 1070–1073.

Binns, K.E. and Salt, T.E., 1995. Excitatory amino acid receptors modulate habituation of the response to visual stimulation of the cat superior colliculus. *Vis. Neurosci.* **12**, 563–571.

Binns, K.E. and Salt, T.E., 1997. Different roles for $GABA_A$ and $GABA_B$ receptors in visual processing in the rat superior colliculus. *J. Physiol. (Lond.)* **504**, 629–639.

Blanks, R.H., Clarke, R.J., Lui, F. *et al.*, 1995. Projections of the lateral terminal accessory optic nucleus of the common marmoset (*Callithrix jacchus*). *J. Comp. Neurol.* **354**, 511–532.

Boussaoud, D., Desimone, R., and Ungerleider, L.G., 1992. Subcortical connections of visual areas MST and FST in macaques. *Vis. Neurosci.* **9**, 291–302.

Bruce, C.J. and Goldberg, M.E., 1985. Primate frontal eye fields. I. Single neurons discharging before saccades. *J. Neurophysiol.* **53**, 603–635.

Bruce, C.J., Goldberg, M.E., Bushnell, M.C. *et al.*, 1985. Primate frontal eye fields. II. Physiological and anatomical correlates of electrically evoked eye movements. *J. Neurophysiol.* **54**, 714–734.

Bruce, C., Desimone, R., and Gross, C.G., 1986. Both striate cortex and the superior colliculus contribute to the visual properties of neurons in the superior temporal polysensory area in the macaque. *J. Neurophysiol.* **5**, 1057–1075.

Bullier, J., Girard, P., and Salin, P.A., 1994. The role of area 17 in the transfer of information to extrastriate visual cortex, in *Primary Visual Cortex in Primates Cerebral Cortex*, Vol. 10 (eds A. Peters and K.S. Rockland), Plenum, New York, pp. 301–330.

Bunt, A.H., Minckler, D.S., and Johanson, G.W., 1977. Demonstration of bilateral projection of the central retina of the monkey with horseradish peroxidase neuronography. *J. Comp. Neurol.* **171**, 619–630.

Büttner, U. and Fuchs, A.F., 1973. Influence of saccadic eye movements on unit activity in simian lateral geniculate and pregeniculate nuclei. *J. Neurophysiol.* **36**, 127–141.

Büttner-Ennever, J.A. and Horn, A.K.E., 1994. Neuroanatomy of saccadic omnipause neurons in nucleus raphe interpositus, in *Four Decades of Seminal Eye Movement Research* (eds T. Brandt, U. Büttner, A.F. Fuchs, D.S. Zee), Thieme Verlag, Stuttgart, pp. 488–495.

Büttner-Ennever, J.A., Cohen, B., Horn, A.K.E. *et al.*, 1996a. Efferent pathways of the nucleus of the optic tract in monkey and their role in eye movements. *J. Comp. Neurol.* **373**, 90–107.

Büttner-Ennever, J.A., Cohen, B., Horn, A.K.E. *et al.*, 1996b. Pretectal projections to the oculomotor complex of the monkey, and their role in eye movements. *J. Comp. Neurol.* **366**, 348–359.

Cajal, S.R. y., 1911. *Histologie du systeme nerveux. Tome II.*, Maloine, Paris, p. 991.

Carpenter, M.B. and Pierson, R.J., 1973. Pretectal region and the papillary light reflex: An anatomical analysis in the monkey. *J. Comp. Neurol.* **149**, 271–300.

Casagrande, V.A., 1994. A third parallel pathway to primate area V1. *Trends Neurosci.* **17**, 305–310.

Chalupa, L.M., 1984. Visual physiology of the mammalian superior colliculus, in *Comparative Neurology of the Optic Tectum* (ed H. Vanegas), Plenum, London, pp. 775–818.

Chalupa, L.M. and Lia, B., 1991. The nasotemporal division of retinal ganglion cells with crossed and uncrossed projections in the fetal rhesus monkey. *J. Neurosci.* **11**, 191–202.

Chalupa, L.M. and Rhoades, 1977. Responses of visual, somatosensory and auditory neurones in the golden hamster superior colliculus. *J. Physiol. (Lond.)* **270**, 595–625.

Clarke, R.J. and Ikeda, H., 1985a. Luminance and darkness detectors in the olivary and posterior pretectal nuclei and their relationship to the pupillary light reflex in the rat. I. Studies with steady luminance levels. *Exp. Brain Res.* **57**, 224–232.

Clarke, R.J. and Ikeda, H., 1985b. Luminance detectors in the olivary pretectal nucleus and their relationship to the pupillary light reflex in the rat. II. Studies using sinusoidal lights. *Exp. Brain Res.* **59**, 83–90.

Clarke, R.J., Blanks, R.H.I., and Giolli, R.A., 2003a. Midbrain connections of the olivary pretectal nucleus in the marmoset (*Callithrix jacchus*): Implications for the pupil light reflex pathway. *Anat. Embryol.* **207**, 149–155.

Clarke, R.J., Zhang, H., and Gamlin, P.D.R., 2003b. Primate pupillary light reflex: Receptive field characteristics of pretectal luminance neurons. *J. Neurophysiol.* **89**, 3168–3178.

Clarke, R.J., Zhang H., and Gamlin, P.D.R., 2003c. Characteristics of the pupillary light reflex in the alert rhesus monkey. *J. Neurophysiol.* **89**, 3179–3189.

Cooper, H.M., Baleydier, C., and Magnin, M., 1990. Macaque accessory optic system. I. Definition of the medial terminal nucleus. *J. Comp. Neurol.* **302**, 394–404.

Clower, D.M., West, R.A., Lynch, J.C. *et al.*, 2001. The inferior parietal lobule is the target of output from the superior colliculus, hippocampus, and cerebellum. *J. Neurosci.* **21**, 6283–6291.

Cooper, H.M. and Magnin, M., 1986. A common mammalian plan of accessory optic system organization revealed in all primates. *Nature* **324**, 457–459.

Cowey, A. and Perry, W.H., 1980. The projection of the fovea to the superior colliculus in rhesus monkeys. *Neuroscience* **5**, 53–61.

Cowey, A. and Stoerig, P., 1991. The neurobiology of blindsight. *Trends Neurosci.* **14**, 140–145.

Cowey, A. and Stoerig, P., 1995. Blindsight in monkeys. *Nature* **373**, 247–249.

Cowey, A., Smith B., and Butter, C. M., 1984. Effects of damage to superior colliculi and pretectum on movement discrimination in rhesus monkeys. *Exp. Brain Res.* **56**, 79–91.

Cowey, A., Stoerig, P., and Bannister, M., 1994. Retinal ganglion cells labelled from the pulvinar nucleus in macaque monkeys. *Neuroscience* **61**, 691–705.

Cucchiaro, J.B., Bickford, M.E., and Sherman, S.M., 1991. A GABAergic projection from the pretectum to the dorsal lateral geniculate nucleus in the cat. *Neuroscience* **41**, 213–226.

Curcio, C.A. and Allen, K.A., 1990. Topography of ganglion cells in human retina. *J. Comp. Neurol.* **300**, 5–25.

Cusick, C.G., 1988. Anatomical organization of the superior colliculus in the monkeys: Corticotectal pathways for visual and visuomotor functions. *Prog. Brain Res.* **75**, 1–15.

Cynader, M. and Berman, N., 1972. Receptive-field organization of monkey superior colliculus. *J. Neurophysiol.* **35**, 187–201.

Dacey, D.M. and Lee, B.B., 1992. The blue-ON opponent pathway in primate retina originates from a distinct bistratified ganglion cell type. *Nature* **367**, 731–735.

Dacey, D.M. and Petersen, M.R., 1992. Dendritic field size and morphology of midget and parasol ganglion cells of the human retina. *Proc. Natl. Acad. Sci. USA* **89**, 9666–9670.

Dacey, D.M., Liao, H.-W., Peterson, B.B., *et al.*, 2005. Melanopsin-expressing ganglion cells in primate retina signal colour and irradiance and project to the LGN. *Nature* **433**, 749–754.

de Monasterio, F.M., 1978. Properties of ganglion cells with atypical receptive field organization in retina of macaques. *J. Neurophysiol.* **41**, 1435–1449.

de Monasterio, F.M. and Gouras, P., 1975. Functional properties of ganglion cells of the rhesus monkey retina. *J. Physiol. (Lond.)* **251**, 167–195.

Desimone, R. and Ungerleider, L.G., 1986. Multiple visual areas in the caudal superior temporal sulcus of the macaque. *J. Comp. Neurol.* **248**, 164–189.

De Valois, R.L., Morgan, H., and Snodderly, D.M., 1974. Psychophysical studies of monkey vision. III. Spatial luminance contrast sensitivity tests of macaque and human observers. *Vision Res.* **14**, 75–81.

Dineen, J.T. and Hendrickson, A., 1983. Overlap of retinal and prestriate cortical pathways in the primate pretectum. *Brain Res.* **278**, 250–254.

Distler, C. and Hoffmann, K.-P., 1989. The pupillary light reflex in normal and innate microstrabismic cats. I. Behavior and receptive-field analysis in the nucleus praetectalis olivaris. *Vis. Neurosci.* **3**, 127–138.

Distler, C. and Hoffmann, K.-P., 1993. Visual receptive field properties in kitten pretectal nucleus of the optic tract and dorsal terminal nucleus of the accessory optic tract. *J. Neurophysiol.* **70**, 814–827.

Distler, C. and Hoffmann, K.-P., 2001. Cortical input to the nucleus of the optic tract and dorsal terminal nucleus (NOT-DTN) in macaques: A retrograde tracing study. *Cereb. Cortex* **11**, 572–580.

Distler, C., Mustari, M.J., and Hoffmann, K.-P., 2002. Cortical projections to the nucleus of the optic tract and dorsal terminal nucleus and to the dorsolateral pontine nucleus in macaques: A dual retrograde tracing study. *J. Comp. Neurol.* **444**, 144–158.

Dodge, R., 1902. Five types of eye movement in the horizontal meridian plane of the field of regard. *Am. J. Physiol.* **8**, 307–329.

Dräger, U.C. and Hubel, D.H., 1975. Responses to visual stimulation and relationship between visual, auditory and somatosensory inputs in mouse superior colliculus. *J. Neurophysiol.* **38**, 690–713.

Dreher, B., Fukada, Y., and Rodieck, R.W., 1976. Identification, classification and anatomical segregation of cells with X-like and Y-like properties in the lateral geniculate nucleus of old-world primates. *J. Physiol. (Lond.)* **258**, 433–452.

Dreher, B., Jervie Sefton, A., Ni, S.Y.K. *et al.*, 1985. The morphology, number, distribution and central projections of class I retinal ganglion cells in albino and hooded rats. *Brain Behav. Evol.* **26**, 10–48.

Duffy, C.J. and Wurtz, R.H., 1991. Sensitivity of MST neurons to optic flow stimuli. I. A continuum of response selectivity to large-field stimuli. *J. Neurophysiol.* **65**, 1329–1345.

Duhamel, J.R., Colby, C.L., and Goldberg, M.E., 1992. The updating of the representation of visual space in parietal cortex by intended eye movements. *Science* **255**, 90–92.

Dürsteller, M.R., Wurtz, R.H., and Newsome, W.T., 1987. Directional pursuit deficits following lesions of the foveal representation within the superior temporal sulcus of the macaque monkey. *J. Neurophysiol.* **57**, 1262–1287.

Eccles, J.C., Linas, R., and Sasaki, K., 1966. Parallel fiber stimulation and the responses induced thereby in the Purkinje cells of the cerebellum. *Exp. Brain Res.* **1**, 17–39.

Edwards, S.B. and Henkel, C.K., 1983. Superior colliculus connections with the extraocular motor nuclei in the cat. *J. Comp. Neurol.* **179**, 451–468.

Feig, S. and Harting, J.K., 1994. Ultrastructural studies of the primate lateral geniculate nucleus: Morphology and spatial relationships of axon terminals arising from the retina, visual cortex (area 17), superior colliculus, parabigeminal nucleus, and pretectum of *Galago crassicaudatus*. *J. Comp. Neurol.* **343**, 17–34.

Feig, S., van Lieshout, D.P., and Harting, J.K., 1992. Ultrastructural studies of retinal, visual cortical (area 17), and parabigeminal terminals within the superior colliculus of *Galago crassicaudatus*. *J. Comp. Neurol.* **319**, 85–99.

Finlay, B.L., Schiller, P.H., and Volman, S.F., 1976. Quantitative studies of single cell properties in monkey striate cortex. IV. Corticotectal cells. *J. Neurophysiol.* **39**, 1352–1361.

Fischer, B. and Boch, R., 1983. Saccadic eye movements after extremely short fixation times in monkey. *Brain Res.* **260**, 21–26.

Fischer, B. and Ramsperger, E., 1984. Human express saccades: Extremely short reaction times of goal directed eye movements. *Exp. Brain Res.* **57**, 191–195.

Fischer, B. and Weber, H., 1993. Express saccades and visual attention. *Behav. Brain Sci.* **16**, 553–557.

Fischer, W.H., Schmidt, M., Stuphorn, V. *et al.*, 1996. Response properties of relay cells in the A-laminae of the cat's dorsal lateral geniculate nucleus of saccades. *Exp. Brain Res.* **110**, 435–445.

Fischer, W.H., Schmidt, M., and Hoffmann, K.-P., 1998. Saccade-induced activity of dorsal lateral geniculate nucleus X- and Y-cells during pharmacological inactivation of the cat pretectum. *Vis. Neurosci.* **15**, 197–210.

Fitzpatrick, D. and Raczkowski, D., 1991. The localization of neurotransmitters in the geniculostriate pathway, in *Neuroanatomy of Visual Pathways and their Development* (eds B. Dreher and S.R. Robinson) *Vision and Visual Dysfunction*, Vol. 3 (series ed J.R. Cronly-Dillon), Macmillan Press, Houndmills, London, pp. 278–301.

Fredericks, C.A., Giolli, R.A., Blanks, R.H. *et al.*, 1988. The human accessory optic system. *Brain Res.* **454**, 116–122.

Fries, W., 1984. Cortical projections to the superior colliculus in the macaque monkey: A retrograde study using horseradish peroxidase. *J. Comp. Neurol.* **230**, 55–76.

Fuchs, A.F., 1967. Saccadic and smooth pursuit eye movements in the monkey. *J. Physiol. (Lond.)* **191**, 609–631.

Fuchs, A.F., Mustari, M.J., Robinson, F.R. *et al.*, 1992. Visual signals in the nucleus of the optic tract and their brain stem destinations. *Ann. N.Y. Acad. Sci.* **656**, 266–276.

Fukishida, K., Kaneko, C.R.S., and Fuchs, A.F., 1992. The neuronal substrate of integration in the oculomotor system. *Prog. Neurobiol.* **39**, 609–639.

Fukuda, Y., Sawai, H., Watanabe, M. *et al.*, 1989. Nasotemporal overlap of crossed and uncrossed retinal ganglion cell projections in the Japanese monkey (*Macaca fuscata*). *J. Neurosci.* **9**, 2353–2373.

Gamlin, P.D.R. and Clarke, R.J., 1995. The pupillary light reflex pathway of the primate. *J. Am. Optom. Assoc.* **66**, 415–418.

Gamlin, P.D.R., Zhang, H., and Clarke, R.J., 1995. Luminance neurons in the pretectal olivary nucleus mediate pupillary light reflex in the rhesus monkey. *Exp. Brain Res.* **106**, 177–180.

Gamlin, P.D.R., Peterson, B.B., and Dacey, D.M., 2001. Physiology and morphology of retinal ganglion cells projecting to the pretectal olivary nucleus of the rhesus monkey. *Invest. Ophthalmol. Vis. Sci.* **42**, 3639.

Gandhi, N.J. and Sparks, D.L., 2003. Changing views of the role of superior colliculus in the control of gaze, in *The Visual Neurosciences*, A Bradford Book (eds L.M. Chalupa and J.S. Werner), MIT Press, Cambridge, MT, USA, pp. 1449–1465.

Ghosh, K.K., Goodchild, A.K., Sefton, A.E. *et al.*, 1996. Morphology of retinal ganglion cells in a New World monkey, the marmoset *Callithrix jacchus*. *J. Comp. Neurol.* **366**, 76–92.

Goldberg, M.E. and Wurtz, R.H., 1972a. Activity of superior colliculus in behaving monkey. I. Visual receptive fields of single neurons. *J. Neurophysiol.* **35**, 542–559.

Goldberg, M.E. and Wurtz, R.H., 1972b. Activity of superior colliculus in behaving monkey. II. Effect of attention on neural responses. *J. Neurophysiol.* **35**, 560–574.

Gomi, H., Shidara, M., Takemura, A. *et al.*, 1998. Temporal firing patterns of Purkinje cells in the cerebellar ventral paraflocculus during ocular following responses in monkey. I. Simple spikes. *J. Neurophysiol.* **80**, 818–831.

Graham, J., 1982. Some topographical connections of the striate cortex with subcortical structures in *Macaca fascicularis*. *Exp. Brain Res.* **47**, 1–14.

Graham, J., Lin, C.S., and Kaas, J.H., 1979. Subcortical projections of six visual cortical areas in the owl monkey, *Aotus trivirgatus*. *J. Comp. Neurol.* **187**, 557–580.

Grasse, K.L. and Cynader, M.S., 1982. Electrophysiology of medial terminal nucleus of accessory optic system in the cat. *J. Neurophysiol.* **48**, 490–504.

Graybiel, A.M., 1978. A satellite system of the superior colliculus: The parabigeminal nucleus and its projections to the superficial collicular layers. *Brain Res.* **145**, 365–374.

Guitton, D., 1992. Control of eye-head coordination during orienting gaze shifts. *Trends Neurosci.* **15**, 174–179.

Harrington, M.E., 1997. The ventral lateral geniculate nucleus and the intergeniculate leaflet: Interrelated structures in the visual and circadian systems. *Neurosci. Biobehav. Rev.* **21**, 705–727.

Harting, J.K., 1977. Descending pathways from the superior colliculus: An autoradiography analysis in the rhesus monkey (*Macaca mulatta*). *J. Comp. Neurol.* **175**, 583–612.

Harting, J.K., Huerta, M.F., Frankfurter, A.J. *et al.*, 1980. Ascending pathways from the monkey superior colliculus: An autoradiographic analysis. *J. Comp. Neurol.* **192**, 853–882.

Harutunian-Kozak, B., Dec, K., and Dreher, B., 1971. Habituation of unitary responses in the superior colliculus of the cat. *Acta Neurobiol. Exp.* **31**, 213–217.

Hashemi-Nezhad, M., Wang, C., Burke, W. *et al.*, 2003. Area 21a of cat visual cortex strongly modulates neuronal activities in the superior colliculus. *J. Physiol. (Lond.)* **550**, 535–552.

Hashikawa, T., van Lieshout, D.P., and Harting, J.K., 1986. Projections from the parabigeminal nucleus to the dorsal geniculate nucleus in the tree shrew (*Tupaia glis*). *J. Comp. Neurol.* **246**, 1–13.

Hattar, S., Liao, H.-W., Takao, M. *et al.*, 2002. Melanopsin-containing retinal ganglion cells: Architecture, projections, and intrinsic photosensitivity. *Science* **295**, 1065–1070.

Hendry, S.H.C. and Reid, R.C., 2000. The koniocellular pathway in primate vision. *Annu. Rev. Neurosci.* **23**, 127–153.

Hikosaka, O. and Wurtz, R.H., 1985. Modification of saccadic eye movements by GABA-related substances. I. Effect of muscimol and bicuculline in monkey superior colliculus. *J. Neurophysiol.* **53**, 266–291.

Hikosaka, O., Sakamoto, M., and Usui, S., 1989. Functional properties of monkey caudate neurons. I. Activities related to saccadic eye movements. *J. Neurophysiol.* **61**, 780–798.

Hoffmann, K.-P., 1973. Conduction velocity in pathways from retina to superior colliculus in the cat: A correlation with receptive field properties. *J. Neurophysiol.* **36**, 409–424.

Hoffmann, K.-P. and Distler, C., 1989. Quantitative analysis of visual receptive fields of neurons in nucleus of the optic tract and dorsal terminal nucleus of the accessory optic tract in macaque monkey. *J. Neurophysiol.* **62**, 416–428.

Hoffmann, K.-P. and Stone, J., 1985. Retinal input to the nucleus of the optic tract of the cat assessed by antidromic activation of ganglion cells. *Exp. Brain Res.* **59**, 395–403.

Hoffmann, K.-P., Distler, C., Erickson, R.G. *et al.*, 1988. Physiological and anatomical identification of the nucleus of the optic tract and dorsal terminal nucleus of the accessory optic tract in monkeys. *Exp. Brain Res.* **69**, 635–644.

Hoffmann, K.-P., Distler, C., and Erickson, R.G., 1991. Functional projections from striate cortex and superior temporal sulcus to the nucleus of the optic tract (NOT) and dorsal terminal nucleus of the accessory optic system (DTN) of macaque monkeys. *J. Comp. Neurol.* **313**, 707–724.

Hoffmann, K.-P., Distler, C., and Ilg, U., 1992. Callosal and superior temporal sulcus contributions to receptive field properties in the macaque monkey's nucleus of the optic tract and dorsal terminal nucleus of the accessory optic tract. *J. Comp. Neurol.* **321**, 150–162.

Hoffmann, K.-P., Bremmer, F., Thiele, A. *et al.*, 2002. Directional asymmetry of neurons in cortical areas MT and MST projecting to the NOT-DTN in macaques. *J. Neurophysiol.* **87**, 2113–2123.

Huerta, M.F. and Harting, J.K., 1983. Sublamination within the superficial gray layer of the squirrel monkey: An analysis of the tectopulvinar projection using anterograde and retrograde transport methods. *Brain Res.* **261**, 119–126.

Huerta, M.F. and Harting, J.K., 1984. The mammalian superior colliculus: Studies of its morphology and connections, in *Comparative Neurology of the Optic Tectum* (ed H. Vanegas), Plenum, London, pp. 687–773.

Huerta, M.F., Krubitzer, L.A., and Kaas, J.H., 1986. Frontal eye field as defined by intracortical microstimulation in squirrel monkeys, owl monkeys, and macaque monkeys. I. Subcortical connections. *J. Comp. Neurol.* **253**, 415–439.

Humphrey, N.K., 1974. Vision in a monkey without striate cortex: A case study. *Perception* **3**, 241–255.

Hutchins, B. and Weber, J.T., 1985. The pretectal complex of the monkey: A reinvestigation of the morphology and retinal terminations. *J. Comp. Neurol.* **232**, 425–442.

Ibbotson, M.R. and Mark, R.F., 1994. Wide-field nondirectional visual units in the pretectum: Do they suppress ocular following of saccade-induced visual stimulation? *J. Neurophysiol.* **72**, 1448–1450.

Ibbotson, M.R., Marotte, L.R., and Mark, R.F., 2002. Investigations into the source of binocular input to the nucleus of the optic tract in an Australian marsupial, the wallaby *Macropus eugenii*. *Exp. Brain Res.* **147**, 80–88.

Ilg, U.J., 1997. Slow eye movements. *Prog. Neurobiol.* **53**, 293–329.

Ilg, U.J. and Hoffmann, K.-P., 1991. Responses of monkey nucleus of the optic tract neurons during pursuit and fixation. *Neurosci. Res.* **12**, 101–110.

Ilg, U.J. and Hoffmann, K.-P., 1993. Functional grouping of the cortico-pretectal projection. *J. Neurophysiol.* **70**, 867–869.

Ilg, U.J. and Hoffmann, K.-P., 1996. Responses of neurons of the nucleus of the optic tract and the dorsal terminal nucleus of the accessory optic tract in the awake monkey. *Eur. J. Neurosci.* **8**, 92–105.

Illing, R.-B., 1996. The mosaic architecture of the superior colliculus. *Prog. Brain Res.* **112**, 17–34.

Isa, T., Endo, T., and Saito, Y., 1998. The visuo-motor pathway in the local circuit of the rat superior colliculus. *J. Neurosci.* **18**, 8496–8504.

Kaas, J.H. and Huerta, M.F., 1988. The subcortical visual system of primates, in *Comparative Primate Biology, Vol. 4, Neurosciences* (ed H.S. Steklis), Alan R. Liss, Inc., New York, pp. 327–391.

Kadoya, S., Wolin, L., and Massopust, L. Jr, 1971a. Collicular unit responses to monochromatic stimulation in squirrel monkey. *Brain Res.* **32**, 251–254.

Kadoya, S., Wolin, L., and Massopust, L. Jr, 1971b. Photically evoked unit activity in the tectum opticum of the squirrel monkey. *J. Comp. Neurol.* **142**, 495–508.

Kaplan, E., 2003. The M, P, and K pathways of the primate visual system, in *The Visual Neurosciences*, A Bradford Book (eds L.M. Chalupa and J.S. Werner), MIT Press, Cambridge, MT, pp. 481–493.

Kawano, K., Shidara, M., and Yamane, S., 1992. Neural activity in dorsolateral pontine nucleus of alert monkey during ocular following responses. *J. Neurophysiol.* **67**, 680–703.

Keller, E.L. and Crandall, W.F., 1983. Neuronal responses to optokinetic stimuli in pontine nuclei of behaving monkey. *J. Neurophysiol.* **49**, 169–187.

Kobayashi, Y., Kawano, K., Takemura, A. *et al.*, 1998. Temporal firing patterns of Purkinje cells in the cerebellar ventral paraflocculus during ocular following responses in monkeys. II. Complex spikes. *J. Neurophysiol.* **80**, 832–848.

Komatsu, H. and Wurtz, R.H., 1988. Relation of cortical areas MT and MST to pursuit eye movements. I. Localization and visual properties of neurons. *J. Neurophysiol.* **60**, 580–603.

Kondo, Y., Takada, M., Kayahara, T. *et al.*, 1992. Single retinal ganglion cells sending axon collaterals to the bilateral superior colliculi: A fluorescent retrograde double-labeling study in the Japanese monkey (*Maccaca fuscata*). *Brain Res.* **597**, 155–161.

Landolt, E., 1891. Nouvelles recherches sur la physiologie des mouvements des yeux. *Archives d'Opthalmologie (Paris)* **11**, 385–395.

Lane, R.H., Allman, J.M., Kaas, J.H. *et al.*, 1973. The visuotopic organization of the superior colliculus of the owl monkey (*Aotus trivirgatus*) and the bush baby (*Galago senegalensis*). *Brain Res.* **60**, 335–349.

Lee, P.H., Helms, M.C., Augustine, G.J. *et al.*, 1997. Role of intrinsic synaptic circuitry in collicular sensorimotor integration. *Proc. Natl. Acad. Sci. USA* **94**, 13299–13304.

Le Gros Clark, W.E., 1933. The medial geniculate body and the nucleus isthmi. *J. Anat. (Lond.)* **67**, 536–548.

Leichnetz, G.R., 1982a. Connections between the frontal eye field and pretectum in the monkey: An anterograde/retrograde study using HRP gel and TMB neurohistochemistry. *J. Comp. Neurol.* **207**, 394–404.

Leichnetz, G.R., 1982b. The medial accessory nucleus of Bechterew: A cell group within the anatomical limits of the rostral oculomotor complex receives a direct prefrontal projection in the monkey. *J. Comp. Neurol.* **210**, 147–151.

Leichnetz, G.R., 1990. Preoccipital cortex receives a differential input from the frontal eye field and projects to the pretectal olivary nucleus and other visuomotor-related structures in the rhesus monkey. *Vis. Neurosci.* **5**, 123–133.

Leichnetz, G.R., 2001. Connections of the medial posterior parietal cortex (area 7 m) in the monkey. *Anat. Rec.* **263**, 215–236.

Lenn, N.J., 1972. An electron microscopic study of accessory optic endings in the rat medial terminal nucleus. *Brain Res.* **43**, 622–628.

Leventhal, A.G., Rodieck, R.W., and Dreher, B., 1981. Retinal ganglion cell classes in the Old World monkey: Morphology and central projections. *Science* **213**, 1139–1142.

Leventhal, A.G., Rodieck, R.W., and Dreher, B., 1985. Central projections of cat retinal ganglion cells. *J. Comp. Neurol.* **237**, 216–226.

Leventhal, A.G., Ault, S.J., and Vitek, D.J., 1988. The nasotemporal division in primate retina: The neural basis of macular sparing and splitting. *Science* **240**, 66–67.

Lisberger, S.G., Miles, F.A., and Zee, D.S., 1984. Signals used to compute errors in monkey vestibuloocular reflex: Role of the flocculus. *J. Neurophysiol.* **52**, 1140–1153.

Livingston, C.A. and Fedder, S.R., 2003. Visual-ocular motor activity in the macaque pregeniculate complex. *J. Neurophysiol.* **90**, 226–244.

Livingston, C.A. and Mustari, M.J., 2000. The anatomical organization of the macaque pregeniculate complex. *Brain Res.* **876**, 166–179.

Lomber, S.G., 2002. Learning to see the trees before the forest: Reversible deactivation of the superior colliculus during learning of local and global visual features. *Proc. Natl. Acad. Sci. USA* **99**, 4049–4054.

Lui, F., Gregory, K.M., Blanks, R.H. *et al.*, 1995. Projections from the visual areas of the cerebral cortex to pretectal nuclear complex, terminal accessory optic nuclei, and superior colliculus in macaque monkey. *J. Comp. Neurol.* **363**, 439–460.

Lünenburger, L., Kleiser, R., Stuphorn, V. *et al.*, 2001. A possible role of the superior colliculus in eye–hand coordination. *Prog. Brain Res.* **134**, 109–125.

Magnin, M., Jeannerod, M., and Putkonen, P.T.S., 1974. Vestibular and saccadic influences on dorsal and ventral nuclei of the lateral geniculate body. *Exp. Brain Res.* **21**, 1–18.

Magoun, H.W., Atlas, D., Hare, W.K. *et al.*, 1935. The afferent pathways of the pupillary light reflex in the monkey. *Brain* **59**, 234–249.

Maioli, M.G., Galletti, C., Squatrito, S. *et al.*, 1984. Projections from the cortex of the superior temporal sulcus to the dorsal lateral geniculate and pregeniculate nuclei in the macaque monkey. *Arch. Ital. Biol.* **122**, 301–309.

Maioli, M.G., Squatrito, S., and Domeniconi, R., 1989. Projections from visual cortical areas of the superior temporal sulcus to the lateral terminal nucleus of the accessory optic system in macaque monkeys. *Brain Res.* **498**, 389–392.

Mark, R.F., James, A.C., and Sheng, X.-M., 1993. Geometry of the representation of the visual field on the superior colliculus of the wallaby (*Macropus eugenii*). I. Normal projection. *J. Comp. Neurol.* **330**, 303–314.

Marrocco, R.T. and Li, R.H., 1977. Monkey superior colliculus: Properties of single cells and their afferent inputs. *J. Neurophysiol.* **40**, 844–860.

Marrocco, R.T., McClurkin, J.W., and Young, R.A., 1981. Spatial properties of superior colliculus cells projecting to the inferior pulvinar and parabigeminal nucleus of the monkey. *Brain Res.* **222**, 150–154.

Maunsell, J.H.R. and Newsome, W.T., 1987. Visual processing in monkey extrastriate cortex. *Annu. Rev. Neurosci.* **10**, 363–401.

Maunsell, J.H.R. and Van Essen, D.C., 1983a. The connections of the middle temporal visual area (MT) and their relationship to a cortical hierarchy in the macaque monkey. *J. Neurosci.* **3**, 2563–2586.

Maunsell, J.H.R. and Van Essen, D.C., 1983b. Functional properties of neurons in middle temporal visual area of the macaque monkey. II. Binocular interactions and sensitivity to binocular disparity. *J. Neurophysiol.* **49**, 1148–1167.

Maunsell, J.H.R. and Van Essen, D.C., 1983c. Functional properties of neurons in middle temporal visual area of the macaque monkey. I. Selectivity for stimulus direction, speed, and orientation. *J. Neurophysiol.* **49**, 1127–1147.

Mays, L.E. and Sparks, D.I., 1980. Dissociation of visual and saccade-related responses in the superior colliculus neurons. *J. Neurophysiol.* **43**, 207–232.

McCrea, R.A., 1988. Neuroanatomy of the oculomotor system: The nucleus prepositus. *Rev. Oculomot. Res.* **2**, 203–223.

Merigan, W.H. and Maunsell, J.H.R., 1993. How parallel are the primate visual pathways? *Annu. Rev. Neurosci.* **16**, 369–402.

Miyashita, M., 1986. Neuronal circuit modifications underlying adaptation of the vestibuloocular reflex, in *Adaptive Processes in Visual and Oculomotor Systems* (eds E.L. Keller and D.S. Zee), Pergamon Press, Oxford, pp. 435–442.

Mize, R., 1992. The organization of GABAergic neurons in the mammalian superior colliculus. *Prog. Brain Res.* **90**, 219–248.

Mohler, C.W. and Wurtz, R.H., 1976. Organization of monkey superior colliculus: Intermediate layer cells discharging before eye movements. *J. Neurophysiol.* **39**, 722–744.

Mohler, C.W. and Wurtz, R.H., 1977. Role of striate cortex and superior colliculus in the visual guidance of saccadic eye movements in monkeys. *J. Neurophysiol.* **40**, 74–94.

Mollon, J.D., 1991. Uses and evolutionary origins of primate colour vision, in *Evolution of the Eye and Visual System* (eds J.R. Cronly-Dillon and R.L. Gregory), *Vision and Visual Dysfunction*, Vol. 2 (series ed J.R. Cronly-Dillon), Macmillan Press, Houndmills, London, pp. 306–319.

Mooney, R.D., Huang, X., and Rhoades, R.W., 1992. Functional influence of interlaminar connections in the hamster's superior colliculus. *J. Neurosci.* **12**, 2417–2432.

Moors, J. and Vendrik, A.J.H., 1979a. Responses of single units in the monkey superior colliculus to stationary flashing stimuli. *Exp. Brain Res.* **35**, 333–347.

Moors, J. and Vendrik, A.J.H., 1979b. Responses of single units in the monkey superior colliculus to moving stimuli. *Exp. Brain Res.* **35**, 349–369.

Munoz, D.P. and Wurtz, R.H., 1993a. Fixation cells in monkey superior colliculus. I. Characteristics of cell discharge. *J. Neurophysiol.* **70**, 559–575.

Munoz, D.P. and Wurtz, R.H., 1993b. Fixation cells in monkey superior colliculus. II. Reversible activation and deactivation. *J. Neurophysiol.* **70**, 576–589.

Mustari, M.J. and Fuchs, A.F., 1989. Response properties of single units in the lateral terminal nucleus of the accessory optic system in the behaving primate. *J. Neurophysiol.* **61**, 1207–1220.

Mustari, M.J. and Fuchs, A.F., 1990. Discharge patterns of neurons in the pretectal nucleus of the optic tract (NOT) in the behaving primate. *J. Neurophysiol.* **64**, 77–90.
Mustari, M.J., Fuchs, A.F., and Wallman, J., 1988a. Response properties of dorsolateral pontine units during smooth pursuit in the rhesus macaque. *J. Neurophysiol.* **60**, 664–686.
Mustari, M.J., Fuchs, A.F., Langer, T.P. *et al.*, 1988b. The role of the primate lateral terminal nucleus in visuomotor behavior. *Prog. Brain Res.* **75**, 121–128.
Mustari, M.J., Fuchs, A.F., Kaneko, C.R.S. *et al.*, 1994. Anatomical connections of the primate pretectal nucleus of the optic tract. *J. Comp. Neurol.* **349**, 111–128.
Mustari, M.J., Fuchs, A.F., and Pong, M., 1997. Response properties of pretectal omnidirectional pause neurons in the behaving primate. *J. Neurophysiol.* **77**, 116–125.
Niimi, K., Kanaseki, T., and Takimoto, T., 1963. The comparative anatomy of the ventral nucleus of the lateral geniculate body in mammals. *J. Comp. Neurol.* **121**, 313–324.
Norden, J.J., Lin, C.S., and Kaas, J.H., 1978. Subcortical projections of the owl monkey (*Aotus trivirgatus*). *Exp. Brain Res.* **32**, 321–334.
Oyster, C.W. and Takahashi, E., 1975. Responses of rabbit superior colliculus to repeated visual stimuli. *J. Neurophysiol.* **38**, 301–312.
Oyster, C.W., Takahashi, E., and Collewijn, H., 1972. Direction-selective retinal ganglion cells and control of optokinetic nystagmus in the rabbit. *Vision Res.* **12**, 183–193.
Pandya, D.N. and Barnes, C., 1987. Architecture and connections of the frontal lobe, in *The Frontal Lobes Revisited* (ed E. Perecman), IRBN Press, New York, pp. 41–72.
Paré, M. and Wurtz, R.H., 1997. Monkey posterior parietal cortex neurons antidromically activated from superior colliculus. *J. Neurophysiol.* **78**, 3493–3497.
Pasik, T., Pasik, P., and Bender, M.B., 1966. The superior colliculi and eye movements. *Arch. Neurol.* **15**, 420–436.
Payne, B.R., 1993. Evidence for visual cortical area homologs in cat and monkey. *Cereb. Cortex* **3**, 1–25.
Payne, B.R., Lomber, S.G., Geeraerts, S. *et al.*, 1996. Reversible visual hemineglect. *Proc. Natl. Acad. Sci. USA* **93**, 290–294.
Perry, V.H. and Cowey, A., 1984. Retinal ganglion cells that project to the superior colliculus and pretectum in the macaque monkey. *Neuroscience* **12**, 1125–1137.
Perry, V.H. and Cowey, A., 1985. The ganglion cell and cone distributions in the monkey's retina: Implications for central magnification factors. *Vision Res.* **25**, 1795–1810.
Perry, V.H., Oehler, R., and Cowey, A., 1984. Retinal ganglion cells that project to the dorsal lateral geniculate nucleus in the macaque monkey. *Neuroscience* **12**, 1101–1123.
Pierson, R.J. and Carpenter, M.B., 1974. Anatomical analysis of papillary reflex pathways in the rhesus monkey. *J. Comp. Neurol.* **158**, 121–144.
Polyak, S., 1941. *The Retina*, University of Chicago Press, Chicago.
Polyak, S., 1957. The pregeniculate gray nucleus and the pupillary reflex pathway, in *The Vertebrate Visual System* (ed H. Klüver), University of Chicago, Chicago, IL, pp. 376–385.
Pong, M. and Fuchs, A.F., 2000a. Characteristics of the pupillary light reflex in the macaque monkey: Metrics. *J. Neurophysiol.* **84**, 953–963.
Pong, M. and Fuchs, A.F., 2000b. Characteristics of the pupillary light reflex in the macaque monkey: Discharge patterns of pretectal neurons. *J. Neurophysiol.* **84**, 964–974.
Price, N.S.C. and Ibbotson, M.R., 2001. Pretectal neurons optimized for the detection of saccade-like movements of the visual image. *J. Neurophysiol.* **85**, 1512–1521.
Provis, J.M., Van Driel, D., Billson, F.A. *et al.*, 1985. Human fetal optic nerve: Overproduction and elimination of retinal axons during development. *J. Comp. Neurol.* **238**, 92–100.
Putkonen, P., Magnin, M., and Jeannerod, M., 1973. Directional responses to head rotation in the neurons from the ventral nucleus of the lateral geniculate body. *Brain Res.* **61**, 407–411.
Raiguel, S., van Hulle, M.M., Xiao, D.K. *et al.*, 1995. Shape and spatial distribution of receptive fields and antagonistic motion surrounds in the middle temporal area (V5) of the macaque. *Eur. J. Neurosci.* **7**, 2064–2082.
Rakic, P. and Riley, K.P., 1983a. Overproduction and elimination of retinal axons in the fetal rhesus monkey. *Science* **219**, 1441–1444.
Rakic, P. and Riley, K.P., 1983b. Regulation of axon number in the primate optic nerve by prenatal binocular competition. *Nature* **305**, 135–137.

Ranson, S.W. and Magoun, H.W., 1933. The central path of the pupilloconstrictor reflex in response to light. *Arch. Neurol. Psychiatry* **30**, 1193–1202.
Rhoades, R.W., Mooney, R.D., and Fish, S.E., 1991. Retinotopic and visuotopic representations in the mammalian superior colliculus, in *Neuroanatomy of Visual Pathways and their Development* (eds B. Dreher and S.R. Robinson), *Vision and Visual Dysfunction*, Vol. 3 (series ed J.R. Cronly-Dillon), Macmillan Press, Houndmills, pp. 150–175.
Ridder, W.H. III and Tomlinson, A., 1997. A comparison of saccadic and blink suppression in normal observers. *Vision Res.* **37**, 3171–3179.
Rizzolatti, G., Buchtel, H.A., Camarda, R. *et al.*, 1980. Neurons with complex visual properties in the superior colliculus of the macaque monkey. *Exp. Brain Res.* **38**, 37–42.
Robinson, D.A., 1972. Eye movements evoked by collicular stimulation in the alert monkey. *Vision Res.* **12**, 1795–1808.
Robinson, D.L. and McClurkin, J.W., 1989. The visual superior colliculus and pulvinar. *Rev. Oculomot. Res.* **3**, 337–360.
Robinson, D.L. and Peterson, S.E., 1992. The pulvinar and visual salience. *Trends Neurosci.* **15**, 127–132.
Rodieck, R.W., 1998. *The First Steps in Seeing*, Sinauer Associates, Inc., Sunderland, MA, USA.
Rodieck, R.W. and Watanabe, M., 1993. Survey of the morphology of macaque retinal ganglion cells that project to the pretectum, superior colliculus and parvicellular laminae of the lateral geniculate nucleus. *J. Comp. Neurol.* **338**, 289–303.
Rodman, H.R., Gross, C.G., and Albright, T.D., 1990. Afferent basis of visual response properties in area MT of the macaque. II. Effects of superior colliculus removal. *J. Neurosci.* **10**, 1154–1164.
Rose, J.E., 1942. The thalamus of the sheep: Cellular and fibrous structure and comparison with pig, rabbit and cat. *J. Comp. Neurol.* **77**, 469–523.
Ross, J., Morrone, M.C., and Burr, D.C., 1997. Compression of visual space before saccades. *Nature* **386**, 598–601.
Rowe, M.H., 1991. Functional organization of the retina, in *Neuroanatomy of Visual Pathways and their Development* (eds B. Dreher and S.R. Robinson), *Vision and Visual Dysfunction*, Vol. 3 (series ed J.R. Cronly-Dillon), Macmillan Press, Houndmills, pp. 1–68.
Saito, H.-A., Yukie, M., Tanaka, K. *et al.*, 1986. Integration of direction signals of image motion in the superior temporal sulcus of the macaque monkey. *J. Neurosci.* **6**, 145–157.
Sawatari, A. and Callaway, E.M., 1996. Convergence of magno- and parvocellular pathways in layer 4B of macaque primary visual cortex. *Nature* **380**, 442–446.
Schiff, D., Cohen, B., and Raphan, T., 1988. Nystagmus induced by stimulation of the nucleus of the optic tract in the monkey. *Exp. Brain Res.* **70**, 1–14.
Schiff, D., Cohen, B., Büttner-Ennever, J.A. *et al.*, 1990. Effects of lesions of the nucleus of the optic tract on optokinetic nysatgmus and after-nystagmus in the monkey. *Exp. Brain Res.* **79**, 225–239.
Schiller, P.H. and Koerner, F., 1971. Discharge characteristics of single units in superior colliculus of the alert rhesus monkey. *J. Neurophysiol.* **34**, 920–936.
Schiller, P.H. and Malpeli, J.G., 1977. Properties of tectal projections of monkey retinal ganglion cells. *J. Neurophysiol.* **40**, 428–445.
Schiller, P.H. and Stryker, M., 1972. Single-unit recording and stimulation in superior colliculus of the alert rhesus monkey. *J. Neurophysiol.* **35**, 915–924.
Schiller, P.H. and Tehovnik, E.J., 2001. Look and see: How the brain moves your eyes about. *Prog. Brain Res.* **134**, 127–142.
Schiller, P.H., Stryker, M., Cynader, M. *et al.*, 1974. Response characteristics of single cells in the monkey colliculus following ablation or cooling of visual cortex. *J. Neurophysiol.* **37**, 181–194.
Schiller, P.H., Malpeli, J.G., and Schein, S.J., 1979. Composition of geniculostriate input of superior colliculus of the rhesus monkey. *J. Neurophysiol.* **42**, 1124–1133.
Schiller, P.H., Sandell, J.H., and Maunsell, J.H., 1987. The effect of frontal eye field and superior colliculus lesions on saccadic latencies in the rhesus monkey. *J. Neurophysiol.* **57**, 1033–1049.
Schiller, P.H., Haushofer, J., and Kendall, G., 2004. An examination of the variables that affect express saccade generation. *Vis. Neurosci.* **21**, 119–127.
Schmidt, M., 1996. Neurons in the cat pretectum that project to the dorsal lateral geniculate nucleus are activated during saccades. *J. Neurophysiol.* **76**, 2907–2918.

Schmidt, M. and Hoffmann, K.-P., 1992. Physiological characterisation of pretectal neurons projecting to the lateral geniculate nucleus in the cat. *Eur. J. Neurosci.* **4**, 318–326.

Schmidt, M., Boller, M., Özen G. *et al.*, 2001. Disinhibition in rat superior colliculus mediated by $GABA_C$ receptors. *J. Neurosci.* **21**, 691–699.

Schneider, G.E., 1969. Two visual systems. *Science* **163**, 895–902.

Schoppmann, A., 1981. Projections from areas 17 and 18 of the visual cortex to the nucleus of the optic tract. *Brain Res.* **223**, 1–17.

Scudder, C.A., Kaneko, C.R., and Fuchs, A.F., 2002. The brainstem burst generator for saccadic eye movements: A modern synthesis. *Exp. Brain Res.* **142**, 439–462.

Sefton, A.J., Dreher, B., and Harvey, A.R., 2004. *Visual System: The Rat Nervous System*, 3rd edn (ed G. Paxinos), Academic Press, San Diego, Chapter 32, pp. 1083–1165.

Segraves, M.A. and Goldberg, M.E., 1987. Functional properties of corticotectal neurons in the monkey's frontal eye field. *J. Neurophysiol.* **58**, 1387–1419.

Sherk, H., 1978. Visual response properties and visual field topography in the cat's parabigeminal nucleus. *Brain Res.* **145**, 375–379.

Sherk, H., 1979. A comparison of visual-response properties in cat's parabigeminal nucleus and superior colliculus. *J. Neurophysiol.* **42**, 1640–1655.

Shidara, M., Kawano, K., Gomi, H. *et al.*, 1993. Inverse-dynamics model eye movement control by Purkinje cells in the cerebellum. *Nature* **365**, 50–52.

Silveira, L.C.L, Yamada, E.S., Perry, V.H. *et al.*, 1994. M and P retinal ganglion cells of diurnal and nocturnal New World monkeys. *Neuroreport* **5**, 2077–2081.

Sincich, L.C., Park, K.F., Wohlgemuth, M.J. *et al.*, 2004. Bypassing V1: A direct geniculate input to area MT. *Nat. Neurosci.* **7**, 1123–1128.

Sommer, M.A. and Wurtz, R.H., 2004a. What the brain stem tells the frontal cortex. I. Oculomotor signals sent from superior colliculus to frontal eye field via mediodorsal thalamus. *J. Neurophysiol.* **91**, 1381–1402.

Sommer, M.A. and Wurtz, R.H., 2004b. What the brain stem tells the frontal cortex. II. Role of the SC-MD-FEF pathway in corollary discharge. *J. Neurophysiol.* **91**, 1403–1423.

Sparks, D.L. and Hartwich-Young, R., 1989. The deep layers of the superior colliculus, in *The Neurobiology of Saccadic Eye Movements* (eds R.H. Wurtz and M.E. Goldberg), Elsevier, Amsterdam, pp. 213–245.

Spatz, W.B. and Tigges, J., 1973. Studies on the visual area MT in primates. II. Projection fibers to subcortical structures. *Brain Res.* **61**, 374–378.

Spatz, W.B., Tigges, J., and Tigges, M., 1970. Subcortical projections, cortical associations, and some intrinsic interlaminar connections of the striate cortex in the squirrel monkey (*Saimiri*). *J. Comp. Neurol.* **140**, 155–174.

Sprague, J.M., 1996. Neural mechanisms of visual orienting responses. *Prog. Brain Res.* **112**, 1–15.

Stanton, G.B., Goldberg, M.E., and Bruce, C.J., 1988a. Frontal eye field efferents in the macaque monkey. I. Subcortical pathways and topography of striatal and thalamic terminal fields. *J. Comp. Neurol.* **271**, 473–492.

Stanton, G.B., Goldberg, M.E., and Bruce, C.J., 1988b. Frontal eye field efferents in the macaque monkey. II. Topography of terminal fields in midbrain and pons. *J. Comp. Neurol.* **271**, 493–506.

Stein, B.E., 1981. Organization of the rodent superior colliculus: Some comparisons with other mammals. *Behav. Brain Res.* **3**, 175–188.

Stein, B.E. and Meredith, M.A., 1991. Functional organization of the superior colliculus, in *The Neural Basis of Visual Function* (ed A.G. Leventhal), *Vision and Visual Dysfunction*, Vol. 4 (series ed J.R. Cronly-Dillon), Macmillan Press, Houndmills, London, pp. 85–110.

Stein, B.E., Jaing, W., Wallace, M.T. *et al.*, 2001. Nonvisual influences on visual-information processing in the superior colliculus. *Prog. Brain Res.* **134**, 143–156.

Stepniewska, I., 2004. The pulvinar complex, in *The Primate Visual System* (eds J.H. Kaas and C.E. Collins), CRC Press, Boca Raton, London, pp. 53–80.

Stoerig, P. and Cowey, A., 1997. Blindsight in man and monkey. *Brain* **120**, 535–559.

Stone, J., Leicester, J., and Sherman, S.M., 1973. The naso-temporal division of the monkey retina. *J. Comp. Neurol.* **150**, 333–348.

Sudkamp, S. and Schmidt, M., 1995. Physiological characterization of pretectal neurons projecting to the lateral posterior-pulvinar complex in the cat. *Eur. J. Neurosci.* **7**, 881–888.

Tamamaki, N., Ulrich, D.J., and Sherman, S.M., 1995. Morphology of physiologically identified retinal X and Y axons in the cat's thalamus and midbrain as revealed by intraaxonal injection of biocytin. *J. Comp. Neurol.* **354**, 583–607.
Tanaka, K., Hikosaka, K., Saito, H. *et al.*, 1986. Analysis of local and wide-field movements in the superior temporal visual areas of the macaque monkey. *J. Neurosci.* **6**, 134–144.
Telkes, I., Distler, C., and Hoffmann, K.-P., 2000. Retinal ganglion cells projecting to the nucleus of the optic tract and the dorsal terminal nucleus of the accessory optic system in macaque monkeys. *Eur. J. Neurosci.* **12**, 2367–2375.
Thach, W.T., 1969. Visual receptive fields of striate cortex neurons in awake monkeys. *J. Neurophysiol.* **32**, 727–742.
Tigges, M. and Tigges, J., 1970. The retinofugal fibers and their terminal nuclei in *Gallago crassicaudatus* (primates). *J. Comp. Neurol.* **138**, 87–102.
Tigges, J., Bos, J., and Tigges, M., 1977. An autoradiographic investigation of the subcortical visual system in chimpanzee. *J. Comp. Neurol.* **172**, 368–380.
Trejo, J.T. and Cicerone, C.M., 1984. Cells in the pretectal olivary nucleus are in the pathway for the direct light reflex of the pupil in the rat. *Brain Res.* **300**, 49–62.
Trejo, L.J., Rand, M.N., and Cicerone, C.M., 1989. Consensual pupillary light reflex in the pigmented rat. *Vision Res.* **29**, 303–307.
Treue, S. and Maunsell, J.H.R., 1996. Attentional modulation of visual motion processing in cortical areas MT and MST. *Nature* **382**, 539–541.
Trevarthen, C.B., 1968. Two mechanisms of vision in primates. *Psychol. Forsch.* **31**, 299–337.
Ungerleider, L.G., Desimone, R., Galkin, T.W. *et al.*, 1984. Subcortical projections of area MT in the macaque. *J. Comp. Neurol.* **223**, 368–386.
Updyke, B.V., 1974. Characteristics of unit responses in superior colliculus of the *Cebus* monkey. *J. Neurophysiol.* **37**, 896–909.
Vaney, D.L., Peichl, L., Wässle, H. *et al.*, 1981. Almost all ganglion cells in the rabbit retina project to the superior colliculus. *Brain Res.* **212**, 447–453.
Vidyasagar, T.R., Kulikowski, J.J., Lipnicki, D.M. *et al.*, 2002. Convergence of parvocellular and magnocellular information channels in the primary visual cortex of the macaque. *Eur. J. Neurosci.* **16**, 945–956.
Volchan, E., deGonzaga Gawryszewski, L., and Rocha-Miranda, C.E., 1982. Visuotopic organization of the superior colliculus of the opossum. *Exp. Brain Res.* **46**, 263–268.
Waespe, W. and Henn, V., 1987. Gaze stabilization in the primate: The interaction of the vestibulo-ocular reflex, optokinetic nystagmus, and smooth pursuit. *Rev. Physiol. Biochem. Pharmacol.* **106**, 37–125.
Wahle, P., Stuphorn, V., Schmidt, M. *et al.*, 1994. LGN-projecting neurons of the cat's pretectum express glutamic acid decarboxylase mRNA. *Eur. J. Neurosci.* **6**, 454–460.
Waleszczyk, W.J., Wang, C., Burke, W. *et al.*, 1999. Velocity response profiles of collicular neurons: Parallel and convergent visual information channels. *Neuroscience* **93**, 1063–1076.
Wall, J.T., Symonds, L.L., and Kaas, J.H., 1982. Cortical and subcortical projections of the middle temporal area (MT) and adjacent cortex in galagos. *J. Comp. Neurol.* **211**, 193–214.
Wallace, S.F., Rosenquist, A.C., and Sprague, J., 1989. Recovery from cortical blindness mediated by destruction of nontectal fibers in the commissure of the superior colliculus in the cat. *J. Comp. Neurol.* **284**, 429–450.
Wässle, H. and Boycott, B.B., 1991. Functional architecture of the mammalian retina. *Physiol. Rev.* **71**, 447–480.
Wässle, H. and Illing, R.-B., 1980. The retinal projection to the superior colliculus in the cat: A quantitative study with HRP. *J. Comp. Neurol.* **190**, 333–356.
Weber, J.T., 1985. Pretectal complex and accessory optic system of primates. *Brain Behav. Evol.* **26**, 117–140.
Weber, J.T. and Giolli, R.A., 1986. The medial terminal nucleus of the monkey: Evidence for a 'complete' accessory optic system. *Brain Res.* **365**, 164–168.
Weber, J.T. and Hutchins, B., 1982. The demonstration of a retinal projection to the medial pretectal nucleus in the domestic cat and the squirrel monkey: An autoradiographic analysis. *Brain Res.* **232**, 181–186.

Weiskrantz, L., 2002. Prime-sight and blindsight. *Conscious. Cogn.* **11**, 568–581.

Weiskrantz, L., Warrington, E.K., Sanderson, M.D. *et al.*, 1974. Visual capacity in hemianopic field following a restricted occipital ablation. *Brain* **97**, 709–728.

Westheimer, G., 1954. Eye movement responses to a horizontally moving visual stimulus. *Arch. Ophthalmol. N.Y.* **52**, 932–941.

Westheimer, G., 1989. History and methodology, in *The Neurobiology of Saccadic Eye Movements* (eds R.H. Wurtz and M.E. Goldberg), Elsevier Science Publishers BV (Biomedical Division), Amsterdam, pp. 3–12.

Westheimer, G. and Blair, S.M., 1974. Unit activity in accessory optic system in alert monkeys. *Invest. Ophthalmol.* **13**, 533–534.

Williams, C., Azzopardi, P., and Cowey, A., 1995. Nasal and temporal retinal ganglion cells projecting to the midbrain: Implications for "Blindsight". *Neuroscience* **65**, 577–586.

Wurtz, R.H., 1996. Vision for the control of movement. *Invest. Ophthalmol. Vis. Sci.* **37**, 2131–2145.

Wurtz, R.H. and Albano, J.E., 1980. Visual-motor function of the primate superior colliculus. *Annu. Rev. Neurosci.* **3**, 189–226.

Wurtz, R.H. and Mohler, C.W., 1976. Organization of monkey superior colliculus: Enhanced visual response of superficial layer cells. *J. Neurophysiol.* **39**, 745–765.

Yabuta, N.H., Sawatari, A., and Callaway, E.M., 2001. Two functional channels from primary visual cortex to dorsal visual cortical areas. *Science* **292**, 297–300.

Yakushin, S.B., Gizi, M., Reisine, H. *et al.*, 2000a. Functions of the nucleus of the optic tract (NOT). I. Adaptation of the gain of the horizontal vestibule-ocular reflex. *Exp. Brain Res.* **131**, 416–432.

Yakushin, S.B., Gizi, M., Reisine, H. *et al.*, 2000b. Functions of the nucleus of the optic tract (NOT). II. Control of ocular pursuit. *Exp. Brain Res.* **131**, 433–447.

Zeki, S.M., 1974. Functional organization of a visual area in the posterior bank of the superior temporal sulcus of the rhesus monkey. *J. Physiol. (Lond.)* **236**, 546–573.

9

The Evolution of Visual Cortex in Primates

Jon H. Kaas

9.1 Introduction

Our current understanding of the evolution of visual cortex in primates is almost completely based on the results of comparative studies of visual cortex organization in extant (existing) primates, as the fossil record is limited to estimates of brain size and surface features seen in the endocasts of skulls. The fossil record does indicate that early primates were rather small mammals who were likely nocturnal (e.g. Martin, 1990; see Kaas and Preuss, 2003 for review). As discussed more fully elsewhere in this book (Chapter 1), the first primates appeared over 65 million years ago (mya), and perhaps over 80 mya. Early primates soon diverged into a strepsirhine branch leading to present-day lemurs, lorises, and galagos, a tarsier line with only a few surviving species in a single family, and the highly diverse branch of anthropoid primates. This line of evolution split into New and Old World monkeys: one branch of Old World monkeys gave rise to apes, with one line of apes giving rise to all modern apes and humans. Currently, there are over 200 extant species of primates.

Obviously, the enormous and daunting task of reconstructing evolution of the differing visual systems in even the major lines of primate evolution remains for the future. Here we outline the likely changes in brain organization from early primates to a few well-studied species of present-day prosimians, tarsiers, and anthropoids. Our approach is cladistic inasmuch as features of visual cortex are compared across taxa, and we presume that frequently shared features are more likely inherited from a common ancestor than

The Primate Visual System: A Comparative Approach Edited by Jan Kremers

independently evolved. However, the usefulness of a cladistic approach is limited since visual cortex organization has been studied in only a few species of primates. Because many species are rare, protected, or unavailable for investigation, this situation is unlikely to change. More likely, we will continue to learn more and more about the visual cortex of a few well-studied species. Fortunately, we can supplement a cladistic approach assuming that information from a few key taxa is especially informative (Kaas, 2002). For example, the fossil record indicates that the brains of many of the extant prosimians, such as galagos, more closely resemble the brains of early primates in shape and size than do the brains of monkeys, apes, and humans, and so, it seems reasonable to base inferences about early primates more on galagos than apes. Galagos seem to have changed less overall, and thus their brains might have retained more features that are primitive. In addition, galago brains have changed little in size, while ape and human brains have greatly enlarged. As major changes in brain size require design changes, because large brains do not function in the same way as small brains (Kaas, 2000), present-day primate brains that are large should differ more than those that are small from the small brains of early primates.

9.2 Features of visual cortex organization that early primates retained from non-primate ancestors

Primates evolved within one of six major clades or superorders of mammals: the Euarchontoglires – a superorder of placental mammals that also includes rodents, lagomorphs, flying lemurs, and tree shrews (Kaas, 2002). We show a schematic of the brain of tree shrew here (Figure 9.1), not because it is primitive and reflects an early stage in the evolution of primate brains as once proposed (Le Gros Clark, 1959), but to show some of the features of visual cortex in tree shrews that are shared by other members of the clade and therefore were likely present in the first primates. Nevertheless, tree shrews resemble primates more than they resemble lagomorphs and rodents in their greater emphasis on vision, expansion of temporal cortex for vision, and semi-arboreal life. Along with flying lemurs, they are more closely related to primates, and unlike flying lemurs, they are available for experimental study. However, early primates were likely nocturnal, while tree shrews are diurnal (Martin, 1990).

As is apparent in Figure 9.1, tree shrews have a primary visual area, V1 or area 17, a second visual area, V2 or area 18, and as many as six or more visual areas occupying much of temporal cortex (Lyon *et al.*, 1998). In the frontal lobe, there might be a poorly developed frontal eye field (FEF) and possibly other visuomotor areas such as the supplementary eye field, but this is uncertain. Microelectrode stimulation studies have only revealed a primary motor area, M1. Overall, well over half of neocortex is devoted to vision.

Some of the features of visual-cortex organization in tree shrews are widely shared by other mammals, as well as by all extant primates, and thus would have been present in early primates. Notably, most or all mammals have a V1 and V2 (e.g. Kaas *et al.*, 1970; Rosa and Krubitzer, 1999).

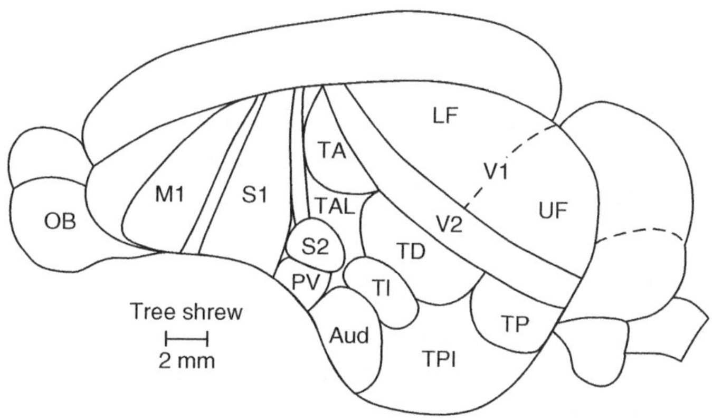

Figure 9.1 A dorsolateral view of the brain of a tree shrew (*Tupaia belangeri*) with proposed visual and other areas of cortex indicated. Primary visual cortex, V1, occupies the caudal pole of the cerebral hemisphere, with the lower visual hemifield (LF) represented rostral to the upper field (UF). V1 is bordered by the second visual area, V2, which also has the LF rostral to the UF. Less well-defined temporal anterior (TA), temporal dorsal (TD), and temporal posterior (TP) visual areas border V2. More lateral visual and multisensory areas include the temporal anterior lateral (TAL) area, the temporal inferior (TI) area, and the temporal posterior inferior (TPI) area. Auditory cortex (AUD) includes a primary area, A1, as well as an adjoining auditory belt. Somatosensory cortex includes a primary area, S1 or area 3b, narrow somatosensory belts just rostral and caudal to S1, a second somatosensory area, S2, and a parietal ventral area, PV. In frontal cortex, which is proportionately small, only a primary motor area, M1, has been identified. The olfactory bulb, OB, marks the front of the brain. See text for references

9.2.1 Area V1

V1 receives the vast majority of projections from the dorsal lateral geniculate nucleus (LGN), but not all (Lyon *et al.*, 2003b). A few LGN neurons commonly project to V2 and adjacent visual areas. LGN inputs to V1 are concentrated in layer 4, with sparse collaterals in layer 6, and a few LGN cells project to superficial layers, largely layer 1 (e.g. Fitzpatrick, 1996; Chapter 6). V1 projects subcortically to the pulvinar in the thalamus (e.g. Lyon *et al.*, 2003a,b), which has several subdivisions, and to visuomotor structures including the superior colliculus (SC) and nuclei in the pons. Layer 6 cells in V1 provide massive feedback connections to the LGN (e.g. Casseday *et al.*, 1979). About half of the ipsilateral cortical connections of V1 are to V2, and the rest distribute to more lateral visual areas (Sesma *et al.*, 1984; Lyon *et al.*, 1998). Nearly all of these cortically projecting cells are in layer 3. Callosally projecting cells are concentrated along the V1/V2 border, but they are not restricted to this border (e.g. Cusick *et al.*, 1984). They originate from both supragranular and infragranular layers. V1 contains a systematic representation of the contralateral visual hemifield with a larger, lateral binocular portion and a smaller, medial monocular portion (Kaas *et al.*, 1972). The center of frontal vision has an expanded representation, while the zero vertical meridian through the center of gaze is represented along the lateral border of V1, with the lower quadrant represented in rostral V1 and the upper quadrant in caudal V1. Part of V1 extends onto the medial wall, but there is no calcarine fissure. Layers in V1 are histologically distinct, but more so in tree shrews than we would hypothesize for early nocturnal primates. Neurons are selective for the orientation of the visual stimuli, and neurons with preferences for different stimulus orientations are systematically arranged in a pinwheel fashion in a number of arrays across cortex (Bosking *et al.*, 1997). Ocular dominance columns do not exist.

9.2.2 Area V2

Most mammals also have the second visual area, V2 (Rosa and Krubitzer, 1999). However, the least shrew (*Cryptotus parva*) does not have a V2 (Catania *et al.*, 1999), apparently a lost feature due to the miniaturization of the brain in this extremely small mammal. It is possible that a few other species will be discovered without a V2, but there is solid evidence for the area in a wide range of species. V2 borders the lateral portion of V1 along a common representation of the vertical meridian. While V1 is shaped like an oval, V2 forms a long-narrow band. Thus, the representation of the contralateral visual hemifield is more distorted in V2 than in V1, so that the representation of paracentral and peripheral vision is split and displaced toward the ends of the long belt, with the lower visual quadrant being represented rostrally and the upper quadrant caudally (Kaas *et al.*, 1972). Positions near or on the zero horizontal meridian relate to the outer border of V2. The main activating input to V2 comes from layer 3 cells in V1, although a small to considerable amount, depending on species, may come from the LGN. V2, in at least many and perhaps most mammals, has a modular structure so that V1 inputs are unevenly distributed in V2, where they alternate with callosal terminations (Sesma *et al.*, 1984; Cusick *et al.*, 1985). Thus, the length of V2 is subdivided into a series of modules across the width of the field differing in densities of V1 or callosal inputs, and possibly in types of V1 inputs. Unfortunately, these modules have not yet been characterized in terms of differences in functional properties, and it is not known if different taxa have similar or quite different modules. However, an alternation of callosal and non-callosal zones is a common feature of V2, and ancestors of primates would have had at least this type of modular organization in V2. V2 provides feedback projections to V1, and projects to regions of extrastriate visual cortex lateral to V2, callosally to V2, V1, and cortex lateral to V2, and to subcortical structures including the pulvinar complex (sometimes called the lateral posterior nucleus or nuclear complex) and the SC (Lyon *et al.*, 1998, 2003b).

Most mammals have additional visual areas in cortex lateral to V2, occupying most of the temporal cortex that is not auditory. Tree shrews, for example, have a series of six visual areas or regions that have been distinguished in cortex lateral to V2 (Lyon *et al.*, 1998), and there might be more. However, not much is known about how visual cortex is divided into areas in most mammals, and proposals for dividing visual cortex into areas in the studied taxa differ considerably, at least in the names of the areas. Thus, it is not at all clear if any of these additional areas existed in the immediate ancestors of primates, and if any of these additional areas have been retained in present-day primates. However, it would be surprising if all the extrastriate visual areas, except for V2, of non-primate mammals evolved independently from those found in primates.

9.3 Features of visual cortex in early primates

Modern primates have been called euprimates to distinguish them from the plesiadaptiforms that are sometimes considered archaic primates (e.g. Block and Boyer, 2002). Here we simply refer to euprimates as primates, allowing that plesiadaptiforms are their recognized sister group. Primates are distinguished from other mammals by having larger brains relative to body size, an enhanced visual system with converged eyes, nails on digits, and grasping hands and feet (Chapter 1; Martin, 1990; Fleagle, 1999). Early primates also had adaptations for leaping as they are thought to have evolved for a nocturnal life of eating fruit and insects in the terminal branches of trees. Plesiadaptiforms had evolved grasping adaptations, but not the forward facing eyes (Block and Boyer, 2002). While plesiadaptiforms were widespread, strepsirhine (prosimian) primates apparently originated in Africa at least 65 mya, and perhaps more than 80 mya. Early primates resembled modern day prosimian primates in many features, especially in brain size relative to body size, and brain shape.

The brain of a modern prosimian primate, a galago, is shown in Figure 9.2. This brain reflects many of the features of visual cortex that were likely present in early prosimian primates. Thus, visual cortex is expanded with an enlarged temporal lobe, and V1 and V2 have rotated to form the caudal pole of the brain, with both of these areas extending onto the medial wall of the cerebral hemisphere and around the ventral surface to enter the calcarine fissure, a fissure that characterizes the brains of all primates. This rotation places the representation of central vision on the dorsolateral surface of the brain, with the lower visual quadrant extending dorsally and medially onto the medial wall and into the calcarine fissure for peripheral vision, and the upper visual quadrant ventrally onto the ventral surface and into the calcarine fissure for peripheral vision. Visual cortex includes much of the temporal lobe and most of posterior parietal cortex, and these regions, together with occipital cortex rostral to V2, contain a number of visual areas, most of which have not been well identified so that many uncertainties remain. Yet, we can estimate the number of visual areas as at least over 10 and perhaps as many as 20, and we can identify some proposed visual areas as well supported by experimental data.

While V1 and V2 are visual areas that were present in early mammals and retained in early primates, both areas had been modified in the time of the first primates. In addition to the rotation of these two areas noted above, both areas received a greatly expanded representation of central vision (Rosa *et al.*, 1997). In addition, most or all extant primates have the modules called *cytochrome blobs* in V1 (dots in Figure 9.2, see Preuss and Kaas, 1996 for review). As these blobs are not found in any of the living close relatives of primates, they likely emerged with or just before the evolution of the first primates. The blobs are irregular patches of tissue in V1 that express a higher level of the metabolic enzyme, cytochrome oxidase (CO), than the surrounding tissue, and they mark modules that contain neurons with different connections and response properties than neurons in interblob portions of V1 (Casagrande and Kaas, 1994). Most notably, blobs receive inputs from the koniocells of the LGN (Chapter 6). One proposal is that neurons in blobs play a special role in color vision, but the prominence of blobs in nocturnal primates suggests that they have a broader role in vision (Casagrande and Kaas, 1994). Nevertheless, blobs

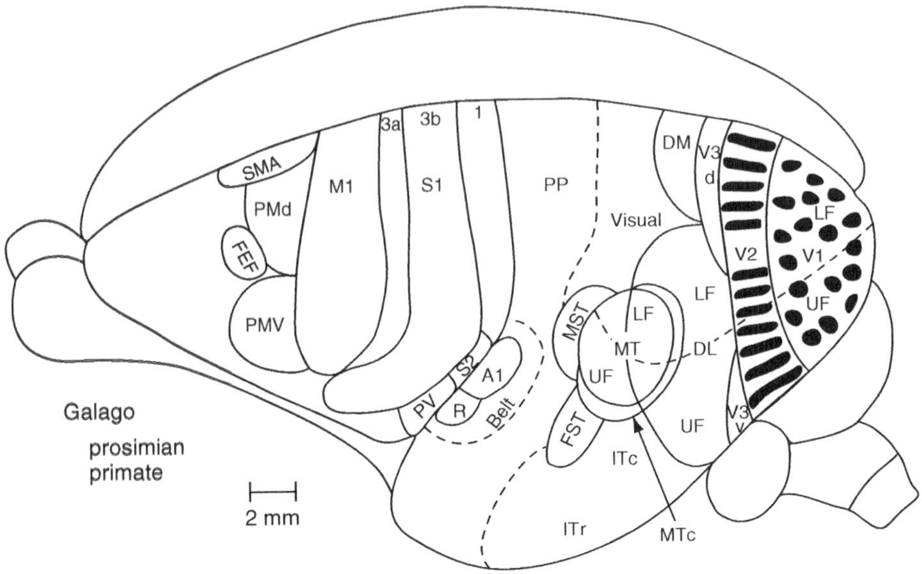

Figure 9.2 A dorsolateral view of the brain of a prosimian galago (*Galago garnetti*, now referred to as *Otolemer garnetti*). V1 and V2 (Figure 9.1) have rotated to occupy the caudal pole of the hemisphere so that the lower field (LF) is represented dorsal to the upper field (UF), with much of paracentral and peripheral vision represented on the medial wall and in the deep calcarine fissure of the medial wall. V1 is characterized by a dot-like array of cytochrome oxidase (CO) dense puffs or blobs (black dots), and V2 is crossed by an alternating pattern of CO dark (black) and CO light bands or stripes. Other well-defined visual areas include the third visual area, V3, which is separated into dorsal (V3d) and ventral (V3v) halves, representing the lower and upper visual quadrants, respectively, and the middle temporal visual area, MT, which represents the lower field dorsally and the central vision caudally. The dorsolateral visual area, DL, with possible rostral and caudal subdivisions, occupies the cortex between MTc and V2, but dorsal and ventral borders are only approximate. Other less well-defined areas include the dorsomedial area, DM, the MT crescent, MTc, the medial superior temporal area, MST, the fundal area of the superior temporal sulcus, FST, which may have dorsal and ventral subdivisions, and caudal and rostral divisions of inferior temporal cortex, ITc and ITr. The posterior half of posterior parietal cortex, PP, including the parietal–occipital junction, is visual, but of uncertain organization. Auditory cortex includes a primary area, A1, a rostral primary-like area, R, an auditory belt of several areas, and a caudally adjoining auditory cortex. Somatosensory cortex includes a primary area, S1 or area 3b, rostrally and caudally adjoining somatosensory belts and laterally adjoining S2 and PV. Motor cortex includes M1, the supplementary motor area, SMA, and dorsal and ventral premotor areas, PMd and PMv. A frontal eye field, FEF, has been identified

emerged as a modular subdivision of V1 in early primates, and this subdivision has been retained in all studied branches of primate evolution.

Another distinguishing feature of V1 in primates is the segregation of projections from the magnocellular (MC) and parvocellular (PC) layers of the LGN in different sublayers of layer 4. In all studied primates, the MC layers project to the outer (external) sublayer and the PC layers project to the inner sublayer of layer 4 of V1 (Casagrande and Kaas, 1994). Note that layer 4 has been variously defined in primates and the common practice of including sublayers external to the MC input in layer 4 is inconsistent with a comparative analysis which places sublayers IVa and IVb of Brodmann in layer 3 (see Casagrande and Kaas, 1994 for review). The important conclusion from the comparative evidence is that from the advent of the first primates, the visual system was specialized to separate MC and PC processing streams, not only at the level of the LGN, but in V1 and beyond (see below).

Many primates also have a segregation of inputs relayed via the LGN from one eye or the other in alternating band-like arrays in layer 4 of V1, the ocular dominance columns (Florence and Kaas, 1992; Horton and Hocking, 1996). Such columns are also found in at least several other mammalian taxa, such as cats, but they are not present in close relatives of primates (tree shrews, rodents, lagomorphs), and thus they appear to have evolved independently in several lines. The functional significance of ocular dominance columns is far from clear, and they may reflect developmental mechanisms selected for reasons unrelated to functional reasons for separating inputs from the two eyes (Kaas and Catania, 2002). Ocular dominance columns have been described in prosimian galagos, and a number of anthropoid primates, but they are variably expressed in some of the smaller New World monkeys (Kaas *et al.*, 1976; Adams and Horton, 2003). Thus, it may be that a tendency to form such columns existed in early primates, and this tendency has been variably expressed. Alternatively, the expression of ocular dominance columns in early primates could have been reduced in some subsequent lines of primate evolution.

A number of other features of V1 in primates likely were retained from non-primate ancestors, although modifications might have occurred. In primates and non-primates such as tree shrews, the local horizontal intrinsic connections of V1 tend to interconnect neurons within nearby modules of neurons of similar functional properties (Bosking *et al.*, 1997). Both primates and tree shrews have groups of neurons selective for visual stimuli of specific orientations, and arrays of such groups are systematically arranged in pinwheel-like structures relative to surface views of V1 (Bosking *et al.*, 1997; Xu *et al.*, 2004a). Callosal connections originate and terminate from locations well within V1 in the tree shrews (Sesma *et al.*, 1984; Cusick *et al.*, 1985; Bosking *et al.*, 2000) and in most mammals as well as prosimian galagos (Cusick *et al.*, 1984; Beck and Kaas, 1994). A modification in galagos is that the longer intrinsic horizontal connections of V1 are more concentrated in the CO blobs (Cusick and Kaas, 1988b).

The second visual area, V2, is the major target of V1 projections in all studied primates (Lyon and Kaas, 2002a,b). The representation of the visual hemifield is split in V2 so that the vertical meridian is along the V1 border, as in other mammals, while nearly all of the outer (rostral) boundary corresponds to the horizontal meridian (e.g. Allman and Kaas, 1974; Rosa *et al.*, 1997). As in other mammals, such as tree shrews (Sesma *et al.*, 1984), the inputs from V1 to V2 are not evenly distributed, but instead are concentrated in band-like arrays across the width of V2. In monkeys, V2 contains a system of CO dense stripes or bands that alternate with CO light bands (see Preuss *et al.*, 1993 for review). The dark CO bands have been divided into two types called *thick* and *thin*, although thickness and thinness do not seem to be reliable distinguishing features (Krubitzer and

Kaas, 1990). In monkeys, the interstripes receive inputs from neurons in the interblobs of V1, the thin stripes receive from both blob and interblob regions, and the thick stripes from a sublayer of cells in V1 (3c by our terminology) that is activated by MC-LGN cell projections to V1 (Xiao and Felleman, 2004). The thick stripes, in turn, project to the middle temporal visual area MT (see below), while the thin and interstripes project to the dorsolateral visual complex, DL (V4) (see following text). Thus, monkeys and other anthropoid primates (Tootell and Taylor, 1995) have a well-organized pattern of modules in V2 that can be distinguished by myelin or CO stains, patterns of connections, and the response properties of neurons in the modules (see Roe, 2004 for review). In addition, at least major aspects of this pattern are expressed in prosimian primates. Most notably, stripe-like arrays of neurons in V2 of galagos project to MT (Krubitzer and Kaas, 1990), while other parts of V2 project to DL (Collins *et al.*, 2001a,b). However, CO stripes and interstripes are only sometimes apparent in V2 of prosimians (see Preuss *et al.*, 1993 for review), and the connections of V1 modules with V2 modules have not yet been established. Thus, it is not apparent if the full anthropoid pattern of modular organization was present in early primates, and what part of this pattern, if any, was inherited from non-primate ancestors. These ancestors apparently lacked CO blobs in V1, CO bands in V2, and many extrastriate visual areas including MT and DL.

The two major targets of V2 projections in primates are the middle temporal visual area, MT, and the dorsolateral visual area, DL (Collins *et al.*, 2001b). While MT is well defined, the dorsal and ventral borders of DL are uncertain. In addition, there is evidence that DL is a complex of two similar areas, with DLr rostral to DLc (Kaas, 2003). As these two divisions of DL are not well defined, there are uncertainties about the existence of caudal DL (DLc) and rostral DL (DLr) in all primate taxa.

MT stands out from surrounding cortex as an area with dense myelination and dense cytochrome oxidase in all investigated primates (see Tootell and Taylor, 1995; Kaas, 1997 for review). The ease with which MT can be identified and localized in various primates has led to a number of studies on the area. To summarize results, MT receives direct feedforward projections from V1, V2, V3, and DL, while projecting to a number of other visual areas including fundus of the superior temporal sulcus (FST), medial superior temporal (MST), and the lateral intraparietal (LIP) region of posterior parietal cortex (e.g. Wall *et al.*, 1982; Maunsell and Van Essen, 1983; Weller *et al.*, 1984; Ungerleider and Desimone, 1986; Krubitzer and Kaas, 1990). MT contains a systematic representation of the contralateral hemifield (e.g. Allman and Kaas, 1971; Allman *et al.*, 1973), with the upper quadrant in ventral or rostroventral MT, and the lower quadrant in dorsal or caudodorsal MT (the representation is rotated somewhat in Old World monkeys). Neurons are highly selective for stimulus orientation and direction of movement (see also Chapter 10), and neurons form pinwheel-like arrays of orientation selective neurons divided into direction of movement domains, in both prosimian (Xu *et al.*, 2004b) and simian (Malonek *et al.*, 1994) primates. In at least some primates, MT has a modular internal organization that segregates neurons selective for processing global or local motion (Born and Tootell, 1992). These modules might relate to the patchy pattern of CO and myelin in MT (Krubitzer and Kaas, 1990) and the patchy pattern of connections with FSTd (Kaas and Morel, 1993), but much is uncertain. What is clear is that none of the close relatives of primates have a visual area with the distinctive, identifying features of MT, so we assume that this visual area evolved with the immediate ancestors of primates

(Kaas and Preuss, 1993; Kaas, 2002). While the origin of MT in primates is uncertain, one speculation is that a cortical area with dense inputs from V1 and V2, located near the middle of the outer border of V2, became displaced into the temporal lobe and acquired the distinctive architectonic and physiological properties of MT (Kaas and Preuss, 1993). As MT is almost exclusively devoted to processing the MC-cell LGN inputs to cortex after relays from V1 and V2, this transformation would likely have emerged with or after the segregation of MC-cell cortical pathways in the immediate ancestors of primates (Chapter 6).

The dorsolateral area, DL, is a visual area or a complex of visual areas that is common to all primates (Kaas, 2003). The major defining feature of DL (also called V4) is a dense and topographic projection pattern from V2 such that both ventral V2 and DL represent the upper visual quadrant, and dorsal V2 and DL represent the lower visual quadrant (Zeki, 1969). In monkeys, DLr appears to have a less dense input from V2, and a more compressed topographic pattern (thus, a smaller representation) compared to DLc: there are other differences in connections and myeloarchitecture as well (Cusick and Kaas, 1988a; Steele *et al.*, 1991; Stepniewska and Kaas, 1996). In prosimians, the architectonic distinctions suggest that DLr and DLc are variably apparent in prosimians (Preuss *et al.*, 1993), and connection patterns with V2 provide only weak evidence for the two divisions (Collins *et al.*, 2001b). DL, especially DLc, provides the major relay of visual information into more ventral parts of the temporal lobe (e.g. Weller and Kaas, 1985, 1987).

There are only a few other visual areas that have been so well identified across a range of primate taxa that we can postulate that these areas were present in early primates. One of these areas is V3, an area about one-half the width of V2 along most of the outer (rostral) border of V2. Until recently, the evidence for V3 was limited, and it was possible to favor other interpretations of visual cortex (e.g. that ventral V3 was another visual area, VP, or that dorsal V3 was part of DM; see Kaas and Lyon, 2001a,b, for review). However, studies of the connection patterns of V1 in galagos (Lyon and Kaas, 2002a), New World monkeys (Lyon and Kaas, 2002a,b), and Old World monkeys (Lyon and Kaas, 2002c) have provided strong evidence for a systematic representation of inputs from V1 to V3 so that the upper quadrant is represented ventrally, and the lower quadrant dorsally, with the horizontal meridian along the V2–V3 border and the vertical meridian along the rostral border of V3, as originally proposed for macaque monkeys (Zeki, 1969). In monkeys, V3 demonstrates a banding pattern of CO dense and CO light stripes that are wider and less obvious than those in V2 (Lyon and Kaas, 2002a,b,c), and optical imaging reveals a clustering of neurons with similar stimulus orientation preferences (Lyon *et al.*, 2002). Other properties of neurons in V3 and the complete connection pattern are not well understood, and it is not known how they vary across primate taxa. As V3 has not been demonstrated in relatives of primates (tree shrews, rodents, lagomorphs), it is likely that V3 evolved with the first primates, and independently in cats (Kaas, 2002).

Two areas with dense interconnections with MT, the MST visual area and the dorsal divisions of the fundal area of the superior temporal sulcus (FSTd), are likely to exist in all or most extant primates because MT connections have been shown with these two regions in several prosimians and a range of species of monkeys (e.g. Maunsell and Van Essen, 1983; Ungerleider and Desimone, 1986; Krubitzer and Kaas, 1990; Kaas and

Morel, 1993; Preuss *et al.*, 1993). As for MT, these areas have no known counterparts in non-primates.

Other areas are also likely to be part of the basic primate plan. These areas include the dorsomedial visual area, DM (Beck and Kaas, 1998a,b, 1999); a target of MT in ventral posterior parietal cortex, VPP (or in macaques, the lateral intraparietal area, LIP); the main target of DL in the temporal lobe (IT or TEO); and the MT crescent (MTc). However, the evidence for most of these additional areas being widespread in primates is limited to that obtained from the connection patterns of better-defined areas, and thus uncertainties remain. Yet, it is clear from the available evidence that early primates had a considerable number of visual areas, on the order of 10–20. Most of these areas emerged with primates, and were retained in most or all extant primates. Undoubtedly, over the next few years, comparative studies will reveal much more about this basic organization of visual cortex in early primates. In addition, there is compelling evidence for the existence of at least one visuomotor area of the frontal lobe in a wide range of primates. Thus, early primates would have had a FEF representing eye movements and projecting to the SC and pons (see also Chapter 8; Huerta *et al.*, 1987; Wu *et al.*, 2000). Inputs from higher-order visual areas provided the guiding sensory inputs to the FEF for these eye movements.

9.4 Visual cortex of tarsiers

The early division of primates, at least 65 mya and perhaps over 80 mya, was into strepsirhine primates including lemurs, galagos, and lorises, and haplorhine primates, which separated into tarsioids and anthropoid primates over 60 mya (Chapter 1). The common ancestor of present-day tarsiers and anthropoids was thought to be a small diurnal primate with increased emphasis on eating fruit, an enhanced visual system, and a reduced olfactory system. The reflecting tapetum on the back of the retina, an adaptation to nocturnal vision in prosimians and many other nocturnal mammals, was lost, and

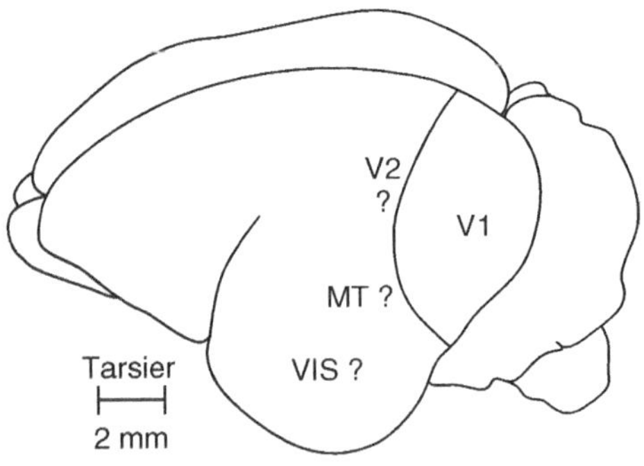

Figure 9.3 A dorsolateral view of the brain of a tarsier (*Tarsius spectrum*). So far, only one visual area, V1 or area 17, has been identified, but other visual areas, including V2 and MT, are likely. An emphasis on visual processing is suggested by the proportionately large V1 and temporal lobe

the retina acquired more cones and a fovea (Ross, 1996). The ancestors of present-day tarsiers returned to a nocturnal niche (Martin, 1990), and adapted by acquiring enormous eyes, while retaining two types of cones and a fovea (Hendrickson *et al.*, 2000). Tarsiers are extremely specialized visual predators that eat no plant food. They are small and survive as only a few remaining species in a single family.

Tarsiers have brains with a proportionally large temporal lobe, and only a trace of a lateral fissure (Figure 9.3). The bulging temporal lobe suggests that a large portion of the brain is devoted to vision, but this is uncertain. What is clear is that primary visual cortex is disproportionately large, and is extremely well differentiated into layers and sublayers (Hassler, 1966; Collins *et al.*, 2001a). Thus, there is an exceptional level of laminar specialization in V1, and much of cortex is occupied by V1 (Stephan, 1984). Unfortunately, not much is known about the rest of visual cortex, largely because tarsiers are rare, protected, and unavailable for experimental study, but also because there have been only a few attempts to examine the cortex histologically, especially in the context of our modern understanding of visual cortex in primates. Thus, it is not yet known if further study will provide architectonic evidence for V2 and MT, areas that have been identified in a wide range of other primates. Because of the specialization of tarsiers as small, nocturnal, visual predators, they might more often reflect specializations of the visual system than primitive features.

9.5 Anthropoid primates

Early anthropoid primates separated from tarsiers over 60 mya, and later diverged into platyrrhine and catarrhine primates (New World and Old World simians) about 40 mya or more. Apes became distinct from Old World monkeys about 25–30 mya (Stewart and Disotell, 1998). Living representatives of these taxa vary considerably in body size, diet, social interaction, and other adaptations, but only New World owl monkeys became nocturnal. Early anthropoids were characterized by larger brains, and an expansion of regions of cortex related to higher-order visual processing in the temporal and posterior parietal cortex. Over 20 visual and visuomotor areas in these regions have been proposed, especially for Old World macaque monkeys (e.g. Lewis and Van Essen, 2000), but in general, these areas are not well established, and it is not certain what areas are widely shared by extant taxa. Estimates of the total number of visual areas are on the order of 30 or more for macaques (Felleman and Van Essen, 1991), and this could even be an underestimate. Macaques have a supplementary eye field as well as a FEF, with both of these frontal lobe areas receiving inputs from a number of higher-order visual and visuomotor areas (Huerta *et al.*, 1987; Huerta and Kaas, 1990). Anthropoids differ from prosimians in that the PC-cell pathway via the LGN to V1 has become so restricted that lesions of V1 cause PC ganglion cells in the retina to degenerate (Weller and Kaas, 1989), suggesting that PC ganglion cells in anthropoid primates have no role other than to activate V1 via the LGN.

Presently, it is not certain if Old World and New World primates have any consistent differences in visual cortex organization, but a number of differences seem likely, given their long, separate evolutionary histories. In addition, some of the smaller New World monkeys may have simpler systems with fewer visual areas than the larger New World monkeys. Initially, early apes were a very successful branch of anthropoid evolution (Stewart and Disotell, 1998), and they generally had larger brains. Little is known about brain organization in extant apes, and this situation is unlikely to change in the near future. However, there is the promise of postmortem studies of cortical architecture. In one such study, histochemical features suggested that the projection of PC-cells of the LGN of Old World monkeys to supragranular layers of V1 is reduced in the ape–human group (Preuss *et al.*, 1999). Thus, we can expect specializations in apes and humans within the visual areas retained from simian ancestors, as well as the possibility of adding areas and modules (see following section).

9.6 Hominid visual cortex

Bipedal hominid primates diverged from apes some 4–5 mya. The hominid line radiated into a number of types, with one branch producing humans some 250 000 years ago. The changes in visual cortex organization that occurred over the course of hominid evolution are largely uncertain. As we are rapidly learning more about how human visual cortex is functionally subdivided into areas, some investigators stress similarities with organizations proposed for macaque monkeys, and others claim major differences. While it seems likely that regions of human visual cortex are specialized for face recognition and object recognition (Gauthier, 2004), as well as other visual and visuomotor abilities, we do not know if the specializations are within widely shared visual areas, or if hominids evolved new visual areas for new abilities. What is certain from the fossil record is that early hominids had chimp-sized brains that were probably not much different from those of the present great apes. The brains of early hominids greatly expanded, especially over the last 2 million years, from around 400–500 cc to 1200–1400 cc of modern humans (Kaas and Preuss, 2003). The sheet of neocortex expanded, going from about 240 cm^2 per hemisphere in hominids with chimp-sized brains to 800 cm^2 per hemisphere in humans. At the same time, the primary areas of cortex (judging from present-day apes and humans) did not expand as much, as great apes have proportionally less neocortex devoted to V1 than monkeys, but proportionally more than humans. This disproportionate expansion of non-primary neocortex suggests that the numbers of non-primary areas increased, a suggestion that awaits experimental verification.

While we are left with uncertainties about the specifics of visual cortex evolution in hominids, some general outcomes are suggested from what is known about scaling problems and how they are solved as brains get bigger (Kaas, 2000). Larger brains and cortical areas have more neurons, making it difficult to maintain proportions of connections between neurons and conduction times over longer connection distances. These problems are addressed by making processing more modular. Cortex should become divided into more areas and areas into modules to reduce connection distances and allow connections to be restricted to specialized streams. In addition, local changes in neural circuits have less impact on the function of larger areas than on small ones, so that smaller areas can be more easily specialized. Thus, it seems reasonable to assume that humans have more visual areas than other extant primates, and that most of these visual areas, especially the larger ones, are modularly organized.

9.7 Conclusions

The results of studies of visual cortex organization in a range of primate and non-primate taxa justify a number of conclusions about the evolution of visual cortex in primates.

1. The immediate ancestors of primates retained two visual areas, V1 and V2, from early mammals. At least V2 was modularly organized with alternating patches of neurons having V1 or callosal connections. Callosal connections extended well into V1, while V1 and V2 projected to several visual areas along the outer border of V2, and perhaps other areas in temporal cortex. Visuomotor areas such as the FEF were either absent or poorly differentiated.
2. Early primates were characterized by an expansion of visual cortex, especially in the temporal lobe. This resulted in a rotation of posterior visual cortex so that V1 and V2 occupied the caudal pole of the hemisphere, while extending from a dorsolateral representation of central vision medially and ventrally to wrap around the medial wall and extend onto the upper and lower banks of the calcarine fissure, a fissure characteristic of all primate brains. V1 and V2 had an expanded representation of central vision, and both areas were modularly organized. V1 had MC-cell and PC-cell geniculate inputs segregated in separate sublayers of layer 4, CO blobs as markers of a modular subdivision, and systematic arrays ('pinwheels') of orientation selective neurons. Ocular dominance columns were weakly segregated in layer 4. V2 had an alternation of three types of band-like modules, but they were not well differentiated in histological structure. These bands differed in types of input from V1 and in targets for feedforward projections. The expanded visual cortex included a number of other visual areas that have been widely retained in extant primates. These include well-identified MT and V3, and DL (and possibly DLr and DLc). Other likely areas of early primates include DM, MST, FST, and VPP (or LIP) area in posterior parietal cortex, and several divisions of inferotemporal cortex. The frontal lobe had an FEF and prefrontal neurons involved in vision. Both MT and V3 had some type of modular organization, and a systematic representation of orientation selective neurons.
3. Tarsiers evolved as highly specialized nocturnal visual predators of insects and small vertebrates. V1 occupies much of the caudal third of the cerebral hemisphere of extant tarsiers, and this V1 is uniquely well differentiated into distinct layers and sublayers. These changes suggest a great dependence on V1 for visual processing, with layers and sublayers having specialized functions. Yet, the temporal lobe is expanded and much of this cortex may contain other visual areas. Unfortunately, almost nothing is known about the organization of extrastriate cortex.
4. Anthropoid primates (monkeys, apes, and humans) have expanded visual systems with a large but uncertain number of visual areas. The numbers may vary across taxa, and include as many as 30–40 areas in some taxa. Little is presently known about the brains of apes, but early hominids had brains about the size of present-day great apes. Subsequent hominids had progressively larger brains so that modern humans have brains of about three times the size of great apes. This increase in size was not accompanied by a proportional increase in the size of V1, suggesting that visual areas retained from early primates did not proportionally expand with the brains, and

new cortical areas emerged. An increase in the modular organization of visual cortex would seem necessary to reduce the scaling problems of a greatly enlarged brain. The results of modern brain-imaging studies have the promise of greatly increasing our understanding of visual cortex organization in humans.

References

Adams, D.L. and Horton, J.C., 2003. Capricious expression of cortical columns in the primate brain. *Nat. Neurosci.* **6**, 113–114.

Allman, J.M. and Kaas, J.H., 1971. A representation of the visual field in the posterior third of the middle temporal gyrus of the owl monkey (*Aotus trivirgatus*). *Brain Res.* **31**, 85–105.

Allman, J.M. and Kaas, J.H., 1974. The organization of the second visual area (VII) in the owl monkey: a second order transformation of the visual field. *Brain Res.* **76**, 247–265.

Allman, J.M., Kaas, J.H., and Lane, R.H., 1973. The middle temporal visual area (MT) in the bush baby (*Galago senegalensis*). *Brain Res.* **57**, 197–202.

Beck, P.D. and Kaas, J.H., 1994. Interhemispheric connections in neonatal owl monkeys (*Aotus trivirgatus*) and galagos (*Galago crassicuadatus*). *Brain Res.* **651**, 57–75.

Beck, P.D. and Kaas, J.H., 1998a. Cortical connections of the dorsomedial visual area (DM) in New World owl monkeys (*Aotus trivirgatus*) and squirrel monkeys (*Saimiri sciureus*). *J. Comp. Neurol.* **400**, 18–24.

Beck, P.D. and Kaas, J.H., 1998b. Cortical connections of the dorsomedial visual area (DM) in prosimian primates. *J. Comp. Neurol.* **398**, 162–178.

Beck, P.D. and Kaas, J.H., 1999. Cortical connections of the dorsomedial visual area in Old World macaque monkeys. *J. Comp. Neurol.* **406**, 487–502.

Block, J.I. and Boyer, D.M., 2002. Grasping primate origins. *Science*, **298**, 1606–1610.

Born, R.T. and Tootell, R.B., 1992. Segregation of global and local motion processing in primate middle temporal visual area. *Nature* **357**, 497–499.

Bosking, W.H., Zhang, Y., Schofield, B. *et al.*, 1997. Orientation selectivity and the arrangement of horizontal connections in tree shrew striate cortex. *J. Neurosci.* **17**, 2112–2127.

Bosking, W.H., Kretz, R., Pucak, M.L. *et al.*, 2000. Functional specificity of callosal connections in tree shrew striate cortex. *J. Neurosci.* **20**, 2346–2359.

Casagrande, V.A. and Kaas, J.H., 1994. The afferent, intrinsic, and efferent connections of primary visual cortex in primates, in *Cerebral Cortex, Vol. 10, Primary Visual Cortex in Primates* (eds A. Peters and K. Rockland), Plenum Press, New York, pp. 201–259.

Casseday, J.H., Jones, D.R., and Diamond, I.T., 1979. Projections from cortex to tectum in the tree shrew, Tupaia glis. *J. Comp. Neurol.* **185**, 253–292.

Catania, K.C., Lyon, D.C., Mock, O.B. *et al.*, 1999. Cortical organization in shrews evidence from five species. *J. Comp. Neurol.* **410**, 55–72.

Collins, C.E., Hendrickson, A.E., and Kaas, J.H., 2001a. Distribution of calbindin, parvalbumin and wisteria in primary visual cortex of tarsier. *Soc. Neurosci. Abs.* **27**, program no. 619.41.

Collins, C.E., Stepniewska, I., and Kaas, J.H., 2001b. Topographic patterns of V2 cortical connections in a prosimian primate. *J. Comp. Neurol.* **431**, 155–167.

Cusick, C.G. and Kaas, J.H., 1988a. Cortical connections of area 18 and dorsolateral visual cortex in squirrel monkeys. *Vis. Neurosci.* **1**, 211–237.

Cusick, C.G. and Kaas, J.H., 1988b. Surface view patterns of intrinsic and extrinsic cortical connections of area 17 in a prosimian primate. *Brain Res.* **458**, 386–388.

Cusick, C.G., Gould, H.J. III, and Kaas, J.H., 1984. Interhemispheric connections of visual cortex of owl monkeys (*Aotus trivirgatus*), marmosets (*Callithrix jacchus*) and galagos (*Galago crassicaudatus*). *J. Comp. Neurol.* **230**, 311–336.

Cusick, C.G., MacAvoy, M.G., and Kaas, J.H., 1985. Interhemispheric connections of cortical sensory areas in tree shrews. *J. Comp. Neurol.* **235**, 111–128.

Felleman, D.J. and Van Essen, D.C., 1991. Distributed hierarchical processing in the primate cerebral cortex. *Cereb. Cortex* **1**, 1–47.

Fitzpatrick, D., 1996. The functional organization of local circuits in visual cortex: insights from the study of tree shrew striate cortex. *Cereb. Cortex* **6**, 329–341.

Fleagle, J.G., 1999. *Primate Adaptation and Evolution*, 2nd edn, Academic Press, New York.

Florence, S.L. and Kaas, J.H., 1992. Ocular dominance columns in area 17 of Old World macaque and Talapoin monkeys: complete reconstructions and quantitative analyses. *Vis. Neurosci.* **8**, 449–462.

Gauthier, I., 2004. Face expertise and category specialization in the human occipitotemporal cortex, in *The Primate Visual System* (eds J.H. Kaas and C.E. Collins), CRC Press, Boca Raton, FL, pp. 289–310.

Hassler, R., 1966. Comparative anatomy of central visual systems in day and night active primates in *Evolution of the Forebrain* (eds R. Hassler and H. Stephan), Georg Thieme, Stuttgart, pp. 419–434.

Hendrickson, A., Djajadi, H.R., Nakamura, L. *et al.*, 2000. Nocturnal Tarsier retina has both short and long/medium-wavelength cones in an unusual topography. *J. Comp. Neurol.* **424**, 718–730.

Horton, J.C. and Hocking, D.R., 1996. Anatomical demonstration of ocular dominance columns in striate cortex on the Squirrel Monkey. *J. Neurosci.* **16**, 5510–5522.

Huerta, M.F. and Kaas, J.H., 1990. The supplementary eye field as defined by intracortical microstimulation: Connections in macaques. *J. Comp. Neurol.* **293**, 299–330.

Huerta, M.F., Krubitzer, L., and Kaas, J.H., 1987. The frontal eye field as defined by intracortical microstimulation in squirrel monkeys, owl monkeys, and macaque monkeys. II. Cortical connections. *J. Comp. Neurol.* **265**, 332–361.

Kaas, J.H., 1997. Theories of visual cortex organization in primates, in *Cerebral Cortex, Vol. 12, Extrastriate cortex in primates* (eds K.S. Rockland, J.H. Kaas, and A. Peters), Plenum Press, New York, pp. 91–125.

Kaas, J.H., 2000. Why is brain size so important: Design problems and solutions as neocortex gets bigger or smaller. *Brain Mind* **1**, 7–23.

Kaas, J.H., 2002. Convergences in the modular and areal organization of the forebrain of mammals: Implications for the reconstruction of forebrain evolution. *Brain Behav. Evol.* **59**, 262–272.

Kaas, J.H., 2003. Early visual areas: V2, V3, DM, DL, and MT, in *The Primate Visual System* (eds J.H. Kaas and C.E. Collins), CRC Press, Boca Raton, pp. 139–159.

Kaas, J.H. and Catania, K.C., 2002. How do features of sensory representations develop? *BioEssays* **24**, 1–10.

Kaas, J.H. and Lyon, D.C., 2001a. Visual cortex organization in primates: theories of V3 and adjoining visual areas. *Prog. Brain Res.* **134**, 285–295.

Kaas, J.H. and Lyon, D.C., 2001b. Connectional and architectonic evidence for dorsal and ventral V3, and dorsomedial visual area in marmoset monkeys. *J. Neurosci.* **21**, 249–261.

Kaas, J.H. and Morel, A., 1993. Connections of visual areas of the upper temporal lobe of owl monkeys: The MT crescent and dorsal and ventral subdivisions of FST. *J. Neurosci.* **13**, 534–546.

Kaas, J.H. and Preuss, T.M., 1993. Archontan affinities as reflected in the visual system, in *Mammal Phylogeny* (eds F. Szalay, M. Novacek, and M. McKenna), Springer-Verlag, New York, pp. 115–128.

Kaas, J.H. and Preuss, T.M., 2003. Human brain evolution, Chapter 45, in *Fundamental Neuroscience*, 2nd edn (ed L.R. Squire), Academic Press, San Diego, pp. 1147–1166.

Kaas, J.H., Hall, W.C., and Diamond, I.T., 1970. Cortical visual area I and II in the hedgehog: The relation between evoked potential maps and architectonic subdivisions. *J. Neurophysiol.* **33**, 595–615.

Kaas, J.H., Hall, W.C., Killackey, H. *et al.*, 1972. Visual cortex of the tree shrew (*Tupaia glis*): Architectonic subdivisions and representations of the visual field. *Brain Res.* **42**, 491–496.

Kaas, J.H., Lin, C.-S., and Casagrande, V.A., 1976. The relay of ipsilateral and contralateral retinal input from the lateral geniculate nucleus to striate cortex in the owl monkey: a transneuronal transport study. *Brain Res.* **106**, 371–378.

Krubitzer, L.H. and Kaas, J.H., 1990. Cortical connections of MT in four species of primates: Areal, modular, and retinotopic patterns. *Vis. Neurosci.* **5**, 165–204.

Le Gros Clark, W.E., 1959. *The Antecedents of Man*, Edinburgh University Press, Edinburgh, UK.

Lewis, V.W. and Van Essen, D.C., 2000. Mapping of architectonic subdivisions in the macaque monkey, with emphasis on parieto-occipital cortex. *J. Comp. Neurol.* **428**, 112–137.

Lyon, D.C. and Kaas, J.H., 2002a. Connectional evidence for dorsal and ventral V3 and other extrastriate areas in the prosimian primates (*Galago garnetti*). *Brain Behav. Evol.* **59**, 114–129.
Lyon, D.C. and Kaas, J.H., 2002b. Evidence from V1 connections for both dorsal and ventral subdivisions of V3 in three species of New World monkeys. *J. Comp. Neurol.* **449**, 281–297.
Lyon, D.C. and Kaas, J.H., 2002c. Evidence for a modified V3 with dorsal and ventral halves in macaque monkeys. *Neuron* **33**, 453–461. *J. Comp. Neurol.* **449**, 281–297.
Lyon, D.C., Jain, N., and Kaas, J.H., 1998. Cortical connections of striate and extrastriate visual areas in the tree shrew. *J. Comp. Neurol.* **401**, 109–128.
Lyon, D.C., Xu, X., Casagrande, V.A. *et al.*, 2002. Optical imaging reveals retinotopic organization of dorsal V3 in New World owl monkeys. *PNAS* **99**, 15735–15742.
Lyon, D.C., Jain, N., and Kaas, J.H., 2003a. The visual pulvinar in tree shrews I. Multiple subdivisions revealed through acetylcholinesterase and Cat-301 chemoarchitecture. *J. Comp. Neurol.* **467**, 593–606.
Lyon, D.C., Jain, N., and Kaas J.H., 2003b. The visual pulvinar in tree shrews II. Projections of four nuclei to areas of visual cortex. *J. Comp. Neurol.* **467**, 607–627.
Malonek, D., Tootell, R.B.H., and Grinvald, A., 1994. Optical imaging reveals the functional architecture of neurons processing shape and motion in owl monkey area MT. *Proc. R. Soc. Lond. B. Biol. Sci.* **258**, 109–119.
Martin, R.D., 1990. *Primate Origins and Evolution: A Phylogenetic Reconstruction*, Princeton University Press, Princeton, N.J.
Maunsell, J.H.R. and Van Essen, D.C., 1983. The connections of the middle temporal visual area (MT) and their relationship to a cortical hierarchy in the macaque monkey. *J. Neurosci.* **3**, 2563–2586.
Preuss, T.M. and Kaas, J.H., 1996. Cytochrome oxidase "blobs" and other characteristics of primary visual cortex in a lemuriform primate (*Cheirogaleus medius*). *Brain Behav. Evol.* **47**, 103–112.
Preuss, T.M., Beck, P.D., and Kaas, J.H., 1993. Areal, modular, and connectional organization of visual cortex in a prosimian primate, the slow loris (*Nycticebus coucang*). *Brain Behav. Evol.* **42**, 237–251.
Preuss, T.M., Qi, H., and Kaas, J.H., 1999. Distinctive compartmental organization of human primary cortex. *PNAS* **96**, 11601–11606.
Roe, A.W., 2004. Modular complexity of area V2 in the macaque monkey, in *The Primate Visual System* (eds J.H. Kaas and C.E.C. Collins), CRC Press, Boca Raton, pp. 109–138.
Rosa, M.G.P. and Krubitzer, L.H., 1999. The evolution of visual cortex: Where is V2? *TINS* **22**, 242–248.
Rosa, M.G.P., Casagrande, V.A., Preuss, T.M. *et al.*, 1997. Visual field representation in striate and prestriate cortices of a prosimian primate (*Galago garnetti*). *J. Neurophysiol.* **77**, 3193–3217.
Ross, C., 1996. Adaptive explanation for the origins of the Anthropoidea (Primates). *Am. J. Primatol.* **40**, 205–230.
Sesma, M.A., Casagrande, V.A., and Kaas, J.H., 1984. Cortical connections of area 17 in tree shrews. *J. Comp. Neurol.* **230**, 337–351.
Steele, G.E., Weller, R.E., and Cusick, C.G., 1991. Cortical connections of the caudal subdivision of the dorsolateral area (V4) in monkeys. *J. Comp. Neurol.* **306**, 495–520.
Stephan, H., 1984. Morphology of the brain in Tarsius, in *Biology of Tarsiers* (ed C. Niemitz), Gustav-Fischer-Berlag, New York, pp. 319–344.
Stepniewska, I. and Kaas, J.H., 1996. Topographic patterns of V2 cortical connections in macaque monkeys. *J. Comp. Neurol.* **371**, 129–152.
Stewart, C.-B. and Disotell, T.R., 1998. Primate evolution – in and out of Africa. *Curr. Biol.* **8**, R582–R588.
Tootell, R.B.H. and Taylor, J.B., 1995. Anatomical evidence for MT and additional cortical visual areas in humans. *Cereb. Cortex* **1**, 39–55.
Ungerleider, L.G. and Desimone, R., 1986. Cortical connections of visual area MT in the macaque. *J. Comp. Neurol.* **348**, 190–222.
Wall, J.T., Symonds, L.L., and Kaas, J.H., 1982. Cortical and subcortical projections of the middle temporal area (MT) and adjacent cortex in Galagos. *J. Comp. Neurol.* **211**, 193–194.
Weller, R.E. and Kaas, J.H., 1985. Cortical projections of the dorsolateral visual area in owl monkeys: The prestriate relay to inferior temporal cortex. *J. Comp. Neurol.* **234**, 35–59.

Weller, R.E. and Kaas, J.H., 1987. Subdivisions and connections of inferior temporal cortex in owl monkeys. *J. Comp. Neurol.* **256**, 137–172.

Weller, R.E. and Kaas, J.H., 1989. Parameters affecting the loss of ganglion cells of the retina following ablations of striate cortex in primates. *Vis. Neurosci.* **3**, 327–349.

Weller, R.E., Wall, J.T., and Kaas, J.H., 1984. Cortical connections of the middle temporal visual area (MT) and the superior temporal cortex in owl monkeys. *J. Comp. Neurol.* **228**, 81–104.

Wu, C.W.-H., Bichot, N.P., and Kaas, J.H., 2000. Converging evidence from microstimulation, architecture, and connections for multiple motor areas in the frontal and cingulate cortex of prosimian primates. *J. Comp. Neurol.* **423**, 140–177.

Xiao, Y. and Felleman, D.J., 2004. Projections from striate cortex (V1) to cytochrome oxidase thin stripes and interstripes of macaque V2. *PNAS* **101**, 7147–7151.

Xu, X., Bosking, W., Sáry, G. *et al.*, 2004a. Functional organization of visual cortex in the nocturnal owl monkey. *J. Neurosci.* **24**, 6237–6247.

Xu, X., Collins, C.E., Kaskan, P.M. *et al.*, 2004b. Optical imaging of visually evoked responses in prosimian primates reveals conserved features of the middle temporal visual area. *PNAS* **101**, 2566–2571.

Zeki, S.M., 1969. Representations of central visual fields in prestriate cortex of monkey. *Brain Res.* **14**, 271–291.

10

The Physiological Basis for Visual Motion Perception and Visually Guided Eye Movements

Uwe J. Ilg, Jan Churan, and *Stefan Schumann*

10.1 Abstract

Events in our physical environment which are transformed by one of our sense organs might trigger two quite different responses of us: Firstly, the neuronal activity elicited by an external event might result in the conscious perception of this event. Secondly, an event in the environment might trigger a goal-directed response, in some circumstances, as for instance in the case of blind-sight patients, even without conscious perception of the event itself. A very basic question is whether these two different responses, i.e. conscious perception as well as generation of a goal-directed motor program, relay on a unique mechanism or on two separate mechanisms. In other words, the question is whether a single brain area or different brain areas are involved in these two processes. These questions will be addressed in this chapter, although its scope will be more focused. Instead of addressing all possible aspects of perception and action, special emphasis will be directed to the processing of visual motion underlying conscious motion perception as well as the generation of adjusted motor programs. Therefore, this chapter represents only a small example of the huge amount of work done in the visual cortex of primates. In addition, the underlying neuronal substrate especially in the rhesus monkey will be discussed in detail. An overview of the techniques that can be used to correlate cortical physiology, visual perception, and visually guided action will be given. It will be shown that in general the processing of visual motion is very similar for action and perception, similar brain areas being involved in both cases.

The Primate Visual System: A Comparative Approach Edited by Jan Kremers

10.2 Processing of visual motion in the primate brain

Whenever the processing of visual motion is addressed, it might be speculated that this processing might depend firstly on the detection of an object and subsequently on its localization at different moments in time. Its velocity could be simply computed as the ratio of difference in space by difference in time (dx/dt). However, it was shown convincingly that motion perception acts independently of object recognition. Moreover, objects that are defined by motion can be recognized ('structure-from-motion', for review see Nakayama, 1985). The motion processing is achieved by elementary motion detectors, either by calculation of a cross-correlation of the spatio-temporal luminance distribution (correlation detectors; Reichardt, 1987) or by the functionally equivalent (van Santen and Sperling, 1985) determination of the spatiot-emporal energy (energy model; Adelson and Bergen, 1985).

In many vertebrate species, motion processing occurs already in the retina. A special, morphologically distinct class of ganglion cells responds directionally selective to visual motion. Clearly, motion signals are generated within retinal circuits (frog: Lettvin *et al.*, 1959; rabbit: from Barlow and Levick, 1965 to Chiao and Masland, 2003; cat: Hoffmann and Schoppmann, 1975). In primates, however, it is more than questionable whether motion processing occurs already in the retina. There is no experimental proof that directionally selective ganglion cells exist in primates. In contrast, there is compelling experimental evidence that the processing of visual motion is a cortical feature in monkeys and humans. In area V1, where the first stage of cortical processing takes place, roughly 20 percent of the neurons respond selectively to the direction of movement of a visual stimulus (Mikami *et al.*, 1986; Hawken *et al.*, 1988; Ilg and Thier, 1996). Most likely, these directionally selective neurons constitute the input to the motion area within the superior temporal sulcus (STS) (Movshon and Newsome, 1996): the middle temporal (MT) area or area V5 (owl monkey: Allman and Kaas, 1971; rhesus monkey: Zeki, 1974). Area MT in turn projects to the next motion area in STS, the middle superior temporal (MST) area (Maunsell and Van Essen, 1983; Ungerleider and Desimone, 1986). It is well established that visual motion processing in primates is intimately related to the neuronal activity observed in these areas (for review, see Albright, 1993).[1] In monkeys, neurons in area MT respond to stochastic motion signals and the strength of the responses increase with the number of coherently moving dots (Britten *et al.*, 1993). Using such stochastic motion signals, a clear relationship between the monkey's behavioral choices and visual responses of area MT was found within single trials (Britten *et al.*, 1996). Some neurons recorded from area MST even responded to inferred motion when the monkeys assumed that an invisible target moves behind an obstacle (Assad and Maunsell, 1995). For area MST, similar neuronal and psychophysical sensitivities were documented (Celebrini and Newsome, 1994). Microstimulation in area MST affected the motion perception of monkeys (Celebrini and Newsome, 1995) as well as the judgments of heading directions (Britten and van Wezel, 1998). More recently, it was shown that microstimulation in area MT affects both the monkey's choices and the speed of

[1] Area MST consists of two representations of the visual field. In most publications, these two subdivisions are named MST-l and MST-d. In brief, neurons in the lateral part of MST (MST-l) are involved in processing visual motion signals related to object motion, whereas neurons in the dorsal part of this area (MST-d) are responsible for the processing of motion signals related to ego motion.

decision-making, i.e. monkeys decided more quickly in favor of the stimulation site's preferred direction (Ditterich *et al.*, 2003). Lesions in areas MT and MST resulted in an impairment of motion perception (Newsome and Paré, 1988; Rudolph and Pasternak, 1999). Finally, using a Glass pattern (Glass, 1969), it was shown that response properties in primate areas MT and MST were similar to the properties of motion perception in humans and monkeys (Krekelberg *et al.*, 2003). Glass patterns do not contain globally coherent motion signals. Instead, they consist only of two frames of random dot pattern and induced motion perception occurs if the two frames were alternatively presented.

In addition to the contributions of areas MT and MST for motion perception, the neuronal activity in these areas was also shown to be related to the execution of smooth pursuit eye movements (SPEMs; Newsome *et al.*, 1988; Thier and Erickson, 1992; Kawano *et al.*, 1994; Ilg and Thier, 2003). Small lesions of these areas yielded an ipsiversive deficit in SPEM (Dursteler and Wurtz, 1988; Yamasaki and Wurtz, 1991). Furthermore, intracortical microstimulation (ICMS) was able to modify ongoing SPEM (Komatsu and Wurtz, 1989; Groh *et al.*, 1997; Born *et al.*, 2000).

It is important to note that the processing of visual motion is definitively not finished at the level of area MST (see below). With respect to the generation of SPEM, it was shown that neurons in the ventral intraparietal (VIP) area in the intraparietal sulcus (Schlack *et al.*, 2003) and in the frontal pursuit area (FPA), a subdivision of the frontal eye field (FEF) (MacAvoy *et al.*, 1991; Tanaka and Lisberger, 2001), were active during the execution of these eye movements. Neurons recorded from the posterior polysensory part of the superior temporal sulcus (STPp) responded to moving stimuli in accordance with the monkeys' perception of motion (Thiele and Hoffmann, 1996). Responsiveness to motion signals was also shown for neurons in area 7A (Siegel and Read, 1997). Finally, sensitivity for motion signals was also documented in areas involved in the processing of vestibular information (PIVC: Gruesser *et al.*, 1990 and area 2v: Buettner and Buettner, 1978).

Synopsis There is convincing evidence that the processing of visual motion in primates is a cortical feature. The first directionally selective responses are recorded from area V1. It is very likely that these directionally selective neurons in area V1 provide the input for areas MT and MST. These areas are massively involved in the processing of visual motion underlying the perception of visual motion as well as the execution of motion-dependent motor programs such as SPEMs.

10.3 Action which depends on motion processing: smooth pursuit eye movements

Since the number of axons in our optic tract is limited, the spatial resolution of our visual system displays a marked anisotropy: only within the central visual field, the fovea, we are able to discriminate up to 60 lines per degree of sight. The spatial resolution sharply decreases with eccentricity. This anisotropy constitutes on the one hand an impressive example of data reduction. On the other hand, it imperatively asks for a very precise motor system which shifts the retinal image of an object of interest into the fovea. If we

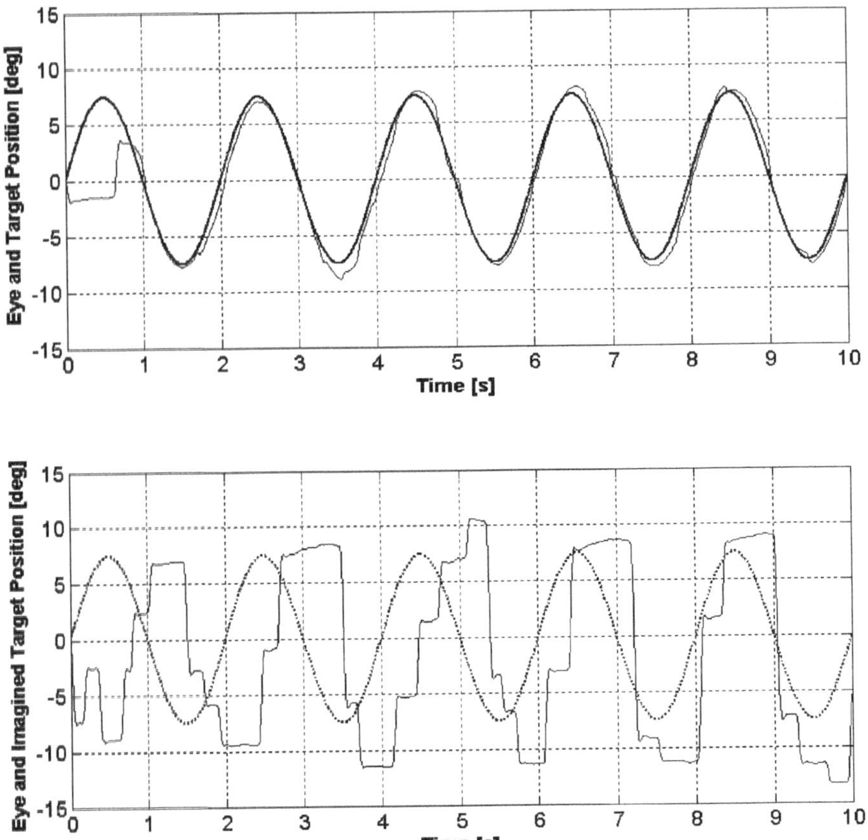

Figure 10.1 The horizontal eye position of a human subject was measured by an infrared eye tracker. The moving target was presented onto a 17 in. computer monitor at a viewing distance of 57 cm. In the upper diagram, the target moved sinusoidally at 0.5 Hz with an amplitude of 7.5° shown by the solid line. In the lower diagram, the monitor was switched off and the subject was instructed to perform the same eye movements as above. The dotted line represents a possible imagined target position. Note the absence of smooth eye movements if the monitor was switched off; only saccades and periods of fixation can be observed. Although the subject performed a series of saccades, he was able to reproduce the periodicity of the target movement

are watching a stationary scene, this task is achieved by fast, ballistic eye movements called *saccades*. If the object of interest moves, we are able to move our eyes at an appropriate velocity yielding a stationary retinal image of the object on the fovea. These eye movements are called smooth pursuit eye movements. In addition to enable the high spatial resolution of foveal vision, SPEM prevents motion blur. Retinal image motion with a velocity as low as 1°/s results in an impairment of grating resolution comparable to two diopters of optical degradation (Carpenter, 1988).

If we ask a subject to move his hand slowly from point A to point B in space, the subject does not have any difficulties in doing so. It seems to be likewise trivial to generate slow or smooth eye movements voluntarily. If a visual target moves on a computer monitor for instance, subjects are able to track the moving target precisely with their eyes. The eye velocity is perfectly matched to target velocity. We asked a subject to perform exactly the same eye movements without a target. We switched the computer monitor off and asked the subject to simply imagine the moving target and to track it. In this condition, the eye movements consisted in a sequence of saccades and fixation periods as illustrated in Figure 10.1. Segments with smooth eye movements were completely absent in this condition. The fact that we are able to perform SPEM only in the presence of a moving target is the reason why these eye movements are frequently used as a biological probe for motion processing. If a subject performs SPEM, it can be taken as indication that successful processing of visual motion has taken place.

Synopsis If an object of interest moves, we are able to move our eyes smoothly at the velocity of the moving target. As a consequence, the image of the target remains stationary on the fovea. Since we are able to generate these eye movements only in the presence of a moving target, they can be used as a biological probe for successful motion processing.

10.4 Comparing motion processing underlying perception and smooth pursuit eye movements

A very simple, but not trivial, question is whether the mechanisms of visual motion processing for perception and for the generation of SPEM use common or separate mechanisms. Influenced by the concept that the cortical visual system consists of two subsystems, a ventral stream responsible for object identification (*what*) and a dorsal stream underlying the localization of objects (*where*) (Ungerleider and Mishkin, 1982), it was proposed that there are also separate visual pathways for perception and action (Goodale and Milner, 1992). Initially, this proposal was supported by the observation that a patient, who had difficulties in reaching appropriately a target, could perform a manual grasping task correctly (Goodale *et al.*, 1991). In other words, the patient's ability to localize the target was normal, whereas her ability to perceive the form of the target and to use this information to appropriately pre-shape the hand was absent. Further experiments addressing the question of separate systems for action and perception focused on the Ebbinghaus illusion: the perceived size of a central circle is influenced by the size of surrounding circles. However, if subjects were asked to grasp for the central circle, the grasping (i.e. the width between thumb and index finger) was not affected by the illusion (Aglioti *et al.*, 1995). The strength of this argument for separate systems for action and perception was reduced by the findings of others showing that perception and grasping could be influenced in a similar manner (Franz *et al.*, 2000). Recently, the question of separate systems for action and perception was also addressed by the analysis of the localization of a target flashed briefly before the execution of a saccade. Here, it was shown that manual pointing and verbal report gave different results: whereas the pointing toward the target briefly presented during the execution of a saccade was correct, the verbal report of the perceived target position reflected a compression of space due to the execution of the saccade (Burr *et al.*, 2001). In summary, there are results from a broad range of experiments supporting the notion of separate systems for perception and action.

To address specifically the question whether separate systems with respect to processing visual motion are used for perception and the generation of SPEM, several experiments following a typical approach were performed in the past: on the one hand, the subject's perception was examined using psychophysical techniques. On the other hand, the identical stimulus was used to elicit eye movements. A very typical stimulus in these studies consisted in a variable ratio of coherently moving dots superimposed onto randomly moving dots displayed within a stationary aperture (Newsome and Paré, 1988). If all dots were moving to the right, the subject clearly perceived rightward movement. If all dots were moving toward the left, the subject perceived leftward motion. If all dots were moving randomly, the perception of the subject was at chance level. The reports of the subjects constituted the *psychometric* function, which is the chance of perceiving the correct direction of motion as a function of the coherence in the moving dots. At the same time, the elicited eye movements could be recorded and analyzed. To determine the *oculometric* function, the direction of the elicited eye movements was determined within a time window of 500 ms as a function of the coherence in the moving dots. To be able to analyze only the smooth eye movements, saccades were eliminated carefully. The obtained *psychometric* and *oculometric* functions are convincingly similar (Krauzlis and Adler, 2001).

The fact that perception and oculomotor control involve the same mechanism was further supported by the observation that *psychometric* and *oculometric* functions could be similarly modified by the subject's anticipation of movement direction. Before the onset of the stimulus, the subjects saw a cue signaling the direction of the subsequent stimulus movement. If the cue predicted rightward movement, both functions were shifted to the left, i.e. a smaller amount of dots moving coherently to the right was necessary to elicit rightward perception as well as rightward eye movement (Krauzlis and Adler, 2001). A common mechanism of visual motion processing for perception and generation of eye movements was also suggested by the similarity of *oculometric* and *psychometric* functions obtained when moving plaids were used (Krauzlis and Stone, 1999), in the case of the directional biases for elongated apertures (Beutter and Stone, 1998), and when the effects of motion coherence (Beutter and Stone, 2000) and of motion integration (Stone *et al.*, 2000) were examined.

Another attempt to reveal whether perception and oculomotor control share a common mechanism of visual motion processing consists in the use of paradoxical motion stimuli. These motion stimuli are called second-order motion to delineate them from first-order motion stimuli (Chubb and Sperling, 1988). Second-order motion cannot be decoded by an elementary motion detector (Reichardt, 1987) nor by the functionally equivalent energy model (Adelson and Bergen, 1985; van Santen and Sperling, 1985). Four different types of motion stimuli are of interest to compare the perception of subjects and their power to elicit SPEM (Figure 10.2): luminance-defined motion (LM), Fourier motion (FM), drift-balanced motion (DBM, Chubb and Sperling, 1988), and theta motion (TM, Zanker, 1993).

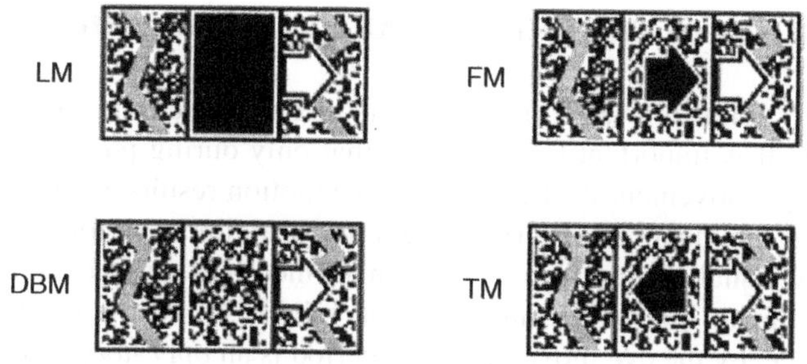

Figure 10.2 A schematic representation of four motion stimuli is shown in space-time diagrams. In case of luminance-defined motion (LM), a rectangle with a different luminance than the background moves toward the right (indicated by the white arrow) across a dynamic random dot pattern (indicated by the grey zig-zag line). In case of Fourier motion (FM), a rectangular area of dots moves coherently. Here, object motion, indicated by the white arrow, and pixel motion, illustrated by the black arrow, are identical. In case of drift-balance motion (DBM), a rectangular aperture moved across the dynamic random dot pattern indicated by white arrow. No pixel motion occurs in this stimulus at all. Finally, in case of theta motion (TM), the pixels within the rectangle move toward the left (black arrow), while the rectangle itself moved to the right (white arrow)

Psychophysical experiments revealed that human subjects (Chubb and Sperling, 1988; Zanker, 1993) and monkeys (Churan and Ilg, 2001) are able to perceive all these motion types correctly. As outlined above, the execution of SPEM is tightly connected to visual motion processing. In line with this notion SPEMs of subjects tracking objects defined by second-order motion followed the direction of the perceived object motion, not the direction of individual moving pixels (Butzer *et al.*, 1997). There is no reason to assume different motion processing underlying the execution of SPEM and the perception of visual motion.

The latency of the initial saccade in the eye movement experiment revealed a strong dependency on the stimulus type: the first-order motion stimulus elicited initial saccades with a latency of 209 ms and the second-order motion stimuli resulted in a saccadic latency of 260 ms (Butzer *et al.*, 1997). In psychophysical experiments, it was shown that second-order motion stimuli have to be presented longer than first-order motion stimuli for successful discrimination (Yo and Wilson, 1992; Derrington *et al.*, 1993). This similarity in the dependence of latency and stimulus duration on the type of motion stimulus also indicates common visual motion processing for perception and action.

Synopsis With respect to the processing of visual motion, there is multiple experimental evidence that the perception of motion and the generation of SPEMs share a common mechanism for motion analysis. The *psychometric* and *oculometric* functions obtained from various visual motion tasks are very similar and reveal similar modifications by expectation. In addition, the processing of second-order motion stimuli reveals similarities in perception and sensorimotor integration.

10.5 Initiation of smooth pursuit eye movements

An important question is how SPEMs are initiated. How is the transition from fixation to pursuit achieved? It is important to emphasize that only during pursuit initiation, before the onset of the eye movement itself, retinal image motion results exclusively from target motion in the environment. This period offers an ideal situation for studies addressing the question how visual motion signals are transformed into eye movement commands.

Since retinal image motion during initiation of SPEM is not yet influenced by a feedback signal, i.e. by the eye movement, this period is an *open-loop* phase in contrast to the *steady-state* phase occurring later during the maintenance of SPEM. During steady-state pursuit, target velocity in space has to be computed by adding the retinal image motion of the target and the executed eye velocity.

If we assume that the retinal image velocity of the target is determined during the open-loop period and transformed into an eye movement program, then eye position would constantly lag target position. Obviously, this does not happen as the typical initiation of a SPEM displayed in Figure 10.3 shows. Approximately 150 ms after the onset of target movement, the eyes started to accelerate, indicated by the increase in eye velocity. At about 300 ms following target onset, the subject performed an initial saccade compensating the lag of eye position. Immediately after the initial saccade, eye velocity was matched to target velocity. Studies on pursuit initiation usually determine the slope

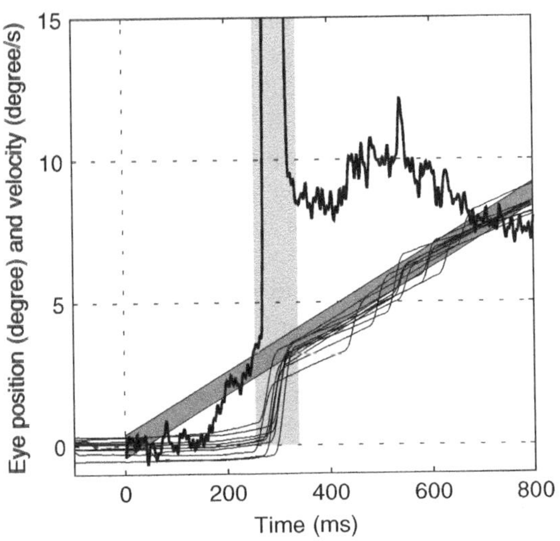

Figure 10.3 Horizontal eye and target position (marked by grey bar) together with the eye velocity of a human subject tracking a small moving target are displayed. Eye positions of individual traces (thin lines) are shown together with the median eye velocity across all shown trials (bold line). The eye position was recorded by means of scleral search coils (Collewijn *et al.*, 1983). As indicated, the target started to move at time 0. Note the stereotyped sequence of pre-saccadic eye acceleration followed by the initial saccade as well as the post-saccadic enhancement. The vertical grey bar gives the time interval in which the initial saccades occurred

of increase in eye velocity (eye acceleration) and the latency of pursuit initiation in addition to the parameters of the initial saccade.

Closer examination of the initial saccade parameters revealed that the amplitude of this saccade was adjusted to the velocity of the target, so that the saccade ended where the target was after its execution, not the target's initial position (Gellman and Carl, 1991). The saccadic system does not only use retinal position error of the target to compute saccade amplitude, but also information about target velocity in space. The presence of this information is also able to explain the phenomenon of *post-saccadic enhancement*, i.e. that the eye velocity before the initial saccade is below target velocity, but ultimately adjusted to target velocity after the saccade.

10.5.1 Latencies and gap paradigm

With respect to the latency of an eye movement, it was shown that a brief (200 ms) temporal gap between the offset of the fixation target and the onset of the new saccade target reduced the saccadic reaction time (e.g. Saslow, 1967; Kalesnykas and Hallett, 2002). These saccades with short latency were called *express saccades* (Fischer and Boch, 1983). To understand why the saccadic reaction time is reduced in the gap paradigm, it is necessary to clarify the different steps involved in the generation of saccades. Obviously, the very first step in this sequence of events is the termination of the fixation process. If the fixation is already terminated as a consequence of the gap, the saccade can be elicited earlier.

Similar to the execution of saccades, the very first step in initiation of SPEM also consists in the termination of fixation. Indeed, similar to saccades a temporal gap between fixation target offset and pursuit target onset also reduced SPEM latency (Merrison and Carpenter, 1995; Knox, 1996; Krauzlis and Miles, 1996a,b). Interestingly, the contrast of fixation and pursuit targets affected the pursuit onset latency, but not the reduction in latency induced by the gap (O'Mullane and Knox, 1999). The similarity of the gap effect on saccadic and pursuit latencies suggests that a common neuronal substrate is involved. It was shown that the activity of neurons in the superior colliculus represents a motor error signal for saccadic and pursuit eye movements (Krauzlis *et al.*, 1997). These neurons increased their activity already during the gap period, before the target was presented. Since this was true for saccades as well as for SPEM (Krauzlis *et al.*, 2002), it is very likely that the superior colliculus is involved in the preparation of saccades as well as in the preparation of SPEM.

In many studies on pursuit initiation, the investigators are interested in avoiding the disturbing initial saccade in order to describe precisely the mechanism of the slow generation of eye velocity. This can easily be achieved if the target does not start to move from the point of fixation, but from a slightly changed position. This paradigm is called step-ramp paradigm. If the size of the target step is adjusted according to the formula

$$\text{target step}[^\circ] = -\text{latency of the subject}[\text{s}] \times \text{target velocity}[^\circ/\text{s}],$$

eye position matches target position asymptotically without an initial saccade (Rashbass, 1961).

10.5.2 Acquisition of motion signals during pursuit initiation

The open-loop period of pursuit is characterized by sequential acquisition of all available visual motion signals, and subsequent transformation of the motion information into a motor command. Pre-saccadic eye acceleration is a function of the eccentricity of the target. Eye acceleration declined if the target started to move in the periphery compared to foveal vision (man: Tychsen and Lisberger, 1986; monkey: Lisberger and Westbrook, 1985). In addition, target movement toward the fovea resulted in higher eye acceleration than movement away from the fovea. To further characterize visual motion processing underlying the initiation of SPEMs, the use of second-order stimuli as explained in section 10.4 (Figure 10.2) turned out to be helpful. It has already been shown that during steady-state pursuit, the direction of SPEM is identical to the direction of perceived object motion (Butzer *et al.*, 1997). We asked whether the parameters of pursuit initiation (onset and acceleration) depend on the order of motion stimuli. Pre-saccadic pursuit initiation depended on the type of motion stimulus. The highest eye acceleration was elicited by FM. DBM resulted in a lower acceleration. In case of TM, the pre-saccadic pursuit was initiated into the direction of pixel motion, which was in contrast to steady-state SPEM (Lindner and Ilg, 2000). However, pursuit latency did not depend on the type of motion stimulus.

It is important to note that the first-order motion components in FM and TM were identical, the only difference being the relation between the first-order motion component and the direction of object motion. If the first-order component was the only source for pursuit initiation, the absolute value of eye acceleration should be identical for TM and FM. But there was a significant difference in the absolute values of eye acceleration – the TM stimulus elicited a significantly smaller acceleration than the FM stimulus. A similar behavior will be presented with respect to the neuronal responses of areas MT and MST in the following text. In addition, if the stimulus consisted in DBM, i.e. without any moving pixels, pre-saccadic pursuit initiation was also observed. Two conclusions can be drawn from these results. First, there are separate systems for decoding first- and second-order motion, and second, these systems interact. Pursuit initiation cannot be described by a simple algebraic summation of the output of both systems.

Luminance-defined stimuli elicited initial saccades with shorter latencies than all other motion-defined stimuli (Butzer *et al.*, 1997). The luminance-defined stimulus could be perceived without further processing of visual-motion information. In support, it was shown that increasing the saliency of the target by a different color compared with the background reduced the pursuit onset latency and increased the initial eye acceleration (Miura *et al.*, 2001).

The area MT in the brain of rhesus monkeys is essential for the visual motion processing underlying the initiation of SPEM (Pack and Born, 2001). If a moving line is viewed through a circular aperture, which blocks the endpoints of the line, only motion perpendicular to line orientation can be observed. Individual neurons recorded from area MT of rhesus monkeys responded initially only to motion perpendicular to the moving line (first 20 ms of the response). Only the late response (>140 ms) signaled the true direction of line motion. Similarly, SPEMs of the monkeys tracking the moving bar changed their direction over time. Initial eye movements were perpendicular to the bar, irrespective of the direction of bar movement, whereas steady-state pursuit was in the true direction of bar movement. A similar change in the tracking direction was reported from human subjects (Masson *et al.*, 2000).

Synopsis In summary, the initiation of SPEM is characterized by a stereotyped sequence of pre-saccadic eye acceleration followed by an initial saccade. The huge increase in eye velocity immediately before and after the initial saccade is achieved by a process called *post-saccadic enhancement*. The open-loop period (i.e. before the onset of eye movement) is characterized by the acquisition of all available motion information.

10.6 Cancelation of self-induced retinal image motion during execution of SPEM

It is a typical but artificial situation in the laboratory when subjects are asked to track a single target moving in front of a dark, homogeneous background. Under natural conditions, objects move in front of a structured background. The presence of a structured background during the execution of SPEM causes two peculiarities: First, we do not perceive the self-induced shift of the retinal image of the background under normal conditions. One might speculate that, similar to saccadic suppression, this might be due to a suppression of motion perception (Bridgeman *et al.*, 1975). However, saccades are very brief, fast eye movements. The costs of visual impairment during saccades are quite small. For SPEM, a similar suppression of vision is not possible since the execution of SPEM lasts much longer and critically depends on the processing of visual motion. With respect to the perception during execution of SPEM, the perception of the stationary surround is not perfect. If one moves a pencil across this text and tracks the tip of the pencil, he will not be able to read the text, but he will perceive the characters moving in the opposite direction to the moving pencil. This apparent background movement is called *Filehne illusion* (Filehne, 1922). A successful attempt to explain why there is only a small misperception of self-induced retinal image motion consists in the *re-afference principle* (von Holst and Mittelstaedt, 1950). According to this principle, our perception of motion does not only depend on retinal image motion, but is also influenced by the executed eye movements via an efference copy signal. In case of a stationary background, the size of the efference copy signal is identical to the self-induced retinal image motion, but with different polarity. Therefore, both signals cancel each other and a stationary background is perceived. In the course of elegant experiments, it was convincingly shown that the gain of the efference copy signal can be modified by artificial retinal image motion during execution of SPEM (Haarmeier and Thier, 1996).

The second peculiarity related to the presence of a structured background during the execution of SPEM consists in the fact that the execution of SPEM is only mildly affected by the background. The effects of a structured background are different during initiation and steady-state pursuit. During pursuit initiation, and before the onset of the eye movement, retinal image motion is only caused by the movement of the target, not yet by an eye movement. Nevertheless, pursuit initiation is quite consistently affected by the presence of a structured background compared to pursuit initiation with a dark, homogeneous background: the initial eye acceleration was reduced and the pursuit latency was increased (Keller and Khan, 1986; Kimmig *et al.*, 1992; Mohrmann and Thier, 1995).

In contrast, during steady-state pursuit there are only marginal influences of the presence of a structured background on pursuit parameters with a huge amount of variations between subjects. Pursuit gain is reduced mildly and compensated by an increase in the number of saccades (Yee *et al.*, 1983; Collewijn and Tamminga, 1984; Ilg and Thier, 1996).

Some years ago, Schwarz addressed the issue of the insensitivity of the pursuit system to self-induced retinal image motion by using brief injections (200 ms) of global motion during various phases of the execution of SPEM. The short duration excluded motion adaptation, which is always present in motion processing. This approach turned out to be very successful for our understanding of the processing of self-induced retinal image motion. A clear asymmetry in the obtained eye velocity profiles became evident: if the background moved in the same direction as the pursuit target, the eye velocity profile was briefly disturbed. If the background moved in opposite direction, no modulation of eye velocity was observed. Note that in natural conditions, i.e. if a target moves in front of a structured background, the direction of self-induced retinal image motion is always opposite to target motion. This explains why the self-induced retinal image motion does not affect the ongoing eye movement (Schwarz and Ilg, 1999; Suehiro *et al.*, 1999; Lindner *et al.*, 2001).

Our data resulting from brief injections of global motion are in agreement with the results from Born *et al.* (2000) who used the above described ICMS in area MT to modulate the ongoing SPEM. Area MT can be divided into regions with wide-field detector characteristics and regions with local-motion detector characteristics (Born and Tootell, 1992). In case of a wide-field detector, the visual response to a moving stimulus increased monotonically with stimulus size. In case of a local-motion detector, the response displayed a maximal value for a specific stimulus size; the response decreased for smaller and bigger stimuli (Born, 2000). Born *et al.* (2000) examined what happened if ICMS was applied at local-motion sites or wide-field sites. They analyzed eye velocity just immediately following the initial saccade (20–60 ms after saccade) and obtained opposite results for both types of motion detectors. ICMS at a local-motion detector site resulted in a shift of eye position in the preferred direction of neurons at the stimulated site. In contrast, ICMS at a wide-field detector site resulted in a shift of eye position in the direction opposite to the preferred direction of neurons at the stimulated site. This effect of ICMS at a wide-field site was identical to the effects of global motion injection (Born *et al.*, 2000).

Synopsis Surprisingly, the presence of a structured background has only marginal effects on steady-state pursuit. Experimental evidence indicates that during the execution of SPEM, the sensitivity for global motion in the direction opposite to target movement (i.e. in the direction of the self-induced retinal image motion) is decreased.

10.7 Smooth pursuit eye movements and motor learning

Many eye movements are executed in response to a stimulus in the environment either as a reflex or at least in a reflex-like manner. Frequently, these stimulus–response loops are subject to adaptation or motor learning. In brief, adaptation or motor learning occurs if the stimulus–response loop generates consistently an error signal. Initially, the error is compensated using retinal information. Subsequently and after successful adaptation, the error is avoided by a modification of the primary response. With respect to SPEM, the transformation of pre-saccadic visual motion information into an eye movement command could be subject to motor learning, as the following experiment shows. The basic paradigm was introduced more than 30 years ago (Barmack, 1970): If a target started to move at a given velocity and changed its velocity after a brief interval (100 ms), post-saccadic eye velocity was determined by the initial target velocity and not by its actual velocity. As a consequence, gaze position deviated from target position. If the paradigm was repeated (300 trials), post-saccadic eye velocity was adjusted to the new target velocity (Kahlon and Lisberger, 1996). It was possible to adapt the eye velocity to an acceleration as well as to a deceleration of the target velocity.

If two moving objects are presented, one acting as target and the other as distractor, pursuit initiation is characterized by vector averaging of both motion signals (Gardner and Lisberger, 2001). The above adaptation of initial eye acceleration offered a possibility to address the question whether this vector averaging occurred rather early or late in the processing of visual motion underlying pursuit initiation. The obtained results clarified that the site of learning was located earlier in processing than the site of averaging (Kahlon and Lisberger, 1999).

With a view to further localize the site of motor learning, we asked whether the motor learning could only be observed during the execution of SPEM, or, alternatively, whether goal-directed hand movements could also be altered as the consequence of motor learning. We determined the accuracy and timing of pointing finger movements from subjects who were instructed to point at a moving target with their index finger. To avoid online modification of the finger movement, the target was switched off when the finger of the subject's hand left the initial start position. During the motor learning, the target started to move for 400 ms at 10°/s and accelerated subsequently to 40°/s. Subjects received positive reinforcement if the finger position was close to the actual target position. As shown in Figure 10.4, the training process actually affected significantly the pointing position of our subjects.

These results suggest that the motor learning dependent changes in the neuronal activity occur somewhere in the sensory pathway common for the execution of SPEM and goal-directed hand movements. This is explained in more detail below. An open but very obvious question at present time asks whether motor learning during the execution of SPEM would also affect hand pointing movements.

Synopsis The open-loop phase during pursuit initiation makes it possible to study specifically how retinal image slip is transformed into a motor command. This initiation can be modulated if the target velocity is consistently changed during the execution of several hundred trials. Initially, the change in target velocity induces a retinal image error which triggers corrective eye movements. In the course of motor learning, the pursuit initiation is adapted appropriately to prevent a build-up of retinal error. Similar effects can be observed during the execution of pointing movements in human subjects. Therefore, the motor learning most likely changes the response properties of neurons in the sensory part of the neuronal substrate underlying the execution of SPEM as well as goal-directed hand movements.

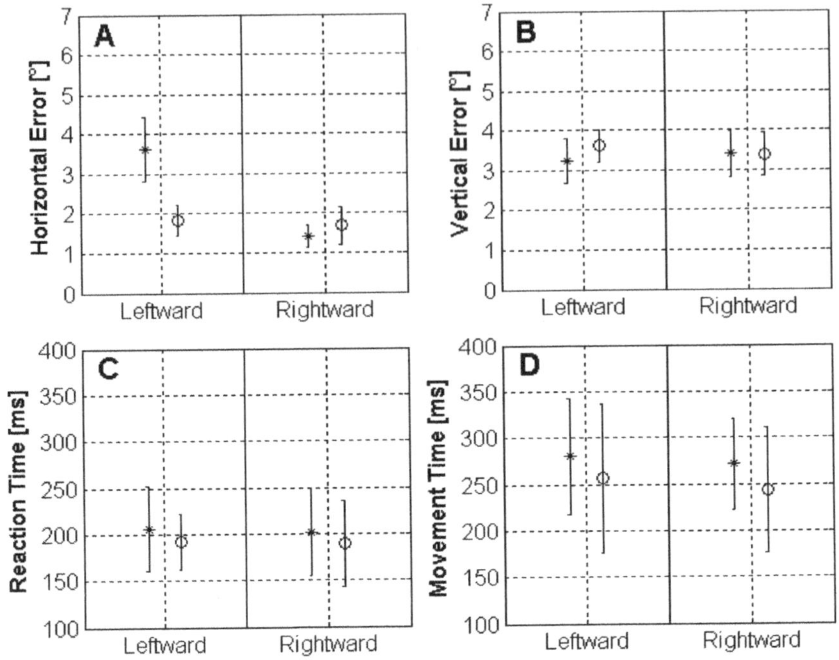

Figure 10.4 Results from pointing movements of six human subjects. Landing position of the subject's finger was determined by means of a touch-screen onto which the visual stimuli were displayed. While the subject had to fixate a stationary target, a second moving target appeared which had to be touched upon color change of the stationary target. To prevent online visual control of the pointing movement, the moving target was switched off as soon as the subject initiated the pointing movement. During the training sessions, the target started to move for 400 ms at 10°/s and subsequently changed velocity to 40°/s. The paradigm was absolutely identical in the pre- and post-training experiments. Only the horizontal pointing error in the direction of the training (leftward, shown in A) was changed as a consequence of the learning; horizontal error in the opposite direction (rightward, also shown in A), vertical error (B), reaction time (C), and movement time (D) remained unchanged by the training procedure. Circles represent pre-training values, asterisks represent post-training values

10.8 Pursuit-related activity and its frame of reference

As mentioned earlier, the execution of SPEM is tightly connected to the processing of visual motion. Elementary motion processing, either in case of correlation detectors (Reichardt, 1987) or in case of the energy model (Adelson and Bergen, 1985; van Santen and Sperling, 1985), is definitively achieved in a retinal frame of reference, which means that the decoding of visual motion relies exclusively upon retinal signals. This does not at all imply that motion processing takes place in the retina. So it might be speculated that SPEM could also be generated within a simple retinal frame of reference. If this assumption is correct, SPEM could be explained as a simple feedback model, with retinal image motion as input and eye movement as output (see Lisberger *et al.*, 1987 for review).

However, some findings are in disagreement with this notion. Firstly, the perception of the moving target during pursuit is not confined to a retinal frame of reference. In case of perfectly adjusted SPEM, the image of the moving target remains stationary on the fovea. Despite the lack of retinal image motion, we perceive the target as moving which cannot result from processing within a retinal frame of reference. Secondly, there is multiple evidence that pursuit is much more driven by the perception of motion than by retinal image motion signals (Steinbach, 1976; Wyatt *et al.*, 1994; Ilg and Thier, 1999). For instance, pursuit can be executed in the expectation of an upcoming target, before the target is visible (see above). Finally, the strongest argument against the retinal frame of reference is elaborated below. The argument consists in the discharge rates of individual neurons in the posterior parietal cortex. The importance of areas MT and MST for the execution of SPEM was documented by single-unit responses (Newsome *et al.*, 1988; Thier and Erickson, 1992; Lisberger and Movshon, 1999), by specific deficits following lesions (Dursteler *et al.*, 1987; Dursteler and Wurtz, 1988; Yamasaki and Wurtz, 1991; see below), and by ICMS in area MT (Groh *et al.*, 1997; Born *et al.*, 2000) and in area MST (Komatsu and Wurtz, 1989; see below).

To answer the important question whether the observed pursuit-related activity can be explained by exclusively visual factors, several studies tried to exclude completely retinal image motion signals. A first step in the experiments was to avoid any background whose self-induced retinal image motion could be the source of pursuit-related activity. Subsequently, to avoid retinal target motion, the pursuit target was either switched off briefly or stabilized electronically on the retina. The pursuit-related activity recorded from area MT dropped as a consequence of this procedure, indicating that area MT acts as a retinal image motion processor (Newsome *et al.*, 1988; Lisberger and Movshon, 1999). In contrast, the activity recorded from area MST remained unchanged (Newsome *et al.*, 1988; Thier and Erickson, 1992). This robustness of discharge rates observed from area MST is taken as indication for the existence of extra-retinal signals affecting neurons in area MST. These extra-retinal signals most likely consisted of an eye movement signal. In contrast, neurons in area MT responded only to visual signals.

The extra-retinal properties were also demonstrated during pursuit of imaginary targets defined by parafoveal cues as shown in Figure 10.5. As already noted, monkeys could be trained to perform SPEM toward an imaginary target (Ilg and Thier, 1999).

During tracking of the imaginary target, there was no visual stimulation of the central visual field where the receptive fields of MST neurons are located. Despite this lack of stimulation, these neurons responded similarly during pursuit of the real and imaginary targets (Ilg and Thier, 2003). This similarity in response is an argument that the MST neurons receive extra-retinal information related to the ongoing eye movement. In conclusion, there is multiple experimental evidence that the discharge rate of neurons in area MST is driven by visual and by extra-retinal signals. The possibility that neurons in area MST use a retinal frame of reference can therefore be excluded. However, the question remains whether the activity codes for eye-in-head, gaze, or target velocity. In order to answer this question, we trained monkeys to execute either isolated eye or combined eye and head movements in response to a moving target.

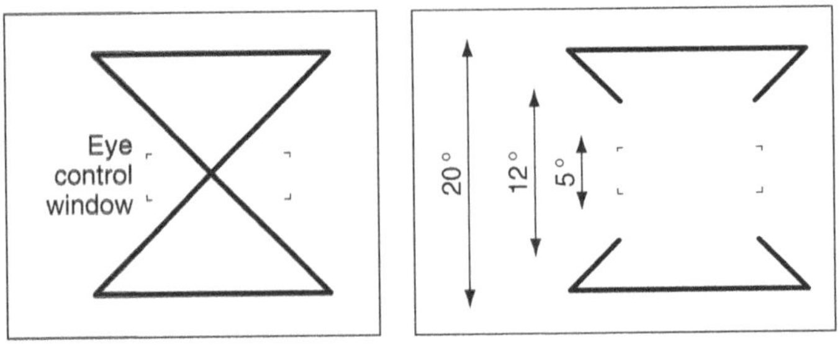

Figure 10.5 Targets to document extra-retinal response properties in the pursuit-related activity of visual-tracking neurons recorded from area MST of rhesus monkeys. Initially, the monkeys were trained to pursue the intersection of diagonals shown at the left. In case of the imaginary targets, the monkeys had to direct their gaze toward the invisible intersection shown at the right. The dimension of the blanked area in the figure was adjusted to the typical size of the receptive fields of the MST neurons

The color of the target signaled whether the monkeys should use eye or head movements for tracking. Figure 10.6 shows the response of a typical neuron recorded from area MST in the two different tracking conditions. Although the eye-in-head movements differed substantially in the two tracking conditions, the discharge rates of the neuron were identical in both cases. The gaze (eye-in-space) movements were also identical in both conditions. So the similarity in discharge rates might be explained as a consequence of identical gaze and resulting identical retinal image motion. However, as shown above, the activity of this neuron remained unchanged during brief removal of the target.

The similarity of responses during eye and head movements excludes the possibility that MST neurons code for eye-in-head movements. However, it is not clear whether the discharge rate coded for gaze or target velocity in space. To solve this question, we employed a combination of vestibular stimulation and a pursuit task. While the monkeys were rotated sinusoidally around the vertical axis at 0.2 Hz, we tracked a target that moved horizontally at 0.33 Hz. The Fourier transformation of the neuronal response revealed that its power density function had a maximum at 0.33 Hz, the frequency of target movement in space. We tested 39 neurons out of the above-mentioned 51 neurons. Finally, linear regressions of the discharge rate versus gaze velocity and of the discharge rate versus target velocity revealed that the activity coded for target motion in space for 34 out of the 39 neurons tested. Our findings suggest that the activity in area MST codes for the trajectory of a target in an external frame of reference. Since the transformation from a retinal frame of reference toward an external reference frame is achieved at this level of cortical processing, this information can be used for the generation of any general goal-directed behavior, not only for SPEM, but also for the perception of the moving target.

Synopsis Although the initial processing of visual motion is doubtless performed within a retinal frame of reference, there is experimental evidence that the motion

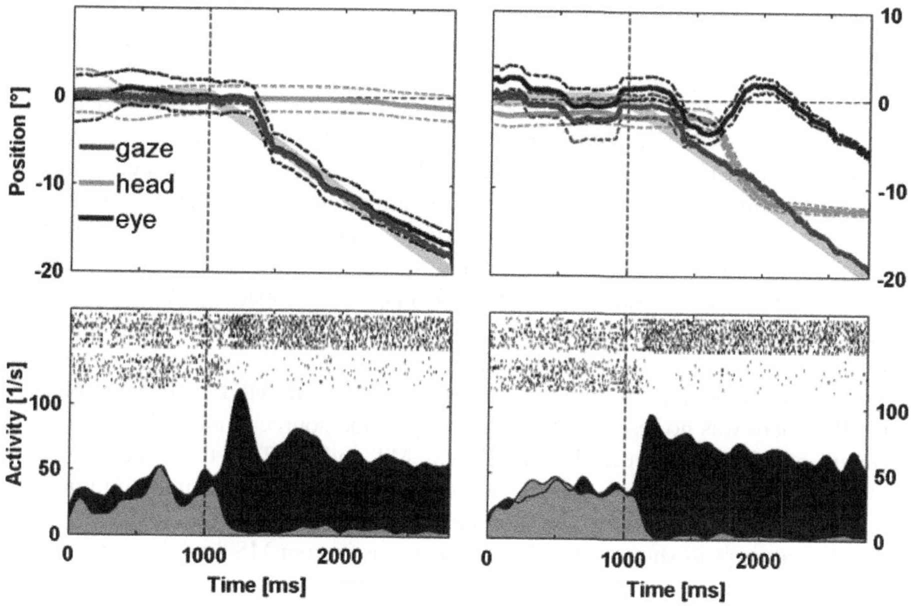

Figure 10.6 Gaze, eye, and head position of a rhesus monkey together with the discharge rate of a typical neuron recorded from area MST (black spike density function and upper raster display gives preferred direction, grey density function and lower raster display represents non-preferred direction) during the execution of two different tracking tasks. On the left, an eye movement trial was requested, the head remained stationary. On the right, the monkey had to execute a head movement trial. The statistical analysis (two-way ANOVA) revealed that the discharge rate was significantly affected by the direction of target movement ($P < 0.0001$), but not by the type of tracking ($P = 0.18$). So although the executed eye-in-head movements were quite different, the response of the neuron was identical in both conditions

processing underlying the execution of SPEMs is performed within an external frame of reference.

10.9 Contributions of area MST to motion perception

So far, the single-unit activity recorded from area MST was presented in connection with the execution of SPEMs. The bottom line was that MST neurons responded to retinal image motion as well as to eye and head movements. To analyze specifically the response properties to various motion signals, it is necessary to prevent the monkeys from pursuing the stimulus. As presented earlier, during the execution of SPEM, the activity elicited by a stimulus could always be the consequence of the executed eye movement. In order to study the responses to motion stimuli, we developed the direction discrimination task (DDT). In brief, while the monkeys had to fixate a stationary target, a moving stimulus was presented. After the presentation, the monkeys had to select one out of two saccade targets thereby reporting the perceived direction of the stimulus.

The results obtained from the pursuit experiments prompted the hypothesis that neurons in area MST might code in a global, modality-independent behavior for object motion in space. In a first experiment, we asked whether MST neurons were activated similarly if the moving stimulus was visual or acoustic. It is important to note that the monkeys had no problem in reporting the direction of the moving acoustic stimulus. To avoid technical problems related to the mechanics of a moving loudspeaker, we applied apparent motion stimuli. Out of a linear arrangement of 48 loudspeakers (spacing of 1°), subsequent speakers were activated (temporal spacing of 50 ms) resulting in an apparent velocity of 20°/s.

However, as indicated by the responses of a typical MST neuron depicted in Figure 10.7, there was no response to the moving sound source at all. We performed this experiment on 32 neurons recorded from areas MT and MST of three monkeys which all gave very clear directionally selective responses to a moving visual stimulus. Not a single neuron gave a response to the moving sound source.

In another attempt to show that individual neurons in area MST might generally code for stimulus motion, we employed second-order motion stimuli as already introduced in

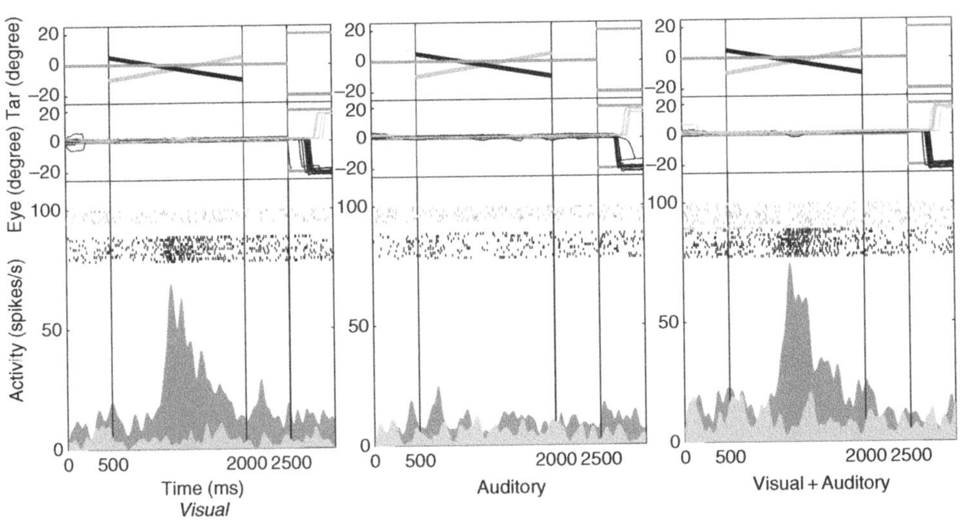

Figure 10.7 The response of a typical MST neuron to a moving visual, auditory, and bimodal stimulus as indicated. The monkey performed a direction discrimination task and reported correctly the direction of stimulus movement by a saccade at the end of the trial. Position of target and eye as well as neuronal activity (raster and spike density function) are shown for each condition. Black traces, rasters, and spike density function represent stimulus and eye movement as well as neuronal activity elicited by stimulus movement in preferred direction, grey informs about the values for the non-preferred direction. Note that there is no excitation to the moving sound source

the DDT. Special emphasis received the TM stimulus since the amount of retinal image motion was identical to the retinal image motion in the FM, but in opposite direction to the perceived movement of the stimulus. We asked whether the responses of individual neurons reflected the perceived direction of stimulus movement or, alternatively, the direction of physical pixel movement.

The neurons recorded from areas MT and MST did not respond to the direction of perceived stimulus movement; instead, they responded to the movement of individual pixels as shown in Figure 10.8. Similar responses showing the apparent inversion of the preferred direction were obtained from 38 neurons recorded from area MT and 68 neurons recorded from area MST of three monkeys. If we used a different type of second-order motion in which the moving object was separated from the background in flicker, a substantial number of neurons maintained their directionality even for the paradoxical motion stimulus (36 percent of the neurons recorded from area MT and 53 percent of the neurons recorded from area MST).

Synopsis Our experimental data do not support the hypothesis that individual neurons in area MST of rhesus monkeys code in a very general and modality-independent manner for motion signals. The neuronal activities recorded from these areas do not parallel the report of the monkeys for all stimuli used in the DDT. These results clearly indicate that area MST is not yet the final stage in the cortical processing of motion stimuli.

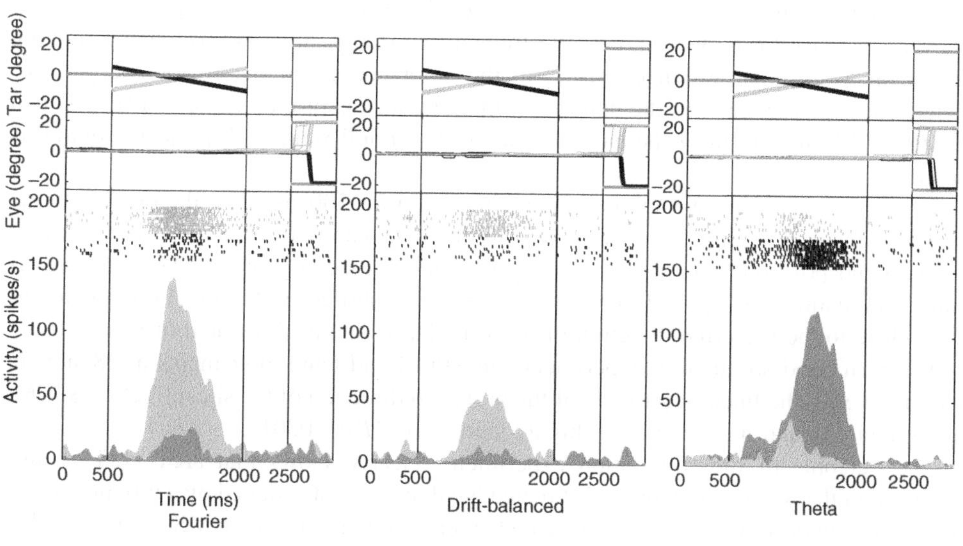

Figure 10.8 Response of a typical MST neuron to first- (Fourier) and second-order motion (drift-balanced and theta motion) as indicated; same conventions as Figure 10.7. Grey denotes stimulus movement in preferred direction as well as eye position traces and neuronal activity. Note that this neuron signaled the movement of the drift-balanced stimulus. In case of the theta stimulus, the neuron apparently inverted its preferred direction (response drawn in black is clearly more pronounced) indicating that the neuron did not code for the direction of stimulus movement, but the physical movement of the individual dots constituting the theta stimulus

10.10 Motion processing for actions other than eye movements

The recordings of single-unit activity from areas MT and MST of rhesus monkeys who either performed the direction discrimination task or performed a tracking task suggest that the observed neuronal activities are causally related to the performance. However, it is quite difficult to prove the existence of this relationship. This is much easier if either the neurons are artificially activated or deactivated. In the first case, we applied current of short 200-μs pulses at a frequency of 200 Hz, usually 80 μA at sites which were previously electrophysiologically explored. On the other hand, we injected up to 5 μl of Muscimol (5 mg/ml) into area MST to produce a transient lesion. Using these two techniques, we were able to document that area MST is indeed important to generate SPEM. As previously described by others (Komatsu and Wurtz, 1989), stimulation yielded an acceleration of SPEM in the preferred direction of the stimulated site. In contrast, stimulation yielded a deceleration if SPEM was performed in the non-preferred direction. Stimulation was only able to modify ongoing SPEM; during fixation the stimulation was not able to affect the eye movements. On the other hand, injections of Muscimol resulted in a decrease in SPEM velocity if the target moved toward the lesion site (ipsiversive) as reported earlier (Dursteler *et al.*, 1987).

Having documented the contributions of area MST for the generation of SPEM, we now asked whether this area is also involved if the monkeys had to perform a hand movement toward a moving target. We trained two rhesus monkeys to touch a moving target on a go-signal which consisted in the color change of the fixation target. We determined the precision of the hand movement by means of a touch-screen (Elo Touch Systems) 33 cm in front of the animal. The visual stimuli were back-projected onto this screen.

In a first experiment, we applied microstimulation for 500 ms while the moving target was presented. The target moved in one of the four cardinal directions. The distance of the hand position from the actual target position (touching error) was measured. If microstimulation was applied, the hand position was shifted in the direction which was very close to the preferred direction of the stimulation site as illustrated in Figure 10.9.

We performed stimulation experiments on SPEM and hand movements at 28 sites in two monkeys. The linear regression of the preferred direction of the site, effect on SPEM, and effect on hand movement was highly significant ($P < 0.0001$).

The outcome of the stimulation experiments clearly shows that area MST is also involved in the organization of goal-directed hand movements elicited by a moving target. To further support this notion, we asked whether deactivation of area MST was also able to modify goal-directed hand movements. In the course of the Muscimol experiments which produced the ipsiversive pursuit deficit as described above, we also performed hand movement experiments. As illustrated in Figure 10.10, the deactivation of area MST affected the hand movements.

If the right hemisphere was impaired, the landing position of the hand was consistent and independent of the direction of target movement shifted toward the left. This shift can be explained since the representation of rightward target movement (i.e. ipsiversive with respect to the lesion site) was missing.

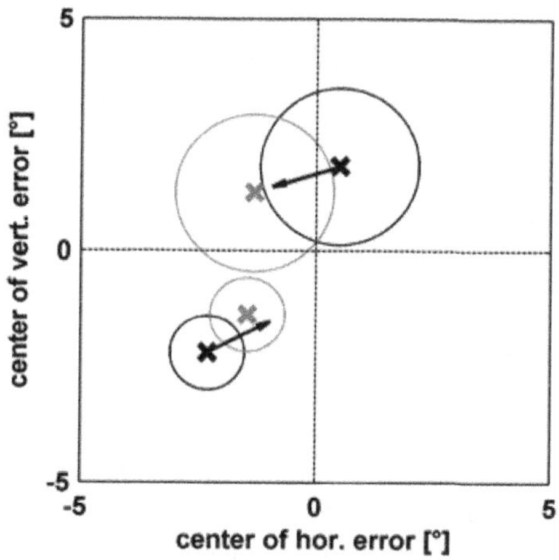

Figure 10.9 Similar to the subject's task shown in Figure 10.4, a rhesus monkey had to point toward a moving target. While the moving stimulus was presented, intracortical microstimulation (ICMS) was applied at a site whose preferred direction was determined previously. The results of two stimulation sites from two different monkeys are shown. Black crosses inform about the pointing error without stimulation based on 160 trials, grey crosses give the pointing error in stimulation trials. The circles represent 50 percent confidence intervals. The arrows inform about the preferred direction of a given site. As indicated, the landing position in stimulated trials was shifted in the direction of the preferred direction as a consequence of the stimulation

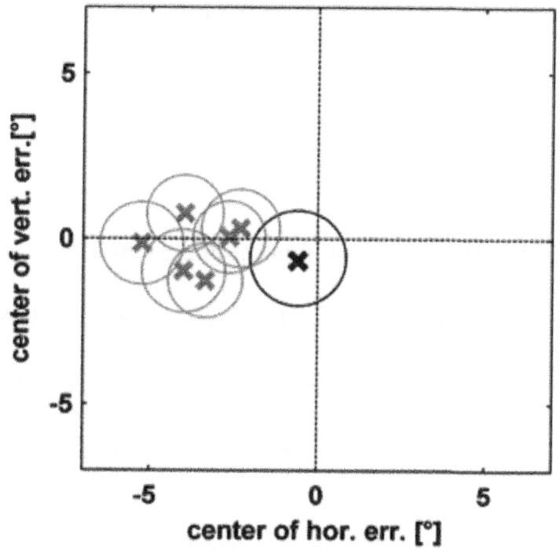

Figure 10.10 Landing position of pointing movements in a rhesus monkey. Different to the results shown in Figure 10.9, here area MST was transiently inactivated by the injection of a small amount of Muscimol. The black cross gives the pre-lesion pointing error of the monkey based on 160 trials, grey crosses inform about the pointing error that resulted from 6 different lesions in area MST of the right hemisphere. For details see Figure 10.9

Synopsis Although the importance of area MST for the generation of SPEM was documented for a long period of time, new experimental data suggest that this area is also involved in the processing of visual motion signals underlying the execution of goal-directed hand movements. So it seems that there is a single cortical machinery decoding the motion signals and providing the input for various motor systems such as the oculomotor or hand movement system.

10.11 Conclusions

With respect to the discoveries related to the neuronal mechanisms underlying the processing of visual motion stimuli, the monkey is an excellent animal model. A huge knowledge how visual motion signals are processed in the brain of rhesus monkeys is available. It seems that similar processing or mechanisms are activated during the perception of visual motion and during the use of motion signals to direct goal-directed behavior. This behavior includes SPEMs as well as goal-directed hand movements elicited by a moving target. The areas MT and MST, located at the conjunction of occipital and parietal cortex, are massively involved in this processing. However, there is also clear evidence that these areas are not the final stages in the processing of visual motion.

Acknowledgement

We are grateful to May Tan who performed the motor learning in pointing movements of human subjects. Financial support was obtained from the German Research Council (DFG, Heisenberg and SFB 550, A3) as well as from the Hermann-and-Lilly-Schilling Foundation.

References

Adelson, E.H. and Bergen, J.R., 1985. Spatiotemporal energy models for the perception of motion. *J. Opt. Soc. Am. A* **2**, 284–299.

Aglioti, S., DeSouza, J.F., and Goodale, M.A., 1995. Size-contrast illusions deceive the eye but not the hand. *Curr. Biol.* **5**, 679–685.

Albright, T.D., 1993. Cortical processing of visual motion. *Rev. Oculomot. Res.* **5**, 177–201.

Allman, J.M. and Kaas, J.H., 1971. Representation of the visual field in striate and adjoining cortex of the owl monkey (*Aotus trivirgatus*). *Brain Res.* **35**, 89–106.

Assad, J.A. and Maunsell, J.H.R., 1995. Neuronal correlates of inferred motion in primate posterior parietal cortex. *Nature* **373**, 518–521.

Barlow, H.B. and Levick, W.R., 1965. The mechanism of directionally selective units in rabbit's retina. *J. Physiol.* **178**, 477–504.

Barmack, N.H., 1970. Modification of eye movements by instantaneous changes in the velocity of visual targets. *Vision Res.* **10**, 1431–1441.

Beutter, B.R. and Stone, L.S., 1998. Human motion perception and smooth eye movements show similar directional biases for elongated apertures. *Vision Res.* **38**, 1273–1286.

Beutter, B.R. and Stone, L.S., 2000. Motion coherence affects human perception and pursuit similarly. *Vis. Neurosci.* **17**, 139–153.

Born, R.T., 2000. Center-surround interactions in the middle temporal visual area of the owl monkey. *J. Neurophysiol.* **84**, 2658–2669.

Born, R.T. and Tootell, R.B., 1992. Segregation of global and local motion processing in primate middle temporal visual area. *Nature* **357**, 497–499.

Born, R.T., Groh, J.M., Zhao, R. *et al.*, 2000. Segregation of object and background motion in visual area MT: Effects of microstimulation on eye movements. *Neuron* **26**, 725–734.

Bridgeman, B., Hendry, D., and Stark, L., 1975. Failure to detect displacement of the visual world during saccadic eye movements. *Vision Res.* **15**, 719–722.

Britten, K.H. and van Wezel, J.A., 1998. Electrical microstimulation of cortical area MST biases heading perception in monkeys. *Nat. Neurosci.* **1**(1), 59–63.

Britten, K.H., Shadlen, M.N., Newsome, W.T. *et al.*, 1993. Responses of neurons in macaque MT to stochastic motion signals. *Vis. Neurosci.* **10**, 1157–1169.

Britten, K.H., Newsome, W.T., Shadlen, M.N. *et al.*, 1996. A relationship between behavioral choice and the visual responses of neurons in macaque MT. *Vis. Neurosci.* **13**, 87–100.

Buettner, U. and Buettner, U.W., 1978. Parietal cortex (2v) neuronal activity in the alert monkey during natural vestibular and optokinetic stimulation. *Brain Res.* **153**, 392–397.

Burr, D.C., Morrone, M.C., and Ross, J., 2001. Separate visual representations for perception and action revealed by saccadic eye movements. *Curr. Biol.* **11**, 798–802.

Butzer, F., Ilg, U.J., and Zanker, J.M., 1997. Smooth-pursuit eye movements elicited by first-order and second-order motion. *Exp. Brain Res.* **115**, 61–70.

Carpenter, R.H.S., 1988. *Movements of the Eye*, Pion Limited, London.

Celebrini, S. and Newsome, W.T., 1994. Neuronal and psychophysical sensitivity to motion signals in extrastriate area MST of the macaque monkey. *J. Neurosci.* **14**(7), 4109–4124.

Celebrini, S. and Newsome, W.T., 1995. Microstimulation of extrastriate area MST influences performance on a direction discrimination task. *J. Neurophysiol.* **73**, 437–448.

Chiao, C.C. and Masland, R.H., 2003 Contextual tuning of direction-selective retinal ganglion cells. *Nat. Neurosci.* **6**, 1251–1252.

Chubb, C. and Sperling, G., 1988. Drift-balanced random stimuli: a general basis for studying non-Fourier motion perception. *J. Opt. Soc. Am. A* **5**, 1986–2007.

Churan, J. and Ilg, U.J., 2001. Processing of second-order motion stimuli in primate middle temporal area and medial superior temporal area. *J. Opt. Soc. Am. A Opt. Image Sci. Vis.* **18**, 2297–2306.

Collewijn, H., Martins, A.J., and Steinman, R.M., 1983. Compensatory eye movements during active and passive head movements: Fast adaptation to changes in visual magnification. *J. Physiol.* **340**, 259–286.

Collewijn, H. and Tamminga, E.P., 1984. Human smooth and saccadic eye movements during voluntary pursuit of different target motions on different backgrounds. *J. Physiol.* **351**, 217–250.

Derrington, A.M., Badcock, D.R., and Henning, G.B., 1993. Discriminating the direction of second-order motion at short stimulus durations. *Vision Res.* **33**, 1785–1794.

Ditterich, J., Mazurek, M.E., and Shadlen, M.N., 2003. Microstimulation of visual cortex affects the speed of perceptual decisions. *Nat. Neurosci.* **6**, 891–898.

Dursteler, M.R. and Wurtz, R.H., 1988. Pursuit and optokinetic deficits following chemical lesions of cortical areas MT and MST. *J. Neurophysiol.* **60**, 940–965.

Dursteler, M.R., Wurtz, R.H., and Newsome, W.T., 1987. Directional pursuit deficits following lesions of the foveal representation within the superior temporal sulcus of the macaque monkey. *J. Neurophysiol.* **57**, 1262–1287.

Filehne, W., 1922. Ueber das optische Wahrnehmen von Bewegungen. *Z. Sinnesphysiol.* **53**, 134–145.

Fischer, B. and Boch, R., 1983. Saccadic eye movements after extremely short reaction times in the monkey. *Brain Res.* **260**, 21–26.

Franz, V.H., Gegenfurtner, K.R., Bulthoff, H.H. *et al.*, 2000. Grasping visual illusions: No evidence for a dissociation between perception and action. *Psychol. Sci.* **11**, 20–25.

Gardner, J.L. and Lisberger, S.G., 2001. Linked target selection for saccadic and smooth pursuit eye movements. *J. Neurosci.* **21**(6), 2075–2084.

Gellman, R.S. and Carl, J.R., 1991. Motion processing for saccadic eye movements in humans. *Exp. Brain Res.* **84**, 660–667.

Glass, L., 1969. Moire effect from random dots. *Nature* **223**, 578–580.
Goodale, M.A. and Milner, A.D., 1992. Separate visual pathways for perception and action. *Trends Neurosci.* **15**, 20–25.
Goodale, M.A., Milner, A.D., Jakobson, L.S. *et al.*, 1991. A neurological dissociation between perceiving objects and grasping them. *Nature* **349**, 154–156.
Groh, J.M., Born, R.T., and Newsome, W.T., 1997. How is a sensory map read out? Effects of microstimulation in visual area MT on saccades and smooth pursuit eye movements. *J. Neurosci.* **17**, 4312–4330.
Gruesser, O.J., Pause, M., and Schreiter, U., 1990. Vestibular neurons in the parieto-insular cortex of monkeys (*Macaca fascicularis*): Visual and neck receptor responses. *J. Physiol.* **430**, 559–583.
Haarmeier, T. and Thier, P., 1996. Modification of the Filehne illusion by conditioning visual stimuli. *Vision Res.* **36**, 741–750.
Hawken, M.J., Parker, A.J., and Lund, J.S., 1988. Laminar organization and contrast sensitivity of direction-selective cells in the striate cortex of the old world monkey. *J. Neurosci.* **8**(10), 3541–3548.
Hoffmann, K.P. and Schoppmann, A., 1975. Retinal input to direction selective cells in the nucleus tractus opticus of the cat. *Brain Res.* **99**, 359–366.
Ilg, U.J. and Thier, P., 1996. Inability of rhesus monkey area V1 to discriminate between self-induced and externally induced retinal image slip. *Eur. J. Neurosci.* **8**, 1156–1166.
Ilg, U.J. and Thier, P., 1999. Eye movements of rhesus monkeys directed towards imaginary targets. *Vision Res.* **39**, 2143–2150.
Ilg, U.J. and Thier, P., 2003. Visual tracking neurons in primate area MST are activated by smooth-pursuit eye movements of an "imaginary" target. *J. Neurophysiol.* **90**, 1489–1502.
Kahlon, M. and Lisberger, S.G., 1996. Coordinate system for learning in the smooth pursuit eye movements of monkeys. *J. Neurosci.* **16**, 7270–7283.
Kahlon, M. and Lisberger, S.G., 1999. Vector averaging occurs downstream from learning in smooth pursuit eye movements of monkeys. *J. Neurosci.* **19**, 9039–9053.
Kalesnykas, R.P. and Hallett, P.E., 2002. Saccadic latency effects of progressively deleting stimulus offsets and onsets. *Vision Res.* **42**, 637–652.
Kawano, K., Shidara, M., Watanabe, Y. *et al.*, 1994. Neural activity in cortical area MST of alert monkey during ocular following responses. *J. Neurophysiol.* **71**(6), 2305–2324.
Keller, E.L. and Khan, N.S., 1986. Smooth-pursuit initiation in the presence of a textured background in monkey. *Vision Res.* **26**, 943–955.
Kimmig, H.G., Miles, F.A., and Schwarz, U., 1992. Effects of stationary textured backgrounds on the initiation of pursuit eye movements in monkeys. *J. Neurophysiol.* **68**, 2147–2164.
Knox, P.C., 1996. The effect of the gap paradigm on the latency of human smooth pursuit of eye movement. *Neuroreport* **7**, 3027–3030.
Komatsu, H. and Wurtz, R.H., 1989. Modulation of pursuit eye movements by stimulation of cortical areas MT and MST. *J. Neurophysiol.* **62**, 31–47.
Krauzlis, R.J. and Miles, F.A., 1996a. Decreases in the latency of smooth pursuit and saccadic eye movements produced by the "gap paradigm" in the monkey. *Vision Res.* **36**, 1973–1985.
Krauzlis, R.J. and Miles, F.A., 1996b. Release of fixation for pursuit and saccades in humans: Evidence for shared inputs acting on different neural substrates. *J. Neurophysiol.* **76**, 2822–2833.
Krauzlis, R.J., Basso, M.A., and Wurtz, R.H., 1997. Shared motor error for multiple eye movements. *Science* **276**, 1693–1695.
Krauzlis, R.J. and Stone, L.S., 1999. Tracking with the mind's eye. *Trends Neurosci.* **22**, 544–550.
Krauzlis, R.J. and Adler, S.A., 2001. Effects of directional expectations on motion perception and pursuit eye movements. *Vis. Neurosci.* **18**, 365–376.
Krauzlis, R.J., Dill, N., and Kornylo, K., 2002. Activity in the primate rostral superior colliculus during the "gap effect" for pursuit and saccades. *Ann. N.Y. Acad. Sci.* **956**, 409–413.
Krekelberg, B., Dannenberg, S., Hoffmann, K.P. *et al.*, 2003. Neural correlates of implied motion. *Nature* **424**, 674–677.
Lettvin, J.Y., Maturana, H.R., McCulloch, W.S. *et al.*, 1959. What the frog's eye tells the frog's brain. *Proc. IRE Nov* **59**, 1940–1951.
Lindner, A. and Ilg, U.J., 2000. Initiation of smooth-pursuit eye movements to first-order and second-order motion stimuli. *Exp. Brain Res.* **133**, 450–456.

Lindner, A., Schwarz, U., and Ilg, U.J., 2001. Cancellation of self-induced retinal image motion during smooth pursuit eye movements. *Vision Res.* **41**, 1685–1694.

Lisberger, S.G. and Westbrook, L.E., 1985. Properties of visual inputs that initiate horizontal smooth pursuit eye movements in monkeys. *J. Neurosci.* **5**, 1662–1673.

Lisberger, S.G. and Movshon, J.A., 1999. Visual motion analysis for pursuit eye movements in area MT of macaque monkeys. *J. Neurosci.* **19**, 2224–2246.

Lisberger, S.G., Morris, E.J., and Tychsen, L., 1987. Visual motion processing and sensory-motor integration for smooth pursuit eye movements. *Annu. Rev. Neurosci.* **10**, 97–129.

MacAvoy, M.G., Gottlieb, J.P., and Bruce, C.J., 1991. Smooth-pursuit eye movement representation in the primate frontal eye field. *Cereb. Cortex* **1**, 95–102.

Masson, G.S., Rybarczyk, Y., Castet, E. *et al.*, 2000. Temporal dynamics of motion integration for the initiation of tracking eye movements at ultra-short latencies. *Vis. Neurosci.* **17**, 753–767.

Maunsell, J.H.R. and Van Essen, D.C., 1983. The connections of the middle temporal visual area (MT) and their relationship to a cortical hierarchy in the macaque monkey. *J. Neurosci.* **3**(12), 2563–2586.

Merrison, A.F. and Carpenter, R.H., 1995. "Express" smooth pursuit. *Vision Res.* **35**, 1459–1462.

Mikami, A., Newsome, W.T., and Wurtz, R.H., 1986. Motion selectivity in macaque visual cortex I. Mechanisms of direction and speed selectivity in extrastriate area MT. *J. Neurophys.* **55**(6), 1309–1327.

Miura, K., Suehiro, K., Yamamoto, M. *et al.*, 2001. Initiation of smooth pursuit in humans. Dependence on target saliency. *Exp. Brain Res.* **141**, 242–249.

Mohrmann, H. and Thier, P., 1995. The influence of structured visual backgrounds on smooth-pursuit initiation, steady-state pursuit and smooth-pursuit termination. *Biol. Cybern.* **73**, 83–93.

Movshon, J.A. and Newsome, W.T., 1996. Visual response properties of striate cortical neurons projecting to area MT in macaque monkey. *J. Neurosci.* **16**(23), 7733–7741.

Nakayama, K., 1985. Biological image motion processing: A review. *Vision Res.* **25**(5), 625–660.

Newsome, W.T. and Paré, E.B., 1988. A selective impairment of motion perception following lesions of the middle temporal visual area MT. *J. Neurosci.* **8**(6), 2201–2211.

Newsome, W.T., Wurtz, R.H., and Komatsu, H., 1988. Relation of cortical areas MT and MST to pursuit eye movements. II. Differentiation of retinal from extraretinal inputs. *J. Neurophysiol.* **60**, 604–620.

O'Mullane, G. and Knox, P.C., 1999. Modification of smooth pursuit initiation by target contrast. *Vision Res.* **39**, 3459–3464.

Pack, C.C. and Born, R.T., 2001. Temporal dynamics of a neural solution to the aperture problem in visual area MT of macaque brain. *Nature* **409**, 1040–1042.

Rashbass, C., 1961. The relationship between saccadic and smooth tracking eye movements. *J. Physiol.* **159**, 326–338.

Reichardt, W., 1987. Evaluation of optical motion information by movement detectors. *J. Comp. Physiol. A* **161**, 533–547.

Rudolph, K. and Pasternak, T., 1999. Transient and permanent deficits in motion perception after lesions of cortical areas MT and MST in the macaque monkey. *cereb. cortex* **9**, 90–100.

Saslow, M.G., 1967. Effects of components of displacement-step stimuli upon latency for saccadic eye movement. *J. Opt. Soc. Am.* **57**, 1024–1029.

Schlack, A., Hoffmann, K.P., and Bremmer, F., 2003. Selectivity of macaque ventral intraparietal area (area VIP) for smooth pursuit eye movements. *J. Physiol.* **551**, 551–561.

Schwarz, U. and Ilg, U.J., 1999. Asymmetry in visual motion processing. *Neuroreport* **10**, 2477–2480.

Siegel, R.M. and Read, H.L., 1997. Analysis of optic flow in the monkey parietal area 7a. *Cereb. Cortex* **7**, 327–346.

Steinbach, M.J., 1976. Pursuing the perceptual rather than the retinal stimulus. *Vision Res.* **16**, 1371–1376.

Stone, L.S., Beutter, B.R., and Lorenceau, J., 2000. Visual motion integration for perception and pursuit. *Perception* **29**, 771–787.

Suehiro, K., Miura, K., Kodaka, Y. *et al.*, 1999. Effects of smooth pursuit eye movement on ocular responses to sudden background motion in humans. *Neurosci. Res.* **35**, 329–338.

Tanaka, M. and Lisberger, S.G., 2001. Regulation of the gain of visually guided smooth-pursuit eye movements by frontal cortex. *Nature* **409**, 191–194.
Thiele, A. and Hoffmann, K.P., 1996. Neuronal activity in MST and STPp, but not MT changes systematically with stimulus-independent decisions. *Neuroreport* **7**, 971–976.
Thier, P. and Erickson, R.G., 1992. Vestibular input to visual-tracking neurons in area MST of awake rhesus monkeys. *Ann. N.Y. Acad. Sci.* **656**, 960–963.
Tychsen, L. and Lisberger, S.G., 1986. Visual motion processing for the initiation of smooth-pursuit eye movements in humans. *J. Neurophysiol.* **56**, 953–968.
Ungerleider, L.G. and Desimone, R., 1986. Cortical connections of visual area MT in the Macaque. *J. Comp. Neurol.* **248**, 190–222.
Ungerleider, L.G. and Mishkin, M., 1982. Two cortical visual systems, in *Analysis of Visual Behavior* (eds D.J. Ingle *et al.*), MIT Press, Cambridge, MA, 549–586.
van Santen, J.P. and Sperling, G., 1985. Elaborated Reichardt detectors. *J. Opt. Soc. Am. A* **2**, 300–321.
von Holst, E. and Mittelstaedt, H., 1950. Das Reafferenzprinzip. *Naturwissenschaften* **37**, 464–476.
Wyatt, H.J., Pola, J., Fortune, B. *et al.*, 1994. Smooth pursuit eye movements with imaginary targets defined by extrafoveal cues. *Vision Res.* **34**, 803–820.
Yamasaki, D.S. and Wurtz, R.H., 1991. Recovery of function after lesions in the superior temporal sulcus in the monkey. *J. Neurophysiol.* **66**, 651–673.
Yee, R.D., Daniels, S.A., Jones, O.W. *et al.*, 1983. Effects of an optokinetic background on pursuit eye movements. *Invest. Ophthalmol. Vis. Sci.* **24**, 1115–1122.
Yo, C. and Wilson, H.R., 1992. Perceived direction of moving two-dimensional patterns depends on duration, contrast and eccentricity. *Vision Res.* **32**, 135–147.
Zanker, J.M., 1993. Theta motion: A paradoxical stimulus to explore higher order motion extraction. *Vision Res.* **33**, 553–569.
Zeki, S.M., 1974. Functional organization of a visual area in the posterior bank of the superior temporal sulcus of the rhesus monkey. *J. Physiol.* **236**, 549–573.

11

Psychophysical Correlates of Identified Physiological Processes

Annette Werner, Joel Pokorny, Vivianne C. Smith, Arne Valberg, Jan Kremers, and *Mark W. Greenlee*

11.1 Introduction

First introduced in 1860 by the German physicist Gustav Fechner in his 'Elemente der Psychophysik' ('Elements of psychophysics'), psychophysics has a long tradition as a tool for the functional investigation of sensory systems (Fechner, 1860). In its original sense it means the study of sensations evoked by physical stimuli, and refers to the methods and models that establish a quantitative relation between subjective perception and the physical stimulus/stimuli that produce that sensation. Over the past 150 years this has provided vision research with many important concepts of signal processing, such as spatial and temporal summation, adaptation and contrast. Furthermore, it uncovered influences from higher-order mechanisms such as attention on visual perception and provided quantitative descriptions of complex visual phenomena, for example, color constancy or the perception of object motion.

The inherent problem in psychophysics is the fact that we cannot measure sensations directly, nor can we communicate our sensations: for example, the redness of an apple may have a different quality between two observers, nevertheless they would both call it red. However, we can measure indirectly sensations or rather the effect on sensations by manipulating stimuli, detection thresholds, or judgments about similarity or non-similarity of stimuli.

The combination of psychophysics and other methods of investigation, such as electrophysiology and, most recently, imaging techniques, has been extremely fruitful for our

The Primate Visual System: A Comparative Approach Edited by Jan Kremers

understanding of how the visual system works. This is true for the system as a whole, as well as its subunits, which can be teased apart and isolated by particular psychophysical techniques.

In this chapter we will describe how psychophysics can be used successfully to characterize the properties of the retino-geniculate (PC, MC, and KC) pathways and how their signals can be modeled in order to account for low-level color computations. At the cortical level, psychophysical phenomena of color and motion perception will be related to the neural activity in identified visual areas and pathways.

11.2 Psychophysical correlates of retino-geniculate pathways

11.2.1 Processing of achromatic information in PC and MC pathways

Here, we consider how psychophysically measured visual function may be related to the activities of single neurons in the primate retino-geniculate pathways. The parvocellular (PC), magnocellular (MC), and koniocellular (KC) pathways are the three major neural pathways conveying information from the retina to the brain (Chapters 5 and 6). The list is not exhaustive, but the cell types of these pathways are the cells usually sampled in retinal and lateral geniculate extra-cellular recordings.

For psychophysicists, the delineation of the chromatic properties of the PC pathways (Chapter 5) first attracted attention, with spectral opponency being of particular interest. Chromatic discrimination was shown to be determined in the PC and KC pathways and chromatic discrimination can be described by models of the activities in these pathways (Zaidi *et al.*, 1992; Smith *et al.*, 2000; see also this chapter). By using equiluminant chromatic modulation, the PC pathway can be isolated in psychophysical studies to a great degree. In these studies, temporal (Kelly, 1974; Swanson *et al.*, 1987; Lee *et al.*, 1990) and spatial (Mullen, 1985; Sekiguchi *et al.*, 1993) tuning show low-pass features. In the temporal domain, however, physiological recordings of single cells in the PC pathway showed much higher temporal resolution than did the psychophysical data (Lee *et al.*, 1990).

It became popular to examine psychophysical functions at equiluminance (Livingstone and Hubel, 1987; Cavanagh, 1991). If thresholds were high for detecting equiluminant stimuli or the percept was diminished, it was concluded that the percept was mediated by the MC pathway. Further, some investigators tended to assume that the MC-pathway processes all luminance stimuli, while the PC-pathway processes only color stimuli. This approach has proved quite dangerous since it ignores the achromatic responsivity of the PC pathway, an issue we consider in this section. Another potentially risky supposition was to assume that the MC-pathway neurons would mediate perception of low-contrast stimuli (Livingstone and Hubel, 1987; Westheimer *et al.*, 1999). This conclusion was based on the large PC- versus MC-pathway difference in contrast gain noted in single-cell physiological response data. However, it is now recognized that summation of the outputs of multiple retinal neurons in the cortex can change the absolute sensitivity of the system without changing the slope of the input contrast gain function (Maunsell, 1992).

The question of which pathway is more sensitive for a given percept or stimulus requires experimental evaluation.

There is a caveat to the use of equiluminance stimulation to silence the MC pathway. Electrophysiological studies indicate that MC-pathway neurons are not completely silenced at equiluminance (Schiller and Colby, 1983; Lee *et al.*, 1989; Kaiser *et al.*, 1990; Dobkins and Albright, 1995); rather they show a frequency-doubled response. Furthermore, individual neurons may have slightly different equiluminance points. The residual response has the potential to be detected in psychophysical threshold measurements under certain stimulus situations that were intended to measure PC-pathway responses (Mollon, 1982; Swanson *et al.*, 1987).

11.2.2 Discrimination of achromatic contrast in the PC and MC pathways

Kaplan and Shapley (1986) pointed out the large difference in the responses of MC- and the PC-pathway cells to changes in stimulus luminance contrast. Presented with a steady luminance stimulus, a cell obtains a constant resting level. A luminance modulation created by drifting a sinusoidal grating across the cell's receptive field, or by temporally modulating a uniform field, elicits a time-locked amplitude response in impulses per second at the Fourier fundamental, with the phase being dependent on the center sign (ON or OFF) of the cell. For the MC pathway, the amplitude response initially shows a very rapid growth with increase in stimulus contrast, but then levels off yielding little change in response amplitude with further increases in contrast. In comparison, the PC-pathway amplitude response shows a steady, almost linear growth with increases in stimulus contrast (see also Chapters 5 and 6).

Pokorny and Smith (1997) developed a set of three psychophysical paradigms that allow measurement of achromatic contrast discrimination in the PC and MC pathways. Contrast thresholds were measured to pulses that were incremented or decremented from the average retinal illuminance. This was done to allow psychophysical differentiation of thresholds mediated by ON and OFF pathways. This work has provided a new view of the function of the achromatic PC pathway. In all three psychophysical paradigms, the stimulus array consisted of four 1° squares with small separations, and the set of four squares appeared in a larger uniform surround.

Paradigm 1, the pulsed-pedestal condition: The observer viewed a uniform field. The stimulus array consisted of four squares that appeared only during the trial period as a pulsed pedestal, with the test square at a higher or a lower retinal illuminance than the other three (Figure 11.1, upper panel). The observer maintained adaptation to the uniform field between trials.

Paradigm 2, the steady-pedestal condition: The four squares were present as a steady pedestal within the surround during the entire protocol and the observer adapted for 1 min to the surround-plus-steady-pedestal stimulus before a measurement began. In this paradigm only the retinal illuminance of the test square changed during the trial period (Figure 11.1, center panel).

Paradigm 3, the pedestal-Δ-pedestal condition: The protocol was identical to the steady-pedestal condition in all respects except that during the trial, the retinal illuminance of all

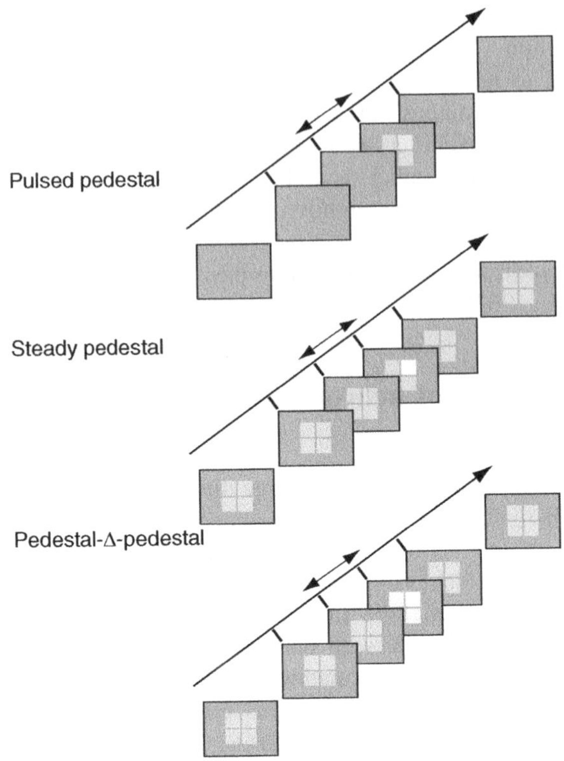

Figure 11.1 Top panel, the pulsed-pedestal condition: The four squares appeared only during the trial period as a pulsed pedestal, with the test square at a higher or a lower retinal illuminance than the other three. Middle panel, the steady-pedestal condition: The four squares were continuously present, the retinal illuminance of the test square changed during a trial. Bottom panel, the pedestal-Δ-pedestal condition: The protocol was identical to the steady-pedestal condition in all respects except that during the trial the retinal illuminance of all the four squares was incremented or decremented (a pedestal-Δ-pedestal), with the test square incremented or decremented by a different amount

the four squares was incremented or decremented (a pedestal-Δ-pedestal), with the test square incremented or decremented by a different amount than the other three squares (Figure 11.1, lower panel).

Figure 11.2 shows the results for each of the three paradigms with a brief (33 ms) pulse presentation. The open squares with a dashed line show results for the pulsed-pedestal paradigm. The thresholds depended on the contrast of the pedestal array and the data are well described by a model fit of a saturation equation (dashed line) that has been used to characterize the contrast response of single cells in the PC pathway (Pokorny and Smith, 1997). The slopes of the fits to the psychophysical data were in the range of slopes for individual PC-pathway cell responses measured by Kaplan and Shapley (1986). There was a discontinuity at the surround retinal illuminance of 115 td (arrow). The data point at 115 td indicates a larger sensitivity than predicted by the model V-shape.

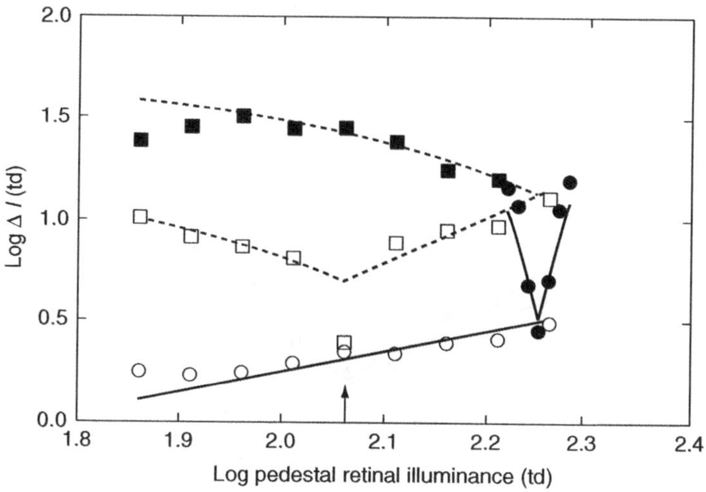

Figure 11.2 Data for the three paradigms with a 33-ms pulse presentation. The open squares show data for the pulsed-pedestal paradigm. The dashed line is a model fit of a saturation equation (Pokorny and Smith, 1997). Open circles show the steady-pedestal paradigm data. The solid line fitting these data had a slope of unity. Closed circles show pedestal-Δ-pedestal data for six small Δ pedestals near 178 td, plotted as a function of total pedestal illuminance. The solid and dashed lines are described in the text (Pokorny and Smith, 1997)

The open circles in Figure 11.2 show data for the steady-pedestal paradigm. Under these stimulus conditions, the thresholds increased monotonically with the pedestal retinal illuminance. The solid line fitting these data had a slope of unity with a single vertical scaling constant. These data were quite different from the pulsed-pedestal data and the result was interpreted to indicate that a different mechanism was mediating the thresholds. The data indicated that this mechanism was yielding thresholds consistent with adaptation to the steady-pedestal luminance and was not affected by the spatio-temporal contrast signals at the edges of the stimulus array. The MC pathway was postulated to mediate the steady-pedestal data for the brief pulse and the MC threshold was raised by saturation in the pulsed-pedestal paradigm.

The closed circles in Figure 11.2 show data for six small Δ-pedestals pulsed from a steady pedestal near 178 td, plotted as a function of total pedestal illuminance. These results showed a very steep V-shape. The solid line fitted to these data was from the saturation equation, and the slopes were in the range of slopes for individual MC-pathway cells measured by Kaplan and Shapley (1986). Closed squares show data for seven larger Δ-pedestals, pedestal steps expected to saturate the MC-pathway response. The data for these pedestals were fitted (dashed line) with the saturation equation and the slopes used for the pulsed-pedestal data. The data and fits indicated a change in pathway sensitivity as the Δ-pedestal increased. Of note, pulsed-pedestal contrasts of 0.02 log unit were sufficient to saturate the MC pathway for a brief period following pulse onset.

The experiment was replicated using a 1.5-s raised cosine stimulus. This is a stimulus designed to have no abrupt transients at onset or offset, which was chosen to favor the PC pathway. Figure 11.3 shows that a V-shape is obtained for the pulsed-pedestal

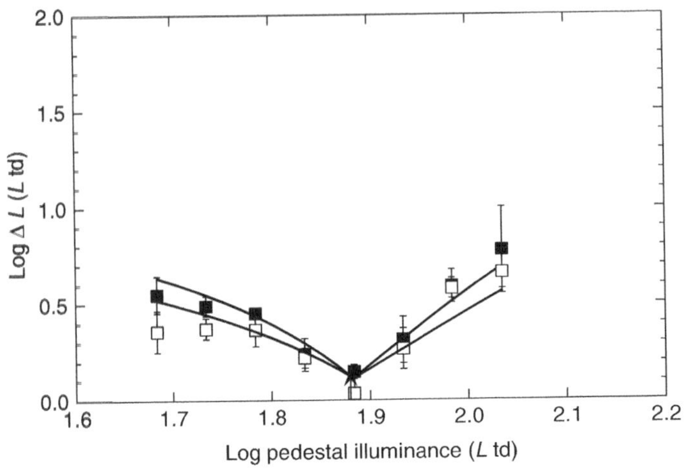

Figure 11.3 Pulsed- and steady-pedestal data for a 1.5-s raised cosine stimulus. The V-shape fits are described in the text (Smith and Pokorny, 2003)

paradigm, with the saturation equation fit revealing the same slope value as for the data fits in Figure 11.2. The steady-pedestal paradigm also yields a V-shape with similar slope values. When the stimulus favors the PC pathway, thresholds are determined by the contrast edges between the pedestal and the background.

11.2.3 Temporal and spatial summation

To examine the temporal and spatial summation properties of the two pathways, data were gathered using the pulsed- and steady-pedestal paradigms for a set of stimuli of four durations (Pokorny and Smith, 1997) and six array sizes (Smith *et al.*, 2001). Varying duration or array size affected only the overall sensitivity without an effect on the curve forms as shown in Figure 11.2. The thresholds are shown in Figure 11.4 as a function of stimulus duration (upper panel) and array size (lower panel); temporal summation is complete within 80 ms for the steady-pedestal paradigm (favoring the MC pathway) and extends beyond 200 ms for the pulsed-pedestal paradigm (favoring the PC pathway). Spatial summation is complete within 1° square for the steady-pedestal paradigm but extends continuously for the pulsed-pedestal paradigm. These data need be accounted for with post-retinal processing.

11.2.4 The temporal recovery from saturation

The data collected using the pulsed-pedestal paradigm saturated the MC pathway at very low pulse contrasts. It was of interest to consider the recovery from saturation. A 145-td pedestal of the four square array was thus presented for a 1000-ms period on the 115-td surround. The test of 25.6-ms duration was then presented before, during, and after the pedestal presentation. The pulse size allowed a log Δ I difference of 0.1 log unit. The data for one observer are shown in Figure 11.5 where the abscissa shows the time from the pedestal onset (stimulus onset asynchrony) and the ordinate shows the threshold in log Δ td (Pokorny *et al.*, 2003). The thresholds for the increment and decrement

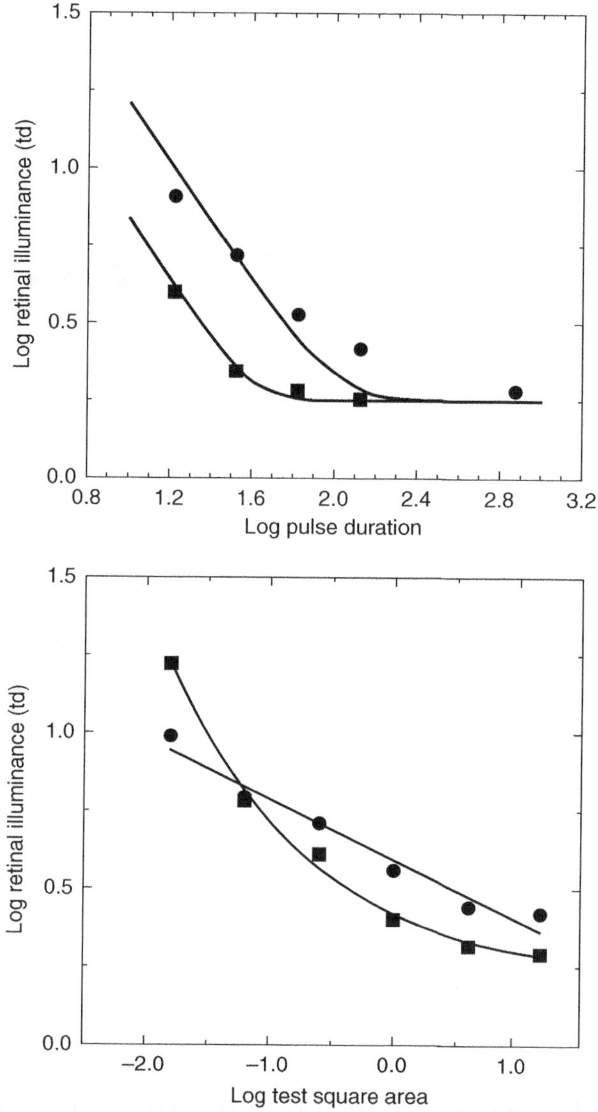

Figure 11.4 Temporal (upper panel) and spatial (lower panel) summation data gathered using the pulsed- and steady-pedestal paradigms. The upper panel includes data for four durations (Pokorny and Smith, 1997); the lower panel has data for six array sizes (Smith *et al.*, 2001)

stimulus conditions are plotted separately. For the increment condition, the thresholds are raised before the pedestal onset but are coincident with the pedestal offset. For the decrement condition, the thresholds are raised at pedestal onset but are raised before the pedestal offset. The threshold increase was about 0.7–0.8 log unit. Both increment and decrement thresholds return to steady state with an exponential decay with a time constant near 25 ms.

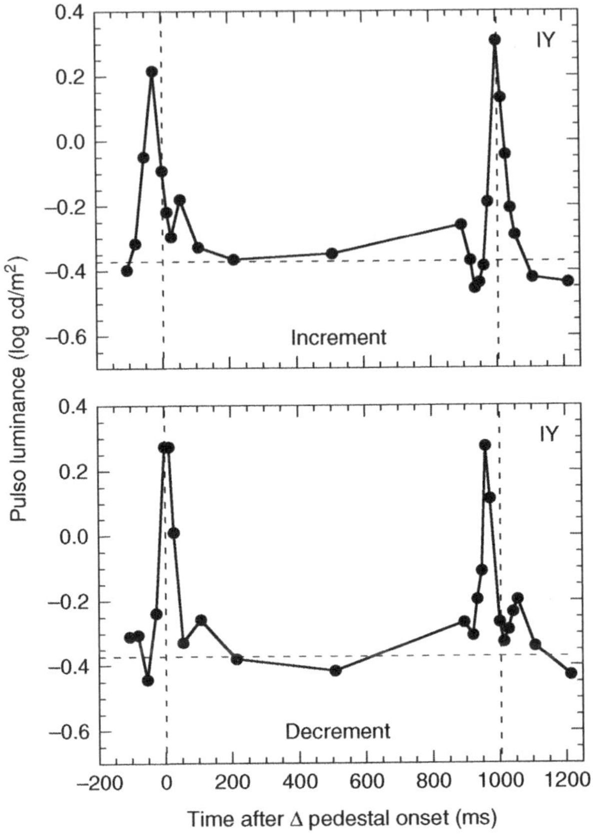

Figure 11.5 The temporal recovery from saturation. Data for a 145-td pedestal of the four square array presented for a 1000-ms period on the 115-td surround. The test of 25.6-ms duration was presented before, during, and after the pedestal. The abscissa shows the time from the pedestal onset and the ordinate shows the threshold in log Δ td. The thresholds for the increment and decrement staircases are plotted separately (Pokorny *et al.*, 2003)

11.2.5 Spatial frequency responses in the MC and PC pathways

Interest developed in assessing the spatial frequency response of the two presumed pathways. The spatial modulation transfer function has been modeled as a sum of four or more underlying spatial frequency channels of varying spatial frequency peaks (Wilson and Bergen, 1979). This paradigm included a large uniform 4° square pedestal and a centrally located test defined by a D6, a sixth derivative of a Gaussian (Swanson *et al.*, 1984). A D6 stimulus presents a spatially localized pattern while having a relatively narrow one-octave spatial frequency bandwidth. As the large uniform pedestal is a source for adaptation for both the PC and the MC pathways, the steady-pedestal paradigm will not allow easy distinction between the two pathways. The pulsed-pedestal paradigm will, however, cause a brief saturation in the MC pathway. A 26.7-ms test was used to look at presumed MC-pathway processing and then a 1-s raised cosine designed to favor presumed PC-pathway processing. Peak spatial frequency was varied in the horizontal direction from 0.25 to 16 cpd, with a fixed Gaussian vertical envelope of 2.5°. The data for a brief 26.7-ms pulse are shown in the upper panel of Figure 11.6. The pulsed-pedestal paradigm yields a band-pass function with a peak at 2 cpd. The steady-pedestal paradigm gives a low-pass function. The two functions show similar sensitivity at 1–4 cpd, and the pulsed-pedestal paradigm shows greater sensitivity at 8 and 16 cpd. The data for a 1-s raised cosine are shown in the lower panel of Figure 11.6. Both pulsed- and steady-pedestal paradigms showed band-pass functions of similar sensitivity. Some overall increase in sensitivity was found compared with the 26.7-ms pulse (lowermost function).

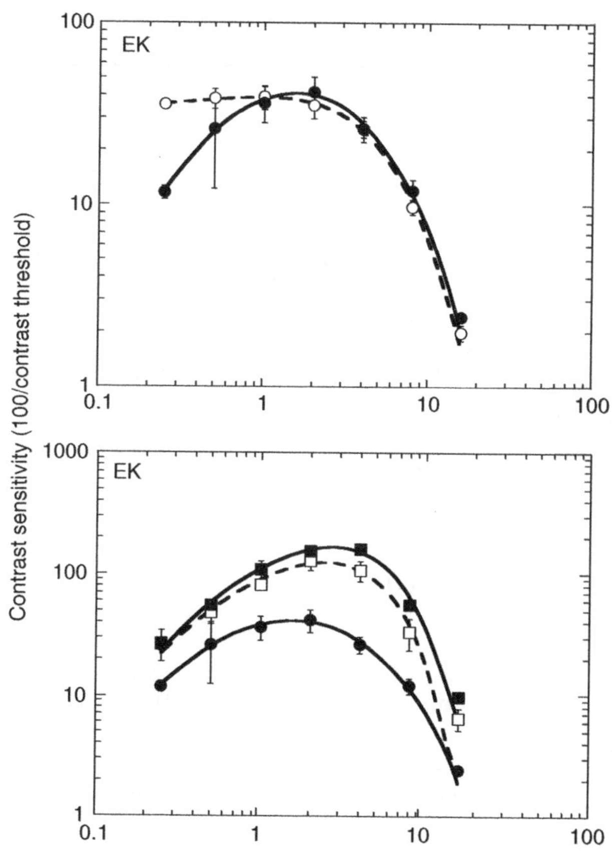

Figure 11.6 Spatial contrast sensitivity functions for pulsed- and steady-pedestal presentation of D6 patterns. Pulsed-pedestal data are shown by open circles and dashed lines, steady-pedestal data by filled circles and solid lines. The upper panel shows data for a presentation duration of 26.7 ms, the bottom panel shows data collected with a 1-s raised cosine (Leonova *et al.*, 2003)

Additionally, however, there was a shift in peak sensitivity to 4 cpd. These results suggested the recruitment of higher spatial frequency channels with increased duration.

11.2.6 Summary

The data from the pedestal paradigm experiments offer insight into the role that the PC pathway plays in achromatic vision. Table 11.1 provides a summary of the response properties of PC- and MC-pathway cells and human observer functions for a variety of experimental manipulations. The results of these studies suggest that psychophysical behavior can be said to mimic cell behavior for stimulus situations where the factors defining performance are retinal; the behavior of cells and observers diverge when post-retinal factors define observer performance.

Table 11.1 Properties of single PC- and MC- pathway cells and psychophysical responses to achromatic stimuli

	PC pathway	MC pathway
Contrast gain		
Cells	Low	High
Psychophysics	Low	High
Pulsed pedestal		
Cells	Depends on contrast	Saturates, recovery within 100 ms
Psychophysics	Depends on contrast	Saturates, recovery within 100 ms
Steady pedestal		
Cells	Less than Weber adaptation	Weber adaptation
Psychophysics	Less than Weber adaptation	Weber adaptation
Critical duration (1000 td)		
Cells (ms)	<50 ms	<50
Psychophysics (ms)	>100 ms	<50
Most effective temporal presentation		
Cells	Extended	Transient
Psychophysics	Extended	Transient
Luminance contrast sensitivity		
Cells	Low	High
Psychophysics	High	High
Spatial summation		
Cells	Local	Local
Psychophysics	Continuous	Power Law
Threshold increment/ decrement identification		
Cells	Yes	Yes
Psychophysics	Yes	No
Spatial modulation transfer function		
Cells	Band-pass, high resolution	Band-pass, high resolution
Psychophysics	Low-pass, limited resolution	Band-pass, high resolution

11.3 Modeling low-level color vision processing

Chromatic information is processed over multiple stages in the visual pathways from retina to the cortex. Color is only indirectly related to the spectral distribution of the light reflected from surfaces or emitted by light sources and the subsequent excitations of cone receptors (see also Chapter 4). Modeling the combination of the receptor signals of L, M, and S cones in three opponent channels (as they were identified anatomically, physiologically – see Chapters 5 and 6 – and psychophysically) is an important first step toward our understanding of color vision.

Here we describe a relatively simple model of some basic color phenomena based upon physiological data. The model starts with the nonlinearity of the cone intensity response and takes differences of cone outputs to form six types of retinal and geniculate opponent cells. Further combinations of outputs of geniculate opponent cell types give a satisfactory model of a variety of data, including color differences and scaling, constant hue perception, the Abney effect, and the Bezold–Brücke phenomenon.

11.3.1 The opponent color code

The MC cells of the primate retina and LGN are the physiological substrate for luminance detection thresholds (Lee *et al.*, 1988). Although MC cells are sensitive to the distinctness of borders between adjacent chromatic areas via a rectifying |M–L| cone-opponent mechanism in their receptive field (Valberg *et al.*, 1992), they are not likely to play a major role in color vision. They mediate the detection of mid-spectral lights (Valberg and Lee, 1989), and they contribute to several other temporal and spatial detection tasks (Lee *et al.*, 1990; Chapter 5).

Cone opponency in retinal ganglion cells is an efficient code for transmitting information from the retina to the brain. Subtracting the signals of one cone type from the signals of another overlapping one removes much of the information common to both cone types, reducing redundancy (Buchsbaum and Gottschalk, 1983; see also Chapter 4). In addition, cone-opponent PC and KC cells in the primate retina and LGN carry chromatic information (but as discussed above, the PC pathway is also involved in supra-threshold achromatic vision). Together, typical responses of ON and OFF PC cells and KC cells to chromatic lights span a 5 log unit range of relative intensity, covering the range commonly encountered in a natural environment.

Modeling the responses of PC cells

PC cells typically combine the responses of two cone types into 'L–M' or 'M–L' opponencies and can be divided into two types, the ON and OFF cells (Wiesel and Hubel, 1966), with an excitatory or inhibitory center response, respectively. The responses, N, of the four opponent classes of PC cells can be modeled by combining the outputs of the cones (given by their membrane potentials, V) in the following way:

$$N_{\mathrm{L-M}} = A_{\mathrm{L}} V_{\mathrm{L}} - A_{\mathrm{M}} V_{\mathrm{M}} (-A_{\mathrm{S}} V_{\mathrm{S}}) + N_o \text{ and } N_{\mathrm{M-L}} = B_{\mathrm{M}} V_{\mathrm{M}} - B_{\mathrm{L}} V_{\mathrm{L}} (+B_{\mathrm{S}} V_{\mathrm{S}}) + N_o$$

where N_o represents the maintained firing activity, and the constants A and B account for the relative weights of the cone inputs (cone excitations having been normalized to equal values for equal energy white). For ON cells $A_{\mathrm{L}} > A_{\mathrm{M}}$ and $B_{\mathrm{M}} > B_{\mathrm{L}}$, and for

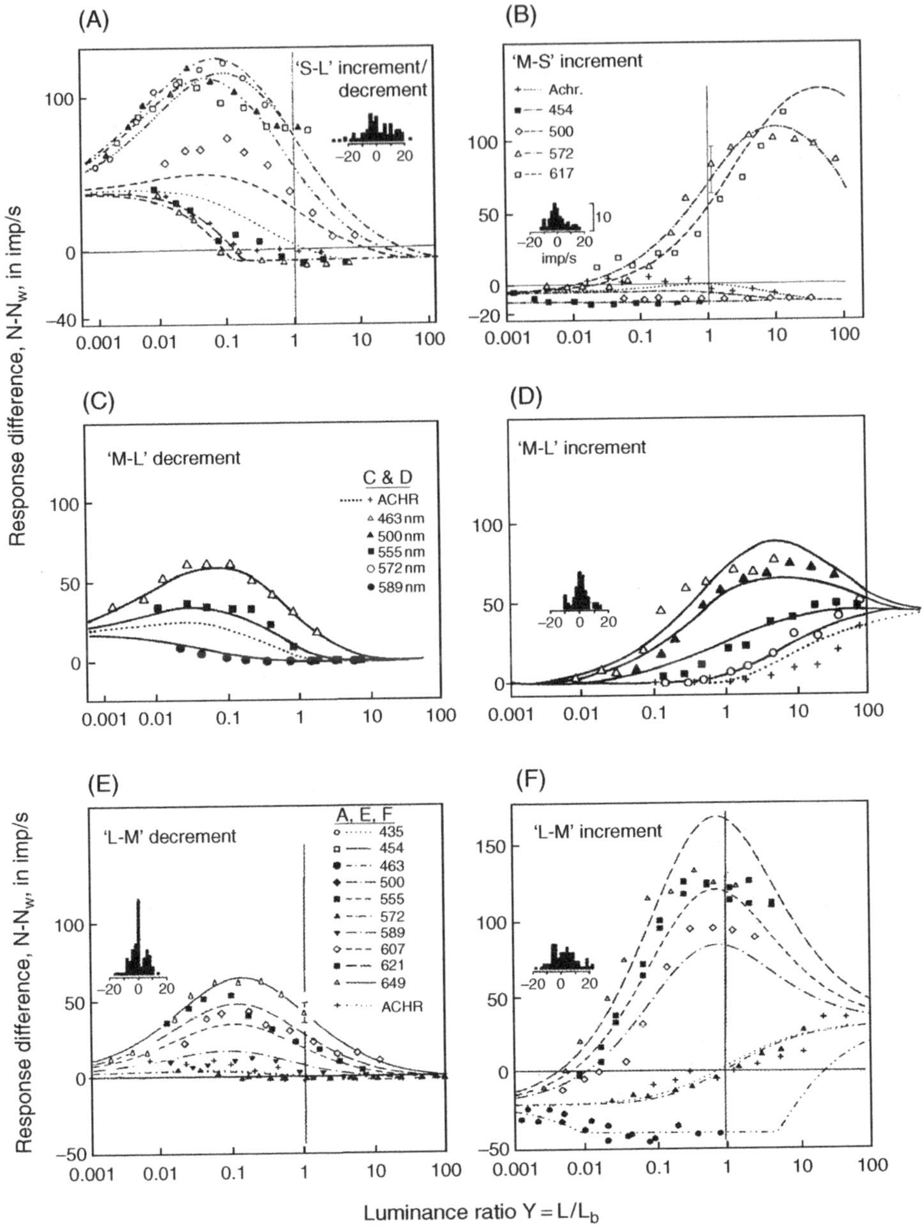

Figure 11.7 Examples of the I-R curves of two KC cells and four PC cells to chromatic and achromatic lights. ON and OFF cells are denoted *Increment* and *Decrement* cells, respectively. The symbols represent the measured firing rate in impulses/s for a given wavelength and luminance ratio between stimulus (4° size, 0.3-s duration) and a white reference (adaptation; same size, 1.2-s duration) field. The curves are derived from a mathematical simulation of these responses, based on a linear combination of receptor potentials. More precisely, the responses of 'M–L' cells can

the more strongly inhibited OFF cells $A_L < A_M$ and $B_M < B_L$ (Derrington *et al.*, 1984; Lee *et al.*, 1987; Valberg *et al.*, 1987). The cone potentials, V, are described by the common nonlinear, hyperbolic function of the cone excitations L, M, and S, being $V_M = (V_{max} \cdot M^n)/(M^n + \sigma_M^n)$ for the M cone. σ_M is the half-saturation constant determined by the state of light adaptation, and the exponent $n = 0.7$. Observe that the same function is used to describe the output of retinal ganglion and LGN cells (Chapter 6). In some instances, an additional inhibitory S-cone input to 'L–M' ON cells and an excitatory S-cone input to 'M–L' ON cells improved the description.

Modeling the response of KC cells

The activity of the two KC cell types (receiving inhibitory or excitatory S-cone input, respectively; Chapter 5) can be modeled by the following combination of cone outputs:

$$N_{M-S} = C_M V_M - C_S V_S(-C_L V_L) + N_o \text{ and } N_{S-L} = D_S V_S - D_L V_L(+D_M V_M) + N_o$$

As for the PC cells, the ratios of the constants (C_M/C_S, D_S/D_L, and D_S/D_M) determine the relative strengths of activation and inhibition. The particular 'Yellow-ON' cell of Figure 11.7B, for instance, could be modeled by an 'M–S' cone combination, but others would need an additional L-cone inhibition (Valberg *et al.*, 1985a, 1986a).

The signals that are transmitted to area V1 by opponent cells are thus directly proportional to the weighted differences of cone potentials. This means that all synaptic processes and interactions in the retina, between photoreceptors, bipolar cells, and ganglion cells, can be lumped together and incorporated in the constants of the equations.

Figure 11.7A–F shows a comparison of recorded data with intensity-response(I-T) curves derived from the model equations (Valberg *et al.*, 1985a, 1986b, 1987). Color stimuli (of 4° size and 0.3-s duration) were exchanged with a white reference or adaptation field (same size and 1.2-s duration), both surrounded by a white adaptation surround. N_W is the response to the repeated white light with a constant luminance $L_b = 100\,\text{cd/m}^2$. Panels A and B are for KC cells (an 'S–L' and an 'M–S' cell), and panels C and D for an ON and an OFF, 'M–L' PC cell, respectively. Similar results for ON and OFF, 'L–M' cells are shown in Figure 11.7E–F. The different symbols represent the responses to different test wavelengths. Nonlinear responses are typical for all cell types within the range of stimulus intensities (of related and unrelated colors; section 11.3.2) encountered in a normal daylight environment. Comparing the responses of OFF and ON PC cells of Figure 11.7C–F we see that for a luminance ratio of stimulus to white light below 1, the responses of the ON cells to white generally increase for luminance *increments*, whereas the OFF cell response increases for luminance *decrements*.

Figure 11.7 *(continued)*
be expressed as $N_{M-L} = B_M V_M - B_L V_L + N_o$, where B_M and B_L are weighting factors that represent the strengths of activation and inhibition by the two opposing cone types. N_o is the empirically determined maintained activity. The small insets within each panel show histograms of the number of deviations between the data points and the mathematical simulation in impulses/s. The rms deviation between data and simulation was normally between 5 and 10 percent of the maximum response of each cell (Valberg, 2005). Reproduced by permission of John Wiley & Sons Ltd

Opponent cells have been demonstrated to adjust their response and sensitivity to steady bright surrounds (Valberg *et al.*, 1985b), and the responses used in the model are representative for those evoked by related colors (with a white surround). In order to understand the different contributions to daylight color vision by ON and OFF cells, the distinction between related and unrelated colors is essential.

11.3.2 Related and unrelated colors

As seen in Figure 11.7, the strongly inhibited OFF cells differentiate between wavelengths for stimuli of low-luminance ratios, as for the related colors. Such colors are perceived on surfaces with reflection factors lower than 1. They are darker than white and contain an amount of induced blackness and are therefore characterized by their lightness (they reflect more or less light). Such colors are the most frequent in a natural environment, where color appearance is strongly dependent on simultaneous contrast. ON cells give spectrally selective responses to higher luminance ratios, albeit with loss of opponency at the highest levels. This qualifies ON cells as candidates for chromatic discrimination of brighter surfaces, e.g. of light sources and other objects emitting light (they are characterized by their brightness and are also called void colors). For a given state of light adaptation, lightness and brightness thus separate colored surfaces into two different domains. Together these domains span four to five decades of relative intensity. The need for more than one coding unit therefore seems natural.

11.3.3 Connecting the model to color perception

As demonstrated in Figure 11.7, responses of opponent cells can be described as proportional to simple differences of cone potentials. In order to establish connections to color perception, we need to go beyond the coding principles in single cells and look for possible ways to combine the outputs of retino-geniculate opponent cells. In the model it is assumed that opponent cells adapt to an extended light surround (the condition of related colors), and that the responses of ON and OFF cells with the same cone opponency are summed. Signals from 'L–M' and 'M–L' cells on the one hand and 'M–S' and 'S–L' cells on the other hand are believed to be handled independently. This leads to the assumption that the perception of a particular hue corresponds to a specific ratio of response magnitudes of independent opponent cells, irrespective of their absolute response (e.g. perceiving a certain orange hue implies that the ratio of the responses of e.g. 'M–S' and 'L–M' cells is constant). Further postulates are that the chromatic strength of a stimulus corresponds to a vector sum of responses of opponent cells, and that the response to achromatic light can be subtracted in a way that allows us to define a pure 'chromatic response'. Let us look at how this model performs on data on color scaling and color appearance.

Color scaling

First, the predictions of the model will be compared with the scaling data of the Munsell color ordering system. In this widely used system, surface colors of painted papers have been ordered by observers according to the dimensions hue, chromatic strength (called 'chroma'), and lightness (called 'value'). Figure 11.8 shows the result for the color chips in a section through the color solid of a constant red hue and an equal spacing

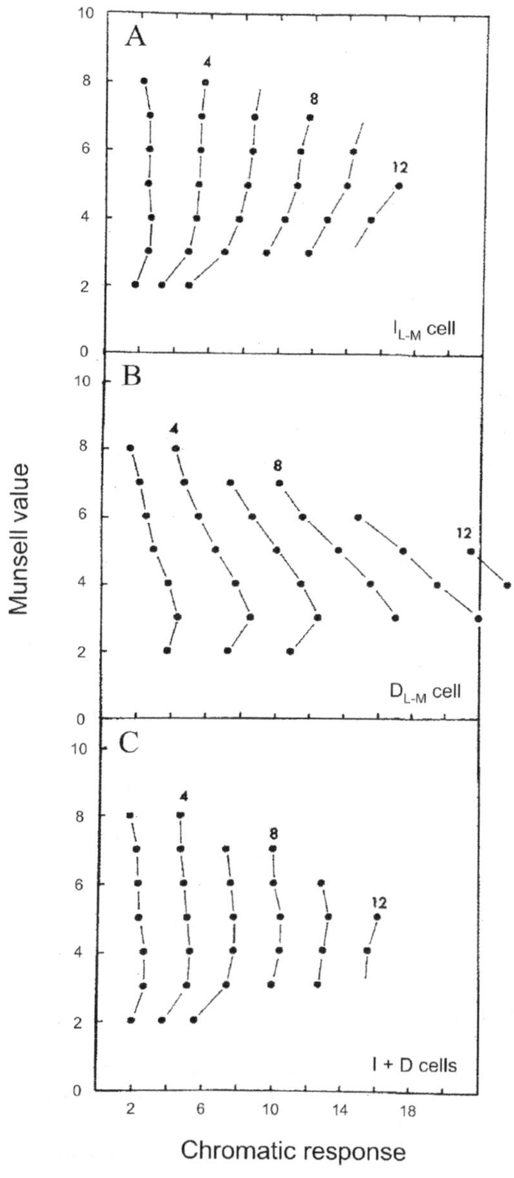

Figure 11.8 (A and B) Model responses of macaque ON and OFF cells (I- and D-cells) to the red stimuli of Munsell hue 5R, varying in chroma and value (lightness). I-cells (A) display a higher firing rate to light colors of high Munsell value, while D-cells (B) respond better to dark colors of low value. The Munsell value axis is a nonlinear luminance ratio scale (×10), and the chromatic response is proportional to the response difference of Figure 10.4. (C) The averaged sum of the responses of A and B gives close to equal steps of chroma for the different lightness levels 3–8 (Valberg, 2005). Reproduced by permission of John Wiley & Sons Ltd

of lightness and chroma. In Figure 11.8A and B are shown the chromatic response amplitude (horizontal axis) of a macaque opponent ON and an OFF cell to all Munsell chips of the red color 5R are shown. Filled circles connected with lines represent different, evenly spaced lightness values of constant chroma. As expected from Figure 11.7, ON cells are more responsive to the lighter colors of high Munsell value (Munsell value 10 corresponds to a luminance ratio $Y = 1.0$ in Figure 11.7), leading to tilted isochroma lines with positive slopes, while OFF cells are more responsive to darker colors of lower value, giving tilted isochromatic lines with negative slopes. The summed and averaged responses of these two types of cells, shown in Figure 11.8C, yield lines of constant chroma responses parallel to the achromatic axis and with approximate equal response increments for the same increments in chroma (Valberg *et al.*, 1986b).

This and similar results for other hues show that the combined responses of $ON_{L-M} + OFF_{L-M}$ and $ON_{M-L} + OFF_{M-L}$ cells to related chromatic colors correlate extremely well with the perception of equal steps of chroma. This combination yields a relatively uniform chroma scale for all lightness values. We therefore hypothesize that such a combination of ON and OFF cell responses might occur at the next level of processing after the LGN, and we use this hypothesis to construct the (p_1, p_2) coordinate system of Figure 11.9.

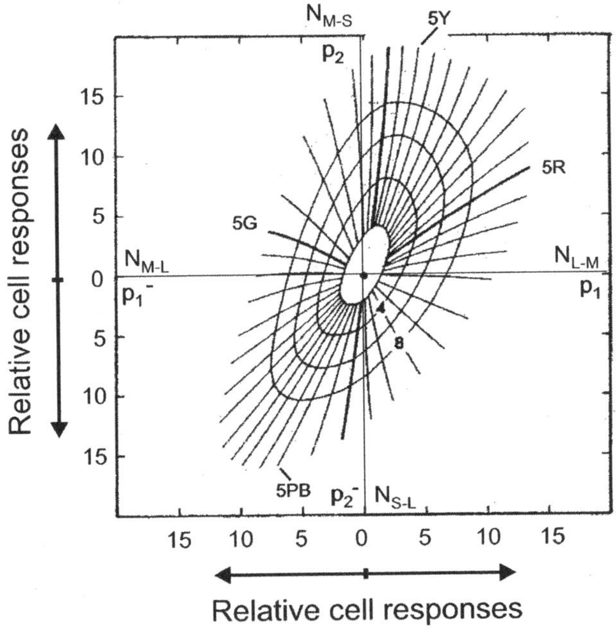

Figure 11.9 The combination of the model chromatic responses, *N*, of the six macaque opponent LGN cell types mentioned in the text to color stimuli of the Munsell color ordering system. The stimuli varied in hue and chroma but had constant lightness (equal to Munsell value 5). The locus of responses to chips of constant hue and varying chroma is approximately a straight line from the origin. Equal chroma steps from the achromatic stimulus (center) around the hue circle are represented by quasi-elliptical loci (Valberg, 2005). Reproduced by permission of John Wiley & Sons Ltd

Constant hues and the Abney effect

In Figure 11.9, the response magnitudes of a combination of representative LGN opponent cells to the Munsell color stimuli of constant chroma have been computed. In a plane of constant lightness, the response magnitudes of the summed ON_{L-M} and OFF_{L-M} model cells to samples of the Munsell color circles are plotted along the positive x-axis (p_1) and that of summed 'M–L' cells for the same chips are represented by the negative x-axis (p_1^-). The response of the 'M–S' cell is plotted along the positive y-axis (p_2), and that of the 'S–L' cell is represented by the negative y-axis (p_2^-). This figure shows only the chromatic responses, i.e. the difference in firing rates between a chromatic stimulus and an achromatic stimulus of the same lightness value (Valberg *et al.*, 1986b; Valberg, 2001).

Stimuli of hues 5Y (yellow), 5R (red), 5PB (blue), and 5G (green) in the Munsell hue system are close to unique hues determined by human subjects. Unique yellow, for instance, is a color that is perceived as neither reddish nor greenish (as is also unique blue), and unique red is neither yellowish nor bluish (Hering, 1920). It is immediately clear from this figure that the cone-opponent directions, represented by the four cardinal coordinate axes p_1 and p_2, cannot individually represent the unique hues. Unique red (5R), for instance, falls nearly midway between the L–M and M–S axes (see also below).

However, a constant Munsell hue line radiating from the white point for every hue around the color circle approximates a straight line. They thus reflect a constant ratio of the responses plotted along the 'cardinal' (p_1, p_2)-opponent directions. The fact that a straight hue line in the diagram of Figure 11.9 is reproduced as a curved line in the CIE (Commission Internationale de L'Eclairage) (x, y) chromaticity diagram gives an obvious explanation of the well-known Abney effect (or Kohlrausch effect). The Abney effect (Abney, 1910) is the term used for the phenomenon that white light added to a colored light of high purity changes the hue of the mixture. Provided luminance is kept constant; even a small amount of white added to saturated unique red will give the mixture a bluish-red appearance (Wyszecki and Stiles, 1982). Increasing amounts of white will intensify the blue shift up to a point after which the blue component diminishes. When white is added to a saturated orange the hue becomes more reddish, and when added to yellow-green it becomes more greenish.

The co-variation demonstrated in Figure 11.9 between a constant response ratio of opponent cells and the perception of *a constant hue*, irrespective of chroma, is strong evidence in favor of the model. It adequately explains the Abney effect as resulting from the response nonlinearity in opponent cells (Valberg *et al.*, 1986b). The quasi-elliptical shapes of the constant chroma loci in Figure 11.9 (chroma 2, 4, 6, and 8 are plotted) show that the equal chroma steps of the Munsell system are not reproduced as equal steps for all hues in the model response. However, allowing for new linear combinations of the cardinal axes p_1 and p_2, giving for instance the F_1 and F_2 coordinates, a more circular shape will result (Valberg *et al.*, 1986b). Because such a linear transformation makes the hue spacing more even, and the chroma scaling for all hues more uniform, without affecting the theoretical explanation of the Abney effect (straight lines will be preserved as straight lines), we use this transformation to explain the Bezold–Brücke phenomenon.

The Bezold–Brücke phenomenon

With increasing luminance, a long-wavelength red light becomes more yellowish and short-wavelength violet light becomes more bluish before they both turn whitish. This phenomenon is known as the Bezold–Brücke phenomenon (Bezold, 1873; Brücke, 1878; Purdy, 1931). The phenomenon can be seen, for instance, when the luminance of a color stimulus spans several log units, as in colored light bulbs where the bright filament appears whitish and the darker colored glass, at some distance away from the filament, has a much higher color strength.

The Abney effect and the Bezold–Brücke phenomenon have long eluded a satisfactory, quantitative explanation. On the basis of what we now know about early levels of color processing, we can provide a plausible physiological account for these phenomena. Since, in the model, chromatic strength (chroma) of a stimulus and its hue are both related in a specific way to the relative activation of opponent cell types, it follows that the model must also predict the change in perceived attributes as a function of stimulus intensity. Does an increase in the luminance of a color stimulus change the response ratios of opponent cells such that the result corresponds with the Bezold–Brücke phenomenon?

Figure 11.10A traces the chromatic responses[1] (the difference in firing rate between a chromatic and an achromatic stimulus of the same luminance) of hypothetical cortical opponent units to color stimuli of the chromaticities and wavelengths given in Figure 11.10B (Valberg, 2005). The responses of the hypothetical cortical units are added as vectors, the radius vector being proportional to chromatic strength, and the orientation of the vectors relative to the coordinate axes being related to hue. For instance, with increasing luminance ratio of the 649-nm stimulus, the resulting response curve makes a leftward turn toward the F_2-axis. This is a consequence of the different dependency on luminance of the units' responses. Since a constant response ratio $F1/F2$ implies a constant hue, the leftward turn is interpreted as a change in the direction of yellow.

The polar plot of Figure 11.10A shows the predictions of the model for how responses to luminance of cortical units combine to yield a representation of chromatic strength and hue for several stimuli of constant chromaticity (Valberg *et al.*, 1991; Valberg, 2005). The curves in the diagrams are projections of the trajectories in 3D space down on a plane. The figure shows that color strength (chroma) initially grows as luminance rises from 0, reaches a maximum for a luminance ratio that is characteristic for each hue, and then decreases again. In accordance with experience, the model predicts the larger hue shifts in the red-orange and violet-green sectors of the color circle, with long-wavelength stimuli becoming more yellowish with increasing intensity and low-luminance violet stimuli turning bluish. Since all curves in Figure 11.10A start at a common point and

[1] Which physiological substrates isolate and process the 'chromatic' and 'achromatic components'? It is still a problem how the achromatic attributes can be linked to the behavior of opponent cells. However, since for every luminance level 'L–M' and 'M–L' cells have rather similar responses to white light, one could imagine that a pure L/M chromatic response can be found as the difference in the responses of these two cell types, and an achromatic percept being associated with stimuli for which this difference is 0 (a sum of the responses of these two cell types might correlate with brightness). The excitatory profile across these color-selective OFF and ON cells probably contains the code for lightness and brightness attributes. Where in the visual pathway such a separation of 'white responses' and 'chromatic responses' would arise, if it takes place at all, is an unsolved issue, although available data suggest that this would not happen before V1.

A relatively wide range of spectral stimuli will evoke a positive response in opponent cells. At the same time, these cells respond to the lightness contrast of chromatic as well as achromatic stimuli (black, grey, and white). The size of the stimulus is also important for their response specificity and magnitude. A PC ON cell, for instance, responds positively for increments in all wavelengths if the stimulus is too small to activate significantly the inhibitory receptive field surround. Therefore, opponent cells of the retina and LGN cannot be associated with a single attribute of the stimulus, such as color. Opponent cells are not color-specific at this level. Only a comparison (probably a cortical process) between responses of different types of cells can determine which stimulus evoked the neural firing, whether it was a luminance contrast, a change in color, a particular stimulus size, etc. At this low level of processing, multiplexing is the rule, meaning that more than one property is signaled along the same channel (Martinez-Oriegas, 1994).

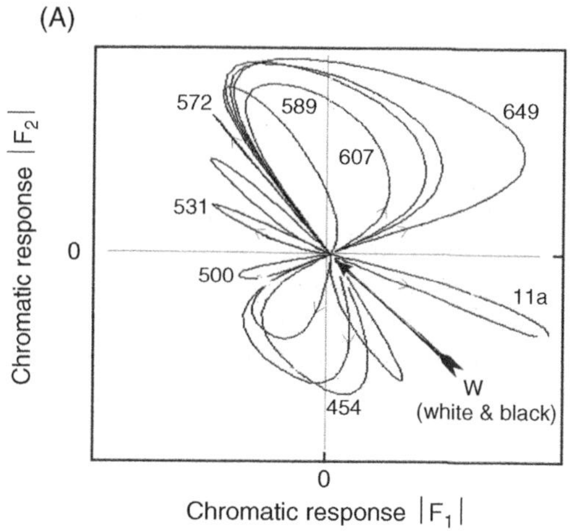

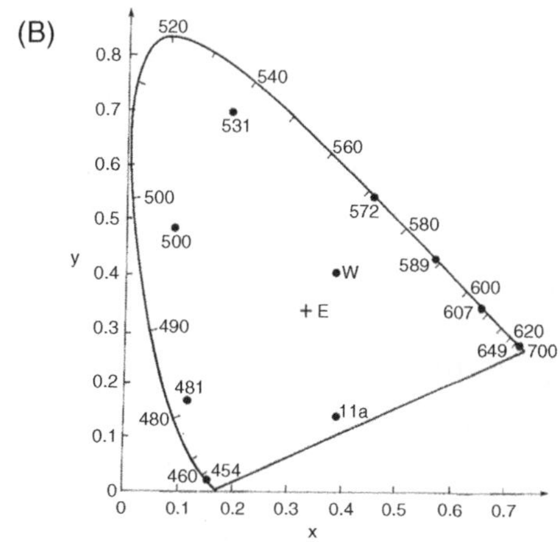

Figure 11.10 (A) Predictions made by the neural color vision model for the changes in perceived hue and color strength (chroma) of eight stimuli of constant chromaticity (shown in B) when luminance ratio is raised from 0 (origin) to high values. The distance from the center of the diagram is proportional to model color strength (chroma), and the radial angle represents hue. The units for cell responses are scaled in such a way that it corresponds to Munsell chroma. Extrapolation beyond the experimental data shows that for extreme high luminance ratios monochromatic stimuli will appear achromatic (Valberg, 2005). Reproduced by permission of John Wiley & Sons Ltd

converge toward the center of the diagram, the model postulates also that chromatic response is lost at very low and very high intensities, and that the stimuli would then appear achromatic. Maximum color strength is largest within the orange-red-purple sector and least within the yellow-green-violet sector.

This overall agreement with perception provides strong evidence that the Abney effect, Bezold–Brücke hue shifts, and the related change of chromatic strength have a common origin in the non-monotonic intensity responses of opponent cells. In addition, the model accounts well for Optical Society of America (OSA) and Munsell color scaling in approximate equidistant color spaces (Valberg *et al.*, 1986b).

11.3.4 Summary

Cortical mechanisms may derive several perceptual attributes of the retinal image by analyzing information in multiple converging and diverging pathways. After the outputs of ON and OFF opponent channels have converged onto cortical cells, probably in the 'blob' regions of V1, they may again be related in new opponent pathways resulting in neurons tuned to chromatic directions other than the cardinal axes (Lennie *et al.*, 1990; Kiper *et al.*, 1997; Wachtler *et al.*, 2003). A more thorough explanation of the model is provided by Valberg (2005).

11.4 Central visual pathways

It is now well established that the higher processing of visual information takes one of two major paths on its way to prefrontal cortex: a ventral path that is primarily involved in the analysis of object-specific features (color, form, and texture) and a dorsal path that extracts spatial coordinates and dynamic trajectories (location in the visual field, motion speed, and direction), which enable the observer to locate and track moving objects (Zeki, 1974, 1978; Van Essen *et al.*, 1981; Ungerleider and Mishkin, 1982; Albright, 1984). In the following we will consider the processing of color and motion within these major streams in the central visual system.

11.4.1 Cortical representation of color

The goal of processing chromatic information in the cortex is to transform the signals from the retino-geniculate pathway as described above in such a way that they become available to the diversity of tasks handled by the visual system: most importantly, it aims at the representation of color as an intrinsic property of an object, independent of the varying illumination or surroundings (the phenomenon of color constancy), by which the object can be reliably identified and distinguished from its background. Secondly, chromatic information contributes to other visual tasks, e.g. scene segmentation or motion perception, whereby the latter does not necessarily result in a conscious color percept. Thus, chromatic processing in the cortex cannot be understood as one single pathway, ending in a single color center, but rather as the sum of several parallel processes, which start in the retina and are completed over several stages in the cortex. The following section will elaborate on the processes leading to conscious color perception whereby the relation between psychophysical phenomena and underlying neural activity will be illuminated.

11.4.2 Representation of hue in the cortex

There are several important transformations of chromatic signals taking place in the visual cortex. The first of these concerns chromatic tuning. As we have seen in section 11.3, neurons in the retino-geniculate-cortico pathway encode chromatic information along cone-opponent axes in color space. It needs to be stressed that these are not identical to our experience of unique hues (Hering-opponent colors, such as unique yellow, blue, green, and red; see above), but rather correspond to the so-called cardinal directions in color space, which have been identified in psychophysical experiments: for example, Krauskopf *et al.* (1982) showed that a sinusoidal modulation of an adaptation light reduces specifically the sensitivity along an [L − M], [S − (L + M)], and [L + M] axis, as compared to intermediate axes in color space.

In our daily experience of color, we can distinguish between millions of different hues and shades, including those of intermediate axes in color space. Furthermore, we experience some hues as unique hues because they do not appear to be mixtures of other colors (e.g. yellow), and we experience extra-spectral colors, i.e. colors that do not correspond to a wavelength region, e.g. purple.

For a long time it has been an open question how and where this transformation takes place. Evidence for the existence of higher-order color mechanisms in the cortex comes from a number of adaptation studies including those on chromatic adaptation/habituation (Krauskopf *et al.*, 1986; Webster and Mollon, 1991, 1994), coherent motion (Krauskopf *et al.*, 1996), and discrimination thresholds along cardinal and non-cardinal directions (Krauskopf and Gegenfurtner, 1992). For example, Webster and Mollon measured changes in color appearance following chromatic adaptation. In their adaptation paradigm post-receptoral channels were specifically desensitized without changing the sensitivity of the receptors. In contrast to the prediction of models based on only three post-receptoral channels, saying that adaptation would only desensitize color appearance for those colors at which the post-receptoral channels are maximally sensitive, they found that color appearance was always distorted away from the direction in color space to which the observer had adapted. Furthermore, Werner *et al.* (2000) observed an asymmetry between the cardinal axes: they examined the changes in the achromatic appearance of a test-patch following chromatic adaptation in different directions in color space. Their results showed faster adaptation toward green as compared with red and faster adaptation toward yellow as compared with blue (Figure 11.11). The fact that this effect could be interocularly transferred suggests a neural substrate of this phenomenon at a stage beyond the binocular combination of chromatic signals in the cortex.

The representation of multiple color directions corresponds to recent findings from single cell recordings which revealed a continuum of tuning directions for isoluminant stimuli in the primary visual cortex (De Valois *et al.*, 2000; Wachtler *et al.*, 2003). It has been suggested that there is at the cortical level a distributed code for chromatic stimuli, whereby each chromaticity is represented by a population of neurons with overlapping tuning curves and different peak sensitivities (Wachtler *et al.*, 2003). In addition, neural clusters were found that encoded red, red-yellow, yellow, green, and blue (Vautin and Dow, 1985; De Valois *et al.*, 2000; Wachtler *et al.*, 2003) and even non-spectral colors, i.e. purple (Dow and Vautin, 1987). How exactly the transformation of LGN signals to cortical color signals occurs is, however, still an open question, since it cannot be explained by simple linear combination of cone signals, as it was described earlier for other color effects (De Valois *et al.*, 2000; Hanazawa *et al.*, 2000).

11.4.3 Mapping of hue in V2

Another feature of our perceptual representation of color is our ability to arrange colors according to their similarity, as it is expressed in color ordering systems. Since the early 17th century, many scientists and painters developed systems for ordering colors according to their appearance, of which the following is only a brief selection. On the basis of his observations of the physical properties of light, Newton (1704) arranged seven so-called principal colors in a circle from red through orange, yellow, green, blue, indigo to violet. Lambert (1772) ordered colors according to their perceptual similarity within a pyramid-formed color space, with dark colors at the base and light colors at the top. The perceptual spacing of the colors was based on a regular change in the ratio of mixed

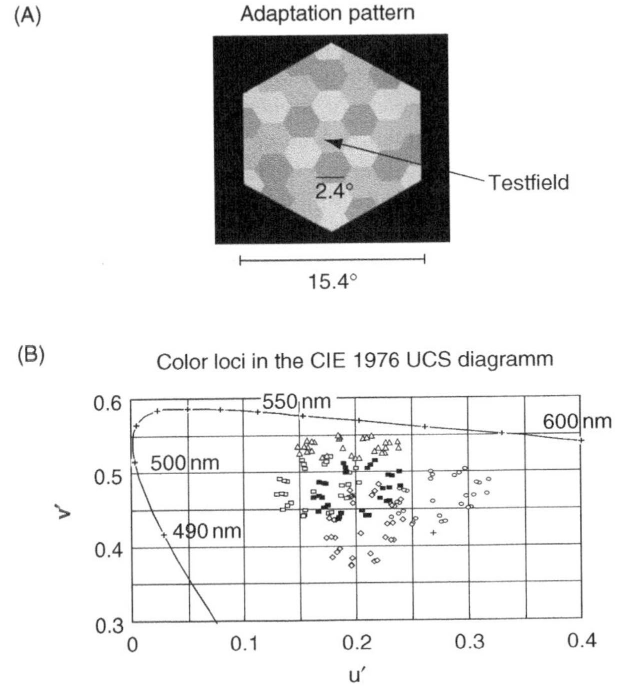

Figure 11.11 (A) Adaptation pattern (complex arrangement). The pattern was heterochromatic, with color loci as indicated in (B). (B) Color loci of the stimuli in the CIE (1976) UCS diagram. The filled symbols (■) represent the stimuli as Munsell papers under D65-illuminant (standard condition), the open symbols show stimuli as Munsell papers under red (O), green (□), blue (◇), and yellow (Δ) illuminant. The locus of the test-patch under standard illuminant, i.e. the locus of D65, is marked by an asterix (★). Note that in order to make the direction of the colorimetric shifts more obvious, the maximal possible colorimetric shifts are depicted. For the actual experiments, the illumination changes produced shifts of 22.88 $\Delta E*_{uv}$ for all color stimuli. (C) Time course of chromatic adaptation. Different symbols refer to the different adaptation illuminants: red (O), green (□), blue (◇), and yellow (Δ). Hundred percent adaptation refers to the maximum possible adaptation shift, i.e. a shift of the color locus of the test-patch in the standard condition to the color locus of the adaptation illuminant. Results of subject AW are shown on the top, those of subject AS on the bottom. Bars indicate the standard deviation (Werner, 2003). Reproduced by permission of Elsevier

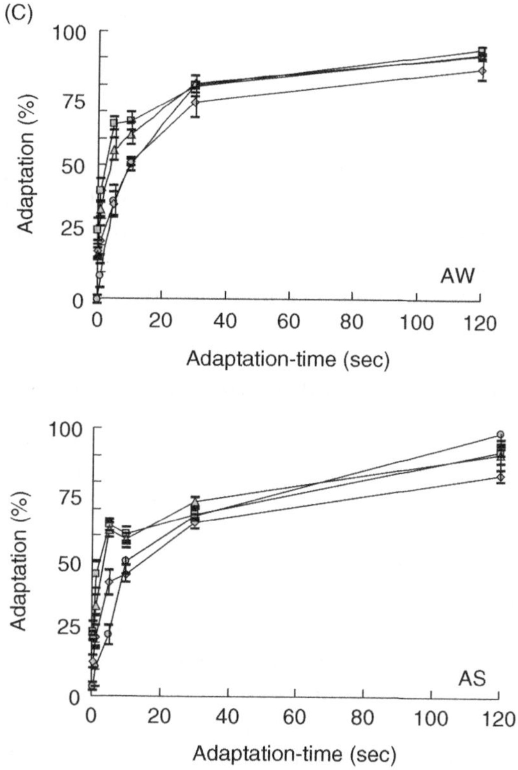

Figure 11.11 *(continued)*

colored waxes. The painter Ph.O. Runge presented colors on the surface of a color globe, with black and white at the two poles. von Goethe (1810), on the other hand, developed a subjective system of complementary colors, in which he also incorporated their emotional effects (e.g. cold, warm). Chevreul (1839) also used a circular order, whereby colors were presented with equal perceptual spacing. J.C. Maxwell (around 1850) was the first to base a color system on the rules of additive color mixing. Expanding the idea of complementary colors, Hering (1920) devised a system where all colors arise from a combination of the opponent colors green and red, blue and yellow, and light and dark. The painter A.H. Munsell classified colors according to their three perceptual dimensions, namely, hue, saturation, and brightness (*Munsell Book of Colors*, 1905/1929); Munsell's color system and its today variants (e.g. the 'HSB' system) are still popular in modern color applications (e.g. printing, color displays). Other recent color appearance systems include the 'OSA Color System' of the Optical Society of America, the 'Swedish Natural Color System', and the 'German Standard Color Chart' (DIN System; see Wyszecki and Stiles, 1982 for a detailed description of modern color ordering systems).

As we have seen in section 11.4.2, an orderly representation of color loci follows already to some degree from an opponent color code. But is this how color is represented also in the brain, i.e. is there something like a 'chromatotopic' map in the cortex?

In the primary visual cortex chromatically sensitive cells are concentrated (although not exclusively) in specific regions, called 'blobs' after their shape. Cells within the same blob tend to share the same chromatic properties (red-green blobs, blue-yellow blobs), and intermediate directions in color space seem to be represented in patches connecting neighboring blobs (Ts'o and Gilbert, 1988). This ordering of color is continued and perhaps even more elaborated, at the following stage, in V2. Recently, optical recordings by Xiao *et al.* (2003) revealed a spatially ordered map of the different hues within the thin stripes, which is reminiscent of the arrangement of colors in the DIN System. Thus, the location within a thin stripe in V2 seems to code the perceived hue, similar to the orientation map in V1. How saturation and brightness, the other two fundamental features of color, are mapped still remains an open question.

11.4.4 Color contrast: chromatic induction and adaptation

Another important feature of color processing in the visual cortex is to bring chromatic signals into a spatial and temporal context: this results in an increasing abstraction of the color code, corresponding to form or motion processing, meaning that the visual signals become less retina-related and instead more scene- and object-related.

The importance of color contrast is twofold: Firstly, it enhances local contrast and thus makes objects more distinguishable from their background (figure ground segmentation), as already recognized by Chevreul (1839) in his 'Law of simultaneous contrast of colors'. Secondly, it constitutes an important computational step toward color constancy: changes of daylight tend to leave the spatial ratios of light reflected from natural surfaces preserved (Nascimento *et al.*, 2002), and therefore the encoding of color as spatial contrast (chromatic and luminance) is a powerful tool to achieve color constancy (Hurlbert and Poggio, 1989; Foster and Nascimento, 1994; Hurlbert and Wolf, 2004).

A prime example of spatial contextual influences is simultaneous color contrast which describes the phenomenon that the perceived color of an area is modified by the presentation of other colors in its vicinity. Typically one color induces the opponent color in an adjacent surface. For example, a grey disk appears greenish when surrounded by a larger red surround. Successive color contrast is the corresponding phenomenon in the temporal domain: the perceived color of an area is modified by the preceding presentation of another color stimulus.

The processes that underlie simultaneous and successive color contrasts are chromatic induction and adaptation. Both are essential mechanisms in color processing, whose investigation uncovers important aspects of the processing of chromatic information. Often a distinction is made on the basis of their time scale, whereby chromatic induction is defined as the class of instantaneous, spatial processes, while adaptation refers to the class of primarily temporal processes (local or spatially extensive). Therefore, chromatic induction is frequently used in connection with an experimental paradigm where only the surround changes and the test is kept constant, whereas chromatic adaptation often refers to an experimental arrangement where background and test change (e.g. as the result of an illumination change). However, one has to keep in mind that chromatic induction and adaptation are extremely difficult to discriminate experimentally because both classes of mechanisms result in the same corresponding shifts of color appearance and both can be spatially extensive processes, either via lateral interactions or – in the case of adaptation – via eye movements.

Chromatic induction and adaptation can be modeled by a combination of processes, including multiplicative gain adjustment and subtractive processes via lateral inhibition (Whittle and Challands, 1969; Jameson and Hurvich, 1972; Walraven, 1976; Shevell, 1978; Geisler, 1981; Hayhoe *et al.*, 1987; Hayhoe and Wenderoth, 1990). These different processes exhibit different temporal properties/time courses, with time constants ranging from a few milliseconds for near-instantaneous processes, to 1–8 s for fast processes, to 20–30 s for slow processes (Fairchild and Lennie, 1992; Fairchild and Reniff, 1995; Rinner and Gegenfurtner, 2000; Werner *et al.*, 2000). Further, there is a multiplicative process of presumably cortical origin, which is instantaneous and accounts for about 60 percent of the adjustments (Rinner and Gegenfurtner, 2000). Figure 11.11 shows typical time courses of the temporal adjustment of color appearance after a change in the overall chromaticity, comparable to an illumination change.

For a long time, chromatic induction and in particular chromatic adaptation were seen as retinal mechanisms (receptor gain control, lateral interactions between the same class of cones; e.g. von Kries, 1904; Evans, 1948; Alpern, 1964; Mollon and Polden, 1975). Also, lateral interactions have been proposed for a second opponent stage (e.g. Jameson and Hurvich, 1964; Shevell, 1978; Guth *et al.*, 1980). More recently, evidence has accumulated for the involvement of cortical stages, as will be outlined in the following paragraphs. Thus, chromatic induction and adaptation are not restricted to one particular level in the visual pathway, but rather continued over several stages in the cortex.

11.4.5 Computation of local color contrast in the cortex

The computation of local cone contrast has been attributed to the function of double opponent cells in the primary visual cortex. These neurons have been identified within the blob regions of the upper layers of V1 (Michael, 1978; Conway, 2001) and are characterized by a combination of spatial and chromatic opponency within their receptive fields.

Double opponency is an especially remarkable step in the processing of cone signals since it allows – in contrast to the simple opponent cells in the LGN – to detect chromatic contrast independently from the presence of luminance contrast. Although their role in chromatic induction has been suspected since a long time (Hurlbert and Poggio, 1989), only recently could this be confirmed experimentally (Conway *et al.*, 2002). Careful measurements of the receptive field properties of double opponent cells showed that these cells do indeed sum linearly their equal responses to a red light in the center and a green light in the surround (at the edge of their receptive field) – just as to be expected from the psychophysical correlate. Furthermore, these cells were also found to respond accordingly to temporal contrast, thus providing also a possible neural substrate for successive color contrast (i.e. chromatic adaptation, see below). The very mechanism that leads to color contrast could potentially also lead to color constancy, since illumination changes affect the receptive field of double opponent cells uniformly, and therefore will be compensated by the antagonistic responses of center and surround.

11.4.6 Contrast adaptation

The encoding of color contrast itself in V1 is subject to adaptation. This refers to the phenomenon that prolonged viewing of colored patterns reduces the perceived intensity of color contrast with the same colors but leaves perception of contrast patterns with other colors undisturbed. This phenomenon contributes to our adjustment to the variety of color contrasts in the natural environment (Webster and Mollon, 1995), thereby supporting color constancy and increasing the efficiency of color coding. Using functional magnetic resonance imaging (fMRI), Engel and Furmanski (2001) showed that the neural correlate of this phenomenon resides in the primary visual cortex, although the underlying computational mechanisms remain unknown.

11.4.7 Large-scale contextual influences

Spatial contextual influences in induction and adaptation can be very extensive: the effect of induction can be measured not only for local but also for remote inducers, over distances up to 10° visual angle (Wachtler *et al.*, 2001; Hurlbert and Wolf, 2004), whereby the influence is exponentially reduced with distance. It can be argued that the visual system employs spatially extensive and complex mechanisms, so that object colors become less dependent on their direct local background, and instead relate more to a global surround, which is less likely to change from scene to scene, at least in natural scenes (so-called grey world assumption).

But double opponent cells code only for local contrast, with receptive fields extending not more than 1°–2° visual angle. Recently, neurons in V1 have been described that show a specific modulation of their chromatic tuning by remote color patches presented as far away as 6° from their classical receptive field centers (Wachtler *et al.*, 2001, 2003). The relation of these cells to double opponent cells is unclear. The size and properties of their receptive fields are reminiscent of so-called silent surround neurons (or neurons with non-classical receptive fields), which are known from higher-order areas (in particular V4), but also from LGN (Chapter 6; Schein and Desimone, 1990; Desimone *et al.*, 1993). Their name stems from the fact that stimuli outside their classical receptive field do not elicit a response unless they are activated together with the neurons center.

In general, contextual influences are more pronounced if the background is patterned ('articulated') than when it is homogeneous. This is true for chromatic induction (Barnes *et al.*, 1999) as well as adaptation. Werner and colleagues (Werner *et al.*, 2000; Werner, 2003a,b) showed that the spectral asymmetries in the time course, as described earlier, are only present if the adaptation pattern is articulated (Figure 11.12). This color-specific asymmetry in the early time course of adaptation has an important implication for the detection of objects against their background in natural scenes: within the typical fixation time in a visual search task (around 5 s), the chromatic difference between the slowly adapting long- or short-wavelength target (typically the chromaticity range of e.g. ripe fruit) and its fast adapting background (typically yellow or green) will be enhanced and thus further facilitate the detection of biologically important objects.

The pronounced effect of articulation on chromatic induction and adaptation has been explained by the combined effect of receptive field size and a temporal modulation of the signals following eye movements, similar to habituation (Zaidi *et al.*, 1992, 1998; Zaidi, 1999). However, it could be demonstrated that the effect of articulation on adaptation cannot be explained by temporal modulation of the chromatic signals alone, but instead requires specific interactions between the chromatic and spatial processing (Werner, 2003a,b; see following section). Furthermore, model calculations for S cone-induced chromatic induction (Monnier and Shevell, 2004) showed that the larger shifts in

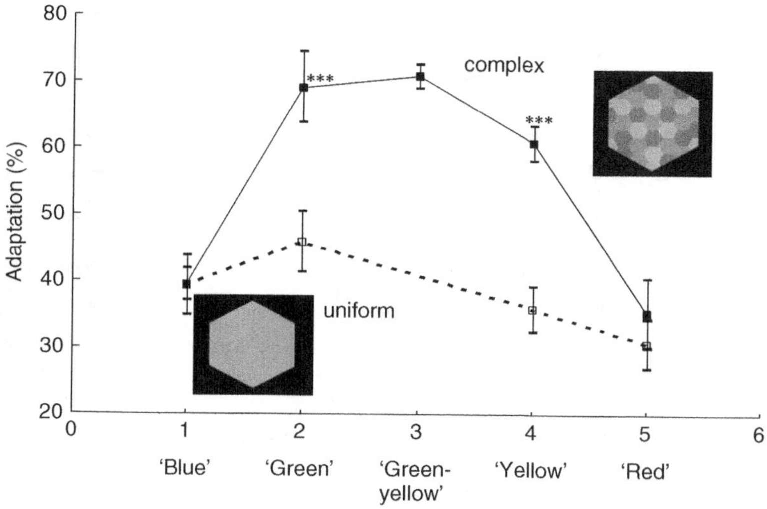

Figure 11.12 The effect of image articulation on chromatic adaptation, demonstrated for different wavelength of the adaptation light. The adaptation lights are ordered along the abscissa according to their corresponding dominant color region. The upper 'curve' shows adaptation using an articulated pattern, the lower 'curve' refers to adaptation using an equivalent uniform field (same mean luminance and chromaticity as the articulated pattern ($L_{mean} = 16.8\,cd/m^2$, $x = 0.316, y = 0.335$)). Asterisks indicate significant differences between the amount of adaptation using the articulated pattern and the uniform pattern (t-test: *** : $p < 0.001$). Results are shown as the average for three subjects (AW, AS, and LR) (Werner, 2003). Reproduced by permission of Elsevier

color appearance observed for patterned surrounds are consistent with the neural receptive field structure of neurons in V1.

11.4.8 Co-processing of color and spatial structure

Contextual effects of induction and adaptation are very specific and depend on the relation between the spatial features (e.g. spatial complexity, spatial scale, and configuration) of the inducing surround and the induced area. Barnes *et al.* (1999) and Hulbert and Wolf (2004) showed that chromatic induction is most pronounced if the spatial scale or chromatic texture of inducers and induced area match; but if they differ, induction is reduced. Similarly, chromatic adaptation is linked to specific spatial features of the adaptation pattern. Middle-wavelength adaptation is most pronounced if the spatial frequency and orientation of the adaptation background and test field match and this can be observed for chromatic as well as luminance structures (Figure 11.13; Werner, 2003a,b). This spatial tuning also includes the depth plane of the adaptation pattern (Shevell and Wei, 2000; Werner, 2003b), although this is not found to the same extent in all experimental paradigms (Shevell and Miller, 1996; Hurlbert and Wolf, 2004).

The spatial characteristics of chromatic adaptation and induction correlate with the so-called multiplexing properties of neurons at different stages in the visual pathway.

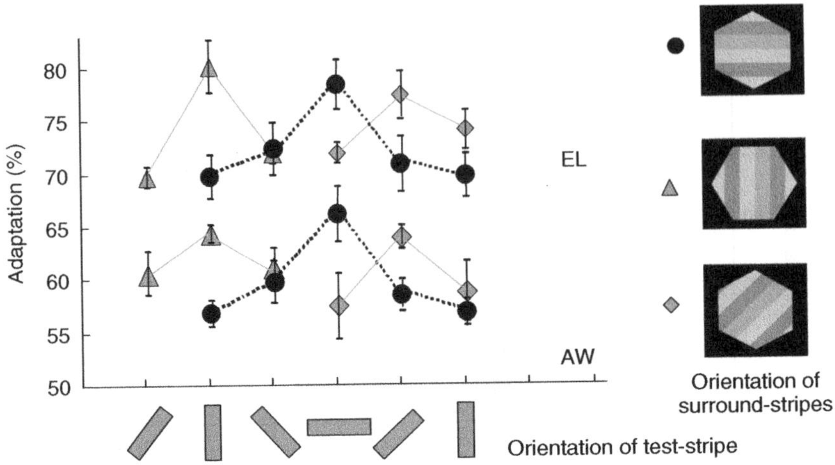

Figure 11.13 Orientation tuning of chromatic adaptation. Adaptation is measured as a function of the relative orientation of the test-stripe, whereby each curve represents a different orientation of the adaptation pattern (background): vertical (grey triangles), horizontal (black dots). Subjects EL (upper curves) and AW (lower curves)

Multiplexing refers to the property of neurons in the visual pathways to respond to more than one stimulus feature, e.g. to color and luminance and/or spatial frequency, orientation, direction, and stereo disparity. For example, Johnson *et al.* (2001) demonstrated that many neurons in the primary visual cortex respond to both luminance and chromatic borders, and additionally are tuned to spatial frequency and orientation, indicating a link between the analysis of color and form in the primary visual cortex. Other physiological experiments have also demonstrated a close cooperation between color and form processing in extrastriate areas (Zeki and Shipp, 1989; Gegenfurtner *et al.*, 1996; Lennie, 1999).

These results are opposed to the classical segregated pathway hypothesis that states that the different sub-modalities of vision (form, color, depth, and motion) are analyzed in parallel (Ungerleider and Mishkin, 1982; Livingstone and Hubel, 1987; Zeki, 1990). With respect to color processing the segregated pathway hypothesis suggests that color is filled into the ready processed form at a late stage (hence 'coloring book model'; Livingstone and Hubel, 1988). Recent results from psychophysics like those reported above for induction and adaptation, as well as findings from physiology and fMRI, challenge this view and indicate instead interlinked processing of different features of an image (Lennie, 1999; Vidyasagar *et al.*, 2002; Gegenfurtner and Kiper, 2003).

11.4.9 Color constancy

Color constancy is a very important property of the color vision system, since it allows us to assign color to an object regardless of changes in the spectral content of the illumination. The term color constancy goes back to von Helmholtz (1896) and since then has been the subject of numerous investigations. The perhaps most impressive and well-known demonstration of color constancy are the 'Mondrian experiments' by Edwin Land. The 'Mondrian', named after the Dutch neo-plasticistic painter Piet Mondriaan (1872–1944), is a complex pattern of multicolored rectangle papers, which is illuminated by the light of three independently controlled projector lamps (one for green, red, and blue light, each). By measuring the light reflected from each of the patches under different illuminations, Land demonstrated that the human perception does not primarily depend on the local light flux emitted from each patch but rather on the ratios between the reflections of neighboring patches, which do not change with an illumination change. On the basis of the concept of constant ratios, Land proposed an algorithm which models color constancy by integrating local contrast signals above a certain threshold over large parts of a visual scene. For the underlying neural 'Retinex Algorithm' refer Land (1964, 1986) and Land and McCann (1971). The principles introduced in the Retinex Algorithm lay still at heart of most modern color constancy algorithms (e.g. lightness algorithms). Experimental proof for such large field computations involving both hemispheres has been demonstrated in psychophysical experiments involving split brain patients (patients whose corpus callosum was cut in order to relieve them of epilepsy attacks; Land *et al.*, 1983). Neurons with correspondingly large receptive fields (16° visual angle) and spanning over both hemispheres have indeed been reported for V4 (Desimone *et al.*, 1993) and, even larger ones, for the infero-temporal (IT) cortex (Komatsu *et al.*, 1992).

Although basic concepts of the underlying mechanisms, multiplicative gain control (von Kries, 1904) and coding the signals as ratios or contrast (Wallach, 1948), have been early recognized, the complexity of the underlying processes is until today not fully understood. Color constancy is often associated with area V4 (in monkey and human cortex), but it needs to be stressed that the phenomenon is the result of an entire series of operations at different stages in the visual pathways, rather than a single mechanism: this includes sensory mechanisms (light and chromatic adaptation, induction, contrast adaptation) as well as cognitive processes (Helmholtz's unconscious inferences), based on previous experience about the physical properties of the real world (e.g. estimating mutual illumination or using specular highlights as an estimate for the illumination; Bloj *et al.*, 1999; Yang and Shevell, 2002) or color memory itself (von Helmholtz, 1896; Jaeger, 1982; Gegenfurtner, 2004).

Although the basic sensory mechanisms for color constancy are already present in the primary visual cortex (MacEvoy and Paradiso, 2001; Wade and Wandell, 2002; Wachtler *et al.*, 2003), Zeki and collaborators have explicitly excluded color constancy at stages earlier than V4, based on early and recent electrophysiological studies (Zeki, 1983a,b; Moutoussis and Zeki, 2002) and more recent fMRI data (Zeki and Marini, 1998; Bartels and Zeki, 2000). It also seems to be clear from clinical studies that lesions in human V4 lead to selective deficits of color constancy and therefore indicate that the operations for color constancy are not completed in V1, but instead involve operations in these higher tier areas (Ruettiger *et al.*, 1999).

11.4.10 Scene segmentation in color constancy

The integration of signals over large parts of the visual field also requires mechanisms that control the extent of the contextual influences. This can be seen if one considers 3D scenes which typically contain multiple illuminations that can vary considerably across the scene because of multiple reflections from different surfaces and the position of an object relative to the light source. For correctly compensating illumination changes in these scenes, it is necessary to identify regions, which are likely to share the same illumination and to restrict contextual influences to those areas. Although this has been recognized as an important mechanism for color constancy by the Gestalt psychologists Katz (1935) and Koffka (1935), scene segmentation in color constancy is still not well understood.

It has been attributed to higher-order, cognitive factors (Bloj *et al.*, 1999; Yang and Shevell, 2002, 2003), grouping factors related to the Gestalt concept (Schirillo and Shevell, 2000), and also sensory mechanisms like the spatial tuning of chromatic adaptation and induction (Werner, 2003a,b; Hurlbert and Wolf, 2004). On the neuronal level, scene segmentation is at least partially achieved in the primary visual cortex (Lamme, 1995; Zipser *et al.*, 1996; Skiera *et al.*, 2000), where a corresponding modulation of neural responses has been shown for color, luminance, motion, or disparity. From the measured delay of the responses (80–100 ms) it is possible that these processes also receive input from higher tier areas, which would allow for the influence of cognitive factors on scene segmentation.

11.4.11 The role of V4 in color processing

The region in and around the fusiform and lingual gyri in the ventro-occipito cortex (V4 after Lueck *et al.*, 1989; Mc Keefry and Zeki, 1997; V8 after Hadjikani *et al.*, 1998; VO after Wandell, 1999) is often associated with a central role in human color processing. It is known since a long time that clinical lesions in humans in or near the fusiform and lingual gyri can result in severe deficits in or even complete loss of color perception ('achromatopsia'; e.g. Verrey, 1888; Damasio *et al.*, 1980; Zeki, 1990; Ruettiger *et al.*, 1999). Subsequently this area has been named V4 after an area in macaque cortex, for which originally a high percentage of color-selective cells was reported (Zeki, 1973; but see Schein *et al.*, 1982). However, it now appears that monkey and human V4 are not truly homologous, since experimental lesions in monkey area V4 do not lead to the same perceptual deficits as in humans: in macaque, even total bilateral ablation of V4 yields only mild deficits in color discrimination and color-ordering tasks which achromatopsic patients cannot perform (Cowey and Heywood, 1995; see also Schiller, 1993; Walsh *et al.*, 1993; Heywood *et al.*, 1995). Future comparative investigations of cortical activity in humans and monkeys with fMRI might shed light on this issue.

Although the exact topography and delimitation of human V4 (hV4) from adjacent cortical areas has become the cause of spirited dispute (Hadjikani *et al.*, 1998; Zeki *et al.*, 1998), there is general consensus about two important properties of color processing in V4: (1) human V4 is not one uniform region, but rather a complex of subunits and satellite areas, each with their own functional specialization (e.g. Howard *et al.*, 1998; Zeki and Marini, 1998); and (2) chromatic stimuli does not activate one particular 'color center', but an entire network of cortical areas, whereby the parts of the network that become mobilized depend on the exact nature of the stimulus. The pattern of co-activations therefore indicates that the hV4 complex plays a key role in linking chromatic signals to memory, attention (Corbetta *et al.*, 1991; Courtney *et al.*, 1998; Beauchamp *et al.*, 1999; Pinsk *et al.*, 2004), and even other modalities as demonstrated by its activation in 'color hearing' synestesia (Nunn *et al.*, 2002). In this phenomenon, individuals experience specific colors when hearing particular words. Together with color activation of the IT cortex, these patterns of activation may play an important role in the assignment of colors to objects (Komatsu *et al.*, 1992; Edwards *et al.*, 2003) and constitute a cognitive level of chromatic processing, as proposed by Zeki and Marini (1998).

11.4.12 Summary

In this section we have followed the processing of chromatic signals that ultimately leads to the perception of color. We have shown that the early cortical areas, instead of merely being distributors for the visual signals, perform major computational steps. Amongst those are the fanning out of the representation of colors from a few cardinal directions into a multitude of hues, and the adjustment of signals with respect to their spatial (and temporal) context, via basic operations like chromatic induction and chromatic adaptation. By doing so, the foundations are laid down at early cortical stages for important visual phenomena, such as color constancy. Their further processing at higher cortical stages is characterized by putting the signals into an even wider spatial context and connecting them via co-activation with hippocampus, IT and frontal cortex, with influences from memory, cognitive processes, and even other sensory inputs, as demonstrated by the phenomenon of synesthesia.

11.5 The cortical representation of motion

The ability to analyze retinal image motion is an important feature of all biological visual systems. Object motion can be used to segment figure from ground, thereby enabling an observer to track with his or her gaze a moving object in a cluttered scene. Retinal image motion can also be evoked by the observer's own eye, head, and body movements, and one important task of the visual system is to separate the different components of image displacement that reflect object or self-motion. The analysis of image motion therefore necessarily involves retinal and extra-retinal sources of information. All biological visual systems have developed elaborated mechanisms for the precise encoding of object motion (Reichardt, 1961; van Santen and Sperling, 1985) and the primate visual system has the most sophisticated cortical architecture devoted to the analysis of visual motion.

Areas in the superior temporal and posterior parietal cortex are involved in the analysis of spatial relationships between objects in the environment and the viewer (Andersen, 1995; Colby, 1998; Andersen *et al.*, 2000). Neurons in the posterior parietal cortex project to the frontal eye fields (FEF) in prefrontal cortex (lateral part of area 6) and, together with information from area MST, are used in the preparation of saccadic and smooth pursuit eye movements (Schiller *et al.*, 1979; Bruce *et al.*, 1985; Lynch, 1987; Krauzlis and Stone, 1999; Tehovnik *et al.*, 2000; see also Chapter 10). Thus, the dorsal visual pathway appears to be specialized for the task of visual motion analysis and oculomotor tracking. fMRI has provided additional information about this processing hierarchy. Areas in the lateral occipitotemporal junction have been identified as the human homolog of the primate areas MT and MST (referred to as V5/V5a or simply MT+; Zeki *et al.*, 1991; Tootell *et al.*, 1995; Heeger *et al.*, 1999).

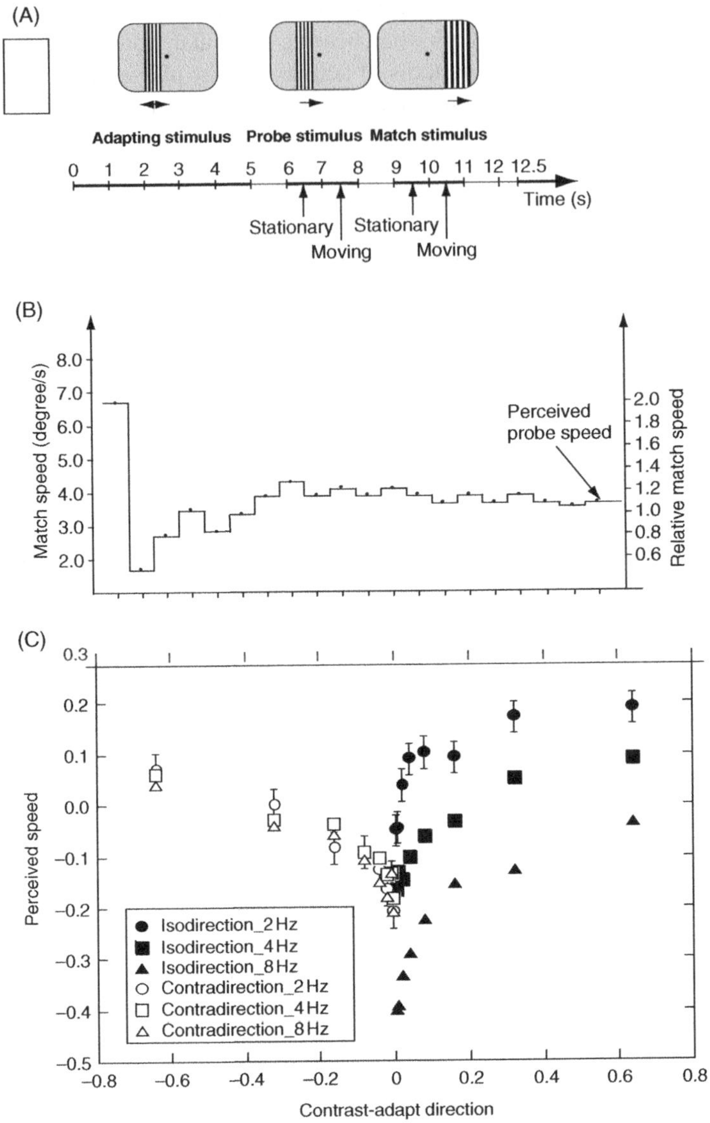

Figure 11.14 The perceived speed of a probe stimulus is shown as a function of probe contrast. (A) Experimental design used by Müller *et al.* (2004) to examine the effect of direction-specific adaptation on the perceived speed of a drifting grating with simultaneous recordings of the motion-onset VEP. (B) Example of one run in which the subject judged the relative speeds of probe and match stimuli. The relative speed of the match stimulus was adjusted over 20 trials to match that of the probe stimulus. (C) The relative perceived speed of the probe stimulus is plotted as a function of the contrast of the probe stimulus. Positive values indicate an increase in perceived speed after adaptation, negative values indicate a decrease in perceived speed. The data points are grand means over the eight observers (Müller *et al.*, 2004). Solid symbols illustrate the mean values for the speed matches after

11.5.1 Motion aftereffects

Psychophysical studies of motion perception have focused on the ability of human observers to detect and discriminate the direction and speed of drifting gratings (Mueller and Greenlee, 1994), random dots (Williams and Sekuler, 1984), and optic flow (Koenderink, 1986). Prolonged viewing of unidirectional motion can induce aftereffects (e.g. the 'waterfall illusion'; Addams, 1834), such that stationary stimuli appear to drift in a direction opposite that of the adaptation direction (Mather *et al.*, 1998). Motion adaptation also affects the perceived speed of moving or flickering stimuli (Mueller and Greenlee, 1994; Müller *et al.*, 2004). An example of the directional selectivity of motion adaptation is given in Figure 11.14. Here the subject viewed an eccentrically presented adapting stimulus that drifted either to the left or to the right at a constant speed. Subsequently the probe stimulus was presented at the same location and subjects were asked to match its perceived speed with that of a subsequently presented match stimulus (presented in the opposite visual hemifield). The findings indicated that motion adaptation has a strong direction and speed tuning. Adaptation effects are most pronounced for stimuli drifting in the same direction with relatively high speeds. The psychophysical results are compared to the motion-onset visual evoked potentials (VEP) acquired in the same subjects in the same sessions (Müller *et al.*, 2004). These phenomena can be best described as the aftereffects of adaptation in directionally selective neurons that respond to the adaptation stimulus and their activity is inhibited after adaptation (Kohn and Movshon, 2003).

11.5.2 First-, second-, and third-order motion

Moving objects are usually associated with luminance contrast contours and the direction and speed of these first-order motion cues are well described by the energy distribution in Fourier space (Adelson and Bergen, 1985; Smith, 1994). However, stimuli can be defined in which motion is induced along virtual contours, such as those found in random flickering textures or contrast varying pedestals (Chubb and Sperling, 1988; Lu and Sperling, 2001; Chapter 10). These motion stimuli have been referred to as second-order (or non-Fourier) motion. In order to correctly detect the direction of motion, the spatio-temporal information has to be assessed over a wider area and integrated to yield a coherent motion signal (Benton and Johnston, 2001). From its onset, psychophysical studies have suggested that first- and second-order motion stimuli could be analyzed by different mechanisms in the visual system. Evidence for this separate processing comes from adaptation studies, where the transfer of adaptation from the one to the other form of motion is minimal (Ledgeway and Smith, 1997). However, recent studies have placed this strict dichotomy into question (Bertone and Faubert, 2003), suggesting a considerable overlap in the ability of motion-sensitive mechanisms to respond to either type of motion.

Figure 11.14 *(continued)*
iso-directional adaptation, open symbols for values obtained after contra-directional adaptation. The different symbols present the findings for the conditions for each of three adaptation speeds (see inset). For the sake of clarity, standard error bars (+1SE) are given only for the slow adaptation condition (2 Hz).

Responses to second-order motion have been explored with fMRI by Smith *et al.* (1998), who compared responses to first- and second-order motion. They showed that most striate and extrastriate regions responded to both types of motion. Selective activation to second-order motion of plaids has been found in V3 using positron emission tomography (PET) (Wenderoth *et al.*, 1999). More recent neuroimaging studies have placed a strict dichotomy of processing for these two types of motion into question. Using fMRI together with motion adaptation, Seiffert *et al.* (2003) and Nishida *et al.* (2003) found comparable responses in retinotopic visual areas and MT+ to first- and second-order motion. Considerable overlap in responses to first-order and two types of higher-order motion was found recently in a PET study (Dupont *et al.*, 2003). Damage to superior temporal and lateral parietal cortex impairs the ability of patients to discriminate the direction and speed of first- and second-order stimuli (Greenlee and Smith, 1997). At least for the patient sample under study in that investigation, the impairments for the two types of motion were highly correlated, suggesting a large degree of overlap between the cortical regions processing these forms of motion.

Recently, a third-order motion analysis has been added to motion systems of first- and second-order (Lu *et al.*, 1999; Seiffert and Cavanagh, 1999). This higher-order system has been attributed an important role for the tracking of object motion, based on the 'saliency' of an object and on attention processes (rather than motion energy as in the case of the first- and second-order systems). This so-called feature tracking system seems to be distinct from the other motion systems by a number of properties: it receives binocular input, is slower (cut-off for temporal frequencies around 3–6 Hz), and is sensitive for isoluminant, chromatic stimuli (Lu and Sperling, 2001). A recent fMRI study identified a region in the intraparietal lobus (IPL) as a neural substrate for the feature tracking system (Claeys *et al.*, 2003).

11.5.3 Coherent motion perception

The ability of human observers to discriminate motion direction has been studied with random dot kinematograms (RDK), where the stimuli are designed such that within a cloud of randomly moving dots, only a small subset of dots drift in a coherent direction (Newsome and Pare, 1988). By varying the number of coherently moving dots in a given RDK stimulus, and by asking subjects to judge the perceived direction of this motion, a psychometric function for motion coherence can be defined. Usually thresholds lie around 5 percent motion coherence, which means that only 25 of a total of 500 random dots need to move in one direction in order for this direction to be reliably perceived. Such perceptual ability demands that the underlying motion-sensitive mechanisms be able to integrate over a wide region of visual space (Benton and Johnston, 2001). Neurons in area MST appear to have just such receptive fields, and the responses of these neurons appear to mediate our sensitivity to these motion fields (Newsome and Pare, 1988; Cook and Maunsell, 2002).

PET has been used to study changes in regional cerebral blood flow (rCBF) evoked when subjects viewed visual motion. Zeki *et al.* (1991), Watson *et al.* (1993), and Dupont *et al.* (1994) found significant responses to random dot motion in the human homolog of area V5/V5a. In a PET study by de Jong *et al.* (1994) subjects viewed optic flow fields with different levels of directional coherence. They found robust responses in the human

V5/V5a complex (MT/MST, also referred to as MT+) in the border region between areas 19 and 37, as well as in area 18 (the human homolog of V3), in the insular cortex, and in areas 19 and 7. Cheng *et al.* (1995) observed activations in occipitotemporal cortex (V5/V5a, BA 19/37) and occipitoparietal areas (V3A, BA 7), and these responses were more pronounced during coherent motion perception compared with random dot motion. This enhancement may be explained by the motion opponency, a feature of neurons that respond selectively to a specific stimulus direction and are inhibited by other neurons tuned to the opposite direction. To test for motion opponency in human MT+ region, Heeger *et al.* (1999) compared BOLD responses to flickering and drifting gratings. They showed that the MT+ region exhibited lower responses to flickering than to drifting gratings, suggesting the presence of 'motion opponency' in area MT+. Single-unit responses in the primate V5 region supported these claims (Heeger *et al.*, 1999). Area V5/MT+ responded somewhat selectively to motion coherence (Rees *et al.*, 2000; Braddick *et al.*, 2001), along with the KO, V3a areas. Singh *et al.* (2000) compared BOLD responses in visual areas to drifting and counterphase flickering gratings. Their results show spatial frequency band-pass tuning functions in V1 and low-pass functions in V2, V3, V3A, and V5, which are in general agreement with single-unit studies (Foster *et al.*, 1985; Levitt *et al.*, 1994; Gegenfurtner *et al.*, 1997). These imaging studies suggest that, similar to that found in single cells in primate area MT/V5, the human MT+ complex contains neuronal populations that are selective to the direction and speed of moving stimuli.

Moving stimuli can be defined by the relative direction or speed of dot clusters, and this kinetic information is detected by the visual system as motion borders. Interestingly, the V5/V5a (MT+) area did not respond selectively to motion borders (Reppas *et al.*, 1997), although it did respond to the global motion in the stimuli. van Oostende *et al.* (1997) describe the kinetic occipital (KO) region in the lateral extrastriate cortex, which responded selectively to kinetic contours (Dupont *et al.*, 1994, 1997; van Oostende *et al.*, 1997; Cornette *et al.*, 1998; Orban *et al.*, 1998). In addition to V5/V5a, several extrastriate areas show selective responses to the direction and speed of visual motion.

11.5.4 Optic flow

Wide-field motion evoked by eye, head, or body movements has been referred to as optic flow (Koenderink, 1986). Cortical responses to optic flow stimuli have been studied by Rutschmann *et al.* (2000), Morrone *et al.* (2000), and Beer *et al.* (2002). The subjects in these studies viewed RDK (random, expansion–contraction, and rotational) motion fields during fMRI measurements. In the Rutschmann *et al.* (2000) study, binocular disparity of the flow fields was also varied in order to obtain responses to movement in 3D space. Little response selectivity was found in the striate and immediate extrastriate regions to disparity in optic flow fields, but area KO responded best to flow fields with disparity gradients. In the PET study by Beer *et al.* (2002) a similar selectivity to optic flow stimuli in area KO/V3b was reported. Morrone *et al.* (2000) found enhanced responses to rotation and expansion, but these differences apparently could only be obtained when the stimuli rapidly alternated between one of two directions, suggesting that these responses might be more related to the processing of sudden direction changes and not to optic flow as such. In summary, the studies reviewed above point to a complex hierarchy of motion-sensitive regions in the human extrastriate cortex. Damage to these regions will necessarily impair the ability of the observer to detect and discriminate complex motion. For example, bilateral occipitoparietal lesions lead to an impairment in the ability to perceive optic flow (Vaina, 1998), as well as to a deficit in the perception of biological motion (Battelli *et al.*, 2003), and could lead to direction-selective losses in motion sensitivity (Blanke *et al.*, 2003).

11.5.5 Pursuit eye movements

Recent electrophysiological and neuroanatomical studies in monkeys suggest that there is a large overlap in the neural control of (small) saccades, fixation, and pursuit (Krauzlis and Stone, 1999). Smooth pursuit allows us to maintain our gaze on a target, despite the fact that the target is in motion. The mechanisms underlying the perception of visual motion and those controlling resultant pursuit appear to be closely related, at least during the initial stages of pursuit programming (Lisberger and Movshon, 1999; see also Chapter 10). Only a few brain-imaging studies have looked at the effects of pursuit eye movements on the responses in visual and oculomotor areas. Kimmig *et al.* (1999) showed that the activity in the FEF was greater during saccades than during pursuit, whereas MT+ (V5/V5a) responded better during pursuit than during saccades. Petit *et al.* (1997) and Rosano *et al.* (2002) found differences in the location of activation in the FEF related to the type of eye movement performed: pursuit was associated with activation in the lateral, saccades with activation in the more medial parts of the FEF. Freitag *et al.* (1998) found that the response in V5/V5a to moving dot stimuli increased during pursuit compared with fixation. Dukelow *et al.* (2001) first isolated area MST by identifying ipsilateral responses to eccentric motion displays and then characterized this area's response to visually guided or self-guided pursuit. The subjects had to pursue either a moving target or an image of their own finger, which they waved in front of their face in the dark. A subsection of the MT/MST complex responded more during the pursuit condition, suggesting its role in the control of pursuit eye movements.

Damage to the MT and MST areas leads to an impairment in pursuit in monkeys (Newsome *et al.*, 1985; Dursteler *et al.*, 1987). Patients with damage in the human homolog of the MT/V5 region demonstrate higher thresholds for discriminating stimulus velocity and impaired working memory for stimulus speed (Greenlee *et al.*, 1995). Kimmig *et al.* (1995) measured in six patients with lesions in the lateral occipitotemporal cortex the pursuit to a sinusoidal target moving on a dark background or over a structured surface. Compared to controls, the patients exhibited a significant reduction in pursuit gain at all but the lowest stimulus speeds (temporal frequency = 0.1 Hz) investigated. The authors speculated that the reduced gain of pursuit could result from the impairment in velocity storage. Accordingly, patients who have difficulty with the velocity discrimination task had difficulty with the pursuit task. The structured backgrounds led to a reduced pursuit gain in both patient and control groups, suggesting that this effect is independent of the effects related to velocity storage. Barton and Sharpe (1998) explored direction-specific impairments in pursuit in a step-ramp task and found evidence for reduced pursuit gain for eye movements into the affected visual hemifield following MT/MST lesions.

11.5.6 Memory for motion

Memory for the speed (Magnussen and Greenlee, 1992) and direction (Blake *et al.*, 1997) of moving stimuli has also been studied. Memory for stimulus speed and direction is very good and discrimination thresholds of 5 percent or less remained constant for delays of up to 30 s. Memory performance declined when masking stimuli were introduced during the retention interval and the masking effect is specific to the relative speed of the test and mask (Magnussen and Greenlee, 1999). This stimulus-selective interference (referred to as memory masking) implies that the capacity of working memory is limited and that this form of sensory memory relies on neurons with response properties similar to those that encode the visual stimuli. Monkeys trained to detect the direction of coherent motion can do so with a precision comparable to that demonstrated by human observers and memory maskers evoke a similar inference (Zaksas *et al.*, 2001; Pasternak and Greenlee, 2005).

Lesions in primate area MT lead to selective loss in motion memory (Newsome and Pare, 1988; see also Huxlin and Pasternak, 2004 for training-induced recovery of motion perception in cats). In monkeys with unilateral MT/MST lesions, motion stimuli were presented in the intact or lesioned hemifield, to assess the role of MT/V5 in the encoding, retention, and retrieval of motion information (Bisley and Pasternak, 2000). MT/MST lesions disrupted the encoding of these stimuli and the retention of direction information for delay periods of a few seconds in duration. Their results suggest that the encoding and the storage of motion information are closely linked: lesions that degrade motion encoding also affect the animal's working memory for motion.

In monkeys with unilateral MT/MST lesions, Bisley and Pasternak (2000) placed either a sample stimulus or a test stimulus in the intact or lesioned hemifield, allowing them to assess the contribution of this area to the encoding, retention, and retrieval of stimulus motion. When the sample stimulus consisted of random dots moving in a broad range of directions (complex motion), MT/MST lesions disrupted both encoding of these stimuli and retention and encoding at a longer delay. However, when the coherent sample (in which all dots moved in the same direction) was placed into the lesioned field, there was no deficit in perceptual thresholds and no additional deficit with a longer delay. This result shows that encoding and storage are closely linked – when the lesion degraded stimulus encoding, retention of that stimulus was also degraded. In the same task, microstimulation of directional columns in area MT disrupted both encoding and retention; it had different effects when delivered at different stages (Bisley *et al.*, 2001).

11.5.7 Biological motion

Biological motion refers to stimulus sequences that arise from complex 3D displacements of body parts including motions of limbs, eyes, and mouths. Based on the original findings of Johansson (1973), coherent object information can be retrieved from sparse dot motion if this motion reflects the real object motion. Striking examples have been presented which indicate that the repetitive motion of a small number of dots can provide reliable information about the gender, build, and even mood of light-point walkers (Troje, 2002). Brain-imaging studies have been conducted to explore the cortical basis of biological motion processing. Using light-point figures (Johansson, 1975) as stimuli, Bonda *et al.* (1996) showed that body movements evoked greater activity in the superior temporal sulcus (STS) region compared with random motion. Puce *et al.* (1998) found that radial gratings evoked a large activation in V5 but little or no activation in the STS region. Viewing the movements of the mouth and eyes evoked activation in the STS but less activation in V5/MT+. Comparison of response patterns in STS to walking patterns and speech movements of the human mouth suggests that this region performs a differential analysis of visual motion and polysensory neurons compare visual and auditory speech-related inputs (Wright *et al.*, 2003). Taken together, these studies suggest the existence of a cortical area in STS that preferentially responds to biological motion.

11.5.8 Summary

In this section we have reviewed psychophysical and brain-imaging studies that explored the cortical mechanisms underlying motion perception and the control of pursuit eye movements. The study of lesions in the motion-sensitive cortical areas MT, MST, and LIP as well as functional imaging studies point to a complex hierarchy of extrastriate visual areas in the human cortex that respond selectively to different aspects of moving stimuli. Together these areas contribute to the analysis of motion-defined boundaries required to segment complex visual scenes into figure and ground. Damage to these areas leads to an impairment in the ability of the observer to extract motion information. Once a moving target has been segmented from a background and selected as a target for attention, the observer has the necessary information required for ocular pursuit. Smooth pursuit is associated with activation in motion-selective visual areas, but also with activation in premotor cortex related to the control of eye movements.

11.6 Conclusion

In this chapter, we demonstrated how psychophysics can help, by measuring and modeling visual performances, to characterize the processing of visual signals along the different stages of the visual pathways. However, only at low levels in the pathway does the behavior of single cells correlate well with psychophysical data; for complex stimuli and tasks, the behavior of cells and observers diverge, indicating more complex networks and influences from higher tier areas in the cortex, including cognition. We have described color and motion processing in the cortex as examples for these multi-area-network characteristics of higher-order information processing.

The challenge of future research will therefore be to identify the function and properties of these networks and their complex cooperation. New approaches in fMRI work will enable us to measure not only the location but also the functional properties of areas and networks. Furthermore, the direct comparison of cortical activity in monkeys and humans will help to identify homologous structures and areas in the cortex. This will help to bring the results from physiology (naturally predominantly from monkeys) and psychophysics (predominantly, but not exclusively, from humans) into closer relation and to understand better the functional properties of cells and networks in the brain. Therefore, reaching the goal may not be so far after all, i.e. 'finding strong and immediate correlations between structure and function of the visual nervous system and the visual performance of the whole organism' (after Gerald Wertheimer, 1990).

References

Abney, W., 1910. On the changes in hue of spectrum colours by dilution with white light. *Proc. R. Soc. Lond. A* **83**, 120–127.

Addams, R., 1834. An account of a peculiar optical phenomena seen after having looked at a moving body. *Lond. Edinb. Philos. Mag. J. Sci.* **5**, 373–374.

Adelson, E.H. and Bergen, J.R., 1985. Spatiotemporal energy models for the perception of motion. *J. Opt. Soc. Am. A* **2**, 284–299.

Albright, T.D., 1984. Direction and orientation selectivity of neurons in visual area MT of macaque. *J. Neurophysiol.* **52**, 1106–1130.

Alpern, M., 1964. Relation between brightness and color contrast. *J. Opt. Soc. Am. A* **54**, 1491–1492.

Andersen, R.A., 1995. Encoding of intention and spatial location in the posterior parietal cortex. *Cereb. Cortex* **5**, 457–469.

Andersen, R.A., Shenoy, K.V., Crowell, J.A. *et al.*, 2000. Neural mechanisms for self-motion perception in area MST. *Int. Rev. Neurobiol.* **44**, 219–233.

Barnes, C.S., Wei, J., and Shevell, S.K., 1999. Chromatic induction with remote chromatic contrast varied in magnitude, spatial frequency, and chromaticity. *Vision Res.* **39**, 3561–3574.

Bartels, A. and Zeki, S., 2000. The architecture of the colour centre in the human visual brain: New results and a review. *Eur. J. Neurosci.* **12**, 172–193.

Barton, J.J. and Sharpe, J.A., 1998. Ocular tracking of step-ramp targets by patients with unilateral cerebral lesions. *Brain* **121**, 1165–1183.

Battelli, L., Cavanagh, P., and Thornton, I.M., 2003. Perception of biological motion in parietal patients. *Neuropsychologia* **41**(13), 1808–1816.

Beauchamp, M.S., Haxby, J.V., Jennings, J.E. *et al.*, 1999. A fMRI version of the Farnsworth–Munsell 100-Hue Test reveals multiple color selective areas in human ventral occipitotemporal cortex. *Cereb. Cortex* **9**, 257–263.

Beer, J., Blakemore, C., Previc, F.H. *et al.*, 2002. Areas of the human brain activated by ambient visual motion, indicating three kinds of self-movement. *Exp. Brain Res.* **143**, 78–88.

Benton, C.P. and Johnston, A., 2001. A new approach to analysing texture-defined motion. *Proc. R. Soc. Lond. B* **268**, 2433–2443.

Bertone, A. and Faubert, J., 2003. How is complex second-order motion processed? *Vision Res.* **43**, 2591–2601.

Bisley, J.W. and Pasternak, T., 2000. The multiple roles of visual cortical areas MT/MST in remembering the direction of visual motion. *Cereb. Cortex* **10**, 1053–1065.

Bisley, J.W., Zaksas, D., and Pasternak, T., 2001. Microstimulation of cortical area MT affects performance on a visual working memory task. *J. Neurophysiol.* **85**, 187–196.

Blake, R., Cepeda, N.J., and Hiris, E., 1997. Memory for visual motion. *J. Exp. Psychol. Hum. Percept. Perform.* **23**, 353–369.

Blanke, O., Landis, T., Mermoud, C. *et al.*, 2003. Direction-selective motion blindness after unilateral posterior brain damage. *Eur. J. Neurosci.* **18**, 709–722.

Bloj, M.G., Kersten, D., and Hurlbert, A.C., 1999. Perception of three-dimensional shape influences colour perception through mutual illumination. *Nature* **402**, 877–879.

Bonda, E., Petrides, M., Ostry, D. *et al.*, 1996. Specific involvement of human parietal systems and the amygdala in the perception of biological motion. *J. Neurosci.* **16**(11), 3737–3744.

Braddick, O.J., O'Brien, J.M., Wattam-Bell, J. *et al.*, 2001. Brain areas sensitive to coherent visual motion. *Perception* **30**, 61–72.

Bruce, C.J., Goldberg, M.E., Stanton, G.B. *et al.*, 1985. Primate frontal eye fields. 2. Physiological and anatomical correlates of electrically evoked eye movements. *J. Neurophysiol.* **54**, 714–734.

Buchsbaum, G. and Gottschalk, A., 1983. Trichromacy, opponent colours coding and optimum colour information transmission in the retina. *Proc. R. Soc. Lond. B* **220**, 89–113.

Cavanagh, P., 1991. Vision at equiluminance, in *Limits of Vision*, Vol. 5 (eds J.J. Kulikowski, V. Walsh, and I.J. Murray), Macmillan, London, pp. 234–250.

Cheng, K., Fujita, H., Kanno, I. *et al.*, 1995. Human cortical regions activated by wide-field visual motion: An $H_2^{15}O$ PET study. *J. Neurophysiol.* **74**, 413–427.

Chevreul, E., 1839. *De la loi du contraste simultané des couleurs et de l'assortiment des object colorés*, Paris.
Chubb, C. and Sperling, G., 1988. Drift-balanced random stimuli: A general basis for studying non-Fourier motion perception. *J. Opt. Soc. Am. A* **5**(11), 1986–2007.
Claeys, K.G., Lindsey, D., Schutter, E.D. *et al.*, 2003. A higher order motion region in human inferior parietal lobule: Evidence from fMRI. *Neuron* **40**, 631–642.
Colby, C.L., 1998. Action-oriented spatial reference frames in cortex. *Neuron* **20**, 15–24.
Conway, B., 2001. Spatial structure of cone inputs to color cells in alert macaque primary visual cortex (V1). *J. Neurophysiol.* **21**, 2768–2783.
Conway, B.R., Hubel, D.H., and Livingstone, M.S., 2002. Color contrast in macaque V1. *Cereb. Cortex* **12**, 915–925.
Cook, E.P. and Maunsell, J.H., 2002. Dynamics of neuronal responses in macaque MT and VIP during motion detection. *Nat. Neurosci.* **5**(10), 985–994.
Corbetta, M., Miezin, F.M., Dobmeyer, S. *et al.*, 1991. Selective and divided attention during visual discriminations of shape, color, and speed: Functional anatomy by positron emission tomography. *J. Neurosci.* **11**, 2383–2402.
Cornette, L., Dupont, P., Rosier, A. *et al.*, 1998. Human brain regions involved in direction discrimination. *J. Neurophysiol.* **79**, 2749–2765.
Courtney, S.M., Petit, L., Haxby, J.V. *et al.*, 1998. The role of prefrontal cortex in working memory: Examining the contents of consciousness. *Philos. Trans. R. Soc. Lond. B Biol. Sci.* **353**, 1819–1828.
Cowey, A. and Heywood, C.A., 1995. There is more to colour than meets the eye. *Behav. Brain Res.* **71**, 89–100.
Damasio, A., Yamada, T., Damasio, H. *et al.*, 1980. Central achromatopsia: Behavioural, anatomic and physiologic aspects. *Neurology* **30**, 1064–1071.
de Jong, B.M., Shipp, S., Skidmore, B. *et al.*, 1994. The cerebral activity related to the visual perception of forward motion in depth. *Brain* **117**, 1039–1054.
Derrington, A.M., Krauskopf, J., and Lennie, P., 1984. Chromatic mechanisms in lateral geniculate nucleus of macaque. *J. Physiol. (Lond.)* **357**, 241–265.
Desimone, R., Moran, J., Schein, S.J. *et al.*, 1993. A role for the corpus callosum in visual area V4 of the macaque. *Vis. Neurosci.* **10**, 159–171.
De Valois, R.L., Cottaris, N.P., Mahon, L.E. *et al.*, 2000. Some transformations of color information from lateral geniculate nucleus to striate cortex. *Proc. Natl. Acad. Sci. USA* **97**, 4997–5002.
Dobkins, K.R. and Albright, T.D., 1995. Behavioral and neural effects of chromatic isoluminance in the primate visual motion system. *Vis. Neurosci.* **12**(2), 321–332.
Dow, B.M. and Vautin, R.G., 1987. Horizontal segregation of color information in the middle layers of foveal striate cortex. *J. Neurophysiol.* **57**, 712–739.
Dukelow, S.P., DeSouza, J.F., Culham, J.C. *et al.*, 2001. Distinguishing subregions of the human MT+ complex using visual fields and pursuit eye movements. *J. Neurophysiol.* **86**, 1991–2000.
Dupont, P., Orban, G.A., De Bruyn, B. *et al.*, 1994. Many areas in the human brain respond to visual motion. *J. Neurophysiol.* **72**, 1420–1424.
Dupont, P., De Bruyn, B., Vandenberghe, R. *et al.*, 1997. The kinetic occipital region in human visual cortex. *Cereb. Cortex* **7**, 283–292.
Dupont, P., Sáry, G., Peuskens, H. *et al.*, 2003. Cerebral regions processing first- and higher-order motion in an opposed-direction discrimination task. *Eur. J. Neurosci.* **17**(7), 1509–1517.
Dursteler, M.R., Wurtz, R.H., and Newsome, W.T., 1987. Directional pursuit deficits following lesions of the foveal representation within the superior temporal sulcus of the macaque monkey. *J. Neurophysiol.* **57**, 1262–1287.
Edwards, R., Xioa, D., Keysers, C. *et al.*, 2003. Color sensitivity of cells responsive to complex stimuli in the temporal cortex. *J. Neurophysiol.* **90**, 1245–1256.
Engel, S.A. and Furmanski, Ch.S., 2001. Selective adaptation to color contrast in human primary visual cortex. *J. Neurosci.* **21**, 3949–3954.
Evans, R.M., 1948. *An Introduction to Color*, John Wiley, New York.
Fairchild, M. and Lennie, P., 1992. Chromatic adaptation to natural and incandescent illuminants. *Vision Res.* **32**, 2077–2085.

Fairchild, M. and Reniff, L., 1995. Time course of chromatic adaptation for color appearance judgments. *J. Opt. Soc. Am. A*, **12**, 824–833.

Fechner, G., 1860. *Elements of Psychophysics* (H.E. Adler, Transl.), Vol. I, Holt, Rinehart, and Winston, New York, 1966.

Foster, D.H. and Nascimento, S.M.C., 1994. Relational colour constancy from invariant cone-excitation ratios. *Proc. R. Soc. Lond. B* **257**, 115–121.

Foster, K.H., Gaska, J.P., Nagler, M. *et al.*, 1985. Spatial and temporal frequency selectivity of neurones in visual cortical areas V1 and V2 of the macaque monkey. *J. Physiol.* **365**, 331–363.

Freitag, P., Greenlee, M.W., Lacina, T. *et al.*, 1998. Effect of eye movements on the magnitude of fMRI responses in extrastriate cortex during visual motion perception. *Exp. Brain Res.* **119**, 409–414.

Gegenfurtner, K., 2004. Cortical mechanisms of colour vision. *Nat. Rev. Neurosci.* **4**, 563–572.

Gegenfurtner, K.R. and Kiper, D.C., 2003. Color vision. *Annu. Rev. Neurosci.* **26**, 181–206.

Gegenfurtner, K.R., Kiper, D.C., and Fenstermaker, S.B., 1996. Processing of colour, form and motion in macaque area V2. *Vis. Neurosci.* **13**, 161–172.

Gegenfurtner, K.R., Kiper, D.C., and Levitt, J.B., 1997. Functional properties of neurons in macaque area V3. *J. Neurophysiol.* **77**, 1906–1923.

Geisler, W.S., 1981. Effects of bleaching and backgrounds on the flash response of the cone system. *J. Physiol. (Lond.)* **312**, 413–434.

Pasternak, T. and Greenlee, M.W., 2005. Working memory in primate sensory systems. *Nat. Rev. Neurosci.* **6**, 97–101.

Greenlee, M.W. and Smith, A.T., 1997. Detection and discrimination of first- and second-order motion in patients with unilateral brain damage. *J. Neurosci.* **17**, 804–818.

Greenlee, M.W., Lang, H.J., Mergner, T. *et al.*, 1995. Visual short-term memory of stimulus velocity in patients with unilateral posterior brain damage. *J. Neurosci.* **15**, 2287–2300.

Guth, S.L., Massof, R.W., and Benzschawel, T., 1980. Vector model for normal and dichromatic color vision. *J. Opt. Soc. Am.* **70**, 197–212.

Hadjikani, N., Liu, A.K., Dale, A.M. *et al.*, 1998. Retinotopy and colour sensitivity in human visual cortical area V8. *Nat. Neurosci.* **1**, 235–241.

Hanazawa, A., Komatsu, H., and Murakami, I., 2000. Neural selectivity for hue and saturation in the primary visual cortex of the monkey. *Eur. J. Neurosci.* **12**, 1753–1763.

Hayhoe, M. and Wenderoth, P., 1990. Adaptation mechanisms in color and brightness, in *From Pigments to Perception. Advances in Understanding Visual Processes. NATO ASI Series A* (eds A. Valberg and B.B. Lee), Plenum Press, New York, pp. 353–367.

Hayhoe, M., Benimoff, N.I., and Hood, D.C., 1987. The time course of multiplicative and subtractive adaptation process. *Vision Res.* **27**, 1981–1996.

Heeger, D.J., Boynton, G.M., Demb, J.B. *et al.*, 1999. Motion opponency in visual cortex. *J. Neurosci.* **19**(16), 7162–7174.

Hering, E., 1920. *Grundzüge der Lehre vom Lichtsinn*, Verlag von Julius Springer, Berlin.

Heywood, C.A., Gaffann, D., and Cowey, A., 1995. Cerebral achromatopsia in monkeys. *Eur. J. Neurosci.* **7**, 1064–1073.

Howard, R., Ffytche, D.H., Barnes, J. *et al.*, 1998. The functional anatomy of imagining and perceiving colour. *Neuroreport* **9**, 1019–1023.

Hurlbert, A. and Poggio, T., 1989. A network for image segmentation using color, in *Neural Information Processing Systems, I* (ed D. Touretzky), Morgan Kaufmann Publishers, Los Altos, CA, pp. 297–303.

Hurlbert, A. and Wolf, K., 2004. Color contrast: A contributory mechanism to color constancy. *Prog. Brain Res.* **144**, 147–160.

Huxlin, K.R. and Pasternak, T., 2004. Training-induced recovery of visual motion perception after extrastriate cortical damage in the adult cat. *Cereb. Cortex* **14**(1), 81–90.

Jaeger, W., 1982. Untersuchungen zu Farbkonstanz und Farbgedächtnis (Investigations concerning colour constancy and colour memory). Heidelberger Akademie der Wissenschaften, mathematisch-naturwissenschaftliche Klasse, 5. Abhandlung.

Jameson, D. and Hurvich, L.M., 1964. Theory of brightness and color contrast in human vision. *Vision Res.* **4**, 135–154.

Jameson, D. and Hurvich, L.M., 1972. Color adaptation: Sensitivity control, contrast, afterimages, in *Handbook of Sensory Physiology* VII/5 (eds D. Jameson and L.M. Hurvich), Springer, Berlin, pp. 568–581.
Johansson, G., 1973. Visual perception of biological motion and a model for its analysis. *Percept. Psychophys.* **14**, 201–211.
Johansson, G., 1975. Visual motion perception. *Sci. Am.* **232**, 76–88.
Johnson, E.N., Hawken, M.J., and Shapley, R., 2001. The spatial transformation of color in the primary visual cortex of macaque monkey. *Nat. Neurosci.*, **4**, 409–416.
Kaiser, P.K., Lee, B.B., Martin, P.R. *et al.*, 1990. The physiological basis of the minimally distinct border demonstrated in the ganglion cells of the macaque retina. *J. Physiol. (Lond.)* **422**, 153–183.
Kaplan, E. and Shapley, R.M., 1986. The primate retina contains two types of ganglion cells, with high and low contrast sensitivity. *Proc. Natl. Acad. Sci. USA* **83**, 2755–2757.
Katz, D., 1935. *The World of Color*, Regan Paul, Trench, Trubner, London.
Kelly, D.H., 1974. Spatio-temporal frequency characteristics of color-vision mechanisms. *J. Opt. Soc. Am.* **64**, 983–990.
Kimmig, H., Pinnow, C., Mergner, T. *et al.*, 1995. Smooth pursuit eye movements in patients with impaired visual motion perception, in *Multisensory Control of Posture* (eds T. Mergner and F. Hlavacka), Plenum Press, New York, pp. 325–330.
Kimmig, H., Greenlee, M.W., Huethe, F. *et al.*, 1999. MR-eyetracker: A new method for eye movement recording in functional magnetic resonance imaging (fMRI). *Exp. Brain Res.* **126**, 443–449.
Kiper, D.C., Fenstemacher, S.B., and Gegenfurtner, K.R., 1997. Chromatic properties of neurons in macaque area V2. *Vis. Neurosci.* **14**, 1061–1072.
Koenderink, J.J., 1986. Optic flow. *Vision Res.* **26**, 161–179.
Koffka, K., 1935. *Principles of Gestalt Psychology*, Hartcourt, Brace & World, New York.
Kohn, A. and Movshon, J.A., 2003. Neuronal adaptation to visual motion in area MT of the macaque. *Neuron* **39**(4), 681–691.
Komatsu, H., Ideura, Y., Kaji, S. *et al.*, 1992. Color selectivity of neurons in the inferior temporal cortex of the awake macaque monkey. *J. Neurosci.* **12**, 408–424.
Krauskopf, J. and Gegenfurtner, K.R., 1992. Color discrimination and adaptation. *Vision Res.* **32**, 2165–2175.
Krauskopf, J., Williams, D.R., and Heeley, D.W., 1982. Cardinal directions in colour space. *Vision Res.* **22**, 1123–1131.
Krauskopf, J., Zaidi, Q., and Mandler, M.B., 1986. Mechanisms of simultaneous color induction. *J. Opt. Soc. Am. A* **3**, 1752–1757.
Krauskopf, J., Wu, H.J., and Farell, B., 1996. Coherence, cardinal directions and higher order color mechanisms. *Vision Res.* **36**, 1235–1245.
Krauzlis, R.J. and Stone, L.S., 1999. Tracking with the mind's eye. *Trends Neurosci.* **22**(12), 544–550.
Lambert, J.H., 1772. Beschreibung einer mit dem Calauschen Wachse ausgemalten Farbpyramide, wo die Mischung jeder Farbe aus Weiss und drey Grundfarben angeordnet, dargelegt und derselben Berechnung und vielfacher Gebrauch angewiesen wird durch J.H. Lampert, Hande und Spener, Berlin.
Lamme, V.A.F., 1995. The neurophysiology of figure-ground segregation in primary visual cortex. *J. Neurosci.* **15**, 1605–1615.
Land, E.H., 1964. The Retinex. *Am. Sci.* **52**, 247.
Land, E.H., 1986. Recent advances in retinex theory. *Vision Res.* **26**, 7–22.
Land, E.H. and McCann, J.J., 1971. Lightness and Retinex Theory, *J. Opt. Soc. Am.* **61**, 1–11.
Land, E.H., Hubel, D.H., Livinstone, M.S. *et al.*, 1983. Colour-generating interactions across the corpus callosum. *Nature* **303**, 616–618.
Ledgeway, T. and Smith, A.T., 1997. Changes in perceived speed following adaptation to first-order and second-order motion. *Vision Res.* **37**, 215–224.
Lee, B.B., Valberg, A., Tigwell, D.A. *et al.*, 1987. An account of responses of spectrally opponent neurones in macaque lateral geniculate nucleus to successive contrast. *Proc. R. Soc. Lond. B* **230**, 293–314.

Lee, B.B., Martin, P.R., and Valberg, A., 1988. The physiological basis of heterochromatic flicker photometry demonstrated in the ganglion cells of the macaque retina. *J. Physiol. (Lond.)* **404**, 323–347.

Lee, B.B., Martin, P.R., and Valberg, A., 1989. Nonlinear summation of M- and L-cone inputs to phasic retinal ganglion cells of the macaque. *J. Neurosci.* **9**, 1433–1442.

Lee, B.B., Pokorny, J., Smith, V.C. *et al.*, 1990. Luminance and chromatic modulation sensitivity of macaque ganglion cells and human observers. *J. Opt. Soc. Am. A* **7**, 2223–2236.

Lennie, P., 1999. Color coding in the cortex, in *Color Vision: From Genes to Perception* (eds K.R. Gegenfurtner and L.T. Sharpe), Cambridge University Press, Cambridge, pp. 235–247.

Lennie, P., Krauskopf, J., Sclar, G., 1990. Chromatic mechanisms in striate cortex of macaque. *J. Neurosci.* **10**, 649–669.

Leonova, A., Pokorny, J., and Smith, V.C., 2003. Spatial frequency processing in inferred PC- and MC-pathways. *Vision Res.* **43**, 2133–2139.

Levitt, J.B., Kiper, D.C. and Movshon, J.A., 1994. Receptive fields and functional architecture of macaque V2. *J. Neurophysiol.* **71**, 2517–2542.

Lisberger, S.G. and Movshon, J.A., 1999. Visual motion analysis for pursuit eye movements in area MT of macaque monkeys. *J. Neurosci.* **19**, 2224–2246.

Livingstone, M.S. and Hubel, D.H., 1987. Psychophysical evidence for separate channels for the perception of form, color, motion and depth. *J. Neurosci.* **7**, 3416–3468.

Livingstone, M.S. and Hubel, D.H., 1988. Segregation of form, color, movement and depth: Anatomy, physiology and perception. *Science* **240**, 740–750.

Lu, Z.L. and Sperling, G., 2001. Three-systems theory of human visual motion perception: Review and update. *J. Opt. Soc. Am. A Opt. Image Sci. Vis.* **18**, 2331–2370.

Lu, Z.L., Lesmes, L.A., and Sperling, G., 1999. The mechanism of isoluminant chromatic motion perception. *PNAS* **96**, 8289–8294.

Lueck, C.J., Zeki, S., Friston, K.J. *et al.*, 1989. The colour centre in the cerebral cortex of man. *Nature* **340**, 386–389.

Lynch, J.C., 1987. Frontal eye field lesions disrupt visual pursuit. *Exp. Brain Res.* **68**, 437–441.

MacEvoy, S.P. and Paradiso, M.A., 2001. Lightness constancy in primary visual cortex. *Proc. Natl. Acad. Sci. USA* **98**, 8827–8831.

Magnussen, S. and Greenlee, M.W., 1992. Retention and disruption of motion information in visual short-term memory. *J. Exp. Psychol. Learn. Mem. Cogn.* **18**, 151–156.

Magnussen, S. and Greenlee, M.W., 1999. The psychophysics of perceptual memory. *Psychol. Res.* **62**, 81–92.

Martinez-Oriegas E., 1994. Chromatic-achromatic multiplexing in human color vision, in *Visual Science and Engineering. Models and Applications,* (ed D.H. Kelly), Marcel Dekker, New York, pp. 117–187.

Mather, G., Verstraten, F., and Anstis, S., 1998. *The Motion Aftereffect: A Modern Perspective*, MIT Press, Boston.

Maunsell, J.H., 1992. Functional visual streams. *Curr. Opin. Neurobiol.* **2**, 506–510.

Michael, C.R., 1978. Color vision mechanisms in monkey striate cortex: Dual-opponent cells with concentric receptive fields. *J. Neurophysiol.* **41**, 572–588.

McKeefry, D.J. and Zeki, S., 1997. The position and topography of the human colour centre as revealed by functional magnetic resonance imaging. *Brain* **120**, 2229–2242.

Mollon, J.D., 1982. Color vision. *Annu. Rev. Psychol.* **33**, 41–85.

Mollon, J.D. and Polden, P.G., 1975. Colour illusion and evidence for interaction between cone mechanisms. *Nature* **258**, 421–422.

Monnier, P. and Shevell, S.K., 2004. Chromatic induction from S-cone patterns. *Vision Res.* **44**, 849–856.

Morrone, M.C., Tosetti, M., Montanaro, D. *et al.*, 2000. A cortical area that responds specifically to optic flow, revealed by fMRI. *Nat. Neurosci.* **3**, 1322–1328.

Moutoussis, K. and Zeki, S., 2002. Response of spectrally selective cells in macaque area V2 to wavelengths and colors. *J. Neurophysiol.* **87**, 2104–2112.

Mueller, R. and Greenlee, M.W., 1994. Effect of contrast and adaptation on the perception of the direction and speed of drifting gratings. *Vision Res.* **34**, 2071–2092.

Mullen, K.T., 1985. The contrast sensitivity of human colour vision to red-green and blue-yellow chromatic gratings. *J. Physiol. (Lond.)* **359**, 381–400.

Müller, R., Göpfert, E., Leineweber, M., *et al.*, 2004. Effect of adaptation direction on the motion VEP and perceived speed of drifting gratings. *Vision Res.* **44**, 2381–2392.

Munsell, A.H., 1905. *A Color Notation.* Boston.

Munsell, A.H., 1929. *Book of Colors: A Revision and Extension of the Atlas of the Munsell Color System.* Baltimore.

Nascimento, S.M., Ferreira, F.P., and Foster, D., 2002. Statistics of spatial cone-excitation ratios in natural scenes. *J. Opt. Soc. Am. A Opt. Image Sci. Vis.* **19**, 1484–1490.

Newsome, W.T. and Pare, E.B., 1988. A selective impairment of motion perception following lesions of the middle temporal visual area (MT). *J. Neurosci.* **8**, 2201–2211.

Newsome, W.T., Wurtz, R.H., Dursteler *et al.*, 1985. Deficits in visual motion processing following ibotenic acid lesions of the middle temporal visual area of the macaque monkey. *J. Neurosci.* **5**, 825–840.

Newton, I., 1704. Opticks: or, a Treatise of the Reflexions, Refractions, Inflexions and Colours of Light. Also two Treatises of the Species and Magnitude of Curvilinear Figures. Printed for Sam, Smith, and Benj. Walford, Printers to the Royal Society, London.

Nishida, S., Sasaki, Y., Murakami, I. *et al.*, 2003. Neuroimaging of direction-selective mechanisms for second-order motion. *J. Neurophysiol.* **90**, 3242–3254.

Nunn, J.A., Gregory, L.J., Brammer, M. *et al.*, 2002. Functional magnetic resonance imaging of synesthesia: activation of V4/V8 by spoken words. *Nat. Neurosci.* **5**, 371–375.

Orban, G.A., Dupont, P., De Bruyn, B. *et al.*, 1998. Human brain activity related to speed discrimination tasks. *Exp. Brain Res.* **122**, 9–22.

Petit, L., Clark, V.P., Ingeholm, J. *et al.*, 1997. Dissociation of saccade-related and pursuit-related activation in human frontal eye fields as revealed by fMRI. *J. Neurophysiol.* **77**, 3386–3390.

Pinsk, M.A., Doniger, G.M., and Kastner, S., 2004. Push-pull mechanisms of selective attention in the human extrastriate cortex. *J. Neurophysiol.* **92**, 622–629.

Pokorny, J. and Smith, V.C., 1997. Psychophysical signatures associated with magnocellular and parvocellular pathway contrast gain. *J. Opt. Soc. Am. A* **14**, 2477–2486.

Pokorny, J., Sun, V.C.W., and Smith, V.C., 2003. Temporal dynamics of early light adaptation. *J. Vis.* **3**, 423–431.

Puce, A., Allison, T., Bentin, S. *et al.*, 1998. Temporal cortex activation in humans viewing eye and mouth movements. *J. Neurosci.* **18**, 2188–2199.

Purdy, D.M., 1931. Spectral hue as a function of intensity. *Am. J. Psychol.* **43**, 541–559.

Rees, G., and Friston, K., and Koch, C., 2000. A direct quantitative relationship between the functional properties of human and macaque V5. *Nat. Neurosci.* **3**, 716–723.

Reichardt, W., 1961. Autocorrelation: a principle for the evaluation of sensory information by the central nervous system, in *Sensory Communication* (ed W.A. Rosenblith), MIT Press, Cambridge, pp. 303–317.

Reppas, J.B., Niyogi, S., Dale, A.M. *et al.*, 1997. Representation of motion boundaries in retinotopic human visual cortical areas. *Nature* **388**, 175–179.

Rinner, O. and Gegenfurtner, K.R., 2000. Time course of chromatic adaptation for color appearance and discrimination. *Vision Res.* **40**, 1813–1826.

Rosano, C., Krisky, C.M., Welling, J.S. *et al.*, 2002. Pursuit and saccadic eye movement subregions in human frontal eye field: A high-resolution fMRI investigation. *Cereb. Cortex* **12**, 107–115.

Ruettiger, L., Braun, D.I., Gegenfurtner, K.R. *et al.*, 1999. Selective color constancy deficits after circumscribed unilateral brain lesions. *J. Neurosci.* **19**, 3094–3106.

Rutschmann, R.M., Schrauf, M., and Greenlee, M.W., 2000. Brain activitation during dichoptic presentation of optic flow stimuli. *Exp. Brain Res.* **134**, 533–537.

Schein, S.J. and Desimone, R., 1990. Spectral properties of V4 neurons in the macaque. *J. Neurosci.* **10**, 3369–3389.

Schein, S.J., Marrocco, R.T., and de Monasterio, F.M., 1982. Is there a high concentration of color-selective cells in area V4 of monkey visual cortex? *J. Neurophysiol.* **47**, 193–213.

Schiller, P., 1993. The effects of V4 and middle temporal (MT) area lesions on visual performance in the rhesus monkey. *Vis. Neurosci.* **10**, 717–746.

Schiller, P.H. and Colby, C.L., 1983. The responses of single cells in the lateral geniculate nucleus of the rhesus monkey to color and luminance contrast. *Vision Res.* **23**, 1631–1641.

Schiller, P.H., True, S.D., and Conway, J.L., 1979. Effects of frontal eye field and superior colliculus ablations on eye movements. *Science* **206**, 590–502.

Schirillo, J.A. and Shevell, S.K., 2000. Role of perceptual organization in chromatic induction. *J. Opt. Soc. Am. A* **17**, 244–254.

Seiffert, A.E. and Cavanagh, P., 1999. Position-based motion perception for color and texture stimuli: Effects of contrast and speed. *Vision Res.* **39**, 4172–4185.

Seiffert, A.E., Somers, D.C., Dale, A.M. *et al.*, 2003. Functional MRI studies of human visual motion perception: Texture, luminance, attention and after-effects. *Cereb. Cortex* **13**, 340–349.

Sekiguchi, N., Williams, D.R., and Brainard, D.H., 1993. Aberration-free measurements of the visibility of isoluminant gratings. *J. Opt. Soc. Am. A* **10**, 2105–2117.

Shevell, S.K., 1978. The dual role of chromatic backgrounds in color perception. *Vision Res.* **18**, 1649–1661.

Shevell, S.K. and Miller, P.R., 1996. Color perception with test and adaptation lights perceived in different depth planes. *Vision Res.* **36**, 949–954.

Shevell, S.K. and Wei, J., 2000. A central mechanism of chromatic contrast. *Vision Res.* **40**, 3173–3180.

Singh, K.D., Smith, A.T., and Greenlee, M.W., 2000. Spatiotemporal frequency and direction sensitivities of human visual areas measured using fMRI. *Neuroimage* **12**, 550–564.

Skiera, G., Petersen, D., Skalej, M. *et al.*, 2000. Correlates of figure-ground segregation in fMRI. *Vision Res.* **40**, 2047–2056.

Smith, A.T., 1994. The detection of second-order motion, in *Visual Detection of Motion* (eds A.T. Smith and R. Snowden), Academic Press, New York.

Smith, V.C. and Pokorny, J., 2003. Psychophysical correlates of parvo- and magnocellular function, in *Normal and Defective Colour Vision* (eds J. Mollon, J. Pokorny, and K. Knoblauch), Oxford University Press, Oxford, pp. 91–107.

Smith, A.T., Greenlee, M.W., Singh, K.D. *et al.*, 1998. The processing of first- and second-order motion in human visual cortex assessed by functional magnetic resonance imaging (fMRI). *J. Neurosci.* **18**, 3816–3830.

Smith, V.C., Pokorny, J., and Sun, H., 2000. Chromatic contrast discrimination: Data and prediction for stimuli varying in L and M cone excitation. *Col. Res. Appl.* **25**, 105–115.

Smith, V.C., Sun, V.C., and Pokorny, J., 2001. Pulse and steady-pedestal contrast discrimination: Effect of spatial parameters. *Vision Res.* **41**, 2079–2088.

Swanson, W.H., Wilson, H.R., and Giese, S.C., 1984. Contrast matching data predicted from contrast thresholds. *Vision Res.* **24**, 63–75.

Swanson, W.H., Ueno, T., Smith, V.C. *et al.*, 1987. Temporal modulation sensitivity and pulse detection thresholds for chromatic and luminance perturbations. *J. Opt. Soc. Am. A* **4**, 1992–2005.

Tehovnik, E.J., Sommer, M.A., Chou, I.H. *et al.*, 2000. Eye fields in the frontal lobes of primates. *Brain Res. Brain Res. Rev.* **32**, 413–448.

Tootell, R.B., Reppas, J.B., Kwong, K.K. *et al.*, 1995. Functional analysis of human MT and related visual cortical areas using magnetic resonance imaging. *J. Neurosci.* **15**, 3215–3230.

Troje, N.F., 2002. Decomposing biological motion: A framework for analysis and synthesis of human gait patterns. *J. Vis.* **2**, 371–387.

Ts'o, D.Y. and Gilbert, C.D., 1988. The organization of chromatic and spatial interactions in the primate striate cortex. *J. Neurosci.* **8**, 1712–1727.

Ungerleider, L.G. and Mishkin, M., 1982. Two cortical visual systems, in *Analysis of Visual Behaviour* (eds D.J. Ingel, M.A. Goodale, and R.J.W. Mansfield), MIT Press, Cambridge, MA, pp. 549–586.

Vaina, L.M., 1998. Complex motion perception and its deficits. *Curr. Opin. Neurobiol.* **8**, 494–502.

Valberg, A., 2001. Unique hues: An old problem for a new generation. *Vision Res.* **41**, 1645–1657.

Valberg, A., 2005. *Light Vision Color*, Wiley & Sons, New York.

Valberg, A. and Lee, B.B., 1989. Detection and discrimination of colour, a comparison of physiological and psychophysical data. *Physica Scripta* **39**, 178–186.

Valberg, A., Tryti, J., and Lee, B.B., 1985a. Computation of responses of opponent-cells in the macaque lateral geniculate nucleus to light stimuli varying in luminance, wavelength and purity. *Institute of Physics Report Series 85-29*, University of Oslo, Oslo, pp. 1–36.

Valberg, A., Lee, B.B., and Tigwell, D.A. *et al.*, 1985b. A simultaneous contrast effect of steady remote surrounds on responses of cells in macaque lateral geniculate nucleus. *Exp. Brain Res.* **58**, 604–608.

Valberg, A., Lee, B.B., and Tigwell, D.A., 1986a. Neurones with strong inhibitory S-cone inputs in the macaque lateral geniculate nucleus. *Vision Res.* **26**, 1061–1064.

Valberg, A., Seim, T., Lee, B.B. *et al.*, 1986b. Reconstruction of equidistant color space from responses of visual neurones of macaques. *J. Opt. Soc. Am. A* **3**, 126–134.

Valberg, A., Lee, B.B., and Tryti, J., 1987. Simulation of responses of spectrally-opponent neurones in the macaque lateral geniculate nucleus to chromatic and achromatic light stimuli. *Vision Res.* **27**, 867–882.

Valberg, A., Lange-Malecki, B., and Seim, T., 1991. Colour changes as a function of luminance contrast. *Perception* **20**, 655–668.

Valberg, A., Lee, B.B., Kaiser, P.K. *et al.*, 1992. Responses of macaque ganglion cells to movement of chromatic borders. *J. Physiol. (Lond.)* **458**, 579–602.

Van Essen, D.C., Maunsell, J.H., and Bixby, J.L., 1981. The middle temporal visual area in the macaque: Myeloarchitecture, connections, functional properties and topographic organization. *J. Comp. Neurol.* **199**, 293–326.

van Oostende, S., Sunaert, S., Van Hecke, P. *et al.*, 1997. The kinetic occipital (KO) region in man: An fMRI study. *Cereb. Cortex* **7**, 690–701.

van Santen, J.P. and Sperling, G., 1985. Elaborated Reichardt detectors. *J. Opt. Soc. Am. A* **2**, 300–321.

Vautin, R.G. and Dow, B.M., 1985. Color cell groups in foveal striate cortex of the behaving macaque. *J. Neurophysiol.* **54**, 273–292.

Verrey, L., 1888. Hemiachromatopsie droite absolue. *Archives de Ophthalmolgie* (Paris) **8**, 289–300.

Vidyasagar, T.R., Kulikowski, J.J., Lipnicki, D.M. *et al.*, 2002. Convergence of parvocellular and magnocellular information channels in the primary visual cortex of the macaque. *Eur. J. Neurosci.* **16**, 945–956.

von Bezold, W., 1873. Ûber das Gesetz der Farbmischung und die physiologischen Grundfarben. *Poggendorffs Annalen der Physik, Leipzig* **150**, 221–247.

von Brücke, E.T., 1878. Ûber einige Empfindungen im Gebiet der Sehnerven. *Sitzungsberichte der Akademie der Wissenschaften in Wien, Mathematisch-Naturwissenschaftliche Klasse, Abteilung 3*, **77**, 39–71.

von Goethe, J.W., 1810. *Zur Farbenlehre*. I. Band (Didaktischer und polemischer Teil), J.G. Kotta, Stuttgart.

von Helmholtz, H., 1896. *Handbuch der Physiologischen Optik*, Voss, Hamburg.

von Kries, J., 1904. Die Gesichtsempfindungen, in Nagels *Handbuch der Physiologie des Menschen*, **3**, 211.

Wachtler, T., Albright, T.D., and Sejnowski T.J., 2001. Nonlocal interactions in color perception: Nonlinear processing of chromatic signals from remote inducers. *Vision Res.* **41**, 1535–1546.

Wachtler, T., Sejnowski, T.J., and Albright, T.D., 2003. Representation of color stimuli in awake macaque primary visual cortex. *Neuron* **37**, 681–691.

Wade, A.R. and Wandell, B.A., 2002. Chromatic light adaptation measured using functional magnetic resonance imaging. *J. Neurosci.* **22**, 8148–8157.

Wallach, H., 1948. Brightness constancy and the nature of achromatic colors. *J. Exp. Psychol.* **38**, 310–324.

Walraven, J., 1976. Discounting the background – the missing link in the explanation of chromatic induction. *Vision Res.* **16**, 289–295.

Walsh, V., Carden, D., Butler, S.R. *et al.*, 1993. The effects of V4 lesions on the visual abilities of macaques: Hue discrimination and colour constancy. *Behav. Brain Res.* **53**, 51–62.

Wandell, B.A., 1999. Computational neuroimaging: Color representations and processing, in *New Cognitive Neuroscience* (ed M. Gazzaniga), MIT Press, Cambridge, MA, pp. 291–304.

Watson, J.D.G., Myers, R., Frackowiak, R.S.J. *et al.*, 1993. Area V5 of the human brain: Evidence from a combined study using positron emission tomography and magnetic resonance imaging. *Cereb. Cortex* **3**, 79–84.

Webster, M.A. and Mollon, J.D., 1991. Changes in colour appearance following post-receptoral adaptation. *Nature* **349**, 235–238.

Webster, M.A. and Mollon, J.D., 1994. The influence of contrast adaptation on color appearance. *Vision Res.* **34**, 1993–2020.

Webster, M.A. and Mollon, J.D., 1995. Colour constancy influenced by contrast adaptation. *Nature* **373**, 694–698.

Wenderoth, P., Watson, J.D., Egan, G.F. *et al.*, 1999. Second order components of moving plaids activate extrastriate cortex: A positron emission tomography study. *Neuroimage* **9**, 227–234.

Werner, A., 2003a. The spatial tuning of chromatic adaptation. *Vision Res.* **43**, 1611–1623.

Werner, A., 2003b. Chromatic adaptation across depth planes. *Perception* **32**, 148.

Werner, A., Sharpe, L.T., and Zrenner, E., 2000. Asymmetries in the time-course of chromatic adaptation and the significance of contrast. *Vision Res.* **40**, 1101–1113.

Westheimer, G., 1990. Relating neural mechanisms to visual perception: Historical and philosophical considerations, in *Visual Perception: The Neurophysiological Foundations* (eds L. Spillmann and J.S. Werner), Academic Press, San Diego, pp. 5–9.

Westheimer, G., Brincat, S., and Wehrhahn, C., 1999. Contrast dependency of foveal spatial functions: Orientation, vernier, separation, blur and displacement discrimination and the tilt and Poggendorff illusions. *Vision Res.* **39**, 1631–1639.

Whittle, P. and Challands, P.D.C., 1969. The effect of background luminance on the brightness of flashes. *Vision Res.* **9**, 1095–1110.

Wiesel, T.N. and Hubel, D.H., 1966. Spatial and chromatic interactions in the lateral geniculate body of the rhesus monkey. *J. Neurophysiol.* **29**, 1115–1156.

Wilson, H.R. and Bergen, J.R., 1979. A four mechanism model for threshold spatial vision. *Vision Res.* **19**, 19–32.

Williams, D.W. and Sekuler, R., 1984. Coherent global motion percepts from stochastic local motions. *Vision Res.* **24**(1), 55–62.

Wright, T.M., Pelphrey, K.A., Allison, T. *et al.*, 2003. Polysensory interactions along lateral temporal regions evoked by audiovisual speech. *Cereb. Cortex* **13**, 1034–1043.

Wyszecki, G. and Stiles, W., 1982. *Color Science*, Wiley, New York, pp. 840–852.

Xiao, Y., Wang, Y., and Felleman, D., 2003. A spatially organized representation of colour in macaque cortical area V2. *Nature* **421**, 536–539.

Yang, J.N. and Shevell, S.K., 2002. Stereo disparity improves colour constancy. *Vision Res.* **42**, 1979–1989.

Yang, J.N. and Shevell, S.K., 2003. Surface color perception under two illuminants: The second illuminant reduces color constancy. *J. Vis.* **3**, 369–379.

Zaidi, Q., 1999. Color and brightness induction: From Mach bands to three dimensional configurations, in *Color Vision: From Genes to Perception* (eds K.R. Gegenfurtner and T.L. Sharpe), Cambridge University Press, Cambridge, pp. 317–343.

Zaidi, Q., Shapiro, A., and Hood, D., 1992. The effect of adaptation on the differential sensitivity of the S-cone color system. *Vision Res.* **32**, 1297–1318.

Zaidi, Q., Spehar, B., and De Bonet, J., 1998. Adaptation to textured chromatic fields. *J. Opt. Soc. Am. A* **15**, 23–31.

Zaksas, D., Bisley, J.W., and Pasternak, T., 2001. Motion information is spatially localized in a visual working-memory task. *J. Neurophysiol.* **86**, 912–921.

Zeki, S.M., 1973. Colour coding in rhesus monkey prestriate cortex. *Brain Res.* **53**, 422–427.

Zeki, S.M., 1974. Functional organization of a visual area in the posterior bank of the superior temporal sulcus of the rhesus monkey. *J. Physiol. (Lond.)* **236**, 549–573.

Zeki, S.M., 1978. Uniformity and diversity of structure and function in rhesus monkey prestriate visual cortex. *J. Physiol.* **277**, 273–290.

Zeki, S.M., 1983a. Colour coding in the cerebral cortex: The reaction of cells in monkey visual cortex to wavelengths and colours. *Neuroscience* **9**, 741–765.

Zeki, S.M., 1983b. Colour coding in the cerebral cortex: The response of wavelength selective and colour coded cells in monkey visual cortex to changes in wavelength composition. *Neuroscience* **9**, 767–781.

Zeki, S.M., 1990. A century of cerebral achromatopsia. *Brain* **113**, 1721–1777.

Zeki, S. and Marini, L., 1998. Three cortical stages of colour processing in the human brain. *Brain* **121**, 1669–1685.

Zeki, S. and Shipp, S., 1989. Modular connections between areas V2 and V4 of macaque visual cortex. *Eur. J. Neurosci.* **1**, 494–506.

Zeki, S., Watson, J.D.G., Lueck, C.J. *et al.*, 1991. A direct demonstration of functional specialization in human visual cortex. *J. Neurosci.* **11**, 641–649.

Zeki, S., McKeefry, D.J., Bartels, A. *et al.*, 1998. Has a new colour area been discovered? *Nat. Neurosci.* **1**, 335.

Zipser, K., Lamme, V.A.F., and Schiller, P.H., 1996. Contextual modulation in primary visual cortex. *J. Neurosci.* **16**, 7376–7389.

Index

The Primate Visual System: A Comparative Approach Edited by Jan Kremers

Printed and bound by CPI Group (UK) Ltd, Croydon, CR0 4YY
16/09/2026
14963473-0003